HANDBOOK OF RELIABILITY ENGINEERING AND MANAGEMENT

Other McGraw-Hill Books of Interest

HANDBOOK OF RELIABILITY ENGINEERING AND MANAGEMENT

W. Grant Ireson
Clyde F. Coombs, Jr.
EDITORS

McGRAW-HILL BOOK COMPANY

New York   St. Louis   San Francisco   Auckland   Bogotá
Hamburg   London   Madrid   Mexico   Milan
Montreal   New Delhi   Panama   Paris
São Paulo   Singapore   Sydney
Tokyo   Toronto

Library of Congress Cataloging-in-Publication Data

Handbook of reliability engineering and management.

Includes index.
Rev. Ed. of: Reliability handbook. 1966.
1. Reliability (Engineering) I. Ireson, William
Grant·   II. Coombs, Clyde F. III. Ireson,
William Grant·   Reliability handbook.
TA169.H36 1988   620'.00452   87-17093
ISBN 0-07-032039-X

This book was originally published in 1966 under the title *Reliability
Handbook*.

234567890 DOC/DOC 89321098

ISBN 0-07-032039-X

The editors for this book were Betty Sun and Dennis Gleason, the
designer was Mark E. Safran, and the production supervisor was
Dianne Walber. This book was set in Times Roman. It was composed
by the McGraw-Hill Book Company Professional & Reference Division composition unit. Printed and bound by R. R. Donnelley & Sons
Company.

CONTENTS

PART 3   ENGINEERING FOR RELIABILITY

APPENDIXES

Appendix A: Tables

Appendix B: Charts

CONTRIBUTORS

William Grant Ireson *Professor Emeritus, Stanford University.* A native of Virginia, Professor Ireson earned B.S. and M.S. degrees from Virginia Polytechnic Institute and Virginia State University and served on its faculty for 7 years. He was Professor of Industrial Engineering at Illinois Tech for three years and joined the Stanford Faculty in 1951. He organized the Industrial Engineering Department and served as its Chairman from its founding until 1975.

Mr. Ireson specializes in quality control, reliability, engineering economic analysis, and engineering education development. He has been a consultant to many private companies, the U.S. Air Force, U.S. Navy, UNIDO, UNESCO, the Ford Foundation, and many government agencies and companies overseas. He has worked in 23 countries.

He is coauthor with E. L. Grant and R. S. Leavenworth of *Principles of Engineering Economy,* coeditor with E. L. Grant of *Handbook of Industrial Engineering and Management,* editor of the *Reliability Handbook,* and author of *Factory Planning and Plant Layout.* He has authored or co-authored over 50 technical papers. He is editor-in-chief of the *Reliability Review,* published by the Reliability Division of ASQC.

Mr. Ireson is a Fellow of both the American Institute of Industrial Engineers and the American Society for Quality Control. He received the Frank and Lillian Gilbreth Award from AIIE, the Eugene L. Grant Award and the Austin J. Bonis Education Award from ASQC, and the Order of Civil Merit Mogryeon Medal from the Government of South Korea. He is a registered professional industrial engineer in California. He is listed in *Who's Who in America* and *Who's Who in Engineering.*

Clyde F. Coombs, Jr. *Hewlett-Packard Company, Palo Alto, California.* A native Californian, Mr. Coombs earned his bachelor's degree in electrical engineering from the University of Utah, and subsequently a master's degree in industrial engineering from Stanford University.

After service in the U.S. Navy, he joined Hewlett-Packard as a manufacturing engineer. He has held a variety of technical and managerial positions in manufacturing, quality assurance, and marketing, and currently is Marketing Manager for the Printed Circuits Division of H-P.

He is the editor-in-chief of the *Printed Circuits Handbook* (three editions), and the *Basic Electronics Instrument Handbook,* all published by McGraw-Hill Book Company. He has authored more than five technical papers published in technical journals.

Mr. Coombs has taught classes in Quality Assurance at Stanford University, and Engineering Economics at San

Francisco State University and the University of Singapore. In 1967 he was named Production Engineer of the Year by the *Electronics Packaging and Production* magazine. He is now a member of the editorial board of that same journal.

John S. Bradin *Associated Services, Huntington, Indiana.* Mr. Bradin has served the electronic industry in various engineering and assurances capacities for over 30 years. He assisted major clients in organizational design and implementation, including formation of a new product assurance department and the establishment of organizational interrelationships for newly established functions, including applied research, systems engineering, product design, and manufacturing for electronic equipment and microelectronic devices.

He has held positions as director of product assurance, reliability assurance manager, program manager, chief engineer, systems engineer, and design engineer for Sparton Electronics Division of Sparton Corp., MEMCOR, and The Louis Allis Company. In addition, Mr. Bradin has directed studies for advanced system concepts, directed field test activities, and led design and development of new electronic products and systems.

Mr. Bradin received his B.S. degree in radio engineering at Tri-State University following service with the U.S. Navy during World War II. He has additional course work through Michigan State University, has produced a large number of audiovisual reports, has given numerous lectures and addresses, and in 1983 participated in an industrial study mission to Japan.

Henry J. Kohoutek *Hewlett-Packard, Fort Collins, Colorado.* Mr. Kohoutek was born in Czechoslovakia and received his academic training at the John Hus University and the Bohemian Technological Institute, both in Prague.

His work experience has been in the computer and electronic instrumentation industries, where he has been involved in all aspects of R&D, including hardware, software, systems, and reliability engineering. He has been granted 18 patents. His managerial assignments span R&D manufacturing and quality assurance. He was responsible for the quality aspects of the Hewlett-Packard VLSI and 9000 computer development programs. At present he is Quality Manager for the Hewlett-Packard Fort Collins Systems Division.

Mr. Kohoutek has published worldwide many papers dealing with a variety of quality issues in the high technology environment and artificial intelligence.

He is active in technical and professional societies such as ASQC, EOQC, IEEE, and AEA.

Arthur A. McGill *Lockheed Missiles and Space Company, Sunnyvale, California.* Arthur A. McGill is Group Engineer responsible for Reliability Problem Assessment at Lockheed Missiles and Space Co. Inc. He attended the University of

Rochester and Casper College and received his B.S. in electrical engineering from the University of Wyoming in 1966 and M.S.E.E. from the University of Santa Clara in 1971. He has taken postgraduate classes in systems management from the University of Southern California.

Mr. McGill joined Lockheed Missiles and Space Co.'s Missile Systems Division in 1966, working on the Polaris missile. He has been at various times responsible for part corrective action and electrical reliability analysis and is currently responsible for reliability assessment of operational missile systems and the Trident II missile. He is a member of IEEE and is a Registered Professional Engineer (Quality) in the state of California. As a result of his work on the Fleet Ballistic Missile program, he received the Department of the Navy Naval Material Command's Reliability, Maintainability and Quality Assurance Award in 1985.

Mr. McGill lives with his wife and three children in Sunnyvale, California. His hobbies are gardening, lapidary (he is president of the Lockheed Gem and Mineral Society), computers, electronics, and home repair.

Robert W. Smiley *Aerojet Strategic Propulsion Company, Sacramento, California.* Mr. Smiley is the Director of Quality Assurance for Aerojet Strategic Propulsion Company in Sacramento, California, which supplies solid rocket motors for Polaris, Minuteman, Peacekeeper, and Midgetman Ballistic Missiles. Bob has spent the last 35 years managing reliability and quality assurance programs for missiles.

His active Navy career spanned from 1943 to 1966. He retired with the rank of captain. In 1951 he was assigned as project officer for the Terrier missile conversion of the USS Boston. In 1954 he transferred to the Navy office at General Dynamics, Pomona, as Quality and Engineering Officer for what is today the Navy's Standard Missile. In 1956, then-Commander Smiley became the Quality and Engineering Officer in the Navy office at Lockheed Sunnyvale, at the beginning of the Polaris program. In 1963 he was the first commanding officer of the Polaris Assembly Facility in Bremerton, Washington.

He left the Navy in 1966 to start his civilian career as the Director of Product Assurance for Lockheed's Missile Division in Sunnyvale, moving laterally in 1972 to become Director of Materiel. In 1977 he was with McDonnell-Douglas Astronautics in St. Louis as Director of Product Assurance for the Harpoon Missile Program. In 1978 he joined Aerojet as Director of Project Assurance for the Tactical Missiles Company. He was transferred to the Strategics Company in 1984.

Mr. Smiley has been very active in ASQC: Named Fellow in 1960; past national Vice President; three terms as Region 6 Director; six years on the Certification Committee; a founder and past chairman of the Reliability Division; fifteen years on the board of directors for the Reliability Symposium; a regular teacher for the ASQC Reliability Engineering course; as well as past chairman of the San Francisco and Bremerton chapters of ASQC.

A graduate of General Motors Institute in Industrial Engineering, he also holds a B.S. and an M.B.A. from the College of Notre Dame, Belmont. ASQC certified him in both Quality and Reliability Engineering, and he has California Professional Engineering licenses in Quality and Industrial Engineering. Currently Vice Chairman of the QA Committee of the Aerospace Industries Association, he is listed in *Engineers of Distinction* and *Who's Who*, and is an active private pilot.

J. Douglas Ekings *Xerox Corporation, Rochester, New York.* Mr. Ekings is presently the manager of the Customer Satisfaction Improvement Team of the Business Products and Systems Group (BP&SG) of Xerox Corporation in Rochester. His responsibility is to support all business groups with quality business strategies. His primary objectives are to improve customer satisfaction and to reduce the cost of quality in all product and service areas. He provides reliability consultation to all business units and develops and implements quality and reliability training programs.

Previously Mr. Ekings was manager of Program and Reliability Assurance, responsible for the development of quality and reliability programs for future product programs. He was also responsible for managing a multinational quality assurance staff whose responsibility was reliability and product safety assurance programs throughout Xerox.

Prior to joining Xerox Corporation he held various managerial positions in reliability engineering and quality assurance with General Dynamics—Electronics Corp., the Martin Company, and Burndy Corp.

Mr. Ekings is a Fellow and Certified Reliability Engineer in the ASQC. He is a Professional Quality Engineer (California) and member of the Executive Committee of the ASQC. He was vice president of ASQC (1983–1985); treasurer (1985–1986); president-elect (1986–1987); and president (1987–1988).

Among the awards he has received are the Austin J. Bonis Education Award (1982), Reliability Education Advancement Award (1983), both from the Reliability Division of ASQC, and the Electronics Division Service Award (1981).

Richard Y. Moss *Hewlett-Packard, Palo Alto, California.* Richard (Dick) Moss received his B.S.E.E. from Princeton University in 1958 and an M.S.E.E. from Stanford University in 1960. He then joined Hewlett-Packard Company in Palo Alto, where from 1960 to 1975 he held various technical and managerial positions in R&D engineering. Since 1975 he has been Hardware Reliability Engineering Manager of the Corporate Quality Department, with responsibility for standards, procedures, and training programs aimed at improving hardward quality and reliability. In this capacity he has traveled to H-P sites worldwide presenting training seminars, auditing performance, and consulting on reliability and quality improvement projects, particularly ESD control. Mr. Moss has authored chapters for three technical reference handbooks, two published by McGraw-Hill and one by Dekker.

Mr. Moss's technical society activities have included numerous presentations for local chapters of ASQC and IEEE, and presentations of papers at the 1978 and 1982 IEEE Electronic Components Conference (ECC); the 1982 ASQC Golden Gate Conference; the 1985 Asia-Pacific EOS/ESD Symposium; the 1985 EOS/ESD Association Tutorial; and the 1986 International Reliability Physics Symposium (IRPS). He has participated in every EOS/ESD Symposium since their inception in 1979, and was a technical session moderator in 1980 and 1982, secretary of the symposium in 1981, vice general chairman in 1983, general chairman in 1984, and chairman of the board in 1985. In addition, he has been a director of the EOS/ESD Association since its founding in 1982 and was president in 1986.

Burton S. Liebesman *Bell Communications Research, Red Bank, New Jersey.* Dr. Liebesman earned the bachelor of science in engineering degree from the United States Naval Academy, and the master of science in electrical engineering and the doctor of philosophy in operations research degrees from New York University. He started as a member of the technical staff at Bell Telephone Laboratories in 1960 and after divestiture of the Bell System became a District Manager in Bell Communications Research. In 1982 he was named a Distinguished Member of Technical Staff at Bell Laboratories. He was an adjunct professor at Rutgers University from 1977 to 1983.

Dr. Liebesman has worked in all phases of quality and reliability assurance during the past fourteen years. His publications have appeared in the *Journal of Quality Technology, the Bell System Technical Journal, the Proceedings of the Reliability and Maintainability Symposium, the Proceedings of the Annual American Society for Quality Control Congress,* and *the Proceedings of the IEEE Global Communications Conference.* He has also contributed to three volumes of *Frontiers in Statistical Quality Control.*

Dr. Liebesman is currently responsible for economic studies of quality and reliability activities, including cost-of-poor-quality studies, quality improvement programs, and warranty analysis procedures. He is also responsible for the coordination of the quality and reliability standards efforts at Bellcore.

Charles R. Miller *United Technologies—Carrier Corporation, Syracuse, New York.* Mr. Miller is the Program Manager for Evaluation Engineering at the Carrier Corporation, where he is responsible for new product reliability evaluation and testing on all types of residential, commercial, and industrial air conditioning, refrigeration, and heating equipment.

His experience includes over ten years in the field of design reliability and product assurance technologies in both electronic and electromechanical systems and products. His previous experience includes ship systems reliability engineering work for the U.S. Naval Sea Systems Command, where he served as Project Reliability Engineer on several new ship designs and systems development projects. He has also been a senior reliability engineer with the General Electric Company, where he performed design reliability and system evaluation work on avionic and shipboard electronic systems and software.

Mr. Miller has served as the chairman of the United Technologies Corporation Interdivisional Committee on Reliability and Maintainability of Systems. He has taught seminars on reliability and evaluation engineering and has written handbooks on computer software reliability for both industry and government. He has authored several papers and articles on the development of effective product reliability programs, with emphasis on the application of proven R&M techniques to commercial product programs.

He holds a B.S. degree from the U.S. Naval Academy.

Kenneth P. LaSala *Air Force Systems Command, Andrews AFB, Washington, D.C.* Mr. LaSala joined the Air Force Systems Command (AFSC) in 1986 as Chief, R&M Division, Product Assurance Engineering Directorate, DSC for Product Assurance & Logistics. As such he is responsible for formulating and implementing AFSC R&M policy, especially with respect to the Air Force R&M 2000 initiative, and conducting assessments of the implementation of the R&M disciplines in AFSC acquisition programs. He is also serving as chairman of the IES ESSEH Environmental Stress Screening Management subcommittee and the IEEE Reliability Society Human Performance Reliability Committee.

Before joining AFSC he served at Headquarters, U.S. Army Materiel Command (HQ AMC) as Chief of Engineering Division, Product Assurance and Testing Directorate. Previously he was assigned to the Office of the Deputy Chief of Naval Material for Reliability, Maintainability and Quality Assurance (DCNM RM&QA). During 1982 to 1983 he served as the Government Resources Coordinator for the DOD Study of Steps Toward Improving Materiel Readiness Posture (DOD R&M Study). He also served as the U.S. representative to the NATO AC/250 Subgroup IX on Reliability and Maintainability. Before that he worked as an R&M engineer in the Undersea Weapons Directorate of the Naval Sea Systems Command.

Mr. LaSala received his B.S. degree in physics at the Rensselaer Polytechnic Institute in 1967 and M.S. in physics from Brown University in 1971. Mr. LaSala is a member of the IEEE Reliability Society and the IES. He has published several papers addressing human-machine reliability and other reliability topics. He is the instructor for Basic Reliability Engineering in the University of Maryland graduate program in reliability and maintainability.

Dwight Q. Bellinger *Division Reliability Manager, TRW Federal Systems Group, McLean, Virginia.* Mr. Bellinger has been with TRW since 1967 in key positions on aerospace, defense, and commercial programs. He implemented reliability and quality programs on a broad range of technologies including the Minuteman missile, rail transportation systems, coal conversion process plants, automobile braking systems,

satellites, power generation and distribution systems, Navy electronic systems, and Navy ship acquisitions. He also conducted reliability research projects for the Department of Energy, the Department of Transportation, and RADC.

Previously Mr. Bellinger worked for Douglas Aircraft, RCA, and IBM, implementing reliability, quality assurance, and failure analysis on the Thor-Delta launch vehicle, the DC-9 and C-5A aircraft, AWACS, the Ballistic Missile Early Warning System (BMEWS), and the SAGE System.

He received his B.A. in statistics from Columbia College and the M.B.A. from Seton Hall University, specializing in operations research. He is a member of IEEE, a senior member of ASQC, and a licensed professional engineer. He has written numerous technical papers in the areas of reliability, system safety, and acquisition management. He was elected chairman of the Engineers' Committee on Three Mile Island, representing sixteen professional societies and reporting to the Congress and the President of the United States.

David L. Burgess *Consultant, Accelerated Analysis, Half Moon Bay, California.* Mr. Burgess was Corporate Reliability Physicist for the Hewlett-Packard Company. He was responsible for designing, developing, and implementing the Failure Analysis Laboratory and supervised all the physics-of-failure analysis conducted by the laboratory.

Previously Mr. Burgess was employed by Fairchild Semiconductor and Monolithic Memories, where he performed and managed reliability engineering and failure analysis functions. He has had over twenty years of experience in failure analysis.

He received the B.E.E. degree from Rensselaer Polytechnic Institute and the M.S.E.E. from San Jose State University. He was general chairman of the International Reliability Physics Symposium in 1983 and is a member of IEEE. Mr. Burgess is coauthor of *Failure and Yield Analysis Handbook.*

John W. Kraus *McDonnell Douglas Astronautics Company.* Mr. Kraus is Staff Manager, Maintainability, Space Station Programs for McDonnell Douglas Astronautics as well as Maintainability Specialist for components. He has had over 35 years of experience in the aerospace industry. He implemented the first successful statistical quality control program at TRW Inc. in the mid-fifties, where he subsequently held such positions as Manager of Product Engineering and Quality Control and Manager of Industrial Engineering.

He later joined the Atomics International Division of North American Aviation, where he directed the establishment of a reliability engineering program and later maintainability and logistics support. He joined McDonnell Douglas (then the Douglas Aircraft Company) in 1966 as Branch Chief, Maintainability Engineering, Effectiveness Engineering Department, where he was responsible for maintainability technology development as well as providing qualified maintainability personnel for the components programs.

Mr. Kraus received a B.S. degree in general engineering

from the Massachusetts Institute of Technology and an M.B.A. from the University of Southern California. He is a senior member of the ASQC and has served on numerous technical and advisory committees for ASQC, the EIA, the AIA, and the American Defense Preparedness Association, of which he is a life member.

Sally Dudley *Hewlett-Packard Company, Palo Alto, California.* Sally Dudley is R&D Manager at Hewlett-Packard's Product Support Division. Her responsibility is to manage the development, implementation, and coordination of maintenance-oriented tools, skill-building programs, and methodologies integral to the delivery of H-P's product support services.

Prior to joining H-P, Sally was a systems engineer for IBM, a systems programmer at Lockheed Missiles and Space Company, and applications manager for Pacific Gas and Electric Co. At PG&E she established an in-house software training program. She joined H-P in 1973 and managed development and support of its order processing system and developed a programmer training program. In 1976 she moved to one of the operational units and managed the installation and maintenance of manufacturing systems.

In 1978 she joined the quality department and managed varied software activities: life-cycle development, quality planning, testing, configuration management, and consulting. She became Quality Manager of Information Networks Division in 1981, where she managed the software quality engineering department, a network test center, hardware materials engineering, environmental testing, and engineering services.

She moved to the corporate offices in 1984 as the company's software quality manager. In that role she provided corporate leadership for the company in improving the quality and reliability of its software products and processes worldwide.

Ms. Dudley earned her B.A. degree in mathematics from Vassar College.

Wayne Nelson *Private Consultant, Schenectady, New York.* Dr. Nelson received his B.S. degree in Physics from Caltech and M.S. (Physics) and Ph.D. (Statistics) degrees from the University of Illinois. He privately consults and teaches for various companies and universities in the United States and abroad and is also employed half time by the General Electric Research and Development Center in Schenectady, N.Y. He consults throughout General Electric on engineering and scientific applications of statistical methods involving planning experiments, accelerated testing, data analysis, sampling, and product life and reliability analysis. He is an adjunct professor at Rensselaer Polytechnic Institute and Union College.

Dr. Nelson has published widely on statistical methods for the analysis of engineering problems. For his publications he received the 1969 Brumbaugh Award, the 1970 Jack Youden Prize, and the 1972 Frank Wilcoxon Prize of the ASQC. He received the ASA Presentation Awards in 1977 and 1979 for outstanding presentations at the Joint Statistical Meetings. In 1981 he received the Dushman Award from G.E. Corporate Research and Development in recognition of his outstanding technical contributions to research and applications on product life data analysis and accelerated testing. He is a Fellow of the American Statistical Association and the American Society for Quality Control.

Dr. Nelson is author of *Applied Life Data Analysis* (John Wiley & Sons, 1982) and is currently completing a book on accelerated testing.

Kailash C. Kapur *Wayne State University, Detroit, Michigan.* Dr. Kapur is a professor in the Department of Industrial Engineering and Operations Research at Wayne State University. He received a bachelor's degree in mechanical engineering with distinction from Delhi University, M.S. degree in operations research and Ph.D. in industrial engineering from the University of California, Berkeley.

Dr. Kapur has worked with General Motors Research Laboratories as a senior research engineer. He was a visiting scholar with Ford and a consultant to the U.S. Army, Tank-Automotive Command. He has served on the Board of Directors of American Supplier Institute, Inc. His consulting work has specialized in design of experiments, the Taguchi methods, design reliability, statistical process control, and quality function deployment.

Dr. Kapur has coauthored with Dr. Lamberson *Reliability in Engineering Design* (John Wiley & Sons) and has written chapters on reliability engineering for several handbooks such as *Industrial Engineering* and *Mechanical Design*. He has published over thirty research papers in technical and research journals. He is a member of ORSA, TIMS, IIE, and ASQC. He is on the editorial board of the IIE Transactions and is a registered professional engineer. He and Dr. Lamberson shared the Allan Chop Technical Advancement Award from the Reliability Division of ASQC in 1987.

PREFACE

The importance of product reliability in the success of a manufacturing concern cannot be overstated. The issue of reliability has developed, and continues to evolve, as one of the key elements of international industrial competition. This is true in all fields, from satellites to instrumentation, to automobiles, to home products, to toys. No longer will the user (consumer) accept products that need frequent repair; there are too many competitive alternatives. In addition, often the user just cannot do without a product. Too much of the quality, and sometimes basic functioning, of the consumer's life depends on the continuing operation of devices to permit dependence on questionable reliability. For example, one usually has no backup for the telephone, the freezer, or the smoke detector. As a result, the issue of reliability has broadened from the area of defense, aerospace, and global communication to become fundamental in all industries. At the same time, there is the realization that periodic repair of a device, and its attendant loss of use, is not an intrinsic and inevitable cost one pays for ownership of such conveniences. A reliable product is expected and at no additional cost. The result of that realization has been one of the causes of a major upheaval in the balance of economic power in the world over the past 20 years.

In the United States, another major issue has brought the question of reliability to the top of management concern; that is, the legal liability incurred when a product malfunctions and causes injury. In the general aviation field, as an example, product liability insurance cost has become one of the biggest elements of product cost for private aircraft, driving companies to limit their participation in this industry or abandon it completely. The general legal assumption is that any manufacturer is under an obligation to provide a reliable product and must literally pay the consequences for not doing so; and a court may define the term *reliable*.

In dealing with this increasing importance of reliability, a new understanding of how to achieve it has also developed. It has become clear that a product designed for continuous, reliable operation and that is also designed to be manufactured and easily maintained is often the least expensive to build and own. It has also become clear that a manufacturing process that is well characterized to create the product, and is in statistical control, not only will be cheaper but will deliver a more predictable, and therefore reliable, result. The Reliability Department alone cannot be expected to compensate for problems best dealt with in design or manufacturing, but it can provide valuable information and leadership toward agreed-upon reliability goals. This cross-functional effort requires a total corporate team commitment to reliability. The critical roles engineering, manufacturing, and service have to play, in addition to the traditional reliability function, must all be focused and balanced by corporate top management.

This book is designed to address this total corporate need for tools to achieve the highest reliability at the least cost; hence the title *Handbook of Reliability Engineering and Management*. In separate and detailed chapters we deal with the management of an industrial organization, its economics, organization, and information flow for different functions to achieve the best total reliability results. Since a product must be designed for reliability, the engineering role is discussed in detail. All of this is supported by the mathematics of reliability not generally available in one volume. We are confident that the result is a reference book that can significantly help an organization turn "reliability" of its product into a competitive advantage both externally with its customers and internally in dealing with total cost of doing business.

We are indebted to the American Society for Quality Control, Milwaukee, Wisconsin, for permission to reproduce Appendixes B.3, B.4, B.5, B.6, and B.7, Charts of Confidence Limits on Reliability, from Lloyd and Liporo's *Reliability: Management, Methods and Mathematics,* copyright 1984.

We are grateful to the Literary Executor of the late Sir Ronald A. Fisher, F.R.S., to Dr. Frank Yates, F.R.S., and to the Longman Group Ltd., London, for permission to reprint data from Table 3 from their book, *Statistical Tables for Biological, Agricultural, and Medical Research* (6th edition, 1974).

W. Grant Ireson
Clyde F. Coombs, Jr.

INTRODUCTION TO RELIABILITY

CHAPTER 1
INTRODUCTION, DEFINITIONS, AND RELATIONSHIPS

W. Grant Ireson
Professor Emeritus, Stanford University

1.1 PURPOSE OF SECTION

This section of the handbook describes in broad terms the entire process by which the quality characteristic known as *reliability* is developed in a product or service. Its purpose is to assist in the development of an effective reliability program plan by:

1. Showing the total spectrum of activities needed
2. Defining the relationships of reliability engineers, quality assurance engineers, production engineers, purchasing, marketing, and management
3. Demonstrating the need for top management understanding and support of a well-planned and -executed reliability program

1.2 CUSTOMER SATISFACTION

Quality and reliability are not free, but poor quality and reliability usually cost much more than good quality and reliability. Warranties, liabilities, recalls, and repairs cost millions of dollars each year because quality and reliability were not given enough emphasis during the design, manufacture, and use stages of product development to attain customer satisfaction. Just as in medicine, the cost of preventing poor quality and reliability is usually much less than the resulting costs of inferior quality and reliability.

Every producer of goods or services knows that the success of a business depends upon the customers' satisfaction with the product or the service, relative to the price charged. This commonly means that the quality of the product or service meets the customers' expectations at a price or cost considered reasonable. The differences in the expectations of different customers account for the fact that products that perform the same general functions are available in widely differing "qualities" and at widely differing prices. These differences are the result of different specifications and the conformance of the product to those specifications.

There are two general concepts of quality as regards specifications:

Quality of design
Quality of conformance

Automobiles offer excellent examples of differences in quality of design. The Rolls Royce, Mercedes, BMW, and Cadillac represent high quality of design when compared with less expensive cars such as the lower-priced Chevrolets, Fords, and Plymouths, and that superiority is represented in the prices charged. All the cars perform the same basic function: provide transportation for the customer. Specification of better materials, tighter tolerances, improved control systems, and many other factors make the higher-priced cars "better" cars.

Quality of conformance deals with how well the actual product conforms to the specifications. Therefore, most of the quality assurance activities deal with the measurement of the quality characteristics specified and procedures and programs to assure that each part that goes into the product complies with all the individual specifications. A product can be produced with total compliance with the specifications, but still be inferior if the specifications are poorly prepared or inferior.

In most cases, the quality of conformance can be determined very quickly by inspection of the parts. Conformance to the specifications may or may not determine how long the product will function properly. The quality of design usually determines how well and how long the product will perform its designated function in an acceptable manner.

1.2.1   Reliability Is Just One Quality Characteristic

Reliability is defined in many different ways, but the most widely accepted definition states that it is the ability or capability of the product to perform the specified function in the designated environment for a minimum length of time or minimum number of cycles or events.

The "life" of an individual product cannot be determined except by running or operating it for the desired time (or number of cycles) or until it fails. Obviously, you cannot wear out all the products to prove that they meet the specifications; so, just as in quality assurance, you must rely on data gained by testing samples of the product. This in turn means that the statements regarding reliability must be in terms of probability of surviving the specified life with satisfactory performance throughout. Thus, reliability is normally stated as one or more of the following:

MTTF	Mean time to failure
MTBF	Mean time between failures
MTBMA	Mean time between (or before) maintenance actions
MTBR	Mean time between (or before) repairs
Θ_0	Mean life in some units such as hours or cycles. Read "theta subzero."
λ	Failure rate in some specified time period. Read "lambda."

Contrary to ordinary quality assurance, reliability assurance requires testing over time, and frequently for very long times, in order to prove beyond a reasonable doubt that the product as designed and *built* will have an acceptable life. Since this will be a probability statement, the confidence level of the results will vary with the amount of testing. To attain a high confidence level of high reliability usually requires that the tests on a number of products be run over several months. The higher the confidence level and the reliability desired, the more testing that will be necessary.

A formal reliability program plan is as necessary as a quality assurance plan because, contrary to the quality assurance situation, it may be months before errors in design, manufacturing, and assembly that adversely affect reliability can be identified. This puts a great premium on

Doing it right the first time!

1.2.2 Reliability Engineering

Reliability engineering has become a recognized profession, and the American Society for Quality Control conducts examinations by which individuals can become Certified Reliability Engineers. The certification examinations usually cover about six subject areas, such as

Definitions

Analytical methods

Various distributions and their uses

Hazard functions

Step stress, life, and demonstration testing

Product life-cycle characteristics

Information regarding the examinations and other requirements, such as experience, for certification can be obtained from the Certification Department, American Society for Quality Control, 310 West Wisconsin Street, Milwaukee, WI 53203. Many of the local sections give short preparatory training programs before the examinations. The exams are given in a large number of locations in the United States as well as in some other countries.

The functions performed by the reliability engineer are probably best described by MIL-STD-785, *Reliability Program for Systems and Equipment Development and Production,* as follows:

> Tasks shall focus on the prevention, detection, and correction of reliability design deficiencies, weak parts, workmanship defects. Reliability engineering shall be an integral part of the item design process, including design changes. The means by which reliability engineering contributes to the design, and the level of authority and constraints on the engineering discipline, shall be identified in the reliability program plan. An efficient reliability program shall stress early investment in reliability engineering tasks to avoid subsequent costs and schedule delays.

The same military standard calls for concurrent reliability accounting.

> Tasks shall focus on the provision of information essential to acquisition, operation, and support management, including properly defined inputs for estimates of operational effectiveness and ownership costs...ensuring...efforts to obtain management data [that] is clearly visible and carefully controlled.

A review of the life cycle of a product will assist in explaining how and where reliability engineering enters into the total process of designing, developing, manufacturing, and marketing a product.

1.3 LIFE CYCLE OF A PRODUCT

Table 1.1 gives a detailed series of steps in the process of designing, manufacturing, and marketing a new product. These steps and the inputs of quality assurance, reliability engineering, and production engineering are described in detail in the following section.

1.3.1 Design Phase

System Definition (Goals).   Reliability engineering and quality assurance usually will have extensive records of past experiences in producing a similar product. Those records will inform the design engineers of the kinds of problems encountered previously and the solutions, if any, that were accomplished. This assists the design engineers in setting realistic objectives for the new product.

Concept.   The design engineers will base the fundamental design concept on their prior experiences, results of research and development activities, and the marketing research regarding the need for the proposed new product. The concepts themselves are founded on the research results and engineering science that support the design engineer's belief that a new product can be designed that will meet the market needs at a price that will provide an attractive profit to the manufacturer.

Preliminary Design.   A design disclosure package is prepared of drawings, specifications, etc., that can be evaluated by all concerned parties. The reliability and quality assurance people will usually be called upon by the design engineers for data inputs and consulting services. The reliability engineering and quality assurance inputs help the design engineers avoid mistakes in the selection of parts, components, manufacturing processes, etc., by pointing out causes of prior difficulties on similar products.

Preliminary Design Review.   A design review committee is usually appointed for any major project. It consists of representatives of the customers (marketing), design engineering, reliability engineering, quality assurance, production engineering, and project management. The preliminary design review (PDR) is carried out after the design disclosure but before any prototypes are built. The purpose is to bring together the knowledge of a number of specialists to analyze the proposed design and to identify any potential problems which may be prevented by design changes.

Remember.   The design engineer(s) have *final* authority for the product design. Quality assurance, reliability engineering, production engineering, and marketing are advisers. In some companies, reliability engineering may have final approval authority.

Redesign.   The design engineers will use the results of the PDR to make changes in the design disclosure package and will present it for a second design review. This step may be repeated two or three times until the design is approved for a prototype.

TABLE 1.1   Life Cycle of a System

Reliability and quality assurance activities at each stage	
Life-cycle stages	Activities
Design phase	
System definition (goals)	R&D data inputs
Concepts for design	
Preliminary design	R&Q consultation, data
Design disclosure (drawings and specs)	
Preliminary design review	Carried out by designers, reliability engineers, quality engineers, production engineers, customer or marketing departments
Redesign: new design disclosure	
Design review repeated (may be repeated several times until design approved)	Same as above
Build prototype	
Prototype testing	Planned and executed by designers, reliability engineers, quality engineers
Design review	Same as above
Redesign	Developed concurrently are quality inspection plans, reliability test program plan, providing of facilities
Production phase	
Release for production (complete drawings and specs)	
Production begins	R&Q perform production tests and verifications; quality sets up change and configuration control system
Continue production program	R&Q continue tests and inspections
Support phase	
Deliver product to customer	R&Q set up field failure report system
Customer starts using product and may find malfunctions and failures	
Field failure, malfunction, and service reports received	Reliability analyzes reports and suggests design changes and corrective actions
Corrective action decided	May require full design review and customer's approval
Corrective action implemented	R&Q follow-up to assure effectiveness
Retirement of product or system	
End of life cycle	

Prototype. One or more prototypes of the design are built by hand, using highly skilled technicians, machinists, etc., to make the prototypes to conform to all the specifications.

Prototype Testing. It is common for reliability engineers to have responsibility for performing all functional tests, but quality assurance will be responsible for ascertaining that the components, etc., conform to specifications. The test results are then presented for a preproduction design review. The same design review committee conducts that review.

Redesign. Prototype testing usually reveals weaknesses, errors, or other desirable changes. Existing prototypes may be modified in order to test the "fix," and another design review may be desirable. Concurrent with the redesign, the reliability engineers will be developing the reliability test program and selecting and procuring necessary test or measuring equipment. Quality assurance will be developing the quality assurance plan, designing the inspection program and forms, and procuring any necessary equipment. Production engineering will be setting up production facilities, obtaining tools, and developing the production plan. All of this takes place in preparation for the next phase.

1.3.2 Production Phase

Release for Production. Complete production drawings, specifications, parts lists, etc., are turned over to production.

Production Begins. The first few production units are submitted to reliability engineering for "production verification tests" and to quality assurance for complete and full inspection against the specifications. Whereas the prototypes were made by highly skilled technicians under very strict control, the production units are the result of production tooling, ordinary workers, and unsorted components and parts. It is to be expected that the quality of conformance and the reliability may have been degraded by the production process unless special provisions have been made in anticipation of this problem. Even when the incoming acceptance procedures and the production personnel have been trained and warned against passing substandard parts, materials, and workmanship, there is still the danger of degradation of reliability.

The inspection and test results are submitted to the design review committee for a complete postproduction design review. As a result of the design review, changes in design, specifications, and production processes may be proposed.

Design Changes. After production has begun, it is necessary for quality assurance to establish a change and configuration control system. This system documents any changes that are made, the serial numbers of the unit on which the change first appeared, and the drawing numbers and/or specifications involved. This is a necessary procedure throughout the life of the product so that spare parts, maintenance procedures, etc., can always be provided for each specific version of the product.

Continue Production. Reliability engineering and quality assurance will continue the test and inspection programs as planned earlier. Results will be recorded and any discrepancies will be brought to the attention of the project manager for corrective action.

1.3.3 Support Phase

Delivered Product. When the customers start using the product, other unanticipated problems may arise. All malfunctions and failures in the customers' hands should be recorded by the service organization and the information fed back to the project manager. Quality assurance and reliability engineering will review all reports and analyze the data to identify causes and propose corrective actions.

Corrective Actions. Decision on corrective action may require a full design review and the customer's approval. Reliability engineering and quality assurance will be responsible for follow-up to see that corrective action is taken and that it is effective.

1.3.4 Retirement Phase

Retirement or Replacement. This is the end of the life cycle, but all the knowledge gained during the design, development, production, and marketing of the product should be recorded so that it can be used later to improve the process for repeat production of the same product or for new products.

1.4 RELIABILITY PROGRAM PLAN

Each company will (or should) develop its own reliability program plan, tailored to the needs of the particular products or services which it produces. Companies with several divisions manufacturing products of significantly different complexity should have different plans for the different lines of products. It is advisable, however, to have only one plan in any given plant so that there will be no confusion regarding which plan applies to which lines of products.

This handbook has been designed to provide how-to information about all of the aspects of designing and operating a reliability program, and in most of the chapters, the subject is addressed from the viewpoint of the production of a complex product, failure of which could cause catastrophic results, such as death or severe injury. For simpler products which do not present serious safety or health hazards, the reliability program can be simplified to provide the necessary protection without wasting money unnecessarily on elaborate programs.

The most common description of a reliability program is that contained in the previously quoted MIL-STD-785. The U.S. Department of Defense has spent many millions of dollars over three decades developing and revising guidelines, standards, and specifications to be used in procurement of military and weapon systems; however, these basic principles are applicable to any production facility. Short descriptions of a large number of these standards and specifications are given in Chap. 5. Each company must tailor the implementation of the generalized program plan to its own specific needs.

MIL-STD-785 lists 17 different tasks that should be considered in the development of a reliability program (see Table 1.2). The first task, 101, says, "The purpose of this task is to develop a reliability program plan which identifies, and ties together, all program management tasks required to accomplish program requirements."

TABLE 1.2 Reliability Program Tasks

Task Section 100: Program Surveillance and Control
Task:
101 Reliability program plan
102 Monitor/control of subcontractors and suppliers
103 Program reviews
104 Failure reporting, analysis, and corrective action system (FRACAS)
105 Failure Review Board (FRB)

Task Section 200: Design and Evaluation
201 Reliability modeling
202 Reliability allocations
203 Reliability predictions
204 Failure modes, effects, and criticality analysis (FMECA)
205 Sneak circuit analysis (SCA)
206 Electronic parts/circuits tolerance analysis
207 Parts program
208 Reliability of critical items
209 Effects of functional testing, storage, handling, packaging, transportation, and maintenance

Task Section 300: Developing and Production Testing
301 Environmental stress screening (ESS)
302 Reliability development/growth test (RDGT) program
303 Reliability qualification test (RQT) program
304 Production reliability acceptance test (PRAT) program

Source: MIL-STD-785B, *Reliability Program for Systems and Equipment Development and Production.*

The following is a list of the functions that are normally assigned to the reliability engineers.

1. Reliability estimation, prediction, and growth plan
2. Participate in all design reviews
3. Reliability apportionment
4. Plan and conduct reliability tests
5. Perform statistical analysis of test data
6. Maintain reliability data system
7. Provide assistance to:
   a. Production
   b.. Quality assurance
   c.. Purchasing
8. Write reliability specifications for purchased items
9. Identify causes of reliability degradation

Each of these functions is briefly explained in the following paragraphs:

1. *Reliability estimation, prediction, and growth plan*:   At each stage of the life cycle, the reliability engineer is expected to use the information gained from tests plus recorded information about the reliability of components, parts, and materials specified in the design to estimate the reliability that can be attained. Tests (as described in Table 1.1) usually show that the estimated reliability has not been attained and that it is necessary to predict what will happen if certain changes are made. In addition, there is a "learning curve" regarding reliability, especially when the design is pushing the state of the art. Therefore, a reliability growth plan is often required for a complex system. The growth plan sets reliability goals in the future, either in terms of time or the number of systems produced. Then the actual test results are compared with the goals in order to make decisions regarding programmatic actions. The mathematical and statistical techniques used to analyze test results and predict future reliability accomplishments are described in Chaps. 18 and 19.

2. *Participate in all design reviews*:   The importance of design reviews cannot be overemphasized. The design review presents the opportunity for all the different viewpoints to be presented and evaluated. If design engineers attempt to obtain the input ideas regarding a design by contacting reliability engineers, quality assurance engineers, production engineers, customer or marketing departments, and management one at a time, there is no chance that the differences can be resolved satisfactorily. Each representative has certain knowledge about reliability consequences that must be weighed against any opposing knowledge from others. The opportunity to present all the reliability considerations at the design stage can save great amounts of money and time (see Chap. 3).

3. *Reliability apportionment*:   Since the reliability of the overall system or product is basically the product of the reliabilities of the individual parts, and since it is easier to attain very high reliability on some kinds of parts than on others, the specification of reliability requirements for the different parts should consider the ease of accomplishment. Lower requirements should be set for difficult parts or components. This allocation of the total reliability among different parts and components is known as apportionment, and is usually the task of reliability engineering.

4. *Plan and conduct reliability tests*:   Most frequently, the reliability test equipment serves for research testing as well as for reliability tests, saving duplication of investment. The reliability engineers are specialists in designing test procedures and setting criteria for equipment.

5. *Perform statistical analysis of test data*:   The reliability engineer is better equipped to interpret the results of tests and to present the findings to the design engineer and project manager. This also works right into the matter of estimating and predicting reliability of a design.

6. *Maintain reliability data system*:   Throughout the product life cycle, reliability information is being collected and must be maintained in a form that can be used in the management of the program. Since the reliability engineer must evaluate and analyze the information in order to make suggestions for changes in design, procedures, processes, etc., the information usually is routed directly to the reliability engineering department where it can be computerized for rapid analysis and for permanent storage into the data base (see Chap. 4).

7. *Provide assistance to*: Production (Chap. 7), quality assurance (Chaps. 8 and 11), and purchasing (Chap. 6).

8. *Write reliability specifications for purchased items*: The reliability of all parts, materials, components, subassemblies, etc., that go into the final product influence the overall reliability. Unreliable supplies can ruin an otherwise good product, and the reliability engineer is the one who is most likely to understand what should be included in the specifications (see Chap. 10).

9. *Identify causes of reliability degradation*: Degradation can occur because of poor parts, improper process control, inadequate tooling, poor workmanship, and many other causes. The reliability engineer must make judgments based on quality assurance information and reliability tests and recommend corrective action.

1.5 SOME SPECIAL RELIABILITY CONSIDERATIONS

While reliability normally is associated with the hardware and its ability to perform the prescribed function, there are some other factors that must be considered as part of the overall reliability program to attain the desired customer satisfaction. They may directly affect the measured performance and/or the probability of functioning for the specified time or cycles. None of these come under the direct control of the reliability engineer, but they should be evaluated by the reliability engineering during the design reviews.

1.5.1 Maintainability and Availability

Practically all electromechanical products require some maintenance to keep them operating in a satisfactory manner. Preventive maintenance is employed to try to prevent unscheduled downtime and interruptions in their use; however, it is practically impossible to prevent all interruptions in service. The ease with which the system or product can be restored to operating condition and the time required are primary considerations in the design stage. The information that has been collected by reliability engineers and quality assurance personnel permits the prediction of frequency of malfunction or failure, and, in turn, the need for planned maintenance and repair actions.

Quantitative measures of maintainability are:

MTTR	Mean time to repair
MTTRS	Mean time to restore system
MTTRF	Mean time to restore function
DLH/MA	Direct labor-hours per maintenance action
TPCR	Total parts cost per removal
PFD	Probability of fault detection

Maintainability demonstration is required on complex systems by most customers. This is usually carried out by industrial engineers using regular repair and maintenance personnel with specialized tools, test equipment, etc., designed for use with the system. Time studies of several trials of each maintenance action are averaged to obtain the expected mean time (see Chap. 15).

Availability is the total time minus the time required for maintenance and repair actions. Obviously, for systems that need to be available full time, there is considerable pressure to reduce the maintenance time and cost of downtime. Typical examples of high-cost systems that justify special attention to maintainability, availability, and cost of downtime are airplanes, large computers, communication systems, and electrical power systems. An unscheduled break in service due to some unexpected part failure can cost thousands of dollars per hour in lost revenues.

Part of the maintainability program requires that a maintenance plan be developed. As shown in Table 1.3, the maintenance plan assures that the provisions for routine and emergency maintenance are considered and provided as part of the overall reliability program and that maintenance data will be analyzed and used in setting up the maintenance plan.

TABLE 1.3 Maintainability Program

Development of a maintainability program involves:

1. Identify phases of program: conceptual, validation, full engineering design, production
2. Tailoring of plan to needs
3. Qualitative requirements
4. Set tasks and classifications

Program surveillance and control tasks:

101 Maintainability program plan
102 Monitor/control of subcontractors and vendors
103 Program reviews:
    Preliminary design reviews (PDR)
    Critical design reviews (CDR)
    Test readiness and others
104 Data collection, analysis, and corrective action system

Design and analysis tasks:

201 Maintainability modeling
202 Maintainability allocation
203 Maintainability prediction
204 Failure modes, effects, and criticality analysis; maintainability information
205 Maintainability analysis
206 Maintainability design criteria
207 Preparation of inputs to the detailed maintainability plan and logistic support analysis (LSA)

Evaluation and test tasks:

301 Maintainability demonstration

Source: MIL-STD-470A, *Maintainability Program for Systems and Equipment.*

1.5.2 Human Factors

Human beings are involved in practically every stage of the life cycle of a product, and over time, humans can and frequently do make mistakes. These mistakes can render an otherwise good design unacceptable. History has recorded a great many costly events that could have been avoided if more consideration had been given to the human factors during the design stage. Also, instructions regarding the performance of duties, or of operation, should be so carefully written that they cannot be misunderstood. Specifically, design engineers and reliability engineers should consider how human factors affect and in turn are affected by:

1. Equipment design characteristics
2. Operational and maintenance procedures
3. Work environment
4. Technical data needed for operation and maintenance
5. Communications
6. Logistics
7. System organization

A massive amount of research has been performed to develop excellent design guidelines for all aspects of human performance. Two of these have been developed by the Department of Defense. They are:

MIL-HDBK-759, *Human Factors Engineering Design for Army Materiel*

MIL-STD-1472, *Human Engineering Design Criteria for Military Systems, Equipment and Facilities*

Consideration of human factors can improve the maintainability of a system, reduce the possibility of detrimental operating procedure, and reduce the chance of damage to the system from improper use. Customers' operators frequently are the cause of malfunctions as a result of inadequate instructions regarding maintenance and operation. Connectors that can be joined in two or more ways (involving polarity, for example) can cause catastrophic damage. Complex corrective actions when something goes wrong may put the operator under such stress that the correct procedure is forgotten or critical steps are omitted. All such possibilities should be recognized and eliminated or reduced by improvements in the design (see Chap. 12).

1.5.3 Cost Effectiveness

Cost effectiveness is not easily defined, but in simple terms it means that the cost of making some improvement in the design, production, or maintenance of a system is exceeded by the prevention of costs resulting from short lives, high maintenance costs, human injuries or deaths, replacement costs, etc. In other words, cost effectiveness means spending some money to prevent the expenditure of a larger amount of money, with due consideration to the time value of money.

Many organizations, including the Department of Defense, look at the "life-cycle costs" of a system. In life-cycle cost analysis, all the expected costs over the entire life of the system are reduced by use of discounted cash flow methods (net present value at some specified discount rate) to obtain the present value of all future costs of the system at a given level and amount of service. Obviously, reliability is probably the most important factor in the development of life-cycle costs.

Reliability can be engineered into the products at the design stage if management is willing to provide the resources as described in this handbook. That does not mean that the product must be "gold-plated," but that the reliability program is tailored to the needs of the product line with due consideration to the consequences of having an inferior product. A breakdown of the costs of quality and reliability can be made as follows:

Controllable costs: The costs of those activities that are planned and included in the quality and reliability budget. These costs include all the activities to assure quality and reliability and all the activities to inspect and test to find out what quality and reliability level has been accomplished.

Resultant costs: These are unplanned costs that result from *not* attaining the required levels of quality and reliability. They include *internal failures*, such as scrap, rework, repairs, etc., within the plant, and *external failures*, which are all the costs incurred from failures and malfunctions after the product is delivered to the customer, field services, replacement, warranties, reduced billings, repairs, liabilities, and loss of reputation.

These costs are frequently categorized under three titles, prevention, appraisal, and failure costs. *Prevention costs* are roughly the same costs that I have described as those budgeted and incurred to assure acceptable quality and reliability. *Appraisal cost* is a term applied to the costs budgeted and incurred to inspect and test the products to ascertain the level of quality and reliability attained. *Failure costs* refer to all of those costs incurred as a result of failures that occur either before or after shipment (internal and external resultant costs).

Reliability engineers should be aware of the effects of their activities on the costs of a system and make a special effort to optimize the overall reliability costs. The design engineers normally are not very concerned about cost matters and strive for perfection regardless of the costs. Reliability, quality assurance, and production engineers can have a great influence on these costs through their participation in the design reviews as well as by providing information to the design engineers (see Chap. 3).

P · A · R · T · 2

MANAGEMENT OF RELIABILITY

CHAPTER 2
ORGANIZING AND MANAGING THE RELIABILITY FUNCTION

John S. Bradin
Quality and Reliability Consultant

2.1 RELIABILITY GOALS AND POLICIES

Organizations require a sense of direction and general guidance to enable them to function effectively. Goals provide the direction and policies provide the guidance. This section first deals with goals in general, then takes up reliability goals in particular. Section 2.2, Reliability Goals, discusses the purposes and uses of goals, who originates them, and how they are applied; it defines related terms, provides criteria for developing goals, and includes a procedure for goal setting. Section 2.3, Reliability Policies, deals with the purposes of policies, their definition, and their use, and provides a list of attributes that policies should possess. The information presented in this section will provide readers with sufficient understanding of goals and policies to enable them to develop their own.

2.2 RELIABILITY GOALS

Goals are used to focus organizational attention toward desired results and to serve as standards against which achievements can be measured. They are established in various ways and at various levels to facilitate achievement of firm requirements. Where requirements are not yet firm, goals serve to specify desired results.

Depending upon the type of business engaged in, product reliability requirements or goals are contractually specified by customers or they are established in house by those who decide what level of reliability the product must have. In the latter case, the reliability level is based on customer and market considerations, as balanced against the cost of reliability and the risk of unreliability. Whichever the case may be, the level is specified as a measurable quantity which, for complex products, must be divided over the major subsystems or units to provide meaningful goals for designers. These goals may be further subdivided down to the lowest level of design responsibility.

Goals are also established in a more general sense to facilitate organizational improvement. These are established at various organizational levels ranging from the division to the department to the functional unit and the multidisciplinary team formed to carry out programs or special-purpose functions. An example of how various activities interrelate in the early stages of a program and how they influence the development of reliability goals is illustrated in Fig. 2.1.

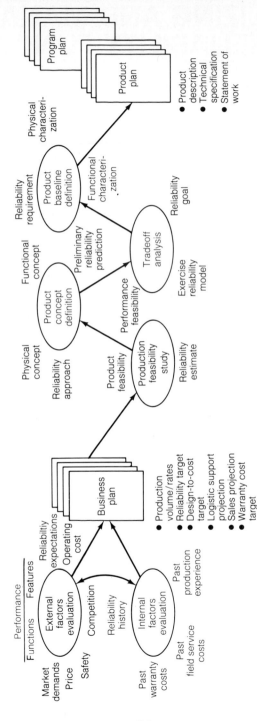

FIG. 2.1 Interactive activities in product planning.

2.2.1 Related Terms

The following related terms are presented in their general relational order:

Organizational purpose: Organizational reason for being.

Organizational mission: Assigned job or business of an organization.

Reliability objective: Sometimes used interchangeably with *reliability goal*, but usually applied on a broader scale with less finality as when coordinating subgoals with higher-level organizational goals. Reliability results desired at intermediate points in a program (e.g., reliability milestones).

Reliability target: A general level of reliability typically proposed in the planning or formative stages of a product development program to indicate a desired result when the degree of attainment is still in question. A tentative value subject to trade-off.

Reliability goal: Quantified reliability level aimed for in early program stages where firm requirements such as formal demonstration tests or acceptance tests are not yet required. Used in conjunction with a threshold value to provide guidance for development programs. Also used to specify desired organizational achievement indirectly related to, but in support of, product-related goals.

Reliability threshold: Limit reliability value below which review action is required. Used to help control technical and cost risks in advanced development programs.

Reliability parameter: A measure of reliability such as mean time between failures, failure rate, probability of survival, or probability of success. Reliability-creating factor of the product development process used as an index of design reliability to facilitate trade-offs in the pursuit of reliability achievement or growth.

2.2.2 Goal Criteria

All goals have certain characteristics in common. The following criteria can be used to assist in developing and arranging goals:

1. *Relevance:*   All goals relate to some higher purpose and should be well suited to the particular needs and circumstances. A good match with needs and circumstances leads to greater success in furthering the higher purpose by directing focus to the right things.

2. *Attainability:*   Goals should be set at levels reasonably attainable within the available time span. Oftentimes, however, a specific level is first desired and then a target date for completion is selected. Whichever is the case, quality, quantity, and time are the interacting variables. For any particular type of goal (quality), goal level (quantity), and completion date (time) are tradable. Other things being equal, the higher the goal, the longer the time to achieve it. However, to maintain interest and commitment, large goals over long periods of time should be avoided. Dividing goals into subgoals over shorter time spans makes them more attainable and individually more satisfying. In all cases, goals should present sufficient challenge to be worthy of pursuit.

3. *Supportability:*   Goal achievement requires wherewithal. The necessary instruments and resources must be available at the time they are needed. It should be determined beforehand what the instruments and resources will be and the extent to which they can or will be provided. Stay with what can be realistically supported.

4. *Compatibility:* Numerous goals are pursued simultaneously within an organization, whether formally stated or not. If established with little or no regard for each other and pursued in isolation, they can result in diverging efforts and possible conflict. Goal setting therefore should be approached from a system perspective to ensure compatibility and to link related goals in a mutually supportable way.

5. *Acceptability:* Goals must be acceptable to those who will be actively involved in their pursuit. Degree of acceptance is influenced by relevance, perceived importance, reasonableness, and desirability of outcome. Where commitment may not be readily obtained, increased acceptance may be brought about through wider participation in the goal-setting process.

6. *Measurability:* Goals provide standards against which performance may be assessed and therefore should be selected for suitability and defined in a way that enables measurement. To make them measurable, goals can be defined qualitatively, quantitatively, and temporally in terms of performance parameters, values, and time scales.

2.2.3   Goal System

Goals are established to urge organizational elements and individuals toward achieving desired ends. Regardless of what level in the organization a goal is set, there are higher purposes to be respected and it is necessary to consider the interrelatedness of goals with respect to each other and their relevance to higher purposes. To be most constructively applied, goals therefore should be systematically arranged in a coherent network. Figure 2.2 illustrates the relationship of objectives with the organizational hierarchy; Fig. 2.3 shows how higher goals are supported by subgoals and that subgoals are negotiable and tradable.

2.2.4   Goal Setting

When working from preestablished firm requirements, reliability goal setting consists of reducing the overall requirements to a series of subgoals, each related in such a way as to reinforce the net outcome. Reliability apportionment is a common, long-standing technique. The goal-setting process is more involved when addressing customer needs and competitive factors in an open-market setting or when developing goals for organizational performance improvement. The following general procedure is useful in establishing organizational performance improvement goals:

1. Clarify what is needed or desired.
2. Review organizational purpose and mission of the organizational unit establishing the goal and the purposes and missions of the units that will be pursuing the goal.
3. Identify key result areas.
4. Determine where the highest payoffs are likely.
5. Select result areas most desirable to pursue.
6. Select goal candidates.
7. Consider resources required to pursue each goal to a successful conclusion.

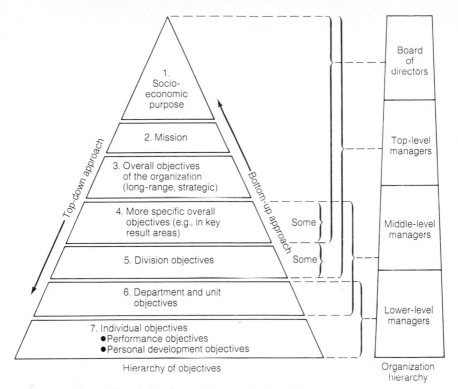

FIG. 2.2 Relationships of objectives and the organizational hierarchy. (*From H. Koontz, C. O'Donnell, H. Weihrich,* Essentials of Management, *4th ed. McGraw-Hill, New York, 1986, p. 99. Reproduced with permission.*)

8. Identify any impediments or risks to goal achievement and determine how they can be overcome or offset.

9. Rank candidate goals according to ease of achievement in combination with degree of payoff.

10. Consider goal interdependencies and adjust goal candidates for maximum coordination and mutual reinforcement.

11. Examine goals for relevance, attainability, supportability, compatibility, acceptability, and measurability.

12. Make final selection of goals and establish deadlines for achievement.

13. Establish measures of success.

14. Develop action plans for key goals. Include provisions for any motivational initiatives needed for obtaining commitment and any provisions for supervisory and management support.

15. Communicate goals in writing.

16. Periodically review progress toward goal achievement and make any adjustments needed to ensure success.

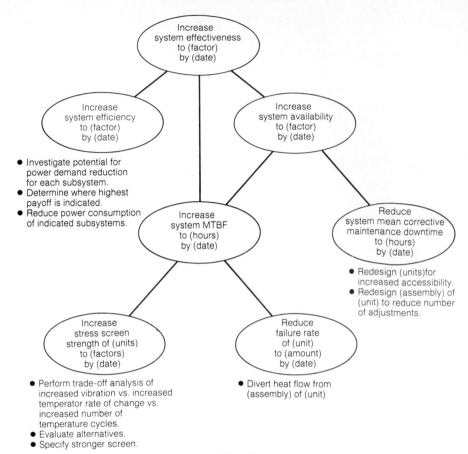

FIG. 2.3 Goal network and supporting activities of the basic levels.

While goal setting is a straightforward rational process which leads up to statements with numerical values and associated due dates, it must be kept in mind that the successful pursuit of goals requires the leadership and active support of management. Once goals are established, managers must see that goals are communicated, are clearly understood, and are accepted through individual commitment. They must ensure that a path toward goals is indicated and that the resources and instruments needed to achieve them are made available.* Indeed, because the goal process is fundamental to the planning and control functions of management, the effectiveness of managers rests to no small extent upon the degree to which they are sensitive to the needs of those pursuing goals and the extent to which they provide the leadership and the material and psychological support needed to achieve the goals.

---

* Miller, James Grier, *Living Systems,* McGraw-Hill Book Company, New York, 1978, p. 652.

2.3   RELIABILITY POLICIES

Organizations periodically face situations that require direction from management to ensure that decisions made and actions taken consistently support the overall mission and goals. Policies provide this direction by defining specific areas of concern and then indicating a desired outcome or specifying the direction decisions and actions should take. In this way policies increase decisiveness by removing uncertainty and thereby reduce the potential for inefficiency, counter-productivity, inappropriate risk taking, or conflict.

While policies are, in a certain sense, related to procedures and rules, they are neither of these. Procedures are precise, step-by-step instructions for required actions in a definite sequence. Rules are requirements to take or not to take specific actions. Both are characterized by a high degree of specialization and rigidity. Neither allows any discretion or deviation. Policies, on the other hand, admit flexibility of judgment within a framework of guidance and direction. The best policies admit the widest possible range of expression and scope of action.

The need for policy is indicated when an organization is faced with decision options that may be precedent-setting. Situations calling for policy include those where:

- Opinions may differ over the best course of action in situations that impact the achievement of overall mission and goals.

- Decisions may have significant consequences beyond the local level at which they are made.

- The choice of action may lead to unnecessary risk, counterproductivity, inefficiency, or conflict.

- Cooperation and reciprocal actions on the part of one organizational element are needed to enable another element to function effectively.

Policies may be established to:

clarify	define	guide	regulate
direct	establish	integrate	authorize
enable	empower	commit	support
provide	admit	ensure	inhibit
prohibit	restrict	disallow	

Reliability policies are needed at a number of levels in the organization to ensure reliability creation, reliability maintenance, and reliability improvement. They are especially needed to smooth the way for key or critical matters. Firms establishing a new reliability function have a compelling need for them; firms engaged in the manufacture and sale of products or systems that can present a significant public safety hazard or can affect national security or national prestige have a critical need for them.

To be effective, reliability policy must issue from high-level management. Management's attitude toward reliability, as expressed through policy, is the most important single ingredient in making reliability engineering and reliability assurance a successful practice in any organization. If the deeds of management in fact support written policies, policies gain credibility and legitimacy and will be respected.

Reliability policy and, ultimately, the final responsibility for the reliability of products and services rests upon the chief executive officer, albeit through successive management levels. The top assurance executive is responsible for pursuing reliability in the manner and to the extent prescribed by general management policy and establishes his or her own policies in support of this function.

2.3.1 Attributes of Policies

Policies, in general, should be:

1. Action-oriented (as contrasted to mere statement of belief)
2. Supportive of organizational goals
3. Consistent with other policies
4. Authoritative, credible, and acceptable at the level of implementation
5. Inclusive to the extent of embracing all aspects of the intended application
6. Specific to the extent of providing unambiguous direction and focus
7. Admissive to the extent of allowing maximum flexibility of choice within the prescribed framework of guidance or direction
8. Concise and readily understandable
9. Relevant to the times and circumstances
10. Stable over relatively long periods of time

2.3.2 Types and Levels of Policies

Policies are established at all levels of the organization, beginning at the corporate level and proceeding through divisional levels and successively finer levels of organizational structure down to departments and functional units. High-level policies such as those at the corporate level are broad and general so as to deal effectively with the broad concerns of top-level management. Corporate reliability policy provides for the establishment and promotion of reliability activity and achievement to fulfill obligations to customers and to society at large. It deals with internal matters pertaining to overall performance and with external matters pertaining to relationships with customers, the community, and involvements with regulatory and other such organizations. It sponsors the reliability function by declaring its intentions to the organization at large and provides for review and evaluation of the overall reliability system.

Divisional policies respond to corporate policies and relate to the more specific issues encountered by departments and functional units. Typically they deal with administration, organizational interrelationships, operating methods, and the maintenance or improvement of organizational performance. Departmental policies deal with situations and conditions more apt to arise on a day-to-day basis. These situations typically include matters relating to suppliers, subcontractors, product design activities, parts and materials, manufacturing, testing, auditing, and reviewing.

2.3.3   Developing and Establishing Policies

The formation of effective and lasting policies requires a comprehensive view of the issues and a full appreciation of the circumstances leading to the need for policy. If the policymaker is not in full possession of the facts and nuances, it is advisable to enlist the views of others who may be deeply involved in the situation and have the breadth of view, knowledge, judgment, and experience to make constructive contributions. It is desirable that those managers and supervisors who would be affected by the policy or who would be expected to carry it out should be considered as potential contributors. Not only can their views be constructive, but their involvement in developing the policy will increase their acceptance and support when the policy goes into effect. While the participative approach is useful, the responsible manager nonetheless must impress the force of office and provide the benefit of experience to create appropriate policy.

It is especially important to the formation of policies for a new reliability function to win the acceptance and cooperation of long-established groups, particularly those that play prominent roles in the organization. Key individuals from those groups brought into the definition and development phases of policy formation can help ensure a well-conceived policy by raising key issues to address.

While the development of many policies is straightforward, some policies are more involved and may require advance planning and study. To assist in such cases, the following outline is presented as a guide:

2.3.4   General Guide for Policy Development

1. State the need. Describe the situation that created the need. Identify who or what is involved, how they are involved, and to what extent.
2. Identify and review any existing policies that relate to the situation.
3. Survey managers and supervisors who will be affected by the new policy. Obtain pros and cons.
4. Determine if a new policy is actually needed or if existing policies should be revised to accommodate the situation.
5. Draft a preliminary policy statement for review and comment by the departments affected. Include purpose and scope.
6. Integrate appropriate suggestions and prepare a revised statement for additional review. Add sections on responsibilities and actions, if appropriate.
7. Check the policy against the attributes presented in Sec. 2.3.1 and prepare a final document.
8. With executive approval and signoff, release the document for distribution.

Policies are communicated by memoranda, letters, instructions, or directives and are included in program plans and various manuals. They are given visibility by way of meetings, workshops, lectures, and training sessions. New policies should be routed to all departments and units affected and acknowledged by signature and date.

Policy statements range in size from simple statements of a paragraph or two in length to comprehensive documents that may include some or all of the following topics:

background	policy
purpose	responsibility
scope	actions
definitions	

While policies are essential to effective and efficient operations, they should be held to the minimum allowed by size of organization, complexity of operations, criticality of processes, management style, and self-responsibility, awareness, and professional level of employees. As guiding forces, they should admit the widest possible latitude for action any given situation allows.

Policies should be reviewed for possible revision or cancellation on a scheduled basis and whenever major changes are made in organization, management, practice, or overall organizational strategy.

2.3.5 Example Policy

Formal policies and their applications vary considerably from industry to industry. Some firms maintain elaborate systems of policies, plans, goals, and methods; others have relatively few. The extremes may be characterized on the one hand by the small entrepreneurial organization functioning in a highly competitive, volatile marketplace, and on the other hand by the very large organization functioning in a highly structured environment such as exists in defense contracting. The small specialty house, functioning "organically," has less need for formal policy because its people have frequent exposure to high-level managers and know their goals, philosophies, ground rules, and views. The employees are well informed about the company's business, are highly self-managed, and see their jobs in the light of the operation as a whole.*

The large defense contractor is, by comparison, mechanistic, compartmentalized, rigorously structured, and functions by way of an elaborate system of controls. The more critical the end use of the product and the more costly and complex the system produced, the greater the need for order, system, rigor of method, and formalization of policies. Because of these widespread differences, no model policy exists that can satisfy the needs of every organization. The example given in this section, therefore, is presented more for structure and style than for specific content.

The following example of corporate policy is for a hypothetical company engaged in the design, development, and production of systems and equipments for critical applications where failures can result in serious safety hazards to the public or extraordinary economic loss to the user.

Policy for Upholding Organizationwide Reliability.

   I. *Background*: Products of the kind produced by the _____ Corporation may, through loss of function or degradation in performance, present potentially serious hazards to public safety or incur extraordinary economic losses to users. The risk of these hazards is diminished by high product reliability, a result attributable not only to the efforts of product designers and other technical specialists who endow the product with an inherent level of

---

* Ibid., p. 649.

reliability but also to the many others throughout the organization who safeguard product reliability by maintaining strict vigilance over operations and processes in offices as well as on the production floor.

II. *Purpose*: The purpose of this policy is to establish a forward-looking attitude whereby individuals throughout the organization more readily anticipate the potential effects of their actions on the reliability of products as ultimately experienced in the hands of customers. The objective is to avert potential reliability problems in their formative stages and thereby statistically reduce the incidence of latent defects to the production process. Success in this area will reduce the burden on the production screening process and materially reduce the chances of undetected problems to pass on into fielded operational systems where the cost of failure can be exorbitant.

III. *Scope*: This policy applies to all direct and supporting organizational functions for all aspects of endeavor from product design through procurement of parts and materials, manufacturing, assembly, shipment, and installation or fielding.

IV. *Policy*: Emphasis is to be placed on forward-looking, anticipative, and preventive activities to reduce the potential for inducing latent reliability problems further "downstream" in the product realization cycle. In particular, attention is to be directed to the following situations:

A. Where economies at local levels are realizable, but the impact on the future reliability of the product is uncertain

B. Where departures or deviations of any kind are perceived as expedients to schedule maintenance or recovery

C. Where there appears to be a demand for decisions under uncertainty

It is firm policy that safety-related reliability is never tradable downward from the established product baselines. Success in achieving the objectives of this policy document requires that all organizational elements be at all times alert to situations that may lead to compromise of product reliability.

V. *Responsibilities:*

A. The Director of Program Management is responsible for ensuring that this policy is applied by all program managers for all programs under their administration.

B. The Director of Engineering is responsible for ensuring that this policy is carried out by all elements of system engineering, design engineering, project engineering, and test engineering throughout all program phases.

C. The Director of Operations is responsible for ensuring that this policy is carried out by the materials function, industrial engineering, manufacturing engineering, manufacturing operations, and all other functions within the department having the potential for affecting product reliability.

D. The Director of Product Assurance is responsible for the dissemination of this policy, for its annual review, and for making recommendations to the Chief Operating Officer concerning its continued relevance, application, and effectiveness.

E. The Director of Employee Relations is responsible for ensuring that all new employees hired for entry into any of the departments identified in parts A through D in the foregoing are made aware of the intent of this policy as a part of their familiarization with the mission, goals, operations, and priorities of the company.

```
_____          _____
Signature, chief operating officer,              Date
_____Corporation
```

2.4 ORGANIZATION AND INTERRELATIONSHIPS

This section deals with organizational functions and forms and their various interrelationships in the pursuit of reliability. A comprehensive list of reliability-related functions is first presented to assist the reader in exploring the many activities within an organization that in one way or another contribute to, or influence, the creation or the preservation of product reliability. By examining all functions in detail, it can be better appreciated how and to what extent they contribute to product reliability. This knowledge can then be put to use to evaluate the overall system and to develop reliability plans most suitable to the needs.

A variety of organizational forms is presented to show where the reliability function can fit into the overall organization and to show the various forms the reliability function itself can take. This will prove useful to those in need of establishing a new reliability function as well as to those interested in considering the possibilities of making structural changes or fine adjustments to an existing system.

This material also points out the importance of developing appropriate selection criteria for evaluating managerial candidates for the new reliability function and why certain personality traits are more important to a start-up operation than to an ongoing, well-established activity. A list of personal attributes is provided to assist in developing the needed personality profile.

Organizational names and individual titles are presented to provide a view of the range and variety of reliability names and titles currently employed by U.S. industry.

Finally, Sec. 2.5, Organizational Forms, provides an outline of how to assess any particular organizational situation and then how to organize a reliability function in response to needs.

2.4.1 Reliability-Related Functions

Customer-experienced reliability of items produced in a corporate environment results from the actions—or inactions—of all organizational elements contributing to or influencing product requirements, the creation of the product, and its state of being thereafter. These organizational elements are distributed throughout a company, being located in groups ranging all the way from marketing departments to shipping and field service departments. Some create reliability, whereas others preserve reliability or make improvements in it. Some impact on reliability

TABLE 2.1   Basic Reliability-Related Functions

1. Corporate planning	11. Design engineering
2. Corporate reliability	12. Components and materials engineering
3. Requirements definition	13. Supplier reliability assurance
4. Program planning and development	14. Reliability information
5. Program management	15. Reliability methods and standards
6. Functional administration	16. Reliability testing and evaluation
7. Technical operations control	17. Production reliability assurance
8. Reliability facilitation	18. Failure analysis and reporting
9. Reliability analysis and statistics	19. Field engineering
10. System engineering	20. Customer service

directly, whereas others influence reliability only indirectly. All, however, have the potential for impacting reliability more or less, in one way or another.

To assist those establishing a new reliability function or reevaluating an existing reliability system, a comprehensive list is presented of technical and managerial functions which in some way may relate to creating, preserving, or influencing reliability. The list is presented to provide the reader with a comprehensive view of reliability-related functions and has been presented in the most general of functional terms to avoid the suggestion of any specific organizational structure or allocation of responsibility.

Not every function listed will be appropriate to any given organization, considering the wide diversity of circumstances and organizational needs. Some companies have large specialized departments focusing in singular areas; others have functional responsibilities broadly distributed. Some are engaged in the relatively fluid business of consumer or industrial products; others are engaged in the more structured business of space vehicles, nuclear power, or defense contracting. Whatever the business or product line, the generic list of functions can serve as a guide to assist in identifying all functional areas directly or indirectly involved in product reliability. In addition, it can serve as a stimulus to thinking not only about how each function affects reliability but also about examining the extent to which interrelationships and interplay between functions contribute to, or detract from, reliability achievement. It is important to consider functions in aggregate as well as individually.

Table 2.1 provides a summary of basic reliability-related functions. The section following provides a brief description of each function, its impact on reliability, and typical activities and subfunctions associated with each.

2.4.2   Activities Associated with Reliability-Related Functions

The following general functions are briefly characterized and their impact on reliability is given to direct the reader's attention to the many organizational elements that contribute to, or in some way influence, product reliability.

1. Corporate planning

Function: Develops strategic plans; establishes broad objectives, policies, and guidelines; ensures that plans at all levels are supportive of organizational purposes and goals; develops initiatives for divisions or strategic business units and functions; provides counsel to, and coordination between, units in developing mutually reinforcing plans.

Impact on reliability: Indirect and long range. Policy, or lack of policy, can influence total organizational perceptions and attitudes toward reliability, quality, and safety. Initiatives from the highest levels of management can favorably influence total organizational performance toward reliability achievement.

a. Corporate goals
b. Corporate policy
c. Product warranty guidelines
d. Product regulations administration
e. Liability prevention programs
f. Quality and productivity improvement programs

2. Corporate reliability

Function: Provides top-level central reliability leadership, advice, and coordination. Develops guidelines and broad initiatives to promote reliability awareness and stimulate reliability achievement at all levels, including the corporation itself.

Impact on reliability: Indirect and long range.
a. Risk assessment guidelines
b. Warranty cost allocation guidelines
c. Product liability program interface
d. International product standards and regulations
e. Organizational development and effectiveness initiatives
f. Reliability system review
g. Cross-divisional reliability communication and coordination
h. Cross-divisional reliability workshops
i. Advanced-technology reliability initiatives
j. Good practices guidelines
k. Error-prevention programs
l. Corporate briefings
m. General counseling on reliability
n. Professional activities coordination

3. Requirements definition

Function: Determines and defines customer needs. For proprietary products sold on the open market, defines all product requirements directly from market needs or from studies of what the market can be persuaded to become interested in. For contracted products, studies customer's requirements and translates them into product-specific requirements.

Impact on reliability: Significant for proprietary products. For contracted products, no impact for accurately interpreted customer requirements, high impact for inaccurate interpretations.

a. Open-market and proprietary products
    (1) Marketing objectives and initiatives
        (a) Marketing plan
        (b) General product characterization
        (c) Price-quality-reliability deliberations

    (2) Product requirements study and definition
       (a) Feasibility studies
       (b) Function and features definition
       (c) Concept definition
       (d) Field use consideration
       (e) Field service objectives
       (f) Cost and warranty studies
       (g) Competitive evaluations
       (h) Warranty and life-cycle cost targets
       (i) Safety impact assessment
       (j) Key safety provisions
       (k) Life, reliability, and maintainability studies
       (l) Life and reliability targets
       (m) Development and production implications
       (n) Agency listing, rating, and approval implications
       (o) Trade-off studies and business and technical risk assessment
       (p) Preliminary reliability requirement
       (q) Baseline system and product specifications
       (r) Subcontractor requirements
       (s) Subcontractor proposal evaluation
    (3) Product plan
       (a) Product description
       (b) Statement of work
       (c) Cost and schedule requirements
       (d) Capital investment summary
  b. Contracted products
    (1) Solicitation study and evaluation
    (2) Proposal and bid responses
    (3) Subcontractor requirements
    (4) Subcontractor proposal evaluation
    (5) Contract negotiations
    (6) Technical requirements translation
    (7) Program requirements package
       (a) Statement of work
       (b) Delivery schedule and funding allocation

4. Program planning and development

Function: Defines program objectives, establishes cost, schedule, performance, and assurances program plans.

Impact on reliability: Moderate to high. Key consideration is the manner in which reliability is scheduled in with respect to the "mainstream" activities and the level of effort allocated to reliability activities.

  a. Requirements review
  b. Reliability history review
  c. Objectives definition
  d. Risk area identification
  e. Priorities development
  f. Work element definition and scheduling
    (1) Work breakdown structure
    (2) Work packages
    (3) Work element scheduling

    g.   Cost element definition and schedule
    h.   Program budget formation
    i.   Key or critical activities definition
    j.   Program organization
    k.   Program coordination provision
    l.   Milestone and program-control-point definition
    m. Program alternatives development
    n.   Task performance measurement criteria
    o.   Program monitoring and evaluation plan
    p.   Reliability growth monitoring plan
    q.   Risk management plan
    r.   Product assurance plan
       (1) Reliability plan
       (2) Quality plan
       (3) Safety plan

5. Program management

Function: Defines, implements, integrates, coordinates, monitors, evaluates, and controls programs.

Impact on reliability: Moderate to high, depending on degree of emphasis on reliability tasks and degree of support in pursuing reliability-related problems to their source, regardless of their origin.

    a.   Risk management provisions
    b.   Program priorities documentation
    c.   Task assignments and dissemination of requirements
    d.   Program integration and coordination
    e.   Work package performance tracking and measurement
    f.   Physical progress evaluation
    g.   Cost, schedule, and performance tracking and control
    h.   Test review board participation
    i.   Reliability growth monitoring
    j.   Product qualification board participation
    k.   Subcontractor and supplier reliability program integration
    l.   Subcontractor and supplier reliability program monitoring
    m. Reliability-critical items and activities status
    n.   Program reviews
    o.   Design reviews
    p.   Program resolution monitoring
    q.   Corrective actions summary reviews
    r.   Status meetings
    s.   Production readiness reviews
    t.   Customer coordination
    u.   Program change impact assessment and adjustment

6. Functional administration

Function: Authorizes, budgets, and administers functional department activities.

Impact on reliability: Moderate to high, depending mainly on budgetary allowances, personnel assignments, resource allocation, staffing levels, and decisions relating to design changes, product qualification, and shipment authorization.

    a. Contract negotiation
    b. Departmental goals and policies
    c. Workload analysis
    d. Workload leveling and personnel assignment
    e. Budget authorization, allocation, and control
    f. Capital expenditure authorization
    g. Work authorization
    h. Resource allocation
    i. Functional organization structuring
    j. Job design
    k. Position descriptions
    l. Personnel qualification standards
    m. Personnel performance standards and appraisal
    n. Workforce requirements forecasts
    o. Personnel requisition approval
    p. Technical data documentation review and approval
    q. Technical proposal review and approval
    r. Design change, waiver, and deviation approval
    s. Product qualification board participation
    t. Product design approval and release
    u. Conditional shipment authorization

7. Technical operations control

Function: Monitors, evaluates, and controls technical operations.

Impact on reliability: Moderate to high, mainly depending on depth and quality of design reviews, critical items control, critical activities control, corrective actions implementation, and decisions relating to design changes and product qualification.

    a. Design criteria review
    b. Design review
    c. Design and redesign effectiveness assessment
    d. Product qualification board participation
    e. Configuration control
    f. Design change impact assessment and compensations
    g. Design change control board participation
    h. Drawings and specifications review and control
    i. Purchase order review
    j. Critical items provisions and control
    k. Software controls review
    l. Test pass-fail criteria review
    m. Test plan approval
    n. Test procedure approval
    o. Reliability growth control
    p. Failure analysis report approval
    q. Failure review board participation
    r. Failed parts and failed hardware control
    s. Corrective action assignment
    t. Corrective action approval, verification, and closeout
    u. Material review board participation
    v. Information feedback network verification
    w. Customer-furnished equipment, material, and tools control

   x. Goal achievement evaluation
   y. Technical data review and approval
   z. Production readiness verification

8. Reliability facilitation

Function: Supports reliability achievement and advancement on a broad scale, basically at the departmental level, by activities that move the overall reliability effort forward on a continuing basis.

Impact on reliability: Mostly indirect and long range.
   a. Interorganization communication and coordination
       (1) Requirements communication
       (2) Reliability awareness programs
       (3) Reliability technology dissemination
       (4) Reliability guidance and consultation
   b. Reliability research and technology advancement
       (1) Improved design techniques
       (2) New or improved anlytical tools
       (3) Efficient test methods
       (4) Advanced statistical methods
   c. Reliability education and training
   d. Reliability certification
   e. Reliability workshops
   f. Reliability improvement studies
   g. Organizational effectiveness analysis and review
   h. Error-prevention programs
   i. Management briefings
   j. Computer utilization enhancements
   k. Professional activities initiatives

9. Reliability analysis and statistics

Function: Performs reliability allocations, predictions, estimates, assessments, and analyses, and applies statistical methods to tests, experiments, and engineering matters.

Impact on reliability: Moderate to high.
   a. Reliability allocation
   b. Reliability prediction
   c. Reliability growth assessment
   d. Reliability growth projection
   e. Reliability qualification test data analysis
   f. Production reliability test data analysis
   g. Field service contract data analysis
   h. Field reliability statistics
   i. Warranty data analysis
   j. Stockpile reliability data analysis
   k. Failure trend analysis
   l. Reliability history data analysis
   m. Cost-of-failure and errors analysis
   n. Cost impact analysis and alerts
   o. Statistical design of experiments
   p. Statistical test plans
   q. Statistical probability theory applications

10. System engineering

Function: Performs studies, analyses, assessments, and design activities at the system level.

Impact on reliability: High.
a. New business activities
b. Operational profile studies
c. Operational environment studies
d. System specification development
e. System design
f. System modeling
g. System effectiveness analyses
h. System reliability considerations and trade-off studies
i. System reliability assessment
j. Baseline system definition
k. Risk modeling
l. Probabilistic risk assessment
m. Risk narratives
n. System event tree analyses
o. System fault tree analyses
p. System failure mode, effects, and criticality analyses
q. Human factors reliability analyses
r. Life-cycle cost studies
s. Design review participation
t. Failure review board participation
u. Integrated logistics support coordination

11. Design engineering

Function: Designs and develops hardware and software. Performs studies analyses, definition, design synthesis, and the preparations required to qualify equipment and component designs for release to production.

Impact on reliability: High.
a. Design specifications development
b. Design criteria development
c. Design synthesis and alternatives assessment
d. Reliability modeling
e. Reliability estimation
f. Fault tree analyses
g. Failure mode, effects, and criticality analyses
h. Tolerance studies
i. Derating and safety margins
j. Sneak circuit analyses
k. Stress analyses
l. Wear and deterioration analyses
m. Reliability–critical item identification and definition
n. Design criticality reduction
o. Design review
p. Failure review board participation
q. Problem identification, definition, and reporting
r. Effects of functional testing, rework, repair, and maintenance
s. Effects of packaging, packing, handling, shipping, and storage
t. Heat management

    u. Fail-safe feature development
    v. Field test and maintainability provisions
    w. Test-analyze-and-fix iterations
    x. Reliability growth initiatives
    y. Software reliability assurance
    z. Design documentation
    aa. Design producibility assurance
    ab. Production readiness implementations

12. Components and materials engineering

Function: Supports the design engineering function by recommending parts and materials, providing application and derating advice, performing component tests and component failure analysis, and interfacing with suppliers.

Impact on reliability: High.
    a. Parts and materials application guidance
    b. Parts and materials specification
    c. Parts and materials selection criteria
    d. Parts derating criteria
    e. Parts availability
    f. Parts qualification
    g. Parts standardization
    h. Parts and materials reliability testing
    i. Parts and materials failure analysis
    j. Parts screening recommendations
    k. Nonstandard parts control
    l. Critical parts and materials identification and control
    m. Design review participation

13. Supplier reliability assurance

Function: Attends to all activities relating to the reliability performance of suppliers and subcontractors and provides for the assurance of reliable incoming parts and materials.

Impact on reliability: High.
    a. Supplier product specifications
    b. Source control drawings
    c. Acceptance test criteria
    d. Acceptance test procedures
    e. Purchase order reliability requirements
    f. Supplier capability surveys
    g. Supplier qualification and selection
    h. Supplier guidance and development
    i. Supplier reliability requirements
    j. Supplier performance monitoring and rating
    k. Supplier problem coordination and resolution
    l. Supplier audits
    m. Supplier product certifications
    n. On-site representation

14. Reliability information

Function: Central repository for reliability data and information. Acquires, processes, stores, and distributes reliability information.

Impact on reliability: Moderate to high, depending on the quality of information collected and disseminated.
  a. Reliability data acquisition and processing
  b. Data reconciliation and validation
  c. Product information data bank
  d. Design activities data bank
  e. Manufacturing activities data bank
  f. Failure rate data bank
  g. Failure experience data base
  h. Reliability data exchange
      (1) GIDEP participation
      (2) Industry coordination
  i. Management reporting

15. Reliability methods and standards

Function: Establishes and disseminates reliability standards, practices, procedures, and related documents.

Impact on reliability: Indirect and long range. Impact can be significant over a lengthy period if standards and practices are not kept relevant and fall into disuse.
  a. Reliability standards
  b. Reliability evaluation procedures
  c. Design reliability guides
  d. Design review checklists
  e. Reliability policy and procedures
  f. Design practice standardization
  g. Reliability information notebook
  h. Supplier and subcontractor reliability guide
  i. Reliability system review procedures

16. Reliability testing and evaluation

Function: Measures and evaluates product reliability level.

Impact on reliability: High.
  a. Integrated test, inspection, and evaluation plans
  b. Reliability test and inspection specifications
  c. Test ground rules and procedures
  d. Pass-fail criteria
  e. Failure detection and localization
  f. Design margin testing
  g. Operational life testing
  h. Environmental development testing
  i. Reliability development growth testing
  j. Environmental stress screening
  k. Reliability qualification testing
  l. Production reliability acceptance testing
  m. Field testing
  n. Test data evaluation
  o. Software reliability verification testing
  p. Test records maintenance
  q. Test reporting
  r. Failure review board participation

17. Production reliability assurance

    Function: Monitors, evaluates, and attends to all aspects of the production system that can degrade reliability. Conducts production reliability tests. Provides for corrective measures and explores opportunities to identify areas that will improve product reliability.

    Impact on reliability: Moderate to high.
    a. Producibility criteria
    b. Producibility analysis and validation
    c. Design review participation
    d. Production readiness implementations
    e. Inspection points identification
    f. Inspection accept-reject criteria
    g. Incoming material reliability assurance
    h. Incoming inspection lot control
    i. Reliability qualification test monitoring
    j. Production reliability acceptance testing
    k. Environmental stress screening trials, plans, and procedures
    l. Environmental stress screening monitoring and reporting
    m. Failure reporting and corrective actions
    n. Failure recurrence control
    o. Perishable material date checks
    p. Failed parts and materials control
    q. Acceptance gauging review
    r. Tooling review
    s. Critical process and items identification and control
    t. Work instruction review
    u. Manufacturing process instruction evaluation
    v. Manufacturing process testing and certification
    w. Manufacturing process control
    x. Production software testing and validation
    y. Production physical environment control
    z. Production operations monitoring and evaluation
    aa. Production lot control
    ab. Packaging, packing, handling, shipping, and storage review
    ac. Problem identification, investigation, and reporting
    ad. Failure review board participation
    ae. Rework practices review
    af. Reliability improvement studies and proposals
    ag. Design change, waiver, and deviation review
    ah. Design change control board participation

18. Failure analysis and reporting

    Function: Locates, identifies, and isolates specific faults and deficiencies in products, parts, and materials and determines the basic underlying cause.

    Impact on reliability: Moderate to high.
    a. Failure analysis
    b. Failure analysis and corrective actions reporting
    c. Failure recurrence reporting
    d. Failure data retention
    e. Failure review board participation
    f. Part physics-of-failure determination

   g. Part physics-of-failure research and studies
   h. Failure analysis technique development
   i. Failed parts and equipment control

19. Field engineering

Function: Conducts field operations, investigations, evaluations, and reviews and provides data and information feedback.

Impact on reliability: High.

   a. Field use environment assessment and reporting
   b. Field performance monitoring, evaluation, and reporting
   c. Field trouble investigation
   d. Field incident reports and failure notices
   e. Failed hardware review
   f. Field failure analysis and reporting
   g. Field repair data feedback
   h. Field service reports and customer information feedback
   i. Warranty repairs and returns data feedback
   j. Stockpile reliability evaluation
   k. Corrective actions recommendations
   l. Design improvement recommendations
   m. Failure review board participation
   n. Field retrofit monitoring

20. Customer service

Function: Attends to all activities relating to product reliability after the customer takes possession of the product. Provides data and information feedback.

Impact on reliability: Moderate to high.

   a. Field failure report verification
   b. Warranty failure investigation and analysis
   c. Warranty failure notices
   d. Replacement or repair actions
   e. Warranty and maintenance contract service
   f. Customer communications and liaison
   g. Marketing and sales support

2.5 ORGANIZATIONAL FORMS

Although there are only a few basic organizational forms from which all others spring, industries are structured in a great many ways. Similar industries employing virtually identical processes and competing in the same market sector may be organized along very different lines at the working unit level. In fact, some corporations have companion divisions with organizational structures unique to themselves.

There are more similarities at the higher corporate levels. Very large companies commonly make their first partitions along major business areas, splitting the organization first into market sectors or groups. From there, a typical company might be organized in divisions, departments, operations, centers, branches, offices, sections, subsections, laboratories, and units. The greater organizational diversity is observed at these levels. These also are the levels where reliability professionals function and it is the reason why so many different reliability forms exist.

To assist the reader in gaining an appreciation for organizational form, several basic forms are first discussed, then a number of specific forms are presented to show the various possibilities. Several "macro" views show total organizations to indicate where the reliability function, or functions, can fit. A series of "micro" views then show the various ways the reliability function itself can be organized.

2.5.1 Basic Organizational Forms

In the general context, organizations can be partitioned along the following lines:

products   programs   processes

customers   projects   geography

functions   technologies

For instance, a functional organization might consist of marketing, engineering, production, quality assurance, and personnel. Each department head would be on the same level and would report directly to the chief operating officer.

A product-oriented structure would have all product managers reporting directly to the chief operating officer. Each product group would be self-sufficient, with its own functional subgroups such as marketing, engineering, etc. Similarly, a customer-oriented structure would be partitioned along customer lines. A defense contractor, for example, might have a major group each for Navy, Army, Air Force, or NASA programs.

Program or project-type structures would be arranged similar to the product-type structure and the customer-related structures in that each would have its own functional subgroups and each would report directly to the chief operating officer. Partitions for technologies, processes, or locations usually are made in conjunction with the others.

While it is perfectly feasible for some firms to organize along these basic lines, it is far more often the practice to employ them in combination. One of the most widely used combining forms is the matrix, wherein all functional department heads as well as the program or project manager report to the chief operating officer. In this manner of functioning, the program or project manager negotiates for needed staff from a pool of professionals in each functional department. Individuals are assigned to projects by their functional manager, yet remain under the direction of their functional managers for policy and administration. While the assigned individuals actually report to two superiors, the potential for conflict is obviated by the fact that the functional manager decides the "who and how" of matters while the program or project manager decides on the "what and when" of things.

A key feature of the matrix arrangement, very common in medium to large organizations, is that people are readily moved in and out of programs as needs in the different phases of a program change. This not only provides high flexibility and adaptability, but the increased mobility also facilitates the interchange of lessons learned from one group to another.

Within each of the foregoing general structures, the organization may be defined along the lines of general functions such as:

policy   support

line   service

staff

Policy functions are directive and managerial. Line functions are primary and contribute directly to creating and marketing the products or services offered by the organization. Staff functions are advisory and supportive and contribute indirectly to major organizational objectives. Support functions are specialized adjuncts to line functions, i.e., components engineering and logistics support. Service functions are general and facilitative, i.e., drafting, document reproductions, and computer services.

2.5.2 Specific Organizational Forms

Figures 2.4 and 2.5 illustrate product assurance organizational forms. Figure 2.4 is representative of a form suitable to medium and large companies and features product assurance directors both at the corporate and divisional levels. The corporate position establishes companywide policy and ensures the maintenance and effectiveness of the product assurance system as it is exercised by each division. In addition to being author and keeper of the product assurance policy manual, the corporate office monitors divisional product assurance performance and conducts annual system reviews.

At the divisional level, the director of product assurance reports directly to the divisional general manager and oversees a broadly based assurance function which, in addition to reliability and maintainability, includes assurances for quality, safety, and producibility. The divisional function is partitioned along three main lines of endeavor: design assurance, manufacturing assurance, and external assurance. The latter has virtual stewardship over customer affairs as they pertain to satisfaction with delivered products and services rendered.

Figure 2.5, somewhat a miniature version of Fig. 2.4, is more suitable to the smaller manufacturer. The product assurance manager reports directly to the general manager as an equal with other top managers such as engineering, manufacturing, marketing, and others similar in function to those found in Fig. 2.4. This function also is broadly based, but because it is smaller it does not differentiate functions as finely as the larger one does. Nevertheless, this smaller organization encompasses all of the key features of the larger configuration.

The organization of Fig. 2.6 is closely related to those shown in Figs. 2.4 and 2.5. Again, the topmost assurance executive is at the corporate level, but in this case the individual is responsible for productivity as well as quality. The significant difference between this form and that of Fig. 2.4 is that this one operates under the identity of quality assurance. The choice of titles is not a matter of "correctness" or effectiveness, either way. One can be as effective as the other. Barring the need for sweeping change, or some other justification, retaining familiar titles may have some virtue. In the case of Fig. 2.6, it may be that the company has a strong tradition in quality and that it practices what may be recognized as "total" quality. The choice of organizational form and name is not only a function of theoretical efficiency but also a function of the people in the company, their perceptions of what quality might or might not mean, and whether or not there is a need to change perceptions and ways of doing things.

Figure 2.7 is "neutral" insofar as being better suited to smaller or larger companies. In this case, the reliability function reports to the support services

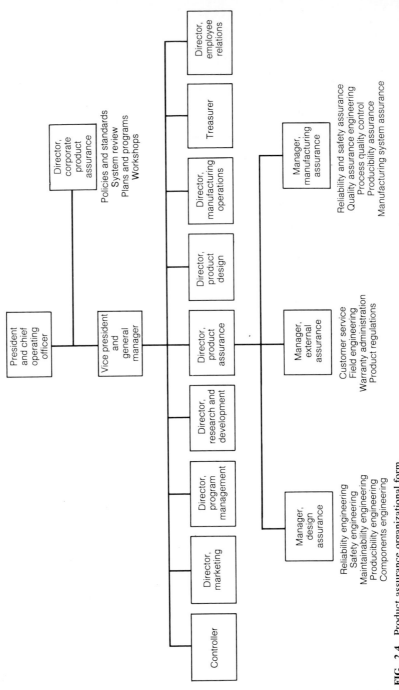

FIG. 2.4 Product assurance organizational form.

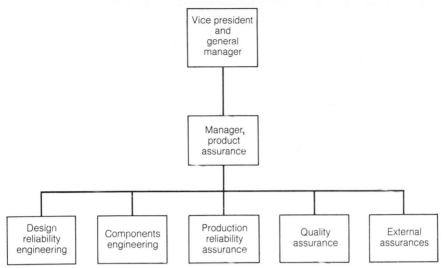

FIG. 2.5   Small group product assurance organizational form.

manager who may be one or two levels from the general manager. This form is entirely suitable for the company that does not have as pressing a demand for the reliability function as, for example, those structured according to Figs. 2.4 to 2.6. In this case, it may be that the products are simple and have sufficient reliability as a natural consequence of materials that have been commonly available for some years and straightforward manufacturing processes requiring little effort to sustain. On the other hand, it may be that the products in fact are quite complex and sophisticated, but the organization is very mature and is staffed with professionals of a very high caliber who utilize reliability principles and practice the reliability discipline on an everyday basis. This is more apt to be the case with companies engaged in designing and producing equipment and systems for critical applications where human lives are at stake or where failures can result in extraordinary loss to users or to society at large. In such cases, the reliability function is more a convenience than a necessity.

Figure 2.8 shows the project organizational form where reliability engineers are assigned to a specific project but take functional direction from their immediate supervisor in their home department. This is the matrix organizational form described under "Basic Organizational Forms." The main advantage of this form is the flexibility it offers, especially to the larger organizations.

Figure 2.9 presents a line-staff form wherein the functional departments have a reliability engineer permanently assigned and policy and guidance are provided by the reliability manager, with day-to-day coordination provided by a reliability coordinator. The coordinator assists by disseminating reliability requirements at the onset of a job, holds regular status meetings with the reliability engineers, provides reliability awareness services, and acts as an overall facilitator. This type of organization is for the smaller companies that might not require the strength inherent in other reliability structures. For this reason, it is important to have a reliability council composed of key upper-level managers to provide a stabilizing influence where differences in opinions may cause stalemates.

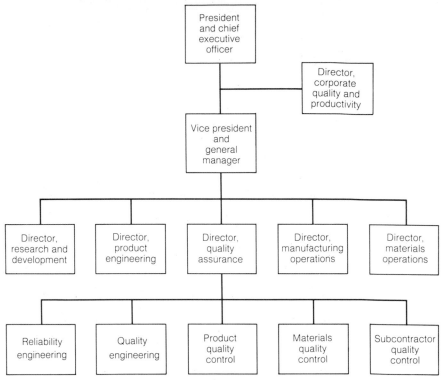

FIG. 2.6 Quality assurance organizational form.

Figure 2.10 represents another approach to reliability achievement when a strong reliability group is not in prospect. This may be the case for companies that do not require high reliability but need a continued reliability presence to ensure the level of achievement desired. Reliability delegates (or representatives) are the prime feature of such an organization. These are not reliability professionals, but key individuals in each significant functional group. These individuals are key in the sense that they are intimately acquainted with the people, the processes, the products, the activities, and the achievements of the functional group they are a normal part of. With briefings, guidance, and support provided by the reliability coordinator, reliability delegates carry out reliability efforts directly within their own home group. They have the advantage of being on the inside and therefore in possession of expert knowledge of the specialty function and the products of their group.

By first providing a reliability goal for each group having a reliability delegate, the delegate then becomes the focal point and the catalyst for helping the group succeed in achieving its goal. For such an approach to work, there must be universal acceptance, particularly by the functional managers. To obtain this acceptance there must be strong support from senior management from the outset. As with the line-staff organization of Fig. 2.9, the reliability coordinator disseminates reliability requirements, holds regular progress meetings, arranges

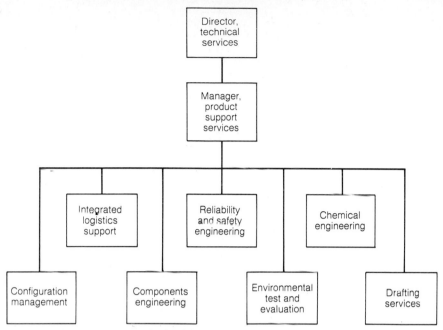

FIG. 2.7 Support group organizational form.

for necessary training and education, and acts as an overall facilitator. A core
group of reliability engineers within the reliability group itself provides specialist
services to the functional groups as needed. A reliability council, responsible to
the chief operating officer, is composed of top-level managers from the operating

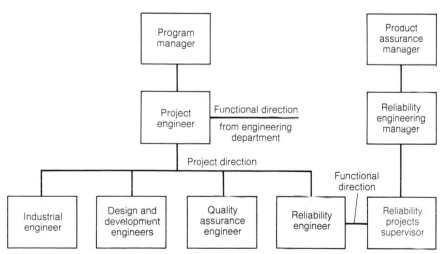

FIG. 2.8 Project organizational form.

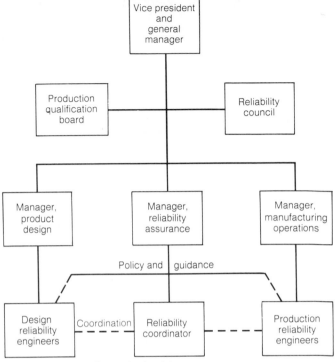

FIG. 2.9 Line-staff organizational form.

groups and the director of reliability assurance and provides general policy and resolves any matters not resolvable at the operating levels.

Figure 2.11 portrays an organization that does not have a separate reliability function but rather has reliability engineers directly in the key functional groups. Reliability engineering is included in the system engineering section as are other specialty functions such as safety engineering and maintainability engineering. Reliability assurance in production is provided by reliability persons within thequality assurance department. This form of organization is typical of a practice quite common in the past whereby quality assurance deals only with production and the quality performance of the design group and others is their own responsibility. As with most any type of organization, what will work in one organization may not in another. The mindset of the participants can have everything to do with the success or failure of the undertaking.

Another small firm organization is given in Fig. 2.12. This form also is typical of the past practice of many companies. Reliability is performed in the engineering department and the quality control department attends to the production effort. This is quite acceptable for many small firms that do not have any special reliability requirements in production, such as reliability acceptance tests, electronics burn-in preconditioning, or environmental stress screening. Many electronics firms today, however, find it necessary to assign reliability engineers to active parts in the production process.

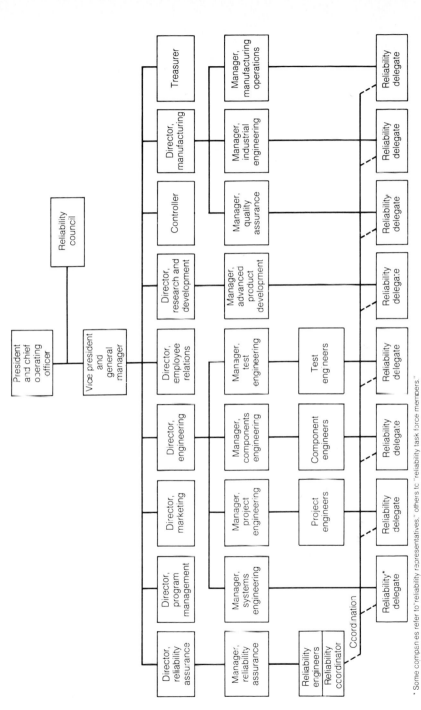

* Some companies refer to "reliability representatives," others to "reliability task force members."

FIG. 2.10 Distributed organizational form.

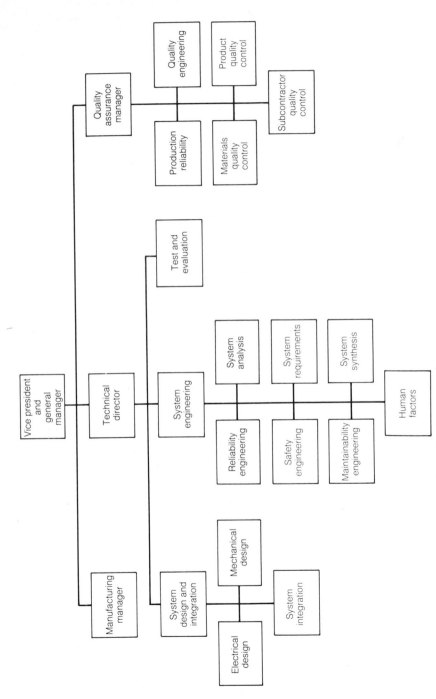

FIG. 2.11 The reliable function in engineering and quality assurance.

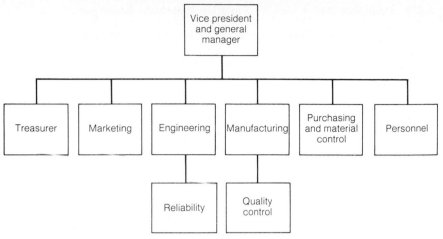

FIG. 2.12 Typical small company organization.

2.6 FIRST MANAGER FOR THE NEW RELIABILITY FUNCTION

Selection criteria for the first manager of the new reliability function differs in a number of respects from those for a stable, well-established activity. The start-up manager should be somewhat of a promoter as well as a manager. A dynamic, entrepreneurial person in this key position at start-up will materially add to the viability and success of the new organization. In a later succession it will be possible to assign a lower key, but competent, individual with a record of steady accomplishment. In any case, the person initially appointed must have the capabilities and management style to match the prevailing conditions and situations and the emphasis at this critical stage of organizational life must be on leadership rather than administration.

To facilitate screening potential candidates, a profile of the "ideal" manager should be developed to accompany the list of otherwise essential technical and managerial qualifications. The profile should include the personality traits senior management desires and approves of for use in creating a guide for screening. Following is a resource list* of personal attributes to assist in developing such a profile. The applicability of any particular attribute in kind or measure will depend entirely upon the particular circumstances. The order of presentation is not necessarily in order of priority and it should not be expected that all the attributes will be required in any specific case.

---

*Several of the attributes listed were developed as a result of reviewing "Personality Profiles" from *Reliability Review*, December 1982 through June 1985.

2.6.1 Desirable Attributes for Start-Up Reliability Managers

1. Strong character. High level of personal integrity and emotional maturity.
2. High achiever. Dynamic, enthusiastic, self-confident.
3. Articulate and persuasive. Clear and forthright communicator. Able to deal with individuals at all organizational levels in a variety of situations.
4. Strong leader. Capable of inspiring associates and taking initiative.
5. Decisive and responsive. Sense of urgency and importance.
6. Team oriented. Respects higher organizational goals. Booster of individuals. Encourages "bottom-up" initiatives. Fosters team spirit.
7. Systems oriented. Comprehends broad issues and sees details in proper context.
8. Efficient and effective. Gets job done to professional standards, within budget, and on time.
9. Diplomatic. Strives for "win-win" resolution of conflict, but exercises candor and humor appropriate to circumstances.
10. Authoritative. Recognized and respected by others. Capable of gaining confidence and winning cooperation of other groups and senior management.
11. Sensitive and attentive. Good listener. Highly approachable.
12. Perceptive and discriminating. Sees beyond appearances to basic underlying issues and root causes of people problems.
13. Ready and available. Takes on demanding jobs without hesitation.
14. Purposeful and steadfast. Resolutely pursues goals and objectives. Maintains sustained effort.
15. Professionally active. Strongly motivated to advance the profession and assist others in their professional development.

CHAPTER 3
ECONOMICS OF RELIABILITY

Henry J. Kohoutek
Product Assurance Manager, Hewlett-Packard Co.

3.1 INTRODUCTION

The practice of engineering involves selecting alternative designs, procedures, plans, and methods that consider time and economy restrictions in their implementation. The satisfaction of the engineer's sense of perfection does not necessarily assure the best alternative. Preferred rational approaches to the selection of alternatives are based on methods of engineering economy that are concerned only with alternatives that already have been established as technically feasible and providing economic analysis of their prospective differences. The word "prospective" indicates that the analysis looks into the future and cannot be viewed as an absolute prediction because of the always-present element of uncertainty associated with time, uncontrolled variations in the value of parameters considered, and imperfections in cost estimation methods. The need for engineering economy comes from the fact that engineers and managers do not work in an economic vacuum but are under strong impact from the processes that managers use to allocate limited resources to produce and distribute various products. Because engineers are basically trained as physical scientists, it is necessary repetitively to stress that economics, which studies these processes, is an empirical social science that provides us with the conceptual framework known as a *theory of choice*. Positive economics, as opposed to normative economics, which is outside our interest, is concerned with questions of facts to which assumptions relating theoretical constructs to real objects are added. Economic models then represent the purely logical aspects of these theories and serve as tools for decision making, behavior or consequence predictions, and testing or refutation of underlying assumptions and propositions based upon economic rather than physical consequences.

The managerial responsibility for the quality of decisions is usually discharged within a framework of

- Clear definition of alternatives
- Identification of aspects common to all alternatives, which then become irrelevant
- Establishing appropriate viewpoints and decision criteria
- Considering consequences and their commeasurability

Given the frequent conflicts of requirements, the different relative position of quality and reliability, as opposed to product performance, project schedule, cost, and their potential impact on business success, the economic aspects of reliability alternatives with their consequences should be studied in detail and well understood. Considering the power and availability of current methods of economic analysis, the principal difficulty in accomplishing this task is in the low availability of valid and detailed reliability cost data and lack of good reliability models. We just do not know well enough the reliability dependencies on design effort, testing, quality control, application environments, etc. Current trends in all major manufacturing industries, emphasizing the importance of quality and its associated attributes, present a significant driving force to alleviate this problem.

A word of caution is in order. Excessive use of reliability cost models can be a pedantic and fruitless exercise. Modeling works best in steady-state conditions, when the state-of-the-art is not being changed. The less quantifiable aspects of industrial processes, e.g., workers' attitudes, levels of standardization, sales practices, are usually left out of the models altogether. And we often fail to recognize that results of our logical arguments, applicable to models, might be irrelevant to the real situation. Experience and common sense help in deciding where to put the effort and how to set realistic expectations.

3.2　RELIABILITY AND VALUE

3.2.1　Introduction

The task of economic science is to devise optimal ways of allocating scarce resources among competing proposals for their use. One of the schemes studied many times in detail, and historically proven as effective in free markets, is based on the price of a given product or service. The concept of price can represent buyers' and sellers' convictions, willingness to pay, personal preferences, efforts to obtain favors, etc., but fundamentally, it refers to the value of the item concerned and its relationships to market situation (described in terms of supply and demand) and to manufacturing cost.

3.2.2　Relative Importance of Reliability, Price, and Performance

The wealth of data about customer behavior, values, beliefs, and attitudes often confirms reliability as the most important product quality attribute, and by that, its impact on the value in exchange, expressed by price, and value in use, expressed, for example, in terms of users' return on investment. Figure 3.1 illustrates two extreme cases. The decisions in the Apollo Space Program were dominated by reliability considerations (weight 0.7 on scale of 1.0) because of the severity of adverse consequences of a failure. Most of the purchase decisions in the area of consumer electronics consider product price much more important than reliability because of small impact of failures, protection by a warranty period, seller's goodwill, etc. The knowledge of relative position of reliability versus other product characteristics, expressed in quantitative weight factors, is important for both formal and practical considerations and can significantly improve the rationality of many decision processes.

3.2.3　Reliability as a Capital Investment

Considering the relationship of reliability to value and its impact on price, product reliability can be analyzed for its attractiveness as a capital investment. A

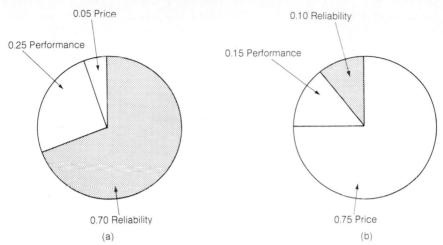

FIG. 3.1 Examples of the relative importance of price, performance, and reliability. (*a*) Space program; (*b*) consumer electronics.

convenient measure of this attractiveness is *return on investment* (ROI). Other measures of capital investment effectiveness can be easily derived from ROI.

The ROI without consideration of the time value of money is an approximation that assumes indefinite life of the project, and the payback method ignores the desired profit or "interest" on the investment. These are approximations that can be used for "quick-and-dirty" analysis of proposed investments in reliability programs to prevent or reduce the number of costly failures of the products. It is highly recommended that the more accurate methods involving the use of present worth or equivalent uniform annual payments and returns be used as described in Grant, Ireson, and Leavenworth cited in the bibliography of this section. Where the required rate of return on the investment is high (to compensate for the uncertainties of the future), these correct methods, as illustrated in the example below, can make great differences in the ROI and the payback period.

- *Payback period* is simply the reciprocal of the simple ROI and is a quick test that, however, ignores total cash flow over time and the time value of money.
- *Benefit/cost ratio* is the ROI multiplied by the expected years of useful life.
- The *net return* is calculated by multiplying the benefit/cost ratio by the investment cost, in our case the cost of a reliability improvement program, and subtracting from it again the investment cost.

Benefit/cost ratio and the net return can be discounted to reflect the time value of money, by multiplying the ROI by the present worth factor, which is found in standard interest tables.

Example. A reliability improvement program under consideration has an expected cost C_R of $50,000 and will avoid annually $N = 250$ failures with average repair cost C_{rep} of $850 each for expected useful life of $L = 8$ years. Let us express the effectiveness of this investment in different measures, considering general and administrative overhead OH $= 30\%$, profit coefficient $P = 10\%$, and 10% discounting schedule defining present worth factor for 8 years of service $F_8 = 5.335$.

Return on investment

$$\text{ROI} = \frac{NC_{\text{rep}}}{C_R(1 + \text{OH})(1 + P)} = \frac{250 \times 850}{50,000(1 + 0.30)(1 + 0.10)} = 2.97$$

Payback period

$$\text{PP} = \frac{1}{\text{ROI}} = \frac{1}{2.97} = 0.336 \text{ years}$$

Benefit/cost ratio

$$\text{B/C} = \text{ROI} \times L = 2.97 \times 8 = 23.76$$

Discounted benefit/cost ratio

$$\text{DB/C} = \text{ROI} \times F_8 = 2.97 \times 5.335 = 15.84$$

Net return

$$\text{NR} = (\text{B/C} \times C_R) - C_R = (23.76 \times \$50,000) - \$50,000 = \$1,138,000$$

Discounted net return

$$\text{DNR} = (\text{DB/C} \times C_R) - C_R = (15.84 \times \$50,000) - \$50,000 = \$742,000$$

All the terms used can be expanded to account for particulars of a given reliability program. With the aid of a computer, additional investigation is possible to assess sensitivities, impact of uncertainties, or compare alternative programs. Similar studies can be made to assess the economic impact of unreliability on product value via performance and availability degradation or increased cost of ownership.

3.2.4 Reliability Impact on Product Positioning

The communicated and perceived reliability levels have a strong impact on the total perceived value of the product and acceptability of its price. The clarity of the communication about product reliability depends objectively on the published data and subjectively on the position of the product reliability in the mind of a prospective buyer. If the market and competitive analyses confirm the importance of reliability in the customer's set of needs, then reliability can form a base for product positioning and can increase its perceived value by taking advantage of some long-term product-independent issues, such as previous product reliability, company image, competitive product weaknesses, and existing goodwill. Successful positioning allows for higher-profit pricing strategy and for more effective advertising.

3.3 RELIABILITY AND COST

3.3.1 Introduction

The study of product reliability impact on cost usually runs into problems of technical nature: cost estimation, lack of data, analysis complexity, optimization methods sensitivity, and so on. A wide variety of economic tools exists to help in solving problems of reliability cost modeling.

3.3.2 Manufacturers' Viewpoint: Cost of Reliability

The concept of *cost of reliability* (COR), the total cost a manufacturer incurs during the design, manufacture, and warranty period of a product of a given reliability, can be developed around the generally accepted notion of *cost of quality* (COQ). The principles of COQ, established in the 1950s, have been verified and found valid in all segments of the manufacturing industry. COQ is applied to measure economic state of quality, to identify opportunities for quality improvement, to verify effectiveness, and to document impact of quality improvement programs. The accounting for COQ tries to identify all the cost items associated with defects in products and processes and then set them in contrast with the cost of doing and staying in business. The classic categorization of COQ applied to COR expresses as unique the costs associated with the following elements:

- *External failure*:   Cost of unreliability during the warranty period, cost of spare parts inventories, cost of failure analysis, etc.
- *Internal failure*:   Yield losses caused by reliability screens and tests, cost of failure-caused manufacturing equipment downtime, cost of redesign for reliability, etc.
- *Reliability appraisal*:   Life testing, environmental ruggedness evaluation, abuse testing, failure data reporting and analysis, reliability modeling, etc.
- *Prevention*:   Design for reliability, reliability standards and guidelines development, customer requirements research, product qualification, design reviews, reliability training, fault-tree analysis, failure modes, effects, and criticality analysis, etc.

Understanding these cost categories is a must for rational planning of reliability assurance resources, environmental and life-testing facilities, training programs, warranty policies, and other services needed for successful reliability programs.

For COR management, the cost information must be further restructured by products, process segments, or departments to identify major contributors and by that, opportunities for improvement. The identified cost levels are usually compared with some measure of revenue or value added to form ratios as management indexes. From the viewpoint of cost management, it is also important to recognize that prevention and appraisal costs are controllable by planning and budgetary mechanisms. Both types of failure cost are, on the other hand, expected or actual results of our inabilities to assure defect-free design, manufacturing, and distribution processes, and to control users' application conditions and environment.

There are also some dangers and pitfalls associated with the use of both COQ and COR. Managers often forget that COQ and COR are dependent variables reflecting successes and failures of the quality and reliability programs. By doing so, they run the risk of degrading reliability levels to achieve short-term cost savings. Other possible dangers are caused by the often difficult identification of defect cause, conflicts coming from cost charges transfer rules, preoccupation with reporting systems, tendencies toward perfectionism, and so on. To prevent these and other pitfalls, managerial prudence and knowledge are required in search for facts, understanding, realistic objectives, priorities with rational execution plans, and progress monitors.

The total COR can be developed similarly as a tool for managing cost and resources associated with a design for reliability, reliability manufacture, and warranty cost reflecting residual unreliability. Because of the different slopes of individual cost curves as functions of increased reliability, the applicability of the concept of cost optimum is self-evident.

In some industries, e.g., semiconductor industry, enough experience-based information has been accumulated to rationally model dependencies of the final product reliability on the number of redesign cycles, levels of screening and testing or, in general, on the level of reliability assurance. This information confirms the intuitive expectation that increase in planned and controlled reliability assurance cost and activities significantly reduces unplanned failure cost. These models, combined with physical failure rate models and warranty cost calculations, allow optimization of the cost of reliability, and by that, allow rational planning of the project, manufacturing, and support resources very early in the design phase [1, 2].

Development of optimal cost vs. reliability strategy starts with a simple formula for total cost of reliability:

$$COR_{total} = CRD + CRM + WC$$

where CRD is the cost of reliability design, modeled by an empirical relationship f_1 among the cost and number of design for reliability cycles, application environment stresses, complexity, and expected general level of quality expressed, e.g., via quality coefficient π_Q. This coefficient, regularly used in MIL-HDBK-217, *Reliability Prediction of Electronic Equipment*, which contains reliability models, reflects the relative impact of reliability program actions on the base reliability defined by the physical nature of components used.

$$CRD = f_1(\pi_Q)$$

The *cost of reliability of manufacturing* (CRM) is a function f_2 of the effectiveness of manufacturing screens, expressed again by the values of the coefficient π_Q and associated fixed and variable costs:

$$CRM = f_2(\pi_Q)$$

The *warranty cost* (WC) is a function of the initial failure rate but also takes into account the effect of the bathtub curve, learning curve, and, of course, repair cost.

These individual expressions allow us to study the total cost of reliability

$$COR_{total} = f_1(\pi_Q) + f_2(\pi_Q) + WC - F(\pi_Q)$$

as a function of the π_Q, and allow us to search for an optimum (see Fig. 3.2) under the conditions of different warranty periods, learning factors, inflation rates, shapes of bathtub curves, etc. The optimum conditions found must be then translated into resources levels and explained in technical terms of reliability engineering. Strategy for reliability, based on minimum total cost and its optimum apportioning among design, manufacturing, and warranty, is generally acceptable to all managers involved.

3.3.3 Users' Viewpoint: Life-Cycle Costing

The birth of the life-cycle costing (LCC) method is traceable to investigations by the Logistics Management Institute for the Assistant Secretary of Defense in the early 1960s. Primary concerns then were the consequences of changing vendors as

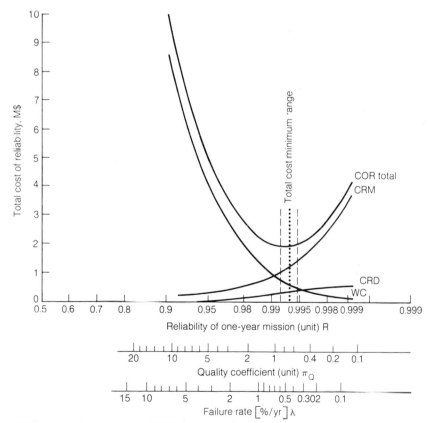

FIG. 3.2 Optimum set of reliability cost curves. (*Adapted from* [2]. *Reproduced with permission.*)

a result of lower bid prices. Since that time, LCC has evolved into a costing discipline, a procurement technique, an acquisition consideration, and a design trade-off tool. LCC requires the identification of all potential system costs through all the phases of the product life cycle: conceptual, development, manufacturing, installation, operation, support, and retirement, which is obviously a very difficult task. During its history, LCC also often tended to degrade into a method for accumulating and reporting cost, but formalization and computerization of models used are changing LCC into an integrated part of the decision support analysis based on relevant, user-oriented concepts of utility and cost.

In application of the LCC methodology, the design trade-offs usually start with attempts to balance acquisition cost with cost of ownership for maximum system capability and affordable total cost. During the planning phase, these trade-offs will propagate through all system levels down to component selection for hardware implementation. This costing and trade-off process, as shown in Fig. 3.3, which illustrates its generalized structure, requires a very close customer-vendor interface.

The customer has to find balance between the contemplated mission requirements and the budgeted resources, while taking into account the internal constraints in available:

- Internal acquisition logistics, reflecting, e.g., status of incoming inspection and testing facilities, installation opportunity, asset management system and procedures
- Support system, which will impact the actual application environment stresses, maintenance strategy, etc.

The vendor, who wants to satisfy the proposed mission requirements, must perform many analyses to gain insight into available implementation alternatives to find the one which guarantees minimum cost. Major constraints for this activity depend on available:

- Tools, skills, information, and resources for accurate analysis
- Creative design implementation alternatives
- Cost-estimating skills and completeness of historical cost data base

This process of finding optimum trade-off for both customer and vendor, internally and between them, must be repeated at different levels of detail throughout the whole acquisition phase. If the vendor is an equipment manufacturer or system integrator, the LCC process is also repeated and refined for each new phase of the product life.

- During the conceptual phase, usually the cost of only one of the proposed alternatives is developed using parametric cost estimating relationships. Cost of other alternatives is calculated from estimated cost differences. When design specifications and implementation strategies are defined, the project enters the validation phase. Here more detailed cost estimates are needed for justification of design-to-cost objectives and need for demonstrable LCC characteristics.
- In the phase of full-scale development, operating and support cost estimates become more accurate because reliability, maintainability, serviceability, and supportability characteristics have been demonstrated. The issues of warranty-

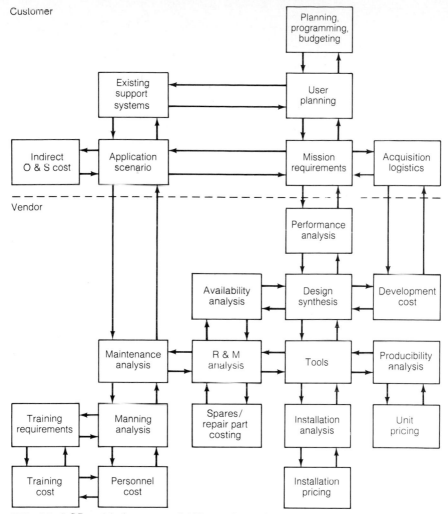

FIG. 3.3 LCC model of system availability. (*Adapted from* [3]. *Reproduced with permission.*)

conditions and pricing are resolved also and form a base for new total LCC reassessment.

- In the life-cycle phases of full-scale manufacturing, delivery, installation, application, and use, the actual cost is measured and stored in cost data bases to allow evaluation of accuracy of previous estimates and improvement of future ones.

The magnitude of the LCC formulation problem and cost estimation difficulties is reflected by the large number of cost-influencing variables, as demonstrated in Fig. 3.4 for cost of availability, and sparsity of clean empirical data. The complexity of

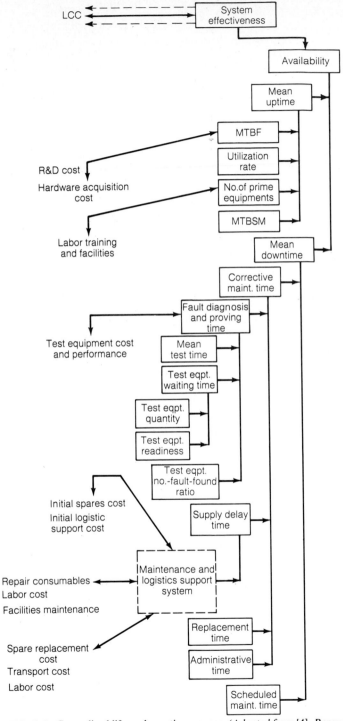

FIG. 3.4 Generalized life-cycle costing process. (*Adapted from* [4]. *Reproduced with permission.*)

the LCC process warrants continued monitoring to provide management with the ability to assess progress. Effective local applications of LCC methodology usually start by acceptance and experimentation with existing systems, many of which were described in past *Proceedings of Annual Reliability and Maintainability Symposia* [3–5].

Constantly increasing requirements for LCC minimization motivate program managers to use computerized methods to optimize cost by varying cost elements; to consider design alternatives, new concepts, or implementation strategies; and to account for uncertainties [6].

In profit-oriented environments, the minimization of some cost is not necessarily the best solution available, so LCC models must be developed to assure strategies for net profit maximization. Profit is a random variable which depends on the amount of up-time logged during the period of use. Key to the calculation of the total life-cycle profit is the profitability characteristic, which relates profit to the operating and support cost and is a function of equipment age and often exhibits the same three life stages (early, useful, and wearout periods) as does the failure rate.

In a majority of practical cases the user is facing decisions in situations much less complex than major military procurement contracts. Because the development and manufacturing costs have been already incurred and are reflected in the equipment purchase price, the user's degrees of freedom in search for LCC minimum are limited only to evaluation of different support and maintenance strategies. The equipment's intrinsic reliability, given the same way as its price, impacts only the cost of maintenance M_C which can be estimated assuming (each per some time period, usually a month or year):

- The cost of spare parts inventory C_{SPI}, reflecting the original manufacturing cost of spare parts C_M and inventory cost rate I_{CR} (as a percentage), including depreciation, interest, handling cost, etc.
- Preventive maintenance cost C_{PM}
- Corrective maintenance cost C_{CM}

$$M_C = C_{SPI} + C_{PM} + C_{CM}$$

$$= C_M I_{CR} + \text{WH}\, \frac{T_R^P + T_T^P}{T_I^P} + \text{WH}\, \frac{\text{MTTR} + T_T^C}{\text{MTBF}}$$

where W = hourly rate of service engineer, including hourly parts cost
H = equipment usage in hours per time period considered (in-use time)
T_R^P = scheduled time for preventive maintenance
T_T^P = expected travel time for preventive maintenance
T_I^P = scheduled preventive maintenance interval, hours
MTTR = mean time to repair
T_T^C = expected travel time for corrective maintenance
MTBF = mean time to failure, expressed in terms of in-use time, not calendar time

For example, typical values of these factors for minicomputer system maintenance in the early 1980s were as follows:

C_M	=	\$5000 at user's site	I_{CR}	=	50% per year
W	=	\$250 per hour	H	=	4000 hours per year
T_R^P	=	0.25 hours	T_T^P	=	0.50 hours
T_I^P	=	2 months	T_T^C	=	1.5 hours
MTTR	=	0.5 hours	MTBF	=	3000 hours

For competitive analysis or comparison and evaluation of alternative strategies, it is customary to express yearly maintenance cost as a percent of the purchase price or to standardize it, e.g., per \$1000 of price.

3.4 PARTICULAR ISSUES

The concepts of total COR and LCC discussed above provide an overall framework for cost optimization and for addressing and studying individual reliability issues, their impact on total cost, revenue, and selection of local implementation alternatives.

3.4.1 Economics of Warranty

In the narrow sense defined by the Uniform Commercial Code, the four basic types of warranties are as follows:

- *Warranties of title and against infringement*, representing that the title is transferred and that the transfer is good, unencumbered, and free from patent, trademark, and similar claims
- *Express warranty*, which is created usually to assure the buyer that delivered products will conform with samples, models, and descriptions used by the seller during the bargaining process
- *Implied warranty of fitness*, reflecting the assumption that the seller knows the end use of the goods by the buyer, who is relying on the seller's skill and judgment
- *Implied warranty of merchantability*, assuring that the goods sold meet the minimum standards of merchantability as stated in the Uniform Commercial Code, paragraph 2-314 (1979)

These warranties are representations of an inducement to purchase for which the law will allow a remedy if they prove to be not true. The key issues invoked by this concept evolve around disclaimers, coverage, limitation period and conditions, consistency, and the required disclosures. In the more general framework of a free market situation, however, warranty is viewed as an element of competitive strategy, and the economic issues associated with pricing, cost accounting, and determination of warranty reserves will dominate management's interest.

From the cost accounting point of view, which also greatly influences company management thinking, warranties usually fall into three basic categories:

- *Reimbursement warranty*, the simplest form, in which manufacturing reimburses the dealership for any charges incurred to repair or replace the defective merchandise.
- *Sales warranty*, in which charges are treated as an ordinary sale except that the revenue account is reduced rather than account receivables being increased.
- *Expense warranty*, the most popular, involves the warranty expenses being merged with other operating costs or cost of goods sold.

The lack of adequate costing figures, a satisfactory base for estimation, the complexity of warranty reporting systems, tax regulations, etc., are often stated as reasons for ignoring warranty cost despite the evidence that they must be controlled.

Warranty, as an expressly stated obligation and responsibility of the seller, can be perceived as an integral part of the satisfaction package intended to promote and encourage sales by reducing potential risks for the buyer. This type of warranty increases the obvious expenses by the cost of promotion, which may be significant. Market analysis is necessary to evaluate the estimated ability of the warranty to increase sales and profits, especially in industries where good warranty experience and repeat sales are not strongly linked or where the rate of return on goodwill assets is limited by insufficient demand. If market and competitive situations allow reasonable freedom in product pricing and length of the warranty period, an optimization model can be developed. In most market situations, we see some evidence of trade-offs between higher price and longer warranty versus lower price and shorter warranty, indicating a strong dependency of the cost optimum on the distribution of failures in the early postsale time period.

A company warranty policy and warranty cost control plans form a reasonable base for setting new product reliability goals as this example illustrates.[*]

Objective.   For the family of random access memory (RAM) parts, propose a generic reliability goal that is consistent with the company goal of a warranty cost of 1.5% of the value added shipped (VAS) at the time of product maturity.

Model Development.   Assuming the independence of failures, the failure rate λ, %/yr, of the component group in question is a simple function of the number of units n of the components in the group

$$\lambda = n\lambda_{comp} \qquad (3.1)$$

where λ_{comp} is a failure rate of an individual component. End product-failure rate λ_p, %/yr, and the component group failure rate can, for the purpose of goal setting, be assumed in the same relation as the total end product-production cost to the cost contribution of the component group.

$$K = \frac{\text{component group associated product cost, \$}}{\text{end product total production cost, \$}} \qquad (3.2)$$

so

---

[*] From Ref. 1. Used by permission.

$$\lambda = K\lambda_p \tag{3.3}$$

The number of failures we expect during the warranty period T, measured in years, and the cost of warranty repair C_R, in dollars, then define the hardware failure-related warranty cost C_{HWR}, in dollars

$$C_{HWR} = \lambda_p T C_R \frac{1}{100} \tag{3.4}$$

$$= \lambda_{comp} \frac{n}{K} \, TC_R \, \frac{1}{100} \tag{3.5}$$

If

$$N = \frac{\text{hardware failure rate warranty cost}}{\text{total warranty cost}} = \frac{C_{HWR}}{C_W} \tag{3.6}$$

then we have an expression for the warranty cost C_W, in dollars, in relation to the component failure rate λ_{comp} as follows:

$$C_W = \frac{1}{N} \lambda_{comp} \frac{n}{K} \, TC_R \, \frac{1}{100} \tag{3.7}$$

In ratio management environment, warranty cost C_W is often expressed in percent of VAS. For an individual end product, the VAS is given by the product price P, \$, so warranty cost index W is

$$W = \frac{C_W}{P} \times 100 \tag{3.8}$$

Assuming in the time of new product maturity the mix of new and old products in warranty to be close to ratio M, unit, of shipment values

$$M = \frac{\text{new product VAS, \$}}{\text{total VAS, \$}} \tag{3.9}$$

we can predict the new level of company warranty cost index

$$W_{com} = MW_{new} + (1 - M)W_{old} \tag{3.10}$$

where W_{old} = the company current warranty cost index based on "old" product's performance, %
W_{new} = is warranty cost index for the "new" product, %, given by

$$W_{new} = \frac{C_W}{P} \times 100 \tag{3.11}$$

In most of the situations, W_{com} is being set as a long-term goal using other means. If values of n, K, T, C_R, N, P, M, and W_{old} are given by current data experience and design consideration, then combining Eqs. (3.7), (3.10), and (3.11), we have

$$W_{com} = \frac{\lambda_{comp} n T C_R M}{PNK} + (1 - M)W_{old} \qquad (3.12)$$

where λ_{comp} is the only unknown value. So the proposed goal for component failure rate λ_{comp} can be derived from the W_{com} goal via rearranged formula (3.12)

$$\lambda_{comp} = \frac{KNP}{nC_R MT} \left[W_{com} - W_{old}(1 - M) \right] \qquad (3.13)$$

As an example, let us assume these values to be given:

Company warranty cost as percent of VAS $\quad W_{old} = 2.0\%$
$W_{com} = 1.5\%$

Warranty periods $\quad T_1 = 3$ mo $= 0.25$ yr on 75% of products, $T_2 = 12$ mo $= 1.0$ yr on 25% of products, so, the average is $.T = 0.4375$ years

Product mix $\quad$ 50% of "old" products
50% of "new" products
$M = 0.50$

Repair cost $\quad C_R = \$500$ per repair
Hardware-related warranty cost $\quad N = 30\%$ of total warranty cost
New product price $\quad P = \$3000$
RAM's contribution $\quad K = 30\%$ of total production cost
Total number of RAMs $\quad n = 12$ in group

Then the annualized failure rate goal required to support the company warranty cost goal can be calculated from:

$$\lambda_{comp} \leq \frac{KNP}{nC_R TM} \left[W_{com} - W_{old}(1 - M) \right]$$
$$= \frac{0.30 \times 0.30 \times 3000}{12 \times 500 \times 0.4375 \times 0.50} \quad [1.5 - 2.0(1 - 0.50)]$$
$$= 0.10\%/yr$$

Result. Assuming duty cycle of 2000 hours per year, the generic failure rate for an individual RAM during the warranty period would not exceed 0.05% per 1000 hours.

3.4.2 Economic Aspects of Product Safety

The close relationship of reliability to issues of product safety is evident from a simple definition of risk:

$$\text{Risk} = \text{probability of a failure} \times \text{exposure} \times \text{consequence}$$

Results obtained by techniques of fault-tree analysis, failure modes and effects analysis, or hazard analysis can be interpreted in terms of negative utility, event probabilities, and severity (criticality), which are standard subjects of cost-benefit studies. The cost factor, strongly dependent on the criticality level, must take into account cost of design for safety, manufacturing for safety, and losses caused by complaints, claims, suits, legal cost, unfavorable publicity, government intervention, etc. The best investments usually result from the lowest cost-benefit ratio, but implementation alternatives must be selected carefully because cost can significantly increase on both probability extremes:

1. low probability requirements could result in overdesign.
2. Accepted high probability of occurrence could be simply perceived as negligence.

A cost-benefit analysis can be complemented by a comparison of the cost of different alternatives to achieve a defined acceptable level of safety. Analytical approach to decision making about safety issues usually follows this simple process:

1. Identify potential hazards and resulting probable accidents associated with current product design.
2. Obtain credible data on accident rates by product over time, e.g., from the National Emergency Injury Surveillance system or insurance companies.
3. Provide a set of alternatives for providing additional increments of safety.
4. Get cost data or cost assessments for these alternatives.
5. Analyze alternatives provided for their effects and cost.

During the analysis it is necessary to follow some practical principles, such as

- Analysis should be a support for judgment, not a substitute for it.
- Analysis must be open and explicit to be useful as a framework for constructive critique and improvement suggestions.
- There is seldom only one single best solution.
- Conclusions from the analysis should be simple.
- Be realistic about improvement implementation prospects.

3.4.3 Economics of Incoming Inspections

The testing objectives may differ from manufacturer to manufacturer (e.g., improved process control, minimized number of field failures, etc.) but are always combined with the basic tendency to minimize production cost. Four basic incoming inspection (I.I.) techniques are available to implement these objectives:

- *100 percent test-stress-retest*:   This comprehensive I.I. technique is the most effective strategy forcing infant failures to appear and failed parts to be removed before they are used in assembly process. The correlation of test results prior to and after burn-in stress often provides important information for vendor's process control and quality improvement. Implementation of this strategy requires investment in both testing and burn-in equipment, indicating its best suitability for high component volumes.
- *100 percent test only*:   One hundred percent testing strategy assures removal of

defective parts delivered in particular batches and, by that, lower assembly rework cost, improved test effectiveness, and total cost of quality. Functional test is only partially effective as a reliability screen.

- *Buying preburned and tested parts*:   This variant of a very effective strategy may bring benefits of the economy of scale because of the expected parts volume differences between vendor's and user's facilities. Sample testing for vendor's performance audit purposes and risk control will increase the total cost without any additional impact in improved quality or reliability.

- *No I.I.*:   This is the best strategy in conditions of good vendor's process control and mutual trust in vendor-user relationships. In less favorable conditions the cost of doing no I.I. will be recognized in process disruptions on subassembly or system levels and, potentially, also in the field.

To determine the most suitable alternative, numerous details of all strategies need to be examined and a composite picture formed. I.I. is often difficult to justify on purely economic terms, especially for low volumes. For high component volumes, the cost-benefit of I.I. is unquestionable in most industrial situations. Also, let us not underestimate the importance of I.I. information feedback to vendors.

   A quick estimate of the break-even component volume can be obtained by this simple reasoning: The break-even point is defined by equality of the cost of not testing and the cost of testing per given period of time, usually one year. Cost of not testing is given by

$$C^{NT} = NR(P^I C^I_R + P^W C^W_R)$$

Cost of testing

$$C^T = P + \frac{NL}{n}$$

where N = number of component units under consideration
R = total fraction defective previously observed or estimated
P^I = fraction defective which fails in-plant
C^I_R = average cost of in-plant repair
P^W = fraction defective which fails during the warranty period
C^W_R = average cost of warranty repair
P = cost of the test equipment
n = number of units tested per hour
L = labor and overhead rate per test hour

So, the break-even volume can be estimated from

$$C^{NT} = C^T$$

$$NR(P^I C^I_R + P^W C^W_R) = P + \frac{N}{n} L$$

$$N = \frac{P}{R(P^I C^I_R + P^W C^W_R) - L/n}$$

3.4.4 Other Topics

For detailed optimization of the complete LCC, the task of reliability management requires a wide variety of economic analyses that address many other diverse topics such as:

- Reliability technology selection
- Manufacturing screening alternatives evaluation
- Optimum maintenance strategies
- ROI in analytical instrumentation

All of them are addressable by the methods of economic evaluation.

3.5 REVIEW OF ECONOMIC TOOLS

3.5.1 Methods of Economic Evaluation

In a majority of manufacturing industries, money will not be allocated to a project or spent unless a good argument is presented to the management that this particular investment will assist in reaching company financial goals. To apply this criterion rationally, technical and economic evaluations of proposed projects are performed. Sometimes these judgments are intuitive, based on experience; in other situations formal detailed analyses are required.

The general concepts forming the framework of sound decisions are quite simple. In the cases of economic decisions, they evolve around notions of profit, growth rate, return on investment, cost-benefit ratio, cash flow, value of money, etc.

The fundamental step in any economic analysis is the selection of decision criteria or figures of merit, which may significantly vary between the private sector (motivated by profit) and public sector (driven by social benefits or possibly by political motives). The strictly economic-benefit-driven situations are much easier to subject to formalized analyses and usually define figure of merit in maximal profitability, maximal ROI, or in minimal time required to recover investment. But even in obvious situations of highly favorable cost-benefit ratios, it is necessary to assess possible changes in assumed conditions to assure stability of expected benefits. We need to understand the risks associated with inflation rates, interest rates, business cycles, errors in cost estimates, effects of obsolescence, depreciation, taxation, and sometimes even social cost.[*]

The basic methods of economic analysis most frequently used are as follows:

Uniform cash flow method requires conversion of all cash flows to a time-adjusted, equivalent, uniform annual amount.

Break-even analysis, suitable for problem solving with incomplete data sets, requires sensitivity analysis to minimize the consequences of errors in estimates.

---

[*] For technical aspects and details of methods of economic analysis see Secs. 3.1 and 3.3.

Present worth method is based on conversion of all cash flows to an equivalent amount discounted to time zero by application of appropriate interest rate formulas.

Rate of return method computes the discounted cash flow rate of the interest return on invested capital by trial and error.

Benefit-cost method, frequently used by federal and state governmental agencies to estimate the economic attractiveness of an investment, computes ratio of probable annual benefits to the equivalent uniform annual cash flow.

Cost effectiveness analysis allows comparison of alternatives on other than solely monetary measures but requires very careful definition of effectiveness, via, e.g., performance, safety, reliability, and relationships to corporate objectives. This presents many difficulties.

Replacement studies, concerned with identification of the most economical time for replacement of existing assets, utilize concept of already incurred (sunk) cost.

Actual implementation of economic analyses and evaluations encounters frequent difficulties in creating consensus on figures of merit, decision criteria and their importance, cost estimation credibility, and unclear power of a wide variety of optimization methods. The major shortcoming of all methods of engineering and managerial economics is their complete disregard of the non–cash flow elements involved (which must be balanced with managerial experience and judgment in the final selection of an alternative).

3.5.2 Cost Estimation Methods

Accurate cost estimation of competing engineering design alternatives is essential for project planning and budgeting decisions by both equipment manufacturer and user. Traditionally, system cost estimates have been prepared using industrial engineering techniques involving detailed studies of necessary operations and materials. These costly estimates, accompanied by volumes of supporting documentation, were subject to frequent extensive revisions in case of even small design changes. Publicized evidence of frequent cost overruns in highly visible government projects indicates their questionable accuracy. These shortcomings resulted in increased interest in statistical and other approaches to cost estimation.

All cost estimating is done by means of analogies and always reflects the future cost of a new system by relating it to some known past experience. Historical cost data incorporate experience with setbacks, design requirements changes, and other difficult to identify and control circumstances in opposition to industrial engineering methods, which tend to be optimistic and not allowing for unforeseen problems. The role of analogy, and the methods of reasoning behind it, is crucial. The art of cost estimation is based on a seven-step process, described, for example, by Barry W. Boehm (see bibliography).

1. Establish objectives for the cost estimating activity to assure support and development of important decision-making information, reevaluate and modify these objectives as the process progresses. Objectives should also help to balance expected accuracy, ratio of absolute to relative estimates, and expected level of conservatism.

2. Assure adequate resources for this miniproject and form a simple project plan.
3. Spell out reliability requirements and document all assumptions.
4. Explore as much detail as feasible to assure good understanding of technical aspects of all parts of the product under consideration in a given phase of its life cycle.
5. For actual estimation use several independent methods and data sources. Most frequently used methods are
    a. Algorithmic estimating models (see below)
    b. Expert judgment, which could reflect also group consensus derived by formalized methods, such as Delphi technique
    c. Analogy based on similarities with past experience, but taking into account impact of inflation, new technologies, productivity growth, etc.
    d. Top-down and bottom-up estimating
6. Compare and iterate estimates to eliminate the optimist-pessimist biases often reflecting a person's roles and incentives. Evaluate the importance of estimates via Pareto analysis, taking into account observed tendencies to overestimate costs related to physically bigger and complex parts of the system.
7. Follow up your estimates with regular comparison with actual cost data collected during the project implementation.

The majority of cost estimating methods is based on the premise that the system cost is in a quantifiable way logically related to some of the system's physical or performance characteristics and, in general, derived from historical cost data by regression analysis.
  The most common forms of estimating algorithms are:

- *Analytical models*

$$\text{Cost} = f(X_1, \ldots, X_n)$$

Where f is some mathematical function relating cost to cost variables $X_1, \ldots, X_n$ correlated with some physical or performance characteristics of the system.

- *Tabular (matrix form) models* provide easy-to-understand, -implement, and -modify relationships which are difficult to express by explicit analytical formulas. Analytical cost models are excellent in describing cost-to-cost estimating relationships. Tabular models are more suitable if cost-to-noncost functions need to be represented. To date, the analytical models used usually contain a small number of variables and are insensitive to many sometimes important factors. Some models take a form of a composite function which can better represent the historical data but at the expense of increased complexity.

Examples of simple models:

1. *Analytical cost estimate model:*   Cost of spare parts procurement support C_{ps}

$$C_{ps} = 0.037 P_{rc}$$

   where P_{rc} is parts repair cost.

2. *Matrix model:*   Relative cost of software qualification testing C_{sq} as a function of required reliability

C_{sq}	Required reliability
$0.55C_N$	Very low
$0.75C_N$	Low
$1.00C_N$	Nominal
$1.25C_N$	High
$1.75C_N$	Very high

where C_N is qualification cost of nominal reliability software product. *Note*: C_N itself is usually a function of product size and complexity, design group skill and experience, user programming language, etc., and can be expressed via analytical models.

The strength of algorithmic models is in their objectivity, repeatability, and computational efficiency in support of families of estimates or sensitivity analysis. They do not handle exceptional conditions and do not compensate for erroneous values of cost variables and model coefficients. Actual experience with cost estimation leads to a conclusion that no single method is substantially superior in all aspects and that strengths and weaknesses of many methods are complementary.

3.5.3 Cost Accounting

To assure a reasonable rate of diffusion of methods of economic analysis to support rational decision making about reliability, a system of actual cost data feedback is necessary to allow evaluation of prediction accuracy, cost estimates, and effectiveness of analytical methods used. In the first-time attempt to identify and measure cost of reliability, it is highly probable that our cost data requirements will not match the established cost accounting system. In this situation the decision to start with a single project study usually prevents deadlock and allows development of information for both reliability improvement program and identification of reliability categories of key importance. Data collection will be manual, necessary forms will be designed separately for this particular study, data compression and interpretation will be done, most probably, on reliability engineers' personal computers.

Only after a demonstrated success of reliability cost management in a single study or project environment can steps be taken to establish a reliability cost accounting system, with the objective of a continuing scoreboard. This continuing scoreboard should be based on a formal reliability cost accounting and reporting system, which parallels the accounting systems required for general financial management and legal purposes. The systematic approach requires at least:

- A list of projects, products, and programs of interest
- A list of departments involved
- A list of accounts where relevant charges are accumulated
- Cost categories (see example in Table 3.1 suitable for LCC model
- Definitions of data entry requirements and formats
- Definitions of data process flow with control points

TABLE 3.1 Examples of Reliability Cost Categories

Prevention costs
  Hourly cost and overhead rates for design engineers, reliability engineers, materials engineers, technicians, test and evaluation personnel
  Hourly cost and overhead rates for reliability screens
  Cost of preventive maintenance program
  Cost of annual reliability training per capital
Appraisal costs
  Hourly cost and overhead rates for reliability evaluation, reliability qualification, reliability demonstration, environmental testing, life testing
  Average cost per part of assembly testing, screening, inspection, auditing, calibration.
  Vendor assurance cost for new component qualification, new vendor qualification, vendor audit
  Cost of test results reports
Internal failure costs
  Hourly cost and overhead rates for troubleshooting and repair, retesting, failure analysis
  Replaced parts costs
  Spare parts inventory cost
  Cost of production changes administration
External failure costs
  Cost to repair a failure
  Service engineering hourly rate and overhead
  Replaced parts costs
  Cost of service kits
  Cost of spare parts inventory
  Cost of failure analysis
  Warranty administration and reporting cost
  Cost of liability insurance

- Formats and frequency of reports and summaries
- Rules of data and results interpretation
- An established base for result comparison and evaluation
- A methodology for cost standards creation and improvement

Collected data must be sorted and compressed to accommodate evaluation from many different viewpoints:

- Product, subassembly, part, component, etc.
- Organization responsibility
- Place and time of occurrence or reporting
- Project or program association

In formatting the data and results, the general preference of graphical representations in forms of tables, Pareto-type distributions, pie charts, trend lines, scattergrams, control charts, etc., is well established. Narratives by specialists or representatives from responsible teams can help in interpretation and in assessing the seriousness of data, especially when reports result in transfer of charges between departments and accounts.

The importance of reliability cost accounting for expense controls of projects or routine activities in every sector is self-evident. But the analysis of data from previous projects, contracts, and economic studies may have longer-term and more fundamental impact on the improvement of managerial decisions, resource allocation, and effectiveness, and by that significantly contribute to the improvement of the business unit's competitive position and probability of success.

3.5.4 Summary

Reliability engineering and management can greatly benefit from prudent application of tools from engineering economics and operations research, especially when facing decisions about costly investments, high risks, or complex situations, such as occur in high-technology environments. The power of these tools, with their rigorous methods and computational accuracy, must be understood in the context of their dependency on quality and relevance of underlying assumptions, historical data, cost estimates, and known unequal treatment of results from physical vs. human sciences. The growing computerization of described methods and availability of computers for daily work, should allow management to concentrate more on the human and strategic aspects of decision making, while being assured of the rigor and accuracy of their technical aspects.

3.6 REFERENCES AND BIBLIOGRAPHY

References

1. Henry J. Kohoutek, "Establishing Reliability Goals for New Technology Products," *1982 Proceedings, Annual Reliability and Maintainability Symposium*, IEEE pp. 460–465.
2. Henry J. Kohoutek, "Development of a Reliability Strategy for new IC Component Family and Process," *Microelectronics and Reliability*, vol. 23, no. 2, 1983, pp. 383–389.
3. Ramesh K. Barasia and T. David Kiang, "Development of a Life Cycle Management Cost Model", *1978 Proceedings, Annual Reliability and Maintainability Symposium*, IEEE, pp. 254–259.4.
4. Donald R. Earles, "LCC—Commercial Application," *1975 Proceedings, Annual Reliability and Maintainability Symposium*, IEEE, pp. 74–85.5.
5. Dr. Hans I. Ebenfelt, "LCC—Defense Application," *1975 Proceedings, Annual Reliability and Maintainability Symposium*, IEEE, pp. 63–73.6.
6. J. T. Henderson, P. E., "A Computerized LCC/ORLA Methodology," *1979 Proceedings, Annual Reliability and Maintainability Symposium*, IEEE, pp. 51–55.

Bibliography

Many papers dealing with the subject of economics of reliability can be found in *Proceedings of the Annual Reliability and Maintainability Symposia* sponsored by IEEE, ASQC, and other professional organizations.

Barry W. Boehm, *Software Engineering Economics*, Prentice-Hall, Englewood Cliffs, N.J., 1981

Armand V. Fiegenbaum, *Total Quality Control*, 3d ed., McGraw-Hill, New York, 1983.

Eugene L. Grant, W. Grant Ireson, and Richard S. Leavenworth, *Principles of Engineering Economy*, 7th ed., John Wiley & Sons, New York, 1982.

Daniel M. Lundvall, "Quality Cost", in J. M. Juran, Frank M. Gryna Jr., and R. S. Bingham Jr. (eds.), *Quality Control Handbook*, 3d ed., McGraw-Hill, New York, 1974, Sec. 5.

Robert L. Mitchess, *Engineering Economics*, John Wiley & Sons, New York, 1980.

Special Issue on Quality Cost, *Quality Progress*, vol. XVI, no. 4, April 1983.

CHAPTER 4
RELIABILITY INFORMATION COLLECTION AND ANALYSIS

W. Grant Ireson
Professor Emeritus, Stanford University

An effective reliability program would be impossible without the collection, recording, analysis, and use of information obtained through the testing and operation of industrial, military, and consumer products. Without the recording, retrieval, and use of information gained through experience with all manner of components, systems, machines, environments, user stresses, and human errors, the production of a reliability product would be like reinventing the wheel each time we want to build a wheelbarrow. The real purpose of developing a formal system of recording, analyzing, and retrieving information gained from our experiences is to enable us to design and build a better, more reliable device without having to repeat all the research, experimentation, design, development, and testing that has previously been carried out to attain the current status of product reliability.

The goal of this chapter is to provide help regarding

1. What information to collect.
2. How to record the information.
3. How to analyze and summarize the information in order to condense great masses of data into simple tables, charts, or diagrams that are easily understood and used by all personnel concerned with quality and reliability.
4. How to file or store the data for easy retrieval.
5. How to use the accumulated knowledge contained in the files.
6. Who should collect what kinds of information.
7. How the data from many different originators should be transmitted to a central analysis and depository unit, and how it should be summarized and indexed for ready reference by all the different users.

Unfortunately, there is no standard method or system by which reliability data are accumulated and recorded. The basic principles are generally applied by most companies and organizations, but each usually designs its own system. The rapid development of computers (and especially the great advances that have been made in micro- and minicomputers) has led most companies and organizations to design a computerized system into which can be fed the data that were contained in manually manipulated files.

Each company should examine all the sources of information, internal and external, and carefully plan a system that will enable it to take full advantage of software programs that are currently available. The cost of developing a complete set of software programs is great, and without the consideration of external sources of reliability data, the company-devised plan may increase the cost of operations for years to come. The list of goals for this chapter is the starting point for the development of a computerized system. Commercially available software can be adapted to the specific needs of the company.

Obviously the needs of a company manufacturing small home appliances will be substantially different from the needs of a company producing microcomputers or automobiles. However, the fundamental principles to be followed in the development of the data system are the same; "Make the shoe to fit the foot." In other words, tailor the system, starting with commercial software for file management, input devices, retrieval methods, and data processing routines, and tailor them to the specific needs and conditions of the company. Some questions that should be answered before selecting either the hardware or the software are:

1. Are specific sources of reliability information relative to my company available? If so, in what form: printed matter, computer disks, computer tapes, microfiche, etc.?

2. What are the formats of the available sources? What problems will I encounter in transposing the data into a format that I can use on my computer?

3. How important are the external data compared with my internally generated data? Will the external data be the major source of information for designers or will they only supplement or verify internally generated information?

4. What is the most economical means for inputting the internal data into the files? Are there standard reporting forms in use, prescribed by customers, standardized for the industry, or individually designed by the company? Are electronic input devices available that are compatible with the computer?

5. Have each of the potential users of reliability data been polled to determine the specific data that should be recorded?

Figure 4.1 provides a comprehensive list of data sources and their users.

4.1 INFORMATION NEEDS TO BE MET

The design, development, production, and maintenance of an acceptably reliable product should *not* be a new adventure each time a new product is developed or each time a modified product is developed from an older product. Each project should be a learning experience which results in complete documentation of all the relevant information so that the learning experience will not be repeated unnecessarily. The information gained in each project can be used in each of the phases, conceptualization, design, production, and support, to help assure that the reliability goals will be attained in a minimum time and with minimum expense. Each of the phases requires certain kinds of information, and the system for

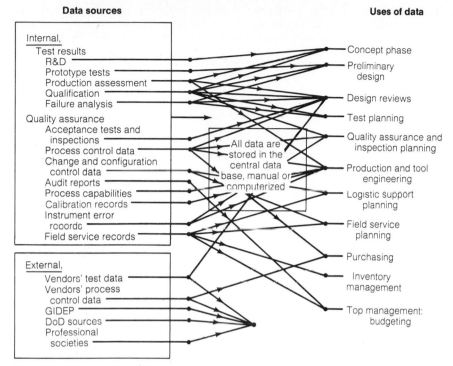

FIG. 4.1   Data sources and uses.

recording, storing, analyzing, and summarizing the information should be designed with the end uses clearly in mind.

The need to record data gained in each project cannot be overemphasized. Too many companies depend on the memories of its employees to provide the data to speed up the development process for a new product, when it is well known that memory is not perfect, details are forgotten, persons with the required knowledge move to other organizations, and the time required to extract the desired information from available persons discourages the responsible persons at every phase from a complete review of previously learned facts.

4.1.1   Conceptualization Phase

The concept for a new or improved product usually is based on scientific facts which the designers believe can be applied through good engineering to accomplish some product goal. Usually many different basic science facts are involved in any product and probably most of them have been applied in the past to similar products, but one or more may never have been tried. The first question, then, is "Do we have any experience in the application of this principle or fact to a successful product?" That information should be readily available.

Most companies develop a line of products and continue over the years to improve them, drop some, and add new ones as the science, technology, and

market demands change. Each product and each improvement involves preliminary designs, prototype testing, and design reviews. The design reviews should
provide permanent records of the experiences gained in product development.
They certainly will contain information about the concepts underlying the
product's design, and the records regarding successes, failures, corrective actions, and design changes. Those histories can be used by the product designer to
guide the development of the product concept by preventing the use of basic
science ideas which have failed in the past and by identifying those that have
resulted in successes. Thus the designer can concentrate on the ideas or concepts
that have not been proved. This type of information is especially relevant because
it represents not only the results of conceptualization and design, but the skills
and knowledge of the personnel. It is better, in that regard, than any similar
information collected by an agency or another company.

Similar information about the concepts can be obtained by reviewing all or
most of the patents on similar products in the U.S. Patent Office. This method can
be effective, but it is usually very expensive, requiring the services of highly
qualified engineers and scientists. Also, it does not provide any information about
the problems encountered in putting the product into production.

Some companies buy one or more units of all competing products, perform an
autopsy, and assess the fine points of the competitors' designs. This method is
helpful in developing ideas for future development, but it, as with the patent
review, is too late. The competitor's product is already on the market.

4.1.2 Design Phase

In the design phase the concept for the product is reduced to specific details
regarding all the components, subassemblies, circuits, mechanical devices, etc. In
this phase the information needs are much more specific and extend well into the
information gained through production and maintenance. The basic principle for
good, reliable designs is to use only proven components and production techniques.

The information required can be summarized briefly as:

1. The reliability of each component under specific use, environments, and
   stresses
2. The effects of changes in stresses, temperature, humidity, pressure, radiation,
   vibration, etc., on the reliability (derating effects and surge effects)
3. The ease or difficulty in diagnosing and repairing failed components
4. Special problems encountered in assembling the components into a complete
   system (danger of damaging components in the assembly operations)
5. Difficulties encountered in the manufacture or fabrication of similar design
   features (how to avoid degradation of the inherent reliability in the production
   processes)
6. Field use problems (kinds of failures or malfunctions resulting from use,
   misuse, and abuses by operators and/or maintenance personnel)

With the best of information systems, the design engineers will nearly always
run into details for which there are no data and no prior experience. However, an

effective information collection, storage, and retrieval system will tend to minimize the problems requiring special testing or qualification during the design stage.

When the designs are extending the state of the art into new frontiers, the number and difficulty of the problems not covered by prior experience will increase. That is what puts the "research" into R&D. In other words, lack of information regarding some aspects of the design concept requires that research be performed to supply the needed data.

4.1.3 Production Phase

Production engineering and processing are always involved in the design reviews at all levels of development, and it needs to have extensive records of prior production projects in order to be able to advise the designers about problems in production that can be anticipated from the design features. The purpose is, of course, to prevent problems at the design stage before the release of the design disclosure to production.

Additionally, production engineering needs to have records that will help it solve production problems during the planning phases for production tooling, inspection, testing, etc. The specific kinds of information the production engineering and processing needs to have include:

1. Process capabilities for every machine tool and process in the plant
2. Accuracy and repeatability of the measuring devices issued to the setup and operating personnel
3. Specific machine operations and processes that have previously been difficult to control (in the statistical quality control sense)
4. Qualification of operating personnel to perform acceptably and to maintain the process standards
5. Control and identification of special materials, components, parts, etc., especially when a visual identification is impossible
6. Accuracy and repeatability of the incoming inspection and acceptance procedures and practices to assure that only "good" components, parts, materials, and subassemblies are submitted to the production facility
7. A record system that assures the identification of the causes of production-induced degradation of reliability.

4.1.4 Support and Maintenance Phase

The support and maintenance phase begins when the product is delivered to the customer and is an extremely important aspect of the maintenance of customer loyalty and repeat sales. It is almost impossible to anticipate all of the stresses that the customer will apply to a product or to anticipate all the different environments to which the product will be subjected, but it is *very* important to try to anticipate these unstated requirements. The founder of one very prominent company in the high-technology field is frequently quoted as saying, "Make it idiotproof!" Data on prior experience will help the designers, production engineers, and maintenance engineers avoid the kinds of difficulties that will result in high frequencies of product failures, high warranty costs, loss of reliability and quality reputation, and loss of customers.

A chapter of this handbook is devoted to the problems of maintenance engineering (Chap. 15), but these are some of the items of information that need to be considered in order to make the product more reliable after it is delivered to the customer:

1. Complete and accurate "change control records" to provide a record of engineering changes, models, components, subassemblies, etc., so that at any time the proper replacement part can be identified and supplied for equipment in the field.

2. Identification of the parts, components, and subassemblies that are most likely to fail, wear out, or degrade the performance, and how to make their identification and replacement easy and fast.

3. Typical abuses the operators apply, and how to derate the components so that they can withstand the extreme stresses applied by the abuses.

4. Records by component, part, and subassembly of failures and malfunctions along with the causes of the failures. The frequencies of such failures indicate the relative importance of solving problems of design, materials, production, maintenance, etc.

5. Records of failure analyses that indicate the necessity to change or upgrade processes or tools. These records are extremely important to the production engineers during the design reviews because they help avoid designs that approach the limits of capability of the tools and processes.

6. Records of the effectiveness of the field service personnel in identifying, correcting, and returning to service the malfunctioning equipment. These records will help to identify the personnel who need additional training, the need for better diagnostic tools, and the effectiveness of the maintenance manuals supplied to the personnel.

7. Records of failures or malfunctions resulting from inadequate, incorrect, or poorly prepared operation and maintenance manuals.

This lengthy listing of the needs for information emphasizes the importance of having a data system that will serve the needs of all the different phases of a product life cycle. These needs, obviously, overlap and are intertwined so that the same information is frequently used by many different persons in the drive to attain high reliability. Therefore, it is imperative that the data collection and storage system be designed so that each responsible person has ready access to the desired information, is not overloaded with unnecessary or trivial data, and can interpret the data to solve his or her immediate problem. Economy dictates that the system be designed to prevent duplication of effort, storage facilities, retrieval facilities, and computer time, and that tasks that can be performed by the computer be programmed so that the personnel can spend time in using the information rather than in trivial tasks such as tabulating, counting, and reading unnecessary data.

4.2 SOURCES OF DATA

4.2.1 Internally Produced Information

Although a great amount of valuable reliability data is available from many external sources, most companies place more importance on their own internally generated information. This stems from the fact that the internally developed information reflects all of the "personality traits" of that organization: personnel capabilities in science, research, development, design, engineering, and production. It also reflects the capabilities of the machines, processes, tools, and management to produce products that comply in all respects to the design specifications. For these reasons, the internally generated data can be used to guide the reliability work with confidence that similar results can be obtained on new products.

Test Results. Because a great many tests are conducted in every quality and reliability program, a test plan should be drawn up for every test to specify the purpose and to assure that there is a record of the environmental conditions, procedures, test equipment, personnel, model and serial number of items tested, and the time and date. The test plan must provide a report form to be completed during the test to show all of the results of the various steps in the test procedure. The test plans and the test results should be in "hard copy" to be filed in the conventional way, but the results should be entered into the computer system as part of the overall test data base for easy future retrieval.

Test results comprise a great portion of the total data system to support reliability programs. These results should include:

Research tests

Prototype tests (see Chaps. 9 and 10)

Environmental tests (see Chaps. 10 and 11)

Development and reliability growth tests (see Chaps. 17, 18, and 19)

Qualification tests (see Chaps. 6, 8, and 19)

Tests on purchased items (see Chap. 6)

Production assessment and production acceptance tests (see Chaps. 7, 8, and 13)

Tests of failed or malfunctioning items (see Chaps. 13 and 14)

Each company must establish its own policy on what kind and how much testing to do and how much and what kind of data to record in its data system. The preceding list of types of tests also refers the reader to various chapters of this handbook in which the authors describe the types of tests needed for more complex products and the kinds and uses of the data generated. Those chapters will be very helpful to the decision maker in determining how extensive the company's testing and data program should be.

Another source of information on test plans and requirements are the several Military Standards and Specifications which deal with many aspects of reliability programs. Chapter 5 provides an annotated bibliography of many of these references.

Quality Control Data.   The quality control group of every company routinely performs inspections and tests on products and equipment in order to maintain the quality standards necessary for reliability products. The results of those inspections and tests are very valuable in design reviews, selection of vendors for components, subassemblies, and supplies, and in advising designers on the problems that may arise in production. Again, it is highly desirable to enter all of those data into a data base so that they can be retrieved quickly and easily. The specific types of information needed are:

1. Incoming inspection and test results identified by component, part number, vendor, and specification
2. In-process quality control data by part number, product, processes, and machines
3. Calibration records of all measuring instruments used in research, testing, and production, including indicators of instrument error, repeatability, accuracy, and frequency of recalibration
4. Results of machine and process capability studies
5. Quality audit records and final test results

Field Support Data.   Practically every manufacturer must maintain some kind of field service facilities to provide repair and maintenance service to its customers. This is one of the most valuable sources of reliability information because it represents the customer's view of the reliability of the product. Every field service or maintenance activity should result in a report to the marketing department and to the reliability organization. The report should provide the following kinds of information:

1. Product, model, and serial number.
2. Part that failed or malfunctioned, part number, serial number (if serialized).
3. Nature of failure. What happened?
4. Cause of failure if determinable on site.
5. Action taken by service personnel.
6. Date and name of service personnel.

It is common to have a standard report form with much of the desired data preprinted in such a way that it can be checked rather than have to write out a lot of words. Even on such a form, there must be space for special remarks by the service person if the preprinted information is not adequate.

For example, many forms have a list of causes of failure ending with "other." Unfortunately, the other block usually gets checked more frequently than the specific blocks because the service person does not take time to determine the cause and the other block is very convenient. Instead of other, there should be a space identified, "If none of the above, explain the cause." When using such a preprinted form, a small notebook should be provided with quick references to the various items printed on the form to help assure that the service person makes a correct diagnosis and knows which items to check under each category.

1. Reporting Activity	2. Report Ser. No.	3. Date Of Trouble	4. Installed In Aircraft/Arresi. Gear/Catapult/Support Equipment Model / Bunc oi Ser. No	5. Aircraft Logbook Time

| System, Set Equipment Or Engine | 6. Model Designation And Model No. | 7. Nomenclature | 8. Serial No. | 9. Time Meter Read./Logbook Time or Events (if applicable) Hour meter / Logbook hrs. / Starts / Landings |

| Unit, Component Accessory, Assembly Or Equipage | 10. Manufacturer's Part No. | 11. Nomenclature | 12. Serial No. | 13. Mfr's Code No. / 14. Contract No. | 15. Time Or Events Hrs. / Starts / Ldg's |

| Subassembly (Electronic) Or Primary Part Failure (Non-electronic) | 16. Manufacturer's Part No. | 17. Nomenclature | 18. Serial No. | 19. Mfr's Code No. | 20. Location (if applicable) |

| Supply Identification Item(s) Returned | 21. Federal Stock Number | 22. (RM, MR copies only) | 23. Quantity | 24. (RM, MR copies only) | 25. (RM, MR copies only) |

| Reason For Report (Check one) | 26. Removal Or Maintenance Action Required As A Result Of: 1 Failure/Suspected Failure Or malfunction 2 Damaged due To improper Maintenance/Operation/Test 3 Damaged or Defective On receipt 4 Damaged Accidentally 5 Scheduled/Directed Removal, high time Overage, excess To requirements | 27. Item overhauled by |

DESCRIPTION OF TROUBLE

(If box 1, 2, 3, or 4 was checked in space 26, complete spaces 28 through 31. If box 5 was checked in space 26, leave spaces 28 through 31 blank.)

28. First Observed/Occurred During

| 1 | Flight operations—Land based | 3 | Pre-flight | 5 | Conditional | 7 | Overhaul/PAR | 9 | Special directed inspection |
| 2 | Flight operations carrier based | 4 | Daily | 6 | Calendar | 8 | Shop maintenance bench test | 10 | Normal operation of support equip., catapults, arresting gear, mirror landing sys. only. |

29. Symptoms— How Discovered Item

D	Incorrect display	I	Low performance	R	Overheating
E	Inoperative	J	Metal in oil	O	Pressure out-of-limits	S	Torque out-of-limits
A	Excessive vibration	F	Interference/Binding	K	Noisy	P	RPM out-of-limits	T	Unstable operation
B	High fuel consumption	G	Intermittent operation	L	None noticed	Q	Surging/Fluctuates	U	Visible defect
C	High oil consumption	H	Leakage	M	Out-of-balance	R	Temperature out of limits	V	Other (Amplify)

30. Part Condition

007	Arced	130	Changed value	201	Distorted/Stretched	750	Missing	585	Sheared
780	Bent	910	Chipped/Nicked	148	Eroded	008	Noisy	196	Shorted/Grounded
135	Binding	999	Circuit defective	250	Frayed/Torn	450	Open	422	Soldering defect
429	Blistered/Peeled	160	Connections defective	001	Gassy	790	Out-of-adjustment	660	Stripped
070	Broken/Cracked	818	Contacts Burned/Pitted	381	Leaking	439	Plugged/Clogged	018	Tested OK – Dirt not work
900	Burned/Burned out	170	Corroded	730	Loose	576	Ruptured/Split/Blown	389	Unknown (Cannot disassemble)
120	Chafed/Galled	200	Dented	004	Low GM or emission	935	Scored	020	Worn—Excessively
								099	Other (Amplify)

31. Cause Of Trouble

A	Design deficiency	D	Faulty overhaul (Quality control)	G	Fluid contamination	J	Operator technique/Adjustment	M	Weather conditions
B	Faulty maintenance (Quality Control)	E	Faulty preservation/Packaging	H	Installation environment (Location in weapons sys.)	K	Other parts primary cause	N	Wrong part installation
C	Faulty manufacturing (Quality Control)	F	Foreign object	I	No failure-replaced to improve sys. performance	L	Undetermined (Cannot disassemble)	O	Other (Amplify)

32. DISPOSITION OR CORRECTIVE ACTION: Select appropriate code(s) from list below and enter in boxes at left to indicate disposition or corrective action taken with respect to each of the items entered in spaces 6, 10, and 16.

Replaced And Returned To Supply
Code Reason

A - Hold 90 days
B - Lack of repair facilities
C - Lack of repair parts
D - Lack of Tech. Pubs
E - Lack of personnel
F - Beyond assigned maintenance level
G - Other—(Defective on receipt, high time, directed removal, excess to requirements, etc.)

Space 6
Space 10
Space 16

Code Corrective Action

H - Used as is
I - Adj./Realign./Serv./Repaired in place
J - Removed-Adj./Realign./Serv./Repaired-reinstalled
K - Removed-repaired made RFI
L - Removed-tested Ok-made RFI
M - Removed-scrapped
N - Surveyed
O - Released for investigation and replaced (Indicate custody in space 35)

33. Maintainability Information

	Hours	Tenths
Man-hours to locate trouble Space 10		
Man-Hours to locate trouble Space 16		
Man-hours to repair/replace/adjust		
Actual time A/C was undergoing repair		
Total time aircraft not flyable due to this malfunction		

ACCESSIBILITY
S Satisfactory
U Unsatisfactory (Amplify)

SPECIAL
1 Frequent trouble item
2 Can be installed wrong (Amplify)

34. Component/Assembly, Subassembly Replaced With
Mfr's Part No.
Serial No.
Mfr's Code No.

35. AMPLIFYING REMARKS (Furnish additional information concerning failure or corrective action not covered above. Do not merely repeat information checked above.) Specify any severe operating conditions, such as hard landings, wheels-up landings, severe maneuvers, etc.)

36. Report Is					Signature	Rank/Rate	Date
0 FUR	1 AMPFUR	2 Urgent AMPFUR	3 Flight Safety AMPFUR	4 Follow up report			

Associated Parts Repaired Or Replaced (Do not list any item reported above)	37. Part No. (Non-electronic parts) Or Part Ref. Designator (Electronic parts)	38. Part Name, Tube Type, Semi Conductor Type Or Description	39. Mfr's Code No.	40. Failure Code (From space 30)	41. Disposition (Code from space 32)	42. Activity Repaired By
						Signature
						Rank/Rate / Date

FAILURE, UNSATISFACTORY OR REMOVAL REPORT
NAVWEPS FORM 13070/3: (10-62)

(Mail this copy to NATSF)

FUR

FIG. 4.2 Example of failure, malfunction, and parts replacement report.

It is also common practice, especially with high-reliability, expensive items, to have the failed part returned to the failure analysis group at the home plant in order to answer the following questions: (1) Can the part be repaired and put back into service? (2) Was the service person's diagnosis correct, and/or is there an underlying cause that should be corrected by a design change, change in production technology, change in supplier, or other corrective action? The returned parts can be sent to the failure analysis laboratory for a detailed autopsy which may lead to important improvements in the design or production (see Chap. 14).

4.2.2   External Information Sources

There is a great amount of important reliability information available to all companies from external sources, and it will save a lot of time, money, and errors if the appropriate information is obtained and used. A brief discussion of the major external sources follows:

Vendors' Inspection and Test Data.   Every company ought to include quality and reliability requirements in every purchase contract for components, supplies, and subassemblies. These contracts should include requirements that the vendor supply, along with the product, the following information:

1. The vendor's written quality control and reliability program plans, which should show how the vendor assures a satisfactory product conforming to the buyer's specifications
2. Quality control records (Shewhart control charts and other process control reports) covering the specific run of the product
3. Inspection reports of the final product, by lot or batch, showing the lot number or identification, the number inspected, and the number of defects or nonconforming units found (and removed)
4. Reliability test documents, showing the test procedure, the environmental conditions, the stresses applied, the type of test equipment, length of test, number tested, number failed, and causes of failure
5. A computation of the estimated reliability for the buyer's specified use, and whether or not the part or item has been "qualified" under one of the Military Standard plans.

In high-value items, the contract may specify that the buyer's quality and reliability engineers can witness the reliability tests and visit the plant to observe the workings of the quality and reliability program plans. This is especially important when the tests are long, expensive, and require very special test facilities. The Department of Defense usually has "resident representatives" stationed in the plants of major contractors to help assure that the products are properly manufactured, inspected, and tested, and to prevent duplication of the inspections and tests by the military agency.

Obviously the demands on the vendors should be tailored to the needs of the customer and the complexity of the product. It is true in most cases that appropriate contractual requirements can prevent duplication of inspections, tests, and failure analyses by the customer. This not only saves money but also helps to assure that there is a clear and complete understanding between the buyer and the seller as to just what the buyer wants.

The uses of the information obtained from vendors are:

1. Evaluation of the vendors' abilities, quality, and reliability in order to select vendors for future orders.
2. Reliability data are used in the conceptualization and design phases in the estimation of the inherent reliability of the proposed product design.
3. Decisions on which components to use in the design. Proven components reduce the time and cost of designing new products.
4. Used by reliability group in planning the reliability test program. Cuts down the investment in test equipment and the cost of running the tests.
5. Used by quality control group in planning the incoming inspection and where and when to inspect in the production line.

Government-Industry Data Exchange Program.   The most comprehensive external data source for all aspects of the design, production, and field support of highly reliable products is the Government-Industry Data Exchange Program (GIDEP). The program is authorized and funded by the U.S. government, but it is open to participation by all companies engaged in producing commercial, off-the-shelf items if the company uses and generates the type of data compiled and distributed through the exchange program. Some government agencies may contractually require participation by companies that produce items for the agencies. The program maintains four data "interchanges":

1. *Engineering data interchange:* Engineering and qualification test reports, nonstandard parts justification data, parts and materials specifications, manufacturing processes, and engineering methodologies and techniques.
2. *Reliability-maintainability data interchange:* Failure rate, failure modes, and replacement rates on parts, components, and materials based on demonstration tests and field performance information.
3. *Metrology data interchange:* Technical data on test systems, calibration systems, test equipment calibration procedures, measurement technologies; designated as a data repository for the National Bureau of Standards.
4. *Failure experience data interchange:* When significant problems are identified regarding parts, components, processes, fluids, materials, safety, and fire hazards, the objective failure information is reported and added to the data base.

In order to make use of GIDEP a company must apply formally to become either a full participant or a partial participant for any one or all four of the interchanges. This means that the company must really participate by providing copies of all relevant reports, test results, etc., to the program in order to receive the benefits. A specific GIDEP representative must be appointed as the contact, and periodic progress reports on GIDEP activity and benefits must be submitted. The reports must not contain "classified" (security) information or proprietary information.

The fully participating company receives the microfilmed data banks, indexes, and all associated documentation. A partial participant receives all program materials and the indexes but not the microfilmed data banks. That company may request a loan of the microfilms of desired reports. Hard copy can be made from the microfilm, or, if the company has remote terminal equipment compatible with the GIDEP Operation Center's computer, it can make direct inquiry and print out the information if desired.

With hundreds of companies involved in all types of products participating, access to the GIDEP interchanges can save any organization great amounts of money and time, as well as prevent serious errors, in the development of high-reliability products. The information is applicable in all phases of the life cycle: conception, design, demonstration, production, and field support. The experience of a great many companies and government agencies has proved over and over that the GIDEP program saves millions of dollars every year.

To obtain complete information about GIDEP and how to become a participant, write:

Director, GIDEP Operations Center
Corona, California   91720

Other Government Data Sources. The Department of Defense has been the primary supporter of research in reliability since the early 1950s and has supported various activities to collect and disseminate data to contractors in order to improve the reliability of military hardware and to reduce costs. The GIDEP program is one of those agencies. Two other agencies under the Department of Defense are:

Technical Information Center   Reliability Analysis Center
Cameron Station   Rome Air Development Center (RBRAC)
Alexandria, VA 22304   Griffis AFB, NY 13441

The Department of Commerce maintains the

National Technical Information Service
5285 Port Royal Road
Springfield, VA 22161

Professional Organizations. The Institute of Electrical and Electronics Engineers, 345 East 47th Street, New York, NY 10017, has developed a great amount of data regarding reliability of electrical hardware of all types.

When using data obtained from external sources, there is always the question of its applicability to the total company environment. A simple example may help clarify this statement.

Environmental stress-test data can be obtained for many components, including derating factors or charts. If you are using those data in designing a product, you must question several points:

1. How much of the variation indicated is the result of measuring instrument error compared with the instrument error in your plant?

2. How precise were the measuring instruments used in the tests compared with the precision of instrumentation in your plant? For data to be truly trustworthy, the instruments must be capable of repeatedly measuring 0.5 or less standard deviations of the characteristic being measured.

3. Did the range of variables in the reported test data cover the range of the same variables that your product will encounter?

4. How many tests, cycles, or operations were carried out to produce the results reported?

5. Do the reported data compare favorably with your internal data?

4.3 DATA STORAGE AND RETRIEVAL METHODS

4.3.1 Manual

In spite of the tremendous advances that have been made in computers, manual methods are still important in the collection, retrieval, analysis, and application of reliability data. Most of the results of tests are first recorded manually on paper forms, and then the appropriate portions of the data are transferred to the computerized data base. Reliability testing programs usually require that specific conditions be observed and recorded in order to validate the test. The procedures must be followed precisely. The results or measurements of each step are then recorded. The plan and procedures are usually printed forms with provisions for the observations to be recorded after each step. See Figs. 8.11, 8.12, and 8.13 in Chap. 8 for some examples of test procedures and a test record sheet. All of this is necessary to assure that there is a complete, accurate, and permanent record of the test that can be used for many purposes.

Failures, malfunctions, or unscheduled maintenance actions require a written report. This report provides important information to the reliability and quality assurance engineers, designers, logistics support, and inventory management as the basis for corrective actions, redesign, production process improvement, and future product design. Most of these failures or unscheduled maintenance actions occur at remote locations and are handled by field service personnel who do not have immediate access to a computer terminal.

A copy of the written report normally accompanies the part, subassembly, or component as it is returned to the home facility for diagnosis, repair, or failure analysis. It helps to reduce the diagnostic time and to assist the repair operation. Another copy of the report is sent to the data center where the failure or malfunction is recorded in a specialized format in the computer data bank.

The design of the report forms is not standard, although some Department of Defense agencies have special forms to facilitate the transfer of the data to the computer data base. Most companies design their own forms to provide just the desired data and avoid excess writing. Some examples of report forms are given here.

Figure 4.2 shows a form that was used by the Bureau of Naval Weapons for aeronautical-material deficiencies. Figure 4.3 shows a form used for electronic equipment failure or replacement by the Navy's Bureau of Ships. Figure 4.4 shows the codes used on Figure 4.3 to indicate the nature of the failure. The list of codes and types of failures helps the service personnel to describe the failure more accurately, and the codes can be used in computerizing the report. Note that other information relative to maintainability, repair time, downtime, etc., is also recorded. A similar Air Force form is shown in Fig. 4.5.

Manually prepared reports are normally filed for future use or reference. Frequently only the most significant information is stored in the computer data

ELECTRONIC EQUIPMENT FAILURE/REPLACEMENT REPORT   DD—787 (PROPOSED)   REPORT BUSHIPS 10550—1

1. DESIGNATION OF SHIP OR STATION

3. TYPE OF REPORT (check one)		4. TIME FAIL. OCCURRED OR MAINT. BEGAN			
1. OPERATIONAL FAILURE	4. STOCK DEFECTIVE	MONTH	DAY	YEAR	TIME
2. PREVENTIVE MAINTENANCE (PoMSEE)	5. REPAIR OF REPLACEABLE UNIT OR PLUG-IN ASSEMBLY				
3. PREVENTIVE MAINTENANCE (NOT PoMSEE)	6. OTHER	5. TIME FAIL. CLEARED OR MAINT. COMPL.			
		MONTH	DAY	YEAR	TIME

2. REPAIRED OR REPORTED BY

NAME	RATE	AFFILIATION
		1. U.S. NAVY 2. CONTRACTOR 3. CIVIL SERVICE

EQUIPMENT

9. FIRST INDICATION OF TROUBLE (check one)		10. OPERATIONAL CONDITION (check one)	11. TIME METER READING
1. INOPERATIVE	5. UNSTABLE OPERATION	1. OUT OF SERVICE	A. HIGH VOLTAGE
2. OUT OF TOLERANCE, LOW	6. NOISE OR VIBRATION	2. OPERATING AT REDUCED CAPABILITY	B. FILAMENT /ELAPSED
3. OUT OF TOLERANCE, HIGH	7. OVERHEATING	3. UNAFFECTED	12. REPAIR TIME MAN-HOURS TENTHS
	8. VISUAL DEFECT		
4. INTERMITTENT OPERATION	9. OTHER, EXPLAIN		

6. MODEL TYPE DESIGNATION

7. EQUIP. SERIAL NO.

8. CONTRACTOR (NAVY CODE OR COMPLETE NAME)

REPLACEMENT DATA

13. LOWEST DESIGNATED UNIT (U) OR SUB-ASSEMBLY (SA)	14. LOWEST DES. U/SA SERIAL NO.	15. REFERENCE DESIGNATION (V-101, C-14, R11, ETC.)	16. FEDERAL STOCK NUMBER	17. MFR. OF REMOVED ITEM	18. TYPE OF FAILURE	19. PRIMARY OR SECOND-ARY FAIL ?	20. CAUSE OF FAILURE	21. DISPOSITION OF REMOVED ITEM	22. REPL. AVAILABLE LOCALLY ?
						P S			Y N
						P S			Y N
						P S			Y N
						P S			Y N
						P S			Y N

23. REPAIR TIME FACTORS

CODE	DAYS	HOURS	TENTHS	CODE	DAYS	HOURS	TENTHS

24. REMARKS (CONTINUE ON REVERSE SIDE IF NECESSARY)

SRA—1

FIG. 4.3 Failure and replacement report for data file input.

4.14

QUICK REFERENCE LISTING OF MOST OFTEN USED FAILURE CODES
(IF PROPER CODE CANNOT BE FOUND, REFER TO ALPHABETICAL LISTING BELOW.)

ELECTRON TUBES		TRANSISTORS AND SEMICONDUCTOR DIODES		PLUG-IN ASSEMBLIES		ELECTRICAL, ELECTRONIC		OTHER COMMON TYPE OF FAILURE CODES — ELECTRO-MECHANICAL, MECHANICAL, CHEMICAL	
CODE	TYPE OF FAILURE	CODE	TYPE OF FAILURE	CODE	TYPE OF FAILURE	CODE	TYPE OF FAILURE	CODE	TYPE OF FAILURE
002	AIR LEAK	741	ALPHA CUT-OFF LOW	035	DRIFTS	007	ARCING, ARCED	710	BEARING FAILURE
007	ARCING, ARCED	744	BACK RESISTANCE LOW	088	GAIN, LOW	080	BURNED OUT	780	BENT
960	BACK RESISTANCE LOW	739	BETA LOW	094	GAIN, NONE	139	CHANGE OF VALUE	040	BINDING, MECHANICAL
001	BROKEN ENVELOPE	743	FALL TIME, EXCESSIVE	360	INTERMITTENT OPERATION	320	HIGH VOLTAGE BREAKDOWN	070	BROKEN
380	GASSY	745	FORWARD RESISTANCE HIGH	387	LOW PERFORMANCE	380	INSULATION BREAKDOWN	090	BRUSHES, IMPROPER TENSION
004	LEAKAGE	742	Ico HIGH	089	MODULATION, LOW	008	LEAKAGE	720	BRUSH FAILURE
131	LOW GM OR EMISSION	737	OPEN, BASE-TO-COLLECTOR	096	MODULATION, NONE	450	NOISY	150	CHATTERING
009	MARGINAL PART REPLACEMENT	735	OPEN, BASE-TO-EMITTER	022	NO OSCILLATION	450	OPEN	160	CONTACTS, CONNECTION DEFECTIVE
053	MICROPHONIC	156	POOR RECOVERY TIME	462	OUTPUT, LOW	082	OPEN, INTERMITTENT	170	CORRODED
003	MISFIRES (THYRATRONS)	734	RISE TIME, EXCESSIVE	255	OUTPUT, NONE	460	OPEN PRIMARY	210	DETENT ACTION POOR
008	NOISY	740	SATURATION RESISTANCE HIGH	258	OVERHEATS	431	OPEN ROTOR	226	EXCESSIVE PLAY
560	OPEN FILAMENT	738	SHORTED, BASE-TO-COLLECTOR	560	POOR REGULATION	470	OPEN SECONDARY	567	HIGH CONTACT RESISTANCE
011	POOR REGULATION	736	SHORTED, BASE-TO-EMITTER	097	SENSITIVITY, POOR	432	OPEN STATOR	700	LOOSE
005	SCREEN DEFECTS (CATHODE RAY)	731	SHORTED, COLLECTOR-TO-EMITTER	091	SENSITIVITY, LOW	453	OPEN SYNCHRO-3	790	OUT OF ADJUSTMENT
006	SHORTED, INTERMITTENT	749	STORAGE TIME, EXCESSIVE	686	UNSTABLE	550	PITTED	570	RUSTY
	SHORTED, PERMANENT					005	SHORTED INTERMITTENT	770	SLIP RING OR COMMUTATOR FAILURE
018	TESTED OK, DID NOT WORK					006	SHORTED PERMANENT	164	SPEED INCORRECT
						620	SHORTED PRIMARY	650	STICKY
						412	SHORTED ROTOR	945	STRUCTURAL FAILURE
						630	SHORTED SECONDARY	020	WORN EXCESSIVELY
						513	SHORTED STATOR		

ALPHABETICAL LISTING

CODE	TYPE OF FAILURE	CODE	TYPE OF FAILURE	CODE	TYPE OF FAILURE	CODE	TYPE OF FAILURE	CODE	TYPE OF FAILURE
002	AIR LEAK	370	JAMMED	733	OPEN, BASE-TO-EMITTER	560	PUNCTURED	613	SHORTED TO GROUND
741	ALPHA CUT-OFF LOW	380	LEAKAGE	450	OPEN, PERMANENT	097	RESPONSE, NONE	640	SLIPPAGE
007	ARCING, ARCED	730	LOOSE	082	OPEN, INTERMITTENT	734	RISE TIME, EXCESSIVE	770	SLIP RING OR COMMUTATOR FAILURE
744	BACK RESISTANCE LOW	013	LOOSE BASE	460	OPEN PRIMARY	570	RUSTY	026	SOLDER JOINT DEFECTIVE
710	BEARING FAILURE	012	LOOSE ELEMENTS	431	OPEN ROTOR	740	SATURATION RESISTANCE HIGH	164	SPEED INCORRECT
780	BENT	400	LOSS OF RESIDUAL MAGNETISM	453	OPEN SECONDARY	935	SCORED	650	STICKY
739	BETA LOW	004	LOW GM OR EMISSION	432	OPEN STATOR	011	SCREEN DEFECTS (CATHODE RAY)	749	STORAGE TIME, EXCESSIVE
226	BRITTLE	387	LOW PERFORMANCE	453	OPEN WINDING	091	SENSITIVITY, LOW	665	STRIPPED
240	BROKEN BASE	225	MANUFACTURER'S DEFECT (EXPLAIN)	099	OTHER, EXPLAIN	738	SHORTED, BASE-TO-COLLECTOR	945	STRUCTURAL FAILURE
745	BROKEN ENVELOPE	131	MARGINAL PART REPLACEMENT	161	OUTPUT INCORRECT	736	SHORTED, BASE-TO-EMITTER	018	TESTED OK, DID NOT WORK
960	BROKEN GLASS	040	MECHANICAL BINDING	462	OUTPUT, LOW	731	SHORTED, COLLECTOR-TO-EMITTER	947	TORN
250	BRUSH FAILURE	009	MICROPHONIC	255	OUTPUT, NONE	005	SHORTED, INTERMITTENT	965	TUNING DRIVE DEFECTIVE
720	BRUSHES, IMPROPER TENSION	053	MISFIRES (THYRATRONS)	790	OUT OF ADJUSTMENT	006	SHORTED, PERMANENT	670	UNBALANCED
080	BURNED OUT	790	MISSING	258	OVERHEATS	620	SHORTED PRIMARY	680	UNSTABLE
130	CHANGE OF VALUE	928	MODULATION, LOW	928	PEELING	412	SHORTED ROTOR	690	VIBRATION EXCESSIVE
120	CHARRED	927	MODULATION, NONE	927	PINCHED	613	SHORTED SECONDARY	700	WEAK ELECTRICALLY
150	CHATTERING	550	NO OSCILLATION	550	PITTED	630	SHORTED STATOR	966	WINDOW SLUG-IN (MAGNETRON)
910	CHIPPED	008	NOISY	010	POOR FOCUS	600	SHORTED TO CASE	020	WORN EXCESSIVELY
180	CLOGGED	022	NOT DETERMINED	156	POOR RECOVERY TIME	610	SHORTED TO FRAME		
160	CONTACTS, CONNECTION DEFECTIVE	920	OPEN, BASE-TO-COLLECTOR	550	POOR REGULATION TIME				
170	CORRODED	737	OPEN, BASE-TO-COLLECTOR	964	POOR SPECTRUM (MAGNETRON)				
190	CRACKED								
200	DENTED								
210	DETENT ACTION POOR								
230	DIRTY								
035	DRIFTS								
226	EXCESSIVE PLAY								
743	FALL TIME, EXCESSIVE								
240	FRAYED								
745	FORWARD RESISTANCE HIGH								
270	FROZEN								
280	FUNGUS EFFECT								
088	GAIN, LOW								
094	GAIN, NONE								
001	GASSY								
790	GROUNDED								
099	HIGH CONTACT RESISTANCE								
567									
320	HIGH VOLTAGE BREAKDOWN								
742	Ico HIGH								
340	INSTALLED IMPROPERLY								
350	INSULATION BREAKDOWN								
360	INTERMITTENT OPERATION								

FIG. 4.4 Examples of codes to identify failed components and types of failures.

A. JOB CONT NO. | B. PRI | C. TIME SPEC REQ | D. WK AREA | E. EST M/H | F. ORIG RPT NO. | G. REPORT NO. 26407-F | H. | I.

1. WEAPON TMS | 2. SERIAL NO. | 3. TIME | 4. WORK CENTER | 5. WORK ORDER NO. | 6. DAY MO YR | 7. WORK UNIT CODE

1A. AGE WUC | 2A. SERIAL NO. | 3A. TIME | 8. ACT TAKEN | 9. WHEN DIS | 10. HOW MAL | 11. UNITS | 12. LABOR HRS | 13. ASST WORK CEN

1B. ENG TM PSN | 2B. SER MOD YR-MFG SER NO. | 3B. TIME | 14. INST ENG TM PSN | 15. SER MOD YR-MFG SER NO. | 16. TIME

1C. ITEM FSC | 2C. PART NO. | 3C. SERIAL NO. | 17. INST ITEM PT NO. | 18. SERIAL NO. | 19. TIME

J. SYMBOL | K. DISCREPANCY | L. CORRECTIVE ACTION

CORRECTED BY-SIGNATURE & GRADE

DISCOVERED BY-SIGNATURE & GRADE | INSPECTED BY-SIG & GRADE | SUPERVISOR-SIG & GRADE

RECORDS ACTION
☐ UNCLEARED DISCREPANCY ☐ REPLACEMENT TIME CHANGE ITEM ☐ DATA TRANSCRIBED TO APPROP RECORDS

DATE TRANSCRIBED | TRANSCRIBED BY-SIGNATURE & GRADE

AFTO FORM JUL 61 211 PREVIOUS EDITIONS OF THIS FORM ARE OBSOLETE. MAINTENANCE DISCREPANCY/PRODUCTION CREDIT RECORD

FIG. 4.5 Example of maintenance report for data input to data base.

base, and at some future time it may be helpful to go back over these reports to extract some other information. As an example, it may be desirable to study the differences in the detection and repair time for some equipment at different field stations or repair centers. The information might be necessary in order to determine if additional training, new test equipment, or better instructions are needed for some service facilities.

4.3.2   Computerized Data System

The design of a computerized data system is a complex task requiring a lot of time of highly qualified programming specialists. At the beginning of this chapter attention was drawn to the needs for information, especially reliability information. A computerized data system should satisfy the needs enumerated in Sec. 4.1 and the needs of management. For example, the analysis of quality and reliability costs (see Chap. 3) requires that defects, failures, and unscheduled repairs be reported not only for reliability analysis purposes but for the accumulation of failure costs or costs of "unquality."

The same input information frequently is needed by several groups, but their uses of the information differ and thus they desire the outputs to be different. Some persons will want only summaries of data, while others will want a more detailed output. This requires that the coding of the information be set up in such a way that it can be retrieved and even analyzed in different ways. To satisfy all of these needs with one central data base becomes a very difficult programming problem that could require several person-years of programming effort.

Many software companies have developed very effective data base management programs that can be used "as is" or can be adapted to the specific reliability needs. These programs may appear to be very expensive, but they are usually more economical than trying to develop your own data management programs.

Many companies find it beneficial to set up several data bases. One system could be just for test data to handle the results of all tests during development and production. One system might be used just for field service failure and malfunction data and analysis. A third system could be set up just to handle the quality and reliability records of the vendors. Another system could be used for in-plant quality control, process control, and process capability studies. Computer storage capacity is very inexpensive now, and the additional storage space required by separate data systems will usually be more than recovered in programming and operating time for single systems.

The inputting of data into the computer(s) can be accomplished in several ways. Computer terminals can be located throughout the plant facility and even in facilities hundreds or thousands of miles away and the data transmitted to the mainframe. The actual input may be by a person typing in the data according to a specific format, or it may be automatically fed into the computer from sensors or detectors on the test equipment, measuring devices, or machines. Facsimiles of handwritten reports can be transmitted to the data center and the data entered manually. In all cases the data must be coded or indexed so that it will be stored in the proper place in the computer.

The rapid introduction of relatively inexpensive powerful personal computers with 512K or more bytes of RAM, 20 to 30 megabytes on hard disks, and high-density floppy disks has resulted in substantially changed thinking about computerization of reliability and quality data bases. Many companies are providing their engineers with desktop computers that can access the mainframe either to extract data, make complex computations, or store data, but with most of the engineer's work being performed on his or her own computer. A large number of software companies are providing a wide range of program packages that will perform most of the manipulative and analysis activities for reliability and quality work. Anyone interested in such programs for personal computers should examine the magazines *Quality,* published by Hitchcock Publishing Company, and *Quality Progress,* published by the ASQC, for the names and addresses of these software companies.

The whole field of computers and software is changing so fast that it is not advisable to try to list computers and software by name. Probably the available hardware and programs will be entirely different within a year. Persons planning new systems should make a thorough investigation into the available systems before contracting for either hardware or software.

A warning is in order. Where several different people need to have access to the same basic data, the main data files should be in the mainframe. Otherwise, there will always be some person(s) who has (have) not received and entered the latest data into his or her computer files. Unknowingly, the analysis will be incomplete.

4.4   GOALS OF DATA COLLECTION AND ANALYSIS

The goal of a data system is to convert the massive amount of information that is accumulated in all of the different divisions of an organization into an organized form that will enable all of the people to obtain the information they need rapidly and in a form that can be used with confidence in performing their assigned duties. There are many ways by which this organized information can be stored and presented to the users. A computer data bank is one way, but it has some disadvantages. In many companies the most commonly used data are printed for inclusion in loose-leaf binders which become the source documents for action.

Some common types of information contained in these binders are:

Derating charts for various standard electronic components

Mean time to failure (MTTF) for various components or parts under different stresses and environments

Process capabilities for the different production equipments and processes (by individual units)

Qualified parts lists with operating characteristics and design criteria

"Design rules"—standard design practices

Standard operating procedures, test procedures, etc.

The basic idea is that once something has been proved acceptable or effective, that information should be made available to every potential user in an easily accessible form whether by a computer or in information binders. This will have many good effects:

It reduces the number of decisions a person has to make and saves time.

It helps to assure that proven parts, processes, and designs will be specified unless the requirements cannot be met with the proven items.

It helps quality assurance and purchasing in obtaining acceptable parts and components from vendors.

It helps to assure that work will be assigned to the machines and processes that are capable of meeting the specifications.

It reduces the time and expense of prototypes, prototype testing, and design reviews, as well as speeds up the release to production.

It improves the capability to estimate the reliability of a product at the design stage.

Obviously the nature of the products and their intended uses will be the major factor in deciding how extensive the data collection and analysis system should be. A company stamping out stainless steel flatware will need a very simple data system, but one producing television sets, video cassette recorders, and video cameras must have a much more complex system to serve its needs. Determine who needs information, what they need, how they will use it, frequency of need, and the desired form of the information before trying to decide what kind of system to install and how extensive it should be. Do not let a salesperson sell you a system that is not appropriate for your immediate and foreseeable needs!

CHAPTER 5
RELIABILITY STANDARDS AND SPECIFICATIONS

Arthur A. McGill
Lockheed Missiles and Space Company

5.1 PURPOSE

The purpose of this chapter is to provide brief descriptions of some of the more commonly referenced reliability and maintainability documents and to assist in their selection and use. Military Standards (MIL-STDs) generally impose requirements and are "what-to-do" documents. Military Handbooks (MIL-HDBKs) are generally "how-to-do-it" documents. Many of the documents are extensive and complete, others are brief and of limited value. Since these documents are not always easy to obtain, some suggested sources are provided.

5.2 SOURCES OF INFORMATION

Commanding Officer
Naval Publications and Forms Center
5801 Tabor Avenue
Philadelphia, PA 19120-5099

Department of Defense
Technical Information Center
Cameron Station
Alexandria, VA 22304

U.S. Department of Commerce
National Technical Information Service
5285 Port Royal Road
Springfield, VA 22161

VSMF Data Control Services
Information Handling Services
Inverness Business Park
15 Inverness Way East
PO Box 1154
Englewood, CO 80150
(303) 790-0600; (800) 525-7052

Information Marketing International
A Ziff-Davis Information Company

13271 Northend Street
Oak Park, MI 48237
(313) 546-6706; (800) 821-3031

GIDEP Operations Center
Corona, CA 91720-5000
(714) 736-4677
Note: The Government-Industry Data Exchange Program (GIDEP) is a primary source of reliability (failure) data used in military programs.

Reliability Analysis Center
Rome Air Development Center (RBRAC)
Griffis AFB, NY 13441-5700
(315) 330-4151
Note: This source is also referred to as RADC and is a primary source of reliability data used in military programs.

The Institute of Electrical and Electronics Engineers (IEEE)
345 East 47th Street
New York, NY 10017-2394
(212) 705-7900

5.3 DESCRIPTIONS OF SELECTED STANDARDS AND SPECIFICATIONS

MIL-STD-105 *Sampling Procedures and Tables for Inspection by Attributes*

This document addresses the subjects of sampling plans; lot size; inspection levels; average quality levels (AQLs); classification of defects; multiple sampling; and normal, tightened, and reduced sampling. For equipments where the sequential method of testing, based on operating time, may not be appropriate, this document, based on the success ratio, can be used. It includes numerous tables showing accept-reject levels and operating characteristic curves for sampling plans. It would help to have sampling theory well in hand to understand the applicability and limitations of this document.

MIL-HDBK-189 *Reliability Growth Management*

This document is designed for both managers and analysts covering everything from simple fundamentals to detailed technical analysis. Included are concepts and principles of reliability growth, advantages of managing reliability growth, and guidelines and procedures to be used to manage reliability growth. It allows the development of a plan that will aid in developing a final system that meets requirements and lowers the life-cycle cost of the fielded system. The document includes sections on benefits, concepts, engineering analysis, and growth models. It contains lots of plots and curves.

MIL-HDBK-217 *Reliability Prediction of Electronic Equipment*

This handbook includes two basic methods for reliability prediction of electronic equipment. The first is a simple method called the parts count reliability prediction technique, using primarily the number of parts of each category with consideration of part quality, environments encountered, and maturity of the production process. The second method is the part stress analysis prediction technique, employing complex models using detailed stress analysis information as well as environment, quality applications, maximum ratings, complexity, temperature, construction, and a number of other application-related factors. The simple method is beneficial in early trade-off studies and situations where the detailed circuit design is unknown. The complex method requires detailed study and analysis which is available when the circuit design has been defined. Samples of each type of calculation are provided. A bibliography on reliability prediction is included.

MIL-HDBK-251 *Reliability/Design Thermal Applications*

This document details approaches to thermal design: methods for the determination of thermal requirements; selection of cooling methods; natural methods of cooling; thermal design for forced air, liquid-cooled, vaporization, and special (heat pipes) cooling systems. Topics covered are the standard hardware program thermal design, installation requirements, thermal evaluation, improving existing designs, and thermal characteristics of parts. Stress analysis methods are emphasized. Many graphs and nomographs are included. There is an excellent bibliography included. This is an excellent handbook.

MIL-HDBK-263 *Electrostatic Discharge Control Handbook for Protection of Electrical and Electronic Parts, Assemblies and Equipment (excluding electrically initiated explosive devices)*

This document includes definitions, causes and effects (including failure mechanisms), charge sources, list and category of electrostatic-sensitive devices by part type, testing, application information, considerations, protective networks, and a bibliography. This is a useful, enlightening document on a problem that is very difficult to demonstrate to management and manufacturing personnel. It is a problem that is difficult to control, difficult to assess, and pervasive. This document details the problem, determination of causes, and identification by failure diagnosis, and shows methods both in design and control to minimize its occurrence.

MIL-HDBK-338 *Electronic Reliability Design Handbook*

This document is virtually a text on reliability. Currently a two-volume set, it discusses the entire subject, heavily emphasizing the reasons for the reliability discipline. It includes general information, referenced documents, definitions, reliability theory, component reliability design considerations, application guidelines, specification control during acquisition, logistic support (storage), failure reporting and analysis, reliability and maintainability theory, reliability specification allocation and prediction, reliability engineering design guidelines, reliability data collection and analysis, demonstration and growth, software reliability, systems reliability engineering, production and use (deployment) reliability and maintainability (R&M), and R&M management considerations. This document is must reading for the reliability professional who has need of detailed explanations and theory.

MIL-STD-454J *Standard General Requirements for Electronic Equipment*

This document establishes the technical baseline for design and construction of electronic equipment for the Department of Defense. It addresses 75 requirements such as brazing, substitutability, reliability, resistors, and casting. It gives numerous references on the subjects addressed, but in itself is not rigorous or extensive in its treatment of subjects. It has some good illustrations for inspection attributes of selected processes such as soldering.

MIL-STD-470A *Maintainability Program Requirements for Systems and Equipment*

This document includes application requirements, tailorable maintainability program tasks, and an appendix with an application matrix and guidance and rationale for task selection. The topics covered are program surveillance and control, design and analysis, modeling, allocations, predictions, failure mode and effects analysis (FMEA), maintainability, design criteria. Each task item includes a purpose, task description, and details to be specified.

MIL-HDBK-472 *Maintainability Prediction*

This document is to familiarize project managers and design engineers with maintainability prediction procedures. It provides the analytic foundation and application details of five prediction methods. Each procedure details applicability, point of application, basic parameters of measure, information required, correlation, and cautionary notes. This document includes equations and data analysis sheets.

IEEE-STD-500-1984 *IEEE Guide to the Collection and Presentation of Electrical, Electronic, Sensing Component, and Mechanical Equipment Reliability Data for Nuclear-Power Generating Stations*

This book includes failure rates, failure rate ranges, failure modes, environmental factors on generic components used (or potentially used) in nuclear-power generating stations. The guide utilized the Delphi procedure to determine failure rate and failure-mode estimates. A data sheet covering each item covered lists failure mode, failure rate, and repair time or out-of-service time (not every item lists every piece of data).

MIL-STD-690B *Failure Rate Sampling Plans and Procedures*

This document provides samples of life-test records, failure rate sampling plans, failure rate tables at 60 and 90 percent confidence levels, and a reliability nomograph. Tables are provided that allow the determination of the probability of qualification of a lot.

MIL-STD-721C *Definition of Terms for Reliability and Maintainability*

This is a list of terms and definitions.

MIL-STD-756B *Reliability Modeling and Prediction*

This document establishes uniform procedures and ground rules for generating mission reliability models for electrical, mechanical, and ordnance equipment. It details the methods for determining service use (life cycle), creation of the reliability block diagram, construction of the mathematical model for computing the item reliability. Some simple explanations on the applicability and suitability of the various prediction sources and methods are included.

MIL-STD-757 *Reliability Evaluation from Demonstration Data*

This document provides procedures for evaluating achieved reliability with the minimum input information necessary. It defines the criteria under which minimum information is gathered. It provides the simple reliability calculations. The standard is not too helpful.

MIL-STD-781C *Reliability Design Qualification and Production Acceptance Tests: Exponential Distribution*

This document covers the requirements and provides details for reliability qualification tests (preproduction) and reliability acceptance tests (production) for equipment with an exponential time-to-failure distribution. It is the primary formal reliability demonstration test method for repairable equipment which operates on a time basis. Test time is stated in multiples of the design mean time between failures (MTBF). The document applies to fixed ground, mobile ground vehicle, shipboard, jet aircraft, turboprop and helicopter, and air-launched weapon equipment. Specifying any two of three parameters, i.e., lower test MTBF, upper test MTBF, or their ratio, given the desired decision risks, determines the test plan to be utilized. Draft versions of MIL-STD-781D indicate that with its issue, the detailed discussion of statistical test plans and environments will be issued separately as MIL-HDBK-781. MIL-STD-781D will be organized on a tailorable task basis coordinated with MIL-STD-785 to discourage indiscriminate blanket applications.

MIL-STD-785B *Reliability Program for Systems and Equipment, Development and Production*

This document provides general requirements and specific tasks for reliability programs. It is of great importance for reliability program planning. It has task descriptions for basic application requirements including sections on program surveillance and control, design and evaluation, development and production testing. An appendix for application guidance for implementation of reliability program requirements is also included. The subsections are in the form of purpose, task description, and details to be specified by the procuring activity. This is a program management document, not a detailed what-to-do document.

MIL-STD-790C *Reliability Assurance Program for Electronic Parts Specifications*

This document establishes the criteria for a reliability assurance program which is to be met by the manufacturer qualifying electronic parts to the specification. Typical topics covered are document submission, organizational structure, test facilities, and failure analysis reports. This document is twelve pages long.

MIL-STD-810D *Environmental Test Methods and Engineering Guidelines*

This document is designed to provide a more careful assessing of the environments to which items are exposed during their life as well as detailing test methods. Included in the numerous types of tests detailed are purpose, environmental effects, guidelines for determining test procedures and test conditions, references, apparatus, preparation for test, procedures, information to be recorded.

Numerous curves on environments are included, e.g., two-wheeled trailer transverse axis frequency vs. power spectral density curve.

MIL-STD-882B *System Safety Program Requirements*

This document provides requirements for developing and implementing a system safety program to identify the hazards of a system and to impose design requirements and management controls to prevent mishaps by eliminating hazards or reducing risks. Twenty-two tasks are defined in the areas of program management and control and design and evaluation. Typical tasks are system safety program plan, preliminary hazard analysis, and software hazard analysis. An appendix is provided to give some rationale and methods for satisfying the requirements previously detailed.

MIL-STD-883C *Test Methods and Procedures for Microelectronics*

This document establishes uniform methods and procedures for testing microelectronic devices. Basic environmental tests and physical (mechanical) and electrical tests (digital or linear) are specified. Also covered are test procedures for failure analysis, limit testing, wafer lot acceptance, and destructive physical analysis. This document gives extensive treatment to the subject of visual defects.

MIL-STD-965 *Parts Control Program*

This document describes two procedures covering the submission, review, and approval of program parts selection lists (PPSL). Typical topics covered are PPSL approval, meetings, parts control board, and Military Parts Control Advisory Group.

MIL-STD-1388-1A *Logistics Support Analysis*

This document details logistic support analysis (LSA) guidelines and requirements. Tasks detail the purpose, task description, task input, and task output. Typical tasks are program planning and control; development of early LSA strategy; planning; program and design reviews; mission hardware, software, and support; system standardization; early fielding analysis; and supportability assessment.

MIL-STD-1472C *Human Engineering Design Criteria for Military Systems, Equipment and Facilities*

This document presents human engineering principles, design criteria, and practices to integrate humans (their requirements) into systems and facilities. This is desired to achieve effectiveness, simplicity, efficiency, reliability, and safety of the system operation, training, and maintenance. This document contains interesting and useful information on items with which humans commonly interface including data and illustrations on visual fields, controls and displays (manual, visual, and audio), physical dimensions and strengths of humans, anthropometry (DOD-HDBK-743 *Anthropometry of US Military Personnel* is a referenced document), ground workspace design requirements, environments, design for maintainability, design for remote handling, hazards, and safety considerations. This document contains extensive figures and tables on human parameters.

MIL-STD-1543A *Reliability Program Requirements for Space and Missile Systems*

This document covers topics such as design for reliability; failure mode, effects, and criticality analysis (FMECA), reliability analysis; modeling and prediction; discrepancy and failure reporting; maximum preacceptance operation; effects of testing, storage, shelf life; packaging, transportation, handling, and maintainability. It gives application guidance and an appendix for FMEA for space and launch vehicle systems. This document is mainly a "what" and definition document.

MIL-STD-1556B *Goverment/Industry Data Exchange Program (GIDEP)*

This document defines the requirements for participation in the GIDEP program, which includes the engineering, failure experience, reliability-maintainability (RMDI), and metrology data interchanges. It is intended to be applied to prime contractors and major subcontractors (who are users of parts) for the government. The RMDI contains failure rate and mode and replacement rate data on parts, components, assemblies, subsystems, and materials based on field performance information and reliability test of equipment, subsystems, and systems. This data interchange also contains reports on theories, methods, techniques, and procedures related to reliability and maintainability practices.

MIL-STD-1574A *System Safety Program for Space and Missile Systems*

This document is a tailored application of MIL-STD-882A for space, missile, and related systems. It defines the management and technical requirements for system safety from concept to the end of the life cycle. It includes numerous system safety definitions such as accident, credible conditions, system safety engineer, etc. Includes the requirements for a system safety program plan such as associate contractor responses, subcontractor responses, software safety analysis, and a list of publications, specifications, and standards.

MIL-STD-1591 *On Aircraft, Fault Diagnosis, Subsystems, Analysis/Synthesis of*

This document provides criteria for conducting trade-off studies to determine optimal design for an aircraft fault diagnosis and isolation system, i.e., on-board built-in test system. It provides a cost model and maintainability labor power model.

MIL-STD-1629A *Procedures for Performing a Failure Mode, Effects, and Criticality Analysis*

This document shows how to perform a FMECA. It details the modeling method, functional block diagrams, defines severity classification and criticality numbers. It provides sample formats for a FMEA, criticality analysis, FMEA and criticality analysis maintainability information sheet, and damage mode and effects analysis sheet. Examples are provided.

MIL-STD-1635(EC) *Reliability Growth Testing*

This document covers the requirements and procedures for reliability development (growth) tests. It details planning requirements, tests to be performed, test conditions, failure recording, analysis, documentation, and corrective action. It details the requirements of the test reviews. An appendix on the Duane reliability growth model and a moving average approach is provided.

MIL-STD-1670 *Environmental Criteria and Guidelines for Air-Launched Weapons*

This document provides acquisition managers with guidelines for realistic test environments for air-launched weapons representing factory-to-target sequences. It covers environmental test criteria. Some good curves are presented, e.g., railroad vibration environments, jet/turboprop and prop cargo area vibration, worldwide humidity environments. It does not cover nuclear environments, electromagnetic, or laser effect environments.

MIL-STD-1679A *Military Standard—Weapon System Software Development*

This document establishes minimum requirements for Department of Defense software (DOD) development. It addresses topics such as criticality of performance, changing operational environments, and life-cycle costs. Data item descriptions (deliverable documents) are described such as code walk through, design walk through, development baseline, documentation, and error-intermittent reports. Topics covered include software quality assurance, configuration management, allowable control structures, command statements, software listings, cross-reference listings, load maps, and trouble reports. Other requirements such as error limits for software acceptance, patch limits, software reviews, and configuration control are covered. This document is thorough, gives good definitions, and gives an excellent plan for good software design practice as well as listing the DOD requirements.

DOD-STD-1686 *Electrostatic Discharge Control Program for Protection of Electrical and Electronic Parts, Assemblies and Equipment (excluding electrically initiated explosive devices)*

This document covers the establishment and implementation of electrostatic discharge control programs for design, test, inspection, servicing, manufacturing, processing, assembling, installation, packaging, labeling, or other handling of electrical or electronic equipment. Refer to DOD-HDBK-263 for "how-to" information.

MIL-STD-2068 (AS) *Reliability Development Tests*

The purpose of this document is to amplify the requirements for reliability development testing. MIL-STD-785 provides the criteria. It differs from MIL-STD-781 in that it does not demonstrate quantitative reliability requirements or the acceptability of hardware. The tests in this document assure that the majority of reliability problems have been resolved.

MIL-STD-2074 (AS) *Failure Classification for Reliability Testing*

This document contains criteria for classification of failures during reliability testing. This classification into relevant or nonrelevant categories allows the proper generation of MTBF reports.

MIL-STD-2080A (AS) *Maintenance Engineering, Planning, and Analysis for Aeronautical Systems, Subsystems Equipment, and Support Systems Equipments*

This document provides requirements and procedures to be applied in performance of maintenance engineering, planning, and analysis. It is to be used in conjunction with MIL-STD-1388. It provides the requirements and tasks. It describes the planning process. Typical topics include relationship of FMEA to preventive maintenance analysis, failure consequences, recommended maintenance analysis types, and R&M analysis requirements.

DOD 4245.7-M *Transition from Development to Production*

This document provides templates (diagrams) showing critical categories and considerations for design, test, production, facilities, logistics, and management during the transition from development to production. It details areas of risk in each category of concern, provides an outline for reducing the risk of the transition, as well as examples of analysis and control methods. Time lines indicating the general time each discipline will need to be implemented are provided. This document includes considerations for software design as well as computer-aided design. This is a what-to-do not a detailed how-to-do-it manual.

MIL-HDBK-46855B *Human Engineering Requirements for Military Systems Equipment and Facilities*

This document details requirements and tasks to be applied during development and acquisition program phases to improve the human interface with equipment and software. Its use should allow achievement of effective and economical utilization of human resources. Topics covered are analysis functions including human performance parameters, equipment capabilities, and task environments; design; test and evaluation; analysis to be performed such as workload analysis, dynamic simulation, and data requirements. Some selected terms are defined such as critical human factors. An application matrix is provided detailing program phases when each task is appropriate. Refer to MIL-STD-1472 for task details.

MIL-STD-52779A *Software Quality Assurance Program Requirements*

This document covers software quality assurance requirements to the extent that it says "go do it" but it isn't a "how to" document. Topics addressed are software QA programs, tools and technical methodologies, computer program design, work certification, documentation, documentation library control, reviews, configuration management, testing, corrective action, and subcontractor control. Other documents would provide insight into what is needed rather than what topics should be covered in a proposal.

LC-78-2 *Storage Reliability Analysis Summary Report* (vol. 1, *Electrical & Electronic Devices;* vol. 2, *Electromechanical Devices;* vol. 3, Hydraulic and Pneumatic Devices;* vol. 4, *Ordnance Devices;* vol. 5, *Optical and Electro-optical Devices*)

This document summarizes analyses on the nonoperating reliability of missile materiel. This document details the failure mechanisms observed on many part types, generally those itemized in MIL-HDBK-217B. Reliability models are included to allow reliability calculation.

NPRD-2 *Nonelectronic Parts Reliability Data, 1981*

This document provides failure rate and failure mode information for mechanical, electromechanical, electrical, pneumatic, hydraulic, and rotating parts. The assumption that the failures of nonelectronic parts follow the exponential distribution has been made because of the virtual absence of data containing individual times or cycles to failure. Generic failure rate tables include environment; application (military or commercial); failure rate; number of records; number failed; and operating hours. A 60 percent confidence interval is used.

RADC-TR-73-248 *Dormancy and Power On-Off Cycling Effects on Electronic (AD-768 619) Equipment and Part Reliability*

This document is the result of two 12-month programs by Martin Marietta. The first was to collect, study, and analyze reliability information and data on dormant military electronic equipment and parts and develop current dormant failure rates, factors, and prediction techniques. The second study was to provide similar information on power on-off cycling. Over 276 billion part-hours of dormancy information was gathered and 118 billion part-cycles of power on-off information was gathered. The power on-off cycling data resulted in limited success.

RADC-TR-75-22 *Nonelectronic Reliability Notebook*

This document contains sections on failure rates and analytical methods. The analytical section addresses applicable statistical methods, reliability prediction, demonstration, and specification. The analytical methods are in a cookbook format and examples are provided. The section on prediction methods includes the methods of Lipson and Kececioglu in applying strength and stress interference methods. The demonstration test section contains instructions for using most of the standard methods. The statistical methods section describes methods for fitting failure distributions, point and interval estimation, tests for outliers, and tests for increasing hazard rates. The prediction tables list environment, failure rate (90 percent confidence limit), environment, and application factors for those environments.

RADC-TR-83-29 *Reliability, Maintainability, and Life Cycle Cost Effects of Commercial Off-the-Shelf Equipment*

This document provides the results of a study to determine the effects of using commercial electronic equipment in a military environment. Some terms such as militarized or best commercial practice are clarified. A computer program listing is included to perform a life-cycle cost study. Results of an industry survey performed as part of the study are included.

RADC-TR-83-72 *The Evolution and Practical Applications of Failure Modes and Effects Analysis*

This document gives a broad, general background in techniques available for failure effects and analysis. Sixteen techniques such as tabular FMEA, matrix FMEA, sneak circuit analysis, fault-tree analysis, and hardware-software interface analysis are discussed. This is a good working document.

RADC-TR-85-91 *Impact of Nonoperating Periods on Equipment Reliability*

This document is designed to provide the nonoperating reliability prediction equivalent of MIL-HDBK-217. The models were derived using empirical data analysis. Details on data sources are provided. Definitions of dormancy, storage, equipment power on-off cycles, etc., are provided and are especially helpful since their definitions are often misunderstood. Nonoperating failure rate models analogous to MIL-HDBK-217-D were developed with examples of calculations. General models included base failure rates, temperature factors, temperature coefficients, quality factors, and on-off cycling factors. Especially useful are failure mechanisms of dormancy and their implications. These considerations can provide solutions to many of the dormancy problems and give insight into critical areas where resources may be expended to improve the product reliability.

TEOOO-AB-GTP-010 *Parts Application and Reliability Information Manual for Navy Electronic Equipment*

This document has as its basic premise that all failures are due to stress. It covers parts derating, part quality, and design for long life. Electrical derating curves and sections on part selection and application for resistors, capacitors, discrete semiconductors, microcircuits, electrical connectors, relays, crystals, switches, filters, and magnetic devices are included. Appendixes include thermal considerations on electronic parts and descriptions of quality and reliability screening levels of standard parts.

CHAPTER 6
RELIABILITY CONSIDERATIONS IN PROCUREMENT

Robert W. Smiley
Aerojet Strategic Propulsion Company

6.1 INTRODUCTION

As much as half of the cost of most manufactured products is in the raw materials and parts or components used in the manufacture, and these are usually procured from suppliers. Both the reliability and cost of the finished product is therefore heavily influenced by the reliability of these supplies. Consequently, it is very important that there be a comprehensive program to control the reliability of suppliers' materials. This program starts with the selection of materials and parts during the design effort, and includes not only the buying process but receiving, storage, issue, and often usage. In this chapter we will discuss only those aspects of the procurement program which have a significant effect on the reliability of the end product.

The dependence of manufacturers on suppliers should be recognized in the formulation of the procurement reliability program. The higher the product reliability desired, the more a continuous dialogue with suppliers is required. The relationship should be built on the understanding that suppliers know more about their products than users do. However, the user must maintain enough technical expertise in each product class to be a "smart" buyer and be able to communicate effectively with suppliers. This communication should especially include reliability and engineering personnel as well as procurement.

The reliability demanded of a supplier should not necessarily be the highest attainable or available, but should always be the appropriate apportioned level, based on the reliability requirements of the end product. The higher the reliability demanded, the higher the cost.

6.2 KEY ELEMENTS OF THE PROCUREMENT SYSTEM

The purpose of a procurement system is to provide parts, components, and materials meeting technical (including reliability) requirements when needed by the manufacturing or logistics operations at a reasonable cost. Cost includes the price paid; shipping, storage, and issue costs; in-house testing and inspection costs; and rework or repair costs to make the items usable.

The procurement system (see Fig. 6.1) includes all of those activities required to

Select suppliers
Secure suppliers' proposals
Budget procurement's resources
Determine material requirements and plan purchases
Procure (including negotiating)
Expedite deliveries
Receive, inspect, and test
Warehouse and issue
Collect and analyze supplier data
Determine and take corrective action at suppliers' facilities

6.2.1 Defining What Is Desired

Engineering drawings and specifications are the usual definition of the material to be purchased, but rarely are these documents complete enough in themselves to define the purchase. Usually the reliability of the part, along with a description of the reliability demonstration required, is included in the engineering definition, along with the details of the part testing and inspection for acceptance. Auxiliary definitions are often required, including:

1. The controls over the supplier's system required to *assure* the continuing reliability and quality of the supplier's parts
2. The special processes and the controls on them (e.g., soldering) which the buyers want performed in a particular specified manner
3. Configuration control, subpart traceability, age controls, and other aspects of a total reliability program

The U.S. government is a very large buyer, and at least three of the government's agencies—Department of Defense, Department of Energy, and the National Aeronautics and Space Administration—have developed and documented, at the taxpayer's expense, comprehensive procurement systems which ensure that material of the required reliability and quality is obtained. Commercial manufacturers can profitably borrow from these systems. Particularly useful are the documents which specify reliability and quality systems and methods; among them, MIL-STD-785, MIL-Q-9858, NASA NPC 250-1, and NASA NPC-200-2 (see Bibliography for titles) are very useful, particularly since they will be recognized and understood by suppliers accustomed to providing high-reliability components, parts, and materials. Many other documents (e.g., the "DOD Soldering Specification," DOD Standard 1866), are also widely used in the high-technology industries and can be applied either verbatim or in a tailored version.

Teamwork between supplier and buyer in developing both the technical and system requirements is important. Suppliers know more about their products than the buyer and should be made an integral part of the effort to define the procurement. Teamwork internal to the company is also essential, and the team should include Engineering and Product Assurance along with Procurement.

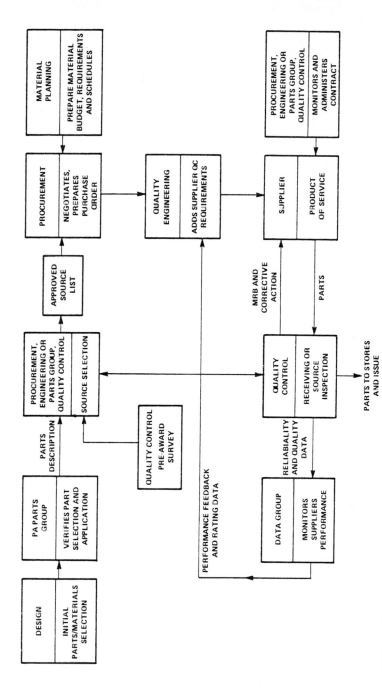

FIG. 6.1 Procurement system for parts and supplier selection and control.

Trade-offs are necessary, since cost or schedule constraints will often preclude satisfying all of the desires of Engineering or Product Assurance.

6.2.2   Selecting Sources

Selection of suppliers is a multistep process which includes some or all of the following:

1. Make-or-buy analysis and decision
2. Request for information and request for quotation
3. Hardware qualification
4. Quotation evaluation
5. Supplier selection

Not all items making up the end product are subject to *make-or-buy* analysis, since company policy frequently dictates either make or buy for certain classes of material, components, or services. However, for those that are not stipulated, a specific make-or-buy committee action should be taken. The committee, chaired by Procurement, should be composed of representatives of Engineering, Manufacturing, and Product Assurance (Reliability and Quality), because many factors must be considered. These include:

The direct cost to manufacture or procure

The time to first need

Delivery rate requirements

Technical support needed

Technical capability required to achieve a successful design or production capability within the budget and time available

The capital needs

The *request for information* (RFI) is the means by which buyers canvass a particular segment of industry to determine which manufacturers are interested in, and capable of, providing the component or material in question. The list would include those who can be reasonably expected, based on past performance, to successfully develop or produce the desired item. Although it is not necessary for the RFI to state all of the requirements, it should describe the more important and unusual ones to preclude suppliers from being involved in a quotation for which they are not qualified. After the returned RFIs are evaluated, the *request for quotation* (RFQ) is the next step. These should be very specific, outlining in detail all of the technical and programmatic requirements for

Reliability	Process control
Reliability assurance	Acceptance test and inspection
Quality control and assurance	Packaging
Configuration management	Preservation and shipping

as well as any special controls required because of the technical nature of the product. If the RFQ includes the requirement for developing a new or significantly modified product, it should also contain the requirements for deliverable software, including:

Design disclosure

Test and inspection plans and procedures

Qualification test plans and procedures

Special process procedures

Engineering data necessary to evaluate the achieved design against the design requirements

If a source competition is to be held among several quoters, the selection criteria should also be specified in detail in the RFQ. Unless all of the requirements are stated, any items requested later in the negotiation or performance phase will undoubtedly raise the price of the item more than if they had been properly detailed in the RFQ.

Hardware qualification will ensure that the actual hardware which the prospective supplier(s) intends to deliver will perform properly in the end product. Qualification can consist of nothing more than a simple ambient test or range through a variety of options to being built into the end product and tested in a complete series of environmental and life tests. The selection is subjective, and made jointly by Product Assurance and Engineering. The details should be in the RFQ so that the supplier knows the testing that the product must pass in order for that supplier to be selected.

Quote evaluation is the process by which Procurement, with Engineering, Product Assurance, and (often) the Program or Project Office, evaluate each supplier's quotation preparatory to making a selection and award. The scope and depth of the evaluation will vary, depending upon the criticality of the item, its cost, and the degree of competition involved. The method of evaluation should have been determined at the time the RFQ went out, and at least the broad basis for competitive selection made known to the quoter, if not the details, including weights to be applied to the various factors. The best evaluation criteria have sometimes been made up by the bidders themselves. Product Assurance must be involved in the evaluation process. The weights assigned for both the evaluation and the final source selection, as well as the tie-breaking criteria, should recognize the importance of reliability and quality relative to cost and schedule. If the quotations are not sufficiently informative, it may be necessary for on-site evaluations to be made at one or more of the bidders' plants by a team comprised of representatives from Procurement, Product Assurance, Engineering, and Manufacturing. This is a check for the relative adequacy of each bidder to produce the item in conformance with the reliability and quality requirements, as well as in the quantity and to the schedule required.

Supplier selection is the final step in this process. A source selection board, consisting of the heads (or lower managers for less important selections) of Procurement, Product Assurance, Engineering, and the Program Office, first set weights for all elements—reliability and quality, performance, schedule, cost, management. It then considers the

Results of the evaluation of the quotations

Reports of the supplier visits

Qualification test results

History of supplier reliability, quality, cost, and schedule performance

Applying the weights previously established, the selection board arrives at a numerical rating for each competitor, which generally determines the selected source (or sources if more than one is to be selected). In the event of tie, often defined as a difference of less than 5 percent in the numerical rating, the tie-breaking criteria are used.

Single or multiple sourcing is a decision that should be a matter of policy. Multiple sourcing is ordinarily selected to provide protection against a single supplier being unable to meet the reliability, performance, or delivery requirements. However, it is an added expense, partly because of the loss of quantity savings and partly because it requires dual administration. Furthermore, dual sourcing results in two populations, differing slightly, which must be accounted for in the reliability tracking system. Furthermore, there is another (psychological) deterrent to its use; when one supplier's product becomes unacceptable, less than maximum effort is applied to correcting the problem because it is believed that the other source will always be available; as a result, although the expense of two sources was incurred, the dual sourcing advantages do not accrue.

6.2.3 Buying and Negotiating

Buying and negotiating can range from simple telephone purchases of off-the-shelf components to negotiated multimillion dollar subcontracts. However, they all involve ordering the desired material from the proper source in the proper quantity to an agreed-to level of reliability and quality, on a desired schedule and at a reasonable price. The final contract or purchase order, not the proposals, determines the work that is to be accomplished and the products to be delivered, and these final contractual agreements are determined at negotiations. It is often possible to find a more cost effective way to achieve a desired product reliability than through the reliability program specified in the RFQ, but this should be worked out between the respective product assurance organizations, not between contract administrators and negotiators.

Most manufacturers' procurement policies, as well as the federal procurement regulations, encourage the use of firm-fixed-price (FFP) contracts, and often require justification for any other form. Although FFP contracts can result in adequate reliability and quality of supplier's product for simple procurements, they also can present a hazard for complex items. This is because FFP subcontractors find one of the easiest places to economize is in the assurance discipline efforts. This is difficult to detect, and the resulting degradation in reliability of the product is generally not immediately obvious. Hence, Product Assurance should serve as the check-and-balance on Procurement's "mandate" to utilize FFP contracts; the buying organization has great latitude in selecting the subcontract form.

One of these alternative forms is the incentive contract (either fixed price or cost plus), which can selectively focus a supplier management's attention. However, a cost-only incentive should be prohibited by Product Assurance when reliability or quality of the product is a significant and difficult factor, since the effect will often be a reduced supplier reliability and quality assurance effort. There should always be an offsetting reliability and quality incentive which is more important to the subcontractor than the cost incentive, i.e., a decrease in profits that is more for a loss in measurable reliability and quality than the increase in profits gained from reduction in costs. And it should be done with the thought in mind that degradation in reliability and quality is more difficult to measure than decreased costs. See Sec. 6.3.1 for further discussion of contract types.

6.2.4 Administering Contracts

There is a long time span between order placement and delivery of material, from a few weeks for simple off-the-shelf components to as long as 18 months for complex, exotic items or materials. Product Assurance and Procurement must continuously monitor the supplier's activities throughout this span, watching progress for reliability, quality, cost, and schedule compliance. If there is development work involved, engineering will also need to monitor, providing guidance and trade-off decisions as the development proceeds. Cost growth must be watched to be sure that cost overruns are not driving reliability and quality cost-cutting. Product Assurance should be directly involved throughout the manufacture, particularly of complex material. This can be achieved by

- Audits of the supplier's compliance with the reliability and quality system requirements
- In-process inspection of the supplier's manufacturing by residents or itinerants
- Review of supplier's in-process test or manufacturing data
- A combination of any or all of these

These actions should be preplanned by Product Assurance for each commodity, and the subcontractor notified in the contract work statement that there will be internal involvement.

Contract administration also includes operating the company's purchase order change system, and the material review board (MRB) system for dispositioning nonconforming material when suppliers request waivers or deviations. These should be highly disciplined systems which ensure positive configuration control; it is equally important that they be operated expeditiously so that suppliers are not unduly delayed, generating unwarranted schedule pressure and a probable degradation of reliability.

Of particular significance in administration of subcontracts is the supplier's corrective action system for reliability and quality problems. This is an important part of the company's overall corrective action system; a simple portion of the system as it relates to suppliers is shown in Fig. 6.1. Problems found in house with a supplier's parts are relayed to the supplier through the buyer, but the Product Assurance corrective action engineers should maintain face-to-face contact with the supplier's quality control personnel until each problem is satisfactorily resolved.

6.2.5 Acceptance and Material Control

The contract administration cycle ends with the acceptance of material, which can be done either at the source, at the destination, or by a combination of actions at both. With reliability-critical material, some testing of the supplies will almost always be performed, usually by the supplier, and sometimes partially or totally duplicated at the receiving end. The higher the required reliability, the more testing that must be performed; often the cost of testing so-called high-reliability parts equals or exceeds the cost of producing the parts.

It is desirable to have suppliers perform acceptance testing (and screening) in their plants, so that only acceptable material is shipped. However, retesting at destination is sometimes advisable to detect any problems which have been caused in shipping. Furthermore, it is prudent to have duplicate test equipment available in the company to allow failure diagnosis of suppliers' parts to be performed as an integral part of failure diagnosis of the end product. An excessive amount of coordination is required if suspect parts must be shipped back to the supplier for analysis during an in-house investigation of an end-product failure.

Care should be taken to handle, stock, and issue material and parts so that their delivered reliability is not degraded. A system of control should operate to ensure compliance with instructions and limitations published jointly by Reliability and Engineering for each item or material which requires special precautions. Sometimes brightly colored labels, e.g., HIGH RELIABILITY PARTS. SPECIAL HANDLING REQUIRED, affixed to containers help to discipline the control system. First-in-first-out (FIFO) material issue procedures should be followed as standard practice to ensure timely discovery of substandard material, to minimize its effect, and to facilitate reliability assessment of the product.

6.3 SELECTED TOPICS

The following seven topics are of special interest to the reliability community and will be discussed in some depth. The topics are

Types of contracts

Supplier reliability programs

Standard parts programs

Upgrading commercial parts

Cost and reliability trade-off studies

Specifying and measuring part reliability

Vendor data and history

6.3.1 Types of Contracts

One of the significant developments in the field of procurement has been an increasing use of precise and quantitative terminology describing reliability requirements in contract and purchase order documents. For purchase of relatively low value off-the-shelf or slightly modified products, fixed-price contracts with these quantitative requirements are often adequate to ensure that parts of adequate reliability are supplied, particularly if the burn-in, screening tests, and reliability demonstrations prerequisite to acceptance are specified in detail. Such reliability requirements are well understood by those suppliers who ordinarily provide so-called high-reliability parts, and there is no need, usually, to utilize other than a fixed-price contract form.

When the purchase is for more complex items, and especially when the contract requires fairly extensive development or redevelopment effort, a fixed-price arrangement may not achieve the desired results. In this situation, the buyer wants the attention of the supplier's management directed to the program, and if, as is often the case, the program does not represent a major part of the supplier's business, management is unlikely to focus on it. A contract form which will help achieve this is the incentive contract, because incentive contract performance directly affects the return on investment and profit on sales figures, which are commonly watched by senior management. Incentives for reliability are not the only ones normally used in incentive contracting. Most contain multiple features, including cost, schedule, performance, and quality. Consequently, the first item for consideration of a reliability incentive is the relative weight with respect to the other parameters. This is usually a matter for negotiation between the manufacturer and customer, but the manufacturer should recommend a weighting that best meets the prime parameters of the contract. If, for example, accuracy is the prime objective of the contract, then accuracy should be given the heaviest weight.

Reliability demonstration tests prescribed by the specification and tests for purposes of determining incentive earnings are not necessarily the same. In some cases there may be little or no resemblance, as, for example, when exceedingly high mean times between failures (MTBF) are involved and it is economically feasible to demonstrate reliability statistically.

With incentive contracting, definition of failure for incentive payment purposes should be established and included in the contract. This may vary somewhat, though not necessarily, from the specification definition of success or failure. For example, if there are incentives for both accuracy and reliability, a determination must be made as to whether excessive drift is one or the other, since the supplier should not suffer a penalty in both areas. Even if the definitions are more arbitrary than scientific, they must be spelled out in detail. It is often easier to define what does *not* constitute a reliability incentive failure. Thus types of failures which do not affect MTBF can be excluded from the MTBF calculations and the incentive criteria. One last consideration is the incentive swing, or the portion of the profit or fee devoted to the incentive. If an incentive is to achieve its basic goal, i.e., obtain management attention, it should represent a significant part of the profit on the contract. Anything less than about 20 percent is unlikely to achieve the purpose. Percentages in the 40 to 50 percent range are not uncommon in successful incentive contracting.

6.3.2 Supplier Reliability Programs

An appropriate reliability program for a supplier can range from as little as collecting failure data on acceptance testing to as much as a complete duplication of the in-house reliability program. Each supplier and product should be analyzed for effect on the reliability of the end product, and an appropriate program should be selected and incorporated in the overall reliability program for the project. The elements of work desired of the supplier should be carefully delineated in the RFQ and in the subcontract work statement.

The selection of various elements for each supplier should be based on the following considerations:

1. Criticality of the reliability of the supplier's component to the reliability of the end product.

2. Amount of development or redevelopment work the supplier must perform to meet the end-product reliability and performance requirements.

3. Complexity of the component, and the ease with which reliability and operability can be ascertained in acceptance testing.

4. History of supplier's past performance in developing and/or manufacturing similar devices.

5. Complexity and performance and reliability requirements of the end product.

6. Budget and other resources available for reliability.

Figure 6.2 is an example of a product assurance plan for a group of suppliers. It is important that the quality control plan and the reliability plan for suppliers be integrated, as in the example, since the work is interdependent.

When the supplier is developing new and unique components that are complex and critical to the reliability of the end product,

1. An apportionment of the overall reliability should be made to the supplier; this should be subject to review and revision as the development progresses, along with the apportionment to all subelements of the product, to allow for optimization of the achieved reliability. The subcontract work statement should allow for such adjustments.

2. The desired reliability demonstration test program should also be delineated from the outset of development.

3. The specific reliability program elements desired of the subcontractor should also be specified (MIL-STD-785 is a useful guide for specifying reliability elements), and the supplier should be asked to prepare for submittal and approval a detailed reliability work plan and schedule.

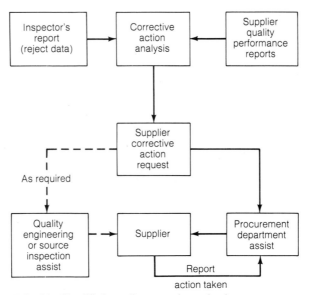

FIG. 6.2 Simplified supplier corrective action loop.

MIL-STD-785 lists the following elements, any or all of which can be required of a subcontractor:

Test requirements for development, qualification, and acceptance

Environmental requirements for equipment design and testing

Components part testing

Maximum preacceptance operation

Integration of government- (customer-) furnished equipment in the system from a reliability point of view

Parts reliability control

Identification of critical items

Supplier and subcontractor reliability programs

Reliability training and indoctrination

Human engineering

Statistical methods

Effects of storage, shelf life, packaging, transportation, handling, and maintenance

Design reviews

Manufacturing control and standards

Failure data collection and analysis, and corrective action

Reliability-demonstration plans (general and specific)

Periodic and final reports

It should be noted that this list implies that subcontractors may in turn be required to assign part or all of the reliability program to their suppliers (see Figure 6.3).

The degree of company involvement in a supplier's development, production, and reliability program is a separate issue and is more dependent upon the probability of the supplier meeting the reliability requirements than on the complexity of the product. The higher that probability, the less direct involvement is required. Involvement can range from nothing more than periodic audits of supplier's conformance to the reliability system, and source inspection of the product, to complete resident offices staffed with reliability, quality, engineering, and procurement personnel. If a meaningful reliability incentive can be devised and incorporated in the subcontract, a lesser degree of company involvement in the supplier's work could be appropriate.

6.3.3 Standard Parts Program

This is the element of an overall reliability program which aims to limit the number of different part sizes, designs, or manufacturers used in a company, project, or equipment design. Standard parts programs are somewhat different than standardization programs, which tend more toward limiting or reducing the number of sizes (e.g., bolt lengths) used in the design or manufacture. Although some of the standardization principles are applied to the standard parts selections, standard parts programs emphasize limiting the number of part numbers and suppliers to maximize the benefit of history and experience with the usage of particular parts.

	RELIABILITY PROGRAM PLAN	CONFIGURATION CONTROL	PART SELECTION APPROVAL	TEST PLAN	NDT PLAN	INSPECTION PLAN	SOURCE INSPECTION	FIRST ARTICLE	Q.C. SYSTEM	INSPECTION SYSTEM	TEST REPORTS	MATERIAL ANALYSIS REPORTS	SPECIAL PROCESS APPROVAL	DISCREPANCY REPORTS	PART AND MATERIAL TRACEABILITY	LOT/BATCH IDENTITY	SERIALIZATION	RELIABILITY DEMONSTRATION
GUIDANCE AND CONTROL SECTION	X	X	X	X		X	X	X	X	X	X		X	X	X	X	X	X
PROPULSION SYSTEM	X	X			X	X	X	X	X			X	X	X	X	X	X	
SEEKER	X	X	X	X		X	X	X	X	X	X		X	X	X	X	X	X
CONTROL SURFACES		X			X	X				X		X			X	X	X	

FIG. 6.3 Sample of supplier reliability and quality control requirements.

6.12

Standard parts are often called approved parts. The advantages of a standard parts program include:

- Can significantly improve equipment reliability by encouraging designers to use only parts with proven reliability and which have been purchased previously from known suppliers
- Reduces the number of different suppliers, providing economies in supplier control activities and permitting better control of fewer suppliers
- Maximizes transfer of good and bad experience with specific suppliers and parts between companies, divisions, projects, or designs
- Provides an administrative mechanism to impose part derating techniques on an equipment design to further improve reliability
- Reduces costs of qualifying new parts or suppliers
- Improves economy of scale by use of larger quantities of fewer parts and enhances company influence over suppliers
- Provides another control over subcontractor design and development activities
- Reduces the number of part specifications which must be prepared and maintained

The disadvantages of a standard parts program include:

- Impedes designers' freedom to select parts which optimize performance, therefore possibly resulting in some reduction in equipment performance by shifting design optimization unnaturally toward part reliability considerations
- May sometimes increase weight or size of parts
- May discourage product improvement efforts by designers
- Effect of a part problem may be greater, since the same part may have multiple uses

One widely used mechanism to implement this program is the "Standard (or Approved) Parts Manual," which is prepared and maintained by parts engineers in Reliability. The manual is usually organized by part types (e.g., semiconductors, connectors) and contains

- Lists of approved parts
- Lists of approved suppliers for each part
- Suppliers' specifications for each part, and/or special company or project specifications
- "Shopping lists" of various screening and burn-in requirements which may be applied to each part to obtain different levels of "goodness"
- Limitations on application of each part, e.g., maximum voltage

Separate from but complementing the manual is a data file containing:

- "Goes-into" listing of using assembly part numbers for each part
- History of quantities purchased and used

- Unit prices paid
- History of acceptance or rejection of each part purchased
- History of all part failures in higher assemblies
- History of all part failures in field usage
- Reference list for part failure diagnosis reports

The system is dynamic in that manuals are updated as history of performance warrants. Furthermore, parts organizations, either on their own or on the request of Engineering or Procurement, independently test new parts as they appear on the market and survey or audit potential new or additional suppliers. This provides a continuing input of new parts or suppliers and ensures keeping up with advances in the state of the art for each part type.

6.3.4 Upgrading Commercial Parts

When the reliability required on an end product requires part reliability higher than the inherent reliability of commercial parts, considerable extra expense is entailed for the additional testing that is necessary. This added expense frequently exceeds the base price of the commercial part. Techniques have been developed to reduce this added expense, generally known as upgrading commercial parts. (Commercial parts are parts bought off the shelf, having had no special reliability testing or special extra controls over the manufacturing process, test, inspection, or materials.)

The reasons for these techniques are the recognition that not every application of a given part stresses the part to the margin of its capability and that only when the reliability of the part in a high-stress application begins to degrade below the part reliability required is it necessary to upgrade the commercial parts. With this premise, only a portion of any lot of commercial parts must have higher reliability than the lot as a whole. The parts having higher reliability can be selected out of the lot by suitable testing or screening. A system of high-reliability identification of the screened parts and of control over the usage of non-high-reliability parts is required to make this technique work.

The screening tests used must be nondestructive to the commercial parts' reliability, since these parts will be used in the low-stress applications. One widely used technique is to monitor the behavior of the important attributes of parts in the lot while the parts are powered and sometimes in a high-temperature environment, and then select for high-reliability application only those parts whose values shift less under time-temperature operation than the mean of the entire population of the lot.

Another method is to power all of the devices in the normal way, and after allowing time for stabilization, to perform an infrared scan of the lot. Those with a radiation pattern different than the norm may be considered to have an inherent defect and are not used in high-stress applications.

A third technique is to buy commercial-grade parts and to perform all of the extra reliability testing in house. This is cost effective if (1) the supplier charges more for the high-reliability version than just the cost of the testing (this occurs frequently with suppliers whose customers are predominantly manufacturers of commercial-grade equipment, and the supplier wants to recover the "inconvenience" costs of a few high-reliability parts), and (2) the company already has the capital investment in high-reliability parts test equipment which is not being fully utilized.

6.3.5   Cost vs. Reliability Trade Studies

The purpose of performing cost vs. reliability trade studies is to quantify and compare the costs of various optional levels of the supplier's parts, or of the supplier's reliability effort, with the expected benefit or incremental benefit from each, so that decisions can be made. Many decisions are necessary to establish the supplier reliability program, many of which can be made better if a trade study is performed. Examples of questions to ask when reaching decisions in which reliability vs. cost trade studies are useful are:

- Should the company reliability program flow down to suppliers? Which elements? To which suppliers?
- Should commercial parts be used? Should they be upgraded? All of them? Some of them? Which ones?
- Should reliability tested parts be used? Which ones, if not all?
- To what confidence level should reliability of suppliers' parts be demonstrated? Which parts?
- Should suppliers provide reliability testing, screening, and demonstration, or should it be performed in house?

Life-cycle cost studies by the military tend to show that improving product reliability improves the life-cycle cost; the studies are not rigorous, however. Similarly, it is generally believed that the higher the reliability of a product, the higher the cost of obtaining it. Accepting these as premises simplifies performance of reliability vs. cost trade studies.

Although gross cost of a given reliability, and the gross benefit, are difficult to quantify, incremental cost and benefit can generally be estimated fairly closely. In procurement, the incremental cost (or price) can be ascertained by asking in an RFQ for different levels of reliability, different levels of demonstration confidence, or different elements of a reliability program. Requesting several different warranty periods is also useful in estimating the incremental cost of different levels of reliability. The change in reliability cost with a change in reliability apportionment to a supplier is another way to cost incremental reliability.

Pricing the benefit of incremental reliability is a matter of converting unreliability to such identifiable costs as the cost of repair or replacement of failed hardware in the factory or in the field, the cost of downtime or loss of use, plus the administrative costs of handling failed hardware. The intangibles (customer dissatisfaction) cannot be reasonably quantified in the usual sense, but a value should be estimated for trade study purposes, a value which will be appropriate in the calculations for the expected near-term gain or loss in sales. (If the company enjoys a good reliability image, this number would be small and relatively insignificant; if the company has a poor reliability reputation, this number should almost be overriding.)

6.3.6 Specifying and Measuring Part Reliability

There are many ways that the desired reliability in suppliers' products can be specified, although some are more appropriate than others for different kinds or classes of product. For each way of specifying, there are one or more appropriate ways of testing to provide the necessary data to measure the attained level. Figure 6.4 categorizes these methods by product types.

In the figure, Type I parts or components are items such as relays or transistors which are to be used in the manufacture of an assembly. Type II are those instruments or equipments which are purchased as complete entities to perform specific functions. Type III relates to large systems, such as a computer mainframe or military or commercial radar.

Reliability Specifications.   Listed below are the commonly used reliability specifications for the types of products shown.

1. *Time to failure (TTF):*   Used when a small sample of the lot of parts can be stressed at a level and for a time long enough for the entire sample to fail. The average or mean of the times to failure is considered the TTF for the lot.

Category	Identity	Useful reliability specifications	Typical test quantity involved	Common characteristic of test
IA	Component parts (samples)	Time to failure (TTF) Stress to failure (STF) Percent failure or success (%F, %S) Failure rate (λ)	Small sample	Destructive
IB	Component parts (lots)	Acceptable performance (a or r) Percent reject (% reject) Stability	Large lots	Nondestructive screening
II	Equipment subassemblies (black boxes)	Mean time between failures (MTBF) Mean time to first failure (MTFF) Mean time to failure (MTTF)	Small quantity; typical of population	To evaluate reliability
III	Systems	Probability of success (R, P_s)	Single	Nondestructive; seldom completely simulates end-use conditions

FIG. 6.4 Categories of reliability specification and testing.

2. *Stress to failure (STF):*   One or more stress factors such as temperature or voltage are increased until an entire sample has failed. The mean value of the STF, and the distribution of failure events about this mean, are used as measures of part reliability of the lot.

3. *Percentage failure or success (%F, %S):* Complementary methods frequently used to specify the reliability of one-shot devices like current-limiting fuses or rifle ammunition.

4. *Failure rate (λ):* Generally applied to component parts and denotes a time-based average rate of failure. Common levels are in fractions of percent of a lot for each 1000 hours of operation, although it can be expressed as 10^6 part-hours per failure, which is one order of magnitude smaller than the 10^5 equivalent of percent per 1000 hours.

5. *Acceptable performance (a or r):*   A measure of the amount key parameters of the parts drift beyond the acceptable limits during a specified mission life.

6. *Percent reject:*   Often used as a measure of reliability, particularly when the population is large and all parts in the lot are tested functionally.

7. *Stability:*   A measure of the change or wander of important parameters of the parts in the lot with time in operation.

8. *Mean time between failures (MTBF):* Denotes the average time between failures for an equipment during the portion of its life cycle in which the exponential failure rule applies, and is used for equipments which can be repaired and returned to use. It may be measured on a single piece of equipment or determined by testing several equipments through several failure incidents and averaging the total operating time for the total number of failures.

9. *Mean time to first failure (MTFF):*   Identical with mean time to failure (MTTF), this is the average time an equipment can be expected to perform before it experiences its first failure, and is especially useful for equipments which will be installed in a use location not accessible for maintenance or repair.

10. *Probability of success (P_s, R):* Denotes that the acceptability of the reliability of a system can be established by a prediction derived from a composite of theoretical analyses and computations based on qualified measurements. These computations generally follow the basic formula

$$P_s = e^{-t/m}$$

where P_s = probability of survival (success)
$\quad t$ = time of test
$\quad m$ = MTBF
$\quad e$ = Naperian base

Measuring Product Reliability.   To measure product reliability usually means testing the product in some manner to obtain time-related failure information. The testing, however, does not necessarily have to be solely for the purpose of measuring the reliability, since ordinary acceptance testing may provide enough time-related failure information to assess the reliability of the product at the same time. However, whatever testing is performed should provide data appropriate to the type of reliability specification used for the product. The following discussion of testing conforms to the categories in Fig. 6.4.

Category I, Part Reliability Testing. This measure is broken into two divisions, IA and IB. Type IA tests are usually performed on samples from lots, whereas type IB tests are usually performed on the entire lot. A reliability testplan will often require both types in order to provide for a complete evaluation of the critical reliability factors. The reliability of parts depends on the values for two types of failure—random catastrophic and degradation. The same tests may provide information for both types, but the data are analyzed differently.

The ideal reliability test for parts is based on the following conditions:

1. The test lot is large.
2. The test lot is expendable and can be tested to failure.
3. The test environmental conditions exactly simulate the final end-use conditions.
4. No time limits are imposed on the test.
5. Complete automatic measurement facilities are available for continuously monitoring the major parameters for each part.
6. The complete life history of each item is obtained and analyzed.

From such a test the following reliability information can be established with a very high degree of confidence:

1. The existence of any early life debugging or burn-in period, its duration and severity being accurately measured.
2. The presence of any flat portion of a life characteristic curve during which the failure rate for the lot is nearly constant and the failure rate events are random in time. The value of this random failure rate would be accurately measured.
3. The individual degradation curves for each of the items would be accurately determined so that final wear-out time is measured to specified acceptance limits.
4. The mean time to wear out for the lot and the distribution about this mean would be measured.

This type of test was used to evaluate the component parts in the American undersea telephone repeater amplifiers. A large percentage of each lot of parts used for this service was expended in comprehensive parts tests. Few projects allow the luxury of this ideal testing, but since the information is needed, simplified tests must be devised to provide approximate measures of reliability.

Measuring random chance failures can be accelerated by running the tests at maximum electrical rating, under temperature, humidity, or vibration environment. If these tests are short, the parts will not be degraded. As large a sample as practical should be tested, and data should be gathered continuously on failure rate to guard against the wear-out period being reached in the testing. (MIL-STD-781, *Reliability Design Qualification and Production Acceptance Tests: Exponential Distribution,* defines wear out as beginning when the failure rate reaches twice the constant value.) The sample size is determined by the length of time allowed for the testing and the confidence desired in the failure rate measurement.

A point estimate at about 60 percent can be determined from:

$$\text{Failure rate} = (f/Nt) \times 10^{-5} \text{ in percent per 1000 hours}$$

where N = the number of parts in the test sample
$\quad t$ = the length of the test in hours
$\quad f$ = the number of failures observed

Figure 6.5 is a family of curves for different values of f matching failure rate against test item hours. It can be seen, by extrapolating the graph, that to measure a failure rate of 0.1 percent per 1000 hours, there can only be one failure in 1 million test item hours, e.g., 100 samples for 10,000 hours each. And 0.1 percent per 1000 hours is not very good reliability for parts destined for complex assemblies. Some relief from this problem can be obtained by using lower confidence levels, as shown in Fig. 6.6; 0.1 percent per 1000 hours could be measured with 10 percent confidence with about 100,000 test item hours, or 100 samples for 1000 hours each.

From these curves, it is clear that the very low failure rates required of parts in complex systems cannot be measured in short reliability acceptance tests. The usual alternative is to substitute an accelerated test on a moderate-sized sample to confirm the uniformity of a high-quality product which has previously proved reliable, or to achieve an accelerated failure rate. Another alternative is to base the total product failure rate on cumulative data obtained over a period of time from tests on many lots.

Accelerated tests for reliability acceptance of parts are widely used.

1. *Derating factor tests:* These tests stress the test sample to the maximum rating of the critical parameter for the part, and acceleration factors are then applied to achieve a probable failure rate which could be applicable at derated conditions. For example, if a test at maximum rated voltage shows the part to have a failure rate of 1 percent per 1000 hours, the following equation will give the approximate failure rate d at a derated application of 20 percent of full rated voltage:

$$d = \frac{\text{full rating}}{(\text{rated voltage/derated voltage})^3}$$

$$= \frac{1.0\%/1000 \text{ hours}}{(V_R/0.2V_R)^3}$$

$$= \frac{1.0}{5^3} = 0.008 \text{ percent per 1000 hours}$$

(The power in the denominator is a function of part type, but third power is typical of electronic parts.)

2. *Step stress:* This is another important form of accelerated test and reveals the uniformity and strength of a product but does not normally yield failure rate data. The step stress test repeatedly employs increased stresses according to a prearranged test plan (see Fig. 6.7). One or more types of stress such as temperature and voltage can be combined in this test with increments of time. After testing at each step or level of stress for the prescribed interval of time, the parameters are measured and the number of rejects or failures is determined. The test is continued according to the plan until the entire sample has failed. A sample size of about 20 to 40 parts is usually adequate. The conditions of environment and electrical stress to be imposed at each step are planned to start at or near the maximum rating for the item being tested and be increased regularly according to the plan until 100 percent failure of the sample results.

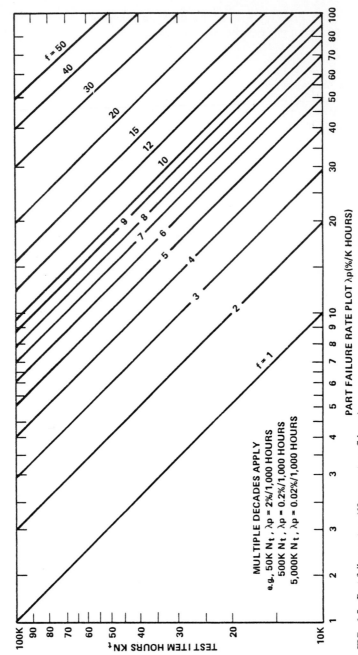

FIG. 6.5 Part failure rate p (60 percent confidence).

6.20

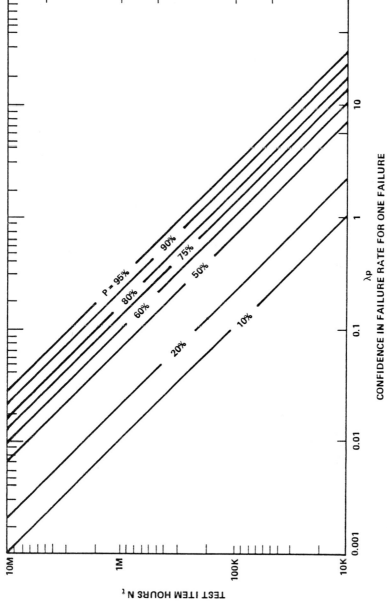

CONFIDENCE IN FAILURE RATE FOR ONE FAILURE

FIG. 6.6 Failure rate of parts, %/1000 hours, for one failure.

6.21

The data, plotted as a density function after smoothing, reveal the strength and uniformity characteristics of the lot.

3. *Degradation tests:*   These tests are used to determine how much the values of key parameters change during power-on cycles. They are used when the design safety factors are not large enough to eliminate possible circuit malfunction-caused by part parameter drift. Most parts have from one to three parameters which must not drift beyond certain limits during the reliability life period; these limits are usually established during the circuit design worst-case analysis. The part degradation tests are used to evaluate the probability of drift beyond the acceptable limits. The tests can be relatively short to predict the ultimate failure point nondestructively.

The degradation test is usually from 1000 to 1500 hours long and is performed on 100 percent of each lot. The parts are cycled with power on and off according to a normal use schedule. The load and temperature conditions are preferably set at the rated maximum for the parts. Repeated measurements of the critical parameters are taken during the tests to determine the part stability and trends of value change measured in the time domain. The exact values for each item at the start and finish of a test are not as important as the degradation path revealed by the repeated measurements, which defines the inherent stability of the parts.

Category II, Equipment Reliability Measuring and Testing.   Measuring equipment reliability involves ascertaining the MTBF (for repairable equipments) or the MTTF (for nonrepairable equipments). The testing generally consists of operational tests performed under simulated end-use conditions with acceptable MTBF and confidence specified. If early or infant mortality failures are removed by a suitable burn-in period, and if wear-out failures are removed by suitable preventive maintenance and replacement, then only chance failures remain. The objective is to measure the chance failure rate over a short period of time and then

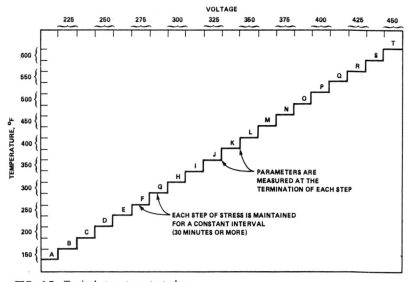

FIG. 6.7 Typical step-stress test plan.

affix to this a probability or measure of confidence that the mean of the short-time sample is typical of a normal life period. One cannot specify an equipment to have a certain MTBF and a certain confidence; one can only specify that an equipment will have a certain inherent MTBF which will be proved to a certain confidence by a specified test. The test measures the most likely value of MTBF, and the amount of statistical data obtained during the test must be evaluated to determine the confidence which can be placed on the measurement. Usually, for reliability acceptance testing, the single-sided description stating the cumulative probability that a measured MTBF is greater than a certain specified minimum value has the greatest usefulness.

In *conventional testing,* equipment is run under specified conditions and the failures counted. Thus if a test of an equipment is run 600 hours, and three failures occur, the point estimate of the MTBF is 600/3 or 200 hours. Referring to Table 6.1, the 90 percent K factors, 3.68 and 0.476, multiply the point estimate to give upper and lower confidence band limits at 736 and 95.2 hours, respectively. An examination of Fig. 6.7 will reveal that the confidence band is quite wide when only enough test time is devoted to allow just a few failures. For the most effective testing, at least 10 failure incidents should be allowed.

The lower-limit half of Table 6.1 can be used to plan equipment reliability acceptance tests to prove with a known confidence that the MTBF is greater than a certain specified figure. However, a simpler method using Table 6.2 will give the same answer. For example, if we want to prove with a 95 percent confidence that the MTBF is greater than 191 hours, and set 10 failures as our acceptance number, the w value at 95 percent confidence for $f = 10$ is 15.70, which when multiplied by the specified 191 hours, gives 3000 equipment hours required. This can be supplied by running one equipment for 3000 hours or, for example, six equipments tested for 500 hours each.

In *sequential reliability acceptance measuring and testing,* the length of the test is not established before the test begins, but depends upon what happens during the test. The sample is tested while subjected to a prescribed environment and duty cycle until the preassigned limitations on the (producer and consumer) risks of making both kinds of wrong decisions based on the cumulative test evidence have been satisfied. The ratio of quantity of failures to the length of test at any test interval is interpreted according to a sequential analysis test plan. Conspicuously good items are accepted quickly, conspicuously bad items are rejected quickly, and items of intermediate reliability require more extensive testing.

The major advantage in using sequential test procedures is that they result in less testing on the average than other procedures when the preassigned limitations on the risks of making both kinds of wrong decisions are the same for both tests. The chief disadvantage is that the test time required to reach a decision cannot be determined prior to testing.

Both risks cannot be limited if a decision is to be made within a reasonable amount of time. Since the consumer's error (of accepting the population when the reliability is less than the specified number of hours) is the desired minimum reliability, it is usually considered to be the more serious and is therefore limiting.

A widely used sequential test plan is found in MIL-STD-781, *Reliability Qualification and Production Approval Tests,* from which Table 6.3 is taken. (It is plotted graphically in Fig. 6.8.) This particular plan is based on consumer and producer risks of 10 percent. In Fig. 6.8, Case A equipment had only 8 failures when the accumulated total test time passed 10.9 times the specified MTBF, and was accepted; Case B reached 14 failures when only 6.8 times the specified MTBF had been reached, and was therefore rejected.

TABLE 6.1 Multiplier K for Two-Sided MTBF Estimate

No. of failures	Upper limit K_U					Lower limit K_L				
	95%	90%	80%	70%	60%	60%	70%	80%	90%	95%
1	28.6	19.2	9.44	6.50	4.48	0.620	0.530	0.434	0.333	0.270
2	9.2	5.62	3.76	3.00	2.43	0.667	0.600	0.515	0.422	0.360
3	4.8	3.68	2.72	2.25	1.95	0.698	0.630	0.565	0.476	0.420
4	3.7	2.92	2.29	1.96	1.74	0.724	0.662	0.598	0.515	0.455
5	3.0	2.54	2.06	1.80	1.62	0.746	0.680	0.625	0.546	0.480
6	2.73	2.30	1.90	1.70	1.54	0.760	0.700	0.645	0.568	0.515
7	2.50	2.13	1.80	1.63	1.48	0.768	0.720	0.667	0.592	0.535
8	2.32	2.01	1.71	1.57	1.43	0.780	0.730	0.680	0.610	0.555
9	2.19	1.92	1.66	1.52	1.40	0.790	0.740	0.690	0.625	0.575
10	2.09	1.84	1.61	1.48	1.37	0.800	0.752	0.704	0.637	0.585
11	2.00	1.78	1.56	1.45	1.35	0.805	0.762	0.714	0.650	0.598
12	1.93	1.73	1.53	1.42	1.33	0.815	0.770	0.720	0.660	0.610
13	1.88	1.69	1.50	1.40	1.31	0.820	0.780	0.730	0.652	0.620
14	1.82	1.65	1.48	1.38	1.30	0.824	0.785	0.736	0.675	0.630
15	1.79	1.62	1.46	1.36	1.28	0.826	0.790	0.746	0.685	0.640
16	1.75	1.59	1.44	1.35	1.27	0.830	0.795	0.750	0.690	0.645
17	1.71	1.57	1.42	1.33	1.26	0.835	0.800	0.760	0.700	0.655
18	1.69	1.54	1.40	1.32	1.25	0.840	0.805	0.765	0.710	0.660
19	1.66	1.52	1.39	1.31	1.24	0.845	0.808	0.767	0.715	0.665
20	1.64	1.51	1.38	1.30	1.23	0.847	0.810	0.768	0.719	0.675
25	1.55	1.44	1.33	1.26	1.21	0.860	0.830	0.790	0.740	0.700
30	1.48	1.39	1.29	1.23	1.18	0.870	0.840	0.806	0.756	0.720
40	1.40	1.32	1.24	1.19	1.16	0.884	0.860	0.826	0.787	0.750
50	1.35	1.28	1.21	1.17	1.14	0.892	0.872	0.847	0.806	0.770
70	1.28	1.23	1.18	1.14	1.11	0.910	0.890	0.860	0.830	0.800
100	1.23	1.19	1.14	1.12	1.09	0.924	0.906	0.880	0.852	0.830
200	1.16	1.13	1.10	1.08	1.06	0.940	0.935	0.916	0.890	0.870
300	1.12	1.10	1.08	1.06	1.05	0.955	0.942	0.930	0.910	0.895
500	1.09	1.08	1.06	1.05	1.04	0.965	0.954	0.942	0.930	0.915

Category III, System Reliability Measuring and Testing. Were it not for the fact that system testing objectives usually encompass considerably more than assessment of the reliability, and for the effect on system reliability of the interactions among subsystems or equipments, system testing could be performed as described in category II for equipments. However, when systems are procured, it is almost universally desired that the operability of each system be demonstrated in an actual or simulated use environment. In an ideal, but rarely realized, situation one or more systems would be available solely to be tested for reliability, in which case the testing methods for equipments would be useful, with the data collected and analyzed to obtain the probability of success P_s reliability measurement.

TABLE 6.2 Multiplier w to Determine Test Time t to Demonstrate MTBF of m or Greater with a Confidence P; $t = wm$, $nt = w$

To number of failures	Multiplier w for confidence P of						
	95%	90%	80%	75%	50%	20%	10%
0 to 1	2.99	2.30	1.61	1.38	0.69	0.223	0.105
2	4.74	3.89	2.99	2.69	1.68	0.824	0.532
3	6.29	5.32	4.28	3.92	2.67	1.530	1.102
4	7.75	6.70	5.51	5.10	3.67	2.297	1.745
5	9.15	8.00	6.72	6.25	4.67	3.089	2.432
6	10.51	9.25	7.90	7.40	5.65	3.903	3.152
7	11.84	10.60	9.07	8.55	6.65	4.733	3.895
8	13.15	11.70	10.23	9.70	7.65	5.576	4.656
9	14.43	13.00	11.38	10.80	8.65	6.428	5.432
10	15.70	14.20	12.52	11.90	9.65	7.289	6.221
11	16.96	15.40	13.65	13.00	10.65	8.160	7.020
12	18.20	16.60	14.77	14.10	11.65	9.031	7.829
13	19.44	17.80	15.90	15.20	12.65	9.91	8.646
14	20.67	18.90	17.01	16.30	13.65	10.794	9.469
15	21.88	20.10	18.12	17.40	14.65	11.682	10.299

Example. To demonstrate m 100 hours with 90% confidence:

$t90,1 = 230$ hours to first failure
$t90,2 = 389$ hours to second failure
$t90,3 = 532$ hours to third failure
$t90,4 = 670$ hours to fourth failure

In the normal situation, where systems are primarily tested for operability, failure occurrence data plotted against operating time of the system will be useful in determining whether the system has passed the infant mortality portion of the bathtub curve (see Fig. 6.9). The data should be smoothed, using a moving average with a base of about three times the desired MTBF, before plotting.

In the design of the system test, special note should be made of the probable system interactions which might affect the reliability of the system. For example, in the design of a reliability measurement test for an airborne radar-based navigation and autopilot system, one would consider the following:

1. *Common power sources*: Voltage surges, poor regulation, conducted radio-frequency interference, loss of power
2. *Radiated interference*: Radio frequency, low frequency, electromagnetic, acoustic noise
3. *Environmental interactions*: Heat, vibration, shock, physical, other
4. *Functional interactions*: Frequency beating, jamming, beam collision, safety
5. *Unspecified conditions*: Environmental, operational, maintenance, transportation and handling, storage, make-ready, checkout

TABLE 6.3 Sequential Test Accept-Reject Criteria

Normalized test time, hours (1)	Total relevant failures for			Normalized test time, hours (1)	Total relevant failures for		
	Reject decision (2)	Accept decision (3)	Continue-test decision (4)		Reject decision (2)	Accept decision (3)	Continue-test decision (4)
0.5	6	...	0–5	16.7	26	...	16–25
1.3	7	...	0–6	17.4	...	16	17–26
2.1	8	...	0–7	17.5	27	...	17–26
2.9	9	...	0–8	18.2	...	17	18–27
3.7	10	...	0–9	18.3	28	...	18–27
4.4	...	0	1–10	19.0	...	18	19–28
4.5	11	...	1–10	19.1	29	...	19–28
5.2	...	1	2–11	19.8	...	19	20–29
5.3	12	...	2–11	19.9	30	...	20–29
6.0	...	2	3–12	20.6	...	20	21–30
6.1	13	...	3–12	20.7	31	...	21–30
6.8	...	3	4–13	21.4	...	21	22–31
6.9	14	...	4–13	21.5	32	...	22–31
7.6	...	4	5–14	22.2	...	22	22–32
7.8	15	...	5–14	22.3	33	...	23–32
8.5	...	5	6–15	23.0	...	23	24–33
8.6	16	...	6–15	23.1	34	...	24–33
9.3	...	6	7–16	23.8	...	24	25–34
9.4	17	...	7–16	24.0	35	...	25–34
10.1	...	7	8–17	24.7	...	25	26–35
10.2	18	...	8–17	24.8	36	...	26–35
10.9	...	8	9–18	25.5	...	26	27–36
11.0	19	...	9–18	25.6	37	...	27–36
11.7	...	9	10–19	26.3	...	27	28–37
11.8	20	...	10–19	26.4	38	...	28–37
12.5	...	10	11–20	27.1	...	28	29–38
12.6	21	...	11–20	27.2	39	...	29–38
13.3	...	11	12–21	27.9	...	29	30–39
13.4	22	...	12–21	28.0	40	...	30–39
14.1	...	12	13–22	28.7	...	30	31–40
14.2	23	...	13–22	29.5	...	31	32–40
14.9	...	13	14–23	30.3	...	32	33–40
15.0	24	...	14–23	31.1	...	33	34–40
15.7	...	14	...	15–24 31.9		34	35–40
15.9	25	...	15–24	32.8	...	35	36–40
16.6	...	15	16–24	33.0	41	40	

Note 1. Column 1 is the total test time (operating time), in hours, accumulated by all equipments in a test group under test divided by the contract specified MTBF, in hours.

Note 2. When the total number of relevant failures accumulated by all equipments in a test group during the corresponding normalized test-time period shown in column 1 (*a*) equals or exceeds the number in column 2, a *reject* decision is made, (*b*) equals or is less than the number in column 3, an *accept* decision is made, (*c*) falls in the range shown in columns 4, a *continue-test* decision is made.

Note 3. A test group may be one or more equipments. A reject or accept decision on a test group is applicable to all equipments of the group from which the test group was selected.

6.26

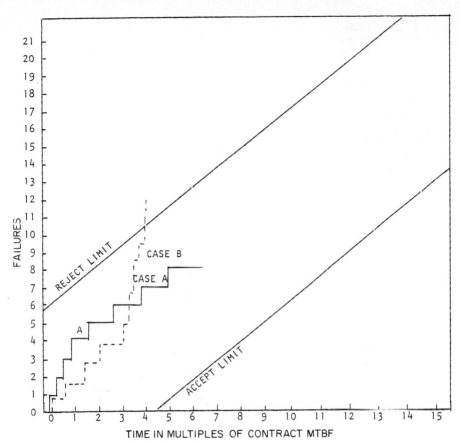

FIG. 6.8   Typical sequential test.

6.3.7   Vendor Data and History

In an ideal world, all decisions, both management and technical, would be based on accurate and complete data, analyzed in depth and presented succinctly andmeaningfully to the decision maker. Although this ideal is rarely met, nevertheless, the data system should strive toward it as a goal.

From a reliability point of view, there are a number of specific *uses* for data, and for which the data system should be designed. They are

Reliability assessment, prediction, and control

Failure recurrence control

Facilitating current and future failure diagnosis

Product, process, and reliability improvement

Supplier evaluation, selection, or reselection

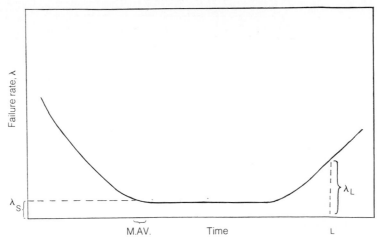

FIG. 6.9 Reliability bathtub curve relations. λ_s = failure rate at the stabilized constant level; M.AV. = moving average base = 3 MTBF = $3/\lambda$; $L = t(\lambda L) = 2\lambda_s$.

There are three basic *sources* of the data required: suppliers' plants, factory, and field. The data required can be conveniently broken into two large categories—background data and operational data.

Background data contain all of the information which should be available to enable intelligent use of the operational data as they are collected. Background data include everything of importance about the product from the development and production programs. They should include

Design disclosure—drawings, specifications, procedures

Engineering studies—parameters, documents, test equipment error analyses, trade studies

Design analyses and reviews—failure mode and effects analyses (FMEAs), worst-case analyses, safety and hazard analyses

Program plans—reliability, quality control, integrated test

Manufacturing plans, flow diagram, shop orders, tool drawings

Inspection plans, flow diagrams, procedures, gauge drawings

Process specifications, procedures

Test plans and procedures—acceptance, reliability, qualification, nondestructive tests (NDT)

Field assembly, repair and test manuals, configuration lists

This background information is particularly useful in interpreting test results to make reliability assessments, and for determining the exact configuration of hardware for failure analysis and control.

Operational data are the specific information regarding precise as-built information, failure and defect data, and time or cycle information. They include:

Equipment operating and test times and cycles

Test data—variables and attributes, from acceptance, reliability assessment or screening, qualification and requalification testing

Inspection data—variables and attributes

Failure analysis reports, including corrective action

Failure and defect data and history

Waivers or deviations granted

As-built configuration lists

Repair, rework, or scrap data

Lot, batch, and serial number records

Results of audits and surveys

Test and inspection success and failure data for a supplier's product should be collected, for at least major suppliers, from the final configuration level in the supplier's plant, through the in-house receiving, in-process, and final assembly levels. If the supplier's product is also tested as part of higher-level assemblies in field delivery or erection testing, these data should also be collected if possible to permit continuous reliability assessment and prediction of the product and can well be the basis for initiating preventive and corrective action when adverse trends are highlighted. In order to support such a use, however, the data system must include continuous analysis of the data as they are received, along with presentation in such a format as to facilitate meaningful trend analysis.

When end-product failures occur which are traceable to a supplier's product, almost all of the background and operational data listed above will be useful in directing the supplier's failure analysis effort and in determining cause and corrective action. In this work, the data system must be extremely flexible, permitting a variety of instant data retrieval, collation, and analysis to be made as needed, and the data presentation for each analysis must be tailored to the immediate need.

For supplier evaluation, selection, or reselection, data which track a supplier's performance against the contract requirement is essential, since a prediction of a supplier's future performance is best based on past performance. For this use, the data system should provide continuing periodic reports of each supplier's performance in some kind of supplier rating format so that relative behavior of competing suppliers can be assessed. The data system output should also trigger corrective action when a supplier's overall performance shows a degradation, e.g., when the defect rate exceeds a predetermined maximum.

BIBLIOGRAPHY

Chapter 5 provides a comprehensive annotated listing of standards, handbooks, and specifications from the Department of Defense. Also included are standards and specifications from other organizations. Many of these items include some aspects of reliability in connection with procurement. Some of the specifications and standards that are directly related to procurement activities are

Specifications

MIL-T-5422, *Testing, Environmental, Aircraft Electronic Equipment*

MIL-Q-9856, *Quality Program Requirements*

MIL-S-52779, *Software Quality Assurance Program Requirements*

Standards

MIL-STD-414, *Sampling Procedures and Tables for Inspection by Variables for Percent Defective*

MIL-STD-690, *Failure Rate Sampling Plans and Procedures*

MIL-STD-721, *Definition of Effectiveness Terms for Reliability, Maintainability, Human Factors and Safety*

MIL-STD-757, *Reliability Evaluation from Demonstration Data*

MIL-STD-781, *Reliability Tests, Exponential Distribution*

MIL-STD-790, *Reliability Assurance Program for Electronic Parts Specifications*

MIL-STD-839, *Parts with Established Reliability Levels, Selection and Use of*

MIL-STD-1556, *Government/Industry Data Exchange Program, Contractor Participation Requirements*

MIL-STD-1635 (EC), *Reliability Growth Testing*

Books

Kapur, K. C. and L. R. Lamberson, *Reliability in Engineering Design*, New York: John Wiley & Sons, 1977.

Lloyd, D. K. and M. Lipow, *Reliability Management, Methods, Mathematics*, 2d ed., published by the authors, 1977.

O'Connor, P. D. T., *Practical Reliability Engineering*, 2d ed., New York: John Wiley & Sons, 1985.

CHAPTER 7
RELIABILITY IN PRODUCTION

Douglas Ekings
Xerox Corporation

7.1 INTRODUCTION

The production process is the essential transition from the product design and development phase to the timely placement of a product with acceptable reliability in the hands of the customer. The objective of the production department during this phase should be to minimize the degradation of the reliability designed into the product. The role of the manufacturing department in the transition process can be described by three basic subphases:

- Product transfer from design to production
- Ongoing reliability activities during production
- Initial introduction of the product to the field

An important additional subphase is the continuous monitoring of field performance with feedback to the product design and development departments. In addition, there must be the implementation of the production reliability function at the proper point with the design and development team, particularly during start-up of production on a new model, to provide the critical linkage from drawing board to field use.

It is typical of industrial enterprises to have strong reliability engineering programs and functional organizations to support the design team. It is just as essential to have the skills of reliability engineering in the production department to provide the continuity and integration of the design and production reliability programs. In addition, communications across departments between members of staff reliability organizations are important in order to develop the policy and procedures to both:

1. Maintain continuity of the production program with the design process
2. Assure a proper transition of responsibility from the line organizations of the design department to the line organizations of the manufacturing department

Experience has also shown that the corporate reliability programs of the design and production of a product only work well if:

1. Supported by senior management
2. There is a common objective to deliver products with acceptable reliability
3. There is an integrated process of reliability activities that provides continuity of effort to the line organizations which is an additional series of functions

7.2 FUNCTIONS OF THE PRODUCTION DEPARTMENT

The production process shown in Fig. 7.1 is typical for most companies, whether large or small. The figure shows the framework of activities that occur during the material acquisition, build-up, assembly, and delivery phases of equipment or subassembly production. The production flow is illustrated in the linkage between the output of the design process and the delivery of product to the customer.

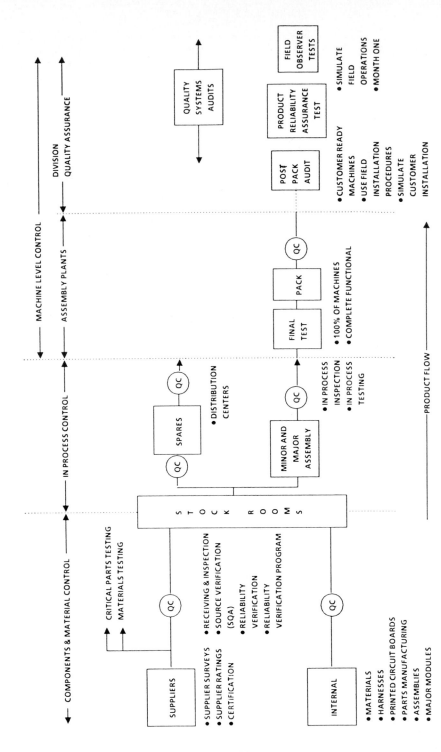

FIG. 7.1 Production process.

This process gives us the framework with which we will show the activities, responsibilities, and integration of the functions of the production reliability program. In order to make clear the reliability activities related to the production phase, a distinction will be made. Reliability engineering activities that relate to the design and development phase are described in other chapters of this handbook. Reliability functions relevant to the production phase are included in the reliability assurance program described here (see Fig. 7.2). The term *reliability assurance* is used here to describe the production reliability activities to denote that:

1. The responsibility for product reliability rests with line management
2. The design determines the reliability potential of the product
3. The principles of quality assurance, which relate to assuring customer satisfaction, are conformed with

7.3 PRODUCTION-RELATED CAUSES OF UNRELIABILITY

The timely and effective implementation of the reliability engineering program during the product development and design phase will provide a level of reliability which will only be degraded through the subsequent production and use phases. There have been many assessments made to quantify the degree of degradation of reliability over the design and manufacturing phases.

An example of this for a complex airborne electronics assembly is shown in Fig. 7.3. We will use 1.0 as the normalized mean time between failures (MTBF) for the "as-designed" reliability. The "as-built" reliability will lessen by a factor to reflect the goodness of the production reliability assurance program. The as-built reliability will further degrade to the "as-maintained" reliability performance in the field.

In Table 7.1 typical causes of unreliability are listed. These hold true for complex electromechanical equipments such as copiers as well as assemblies which are 100 percent electronic. By examination of these categories it should be

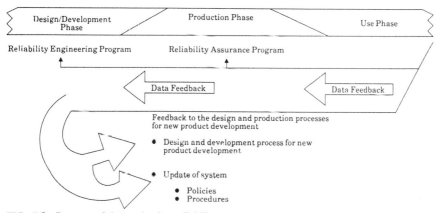

FIG. 7.2 Purpose of the production reliability program.

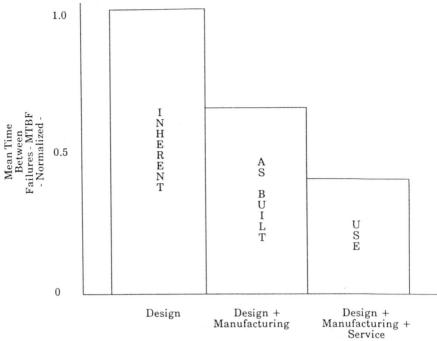

FIG. 7.3 Effects of the manufacturing and use environment on inherent reliability (MTBF).

obvious that:

1. All portions of a company have a role to play.
2. No one organization has 100 percent influence on any one of these categories.
3. Production (Manufacturing) has direct or strong influence in several major categories to a product that meets design objectives.

The better that the failure mechanisms in material, personnel, methods, and procedures are understood, the more effective production will be in achieving reliability objectives. This is the purpose of the production reliability program. Another way of classifying sources of product unreliability is listed in Table 7.2.

It is extremely important, particularly when talking about relative contributions of unreliability, to use the classic bathtub curve (Fig. 7.4). The use here will

TABLE 7.1 Typical Causes of Unreliability

Component selection	Substandard materials*
Component quality*	Environmental stresses
Circuits	Packaging
System interfaces	Material handling*
Workmanship*	Subsystem interaction
Inadequate processes*	

* Controlled or influenced by manufacturing.

TABLE 7.2 Classifying Sources of Product Unreliability

Distribution of causes, %	Source
20–40	Design and development
40–65	Quality of components
15–20	Quality of workmanship

be to direct focus on the causes of unreliability introduced by the manufacturing department, as well as to illustrate how the manufacturing department can degrade the inherent design reliability.

The bathtub curve is divided into three regions. Region I is typically the portion of the curve that has the largest percent contribution from manufacturing unreliability. The failures formed here are referred to as "infant mortality failures." A better description of these failures is "manufacturing leakage." A typical listing of these failures is shown in Table 7.3. By inspection it can be seen that a product will experience infant mortality failures because:

1. There are no well-defined manufacturing processes and procedures.

2. There is not a system to enforce conformance to procedures that do exist.

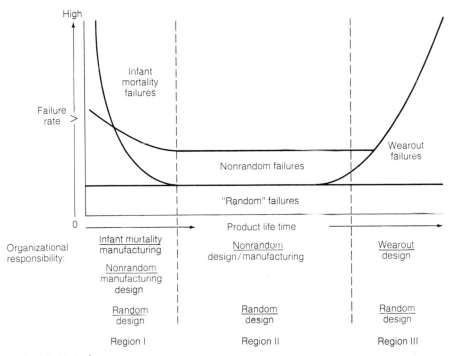

FIG. 7.4 Bathtub curve.

TABLE 7.3 Infant (Region I) Mortality Failure Causes

1. Inadequate test specifications—components or engineering	9. Improper use procedures
2. Inadequate quality control	10. Power surges
3. Inadequate manufacturing processes or tooling—unstable	11. Inadequate marketing on installation environment
4. Inadequate materials	12. Inadequate technical representative training
5. Improper handling or packaging	13. Incomplete final test
6. Marginal components	14. End runaround final test
7. Overstressed components	15. Subsystem interactions
8. Improper setup or installation	

3. There is not a process to detect and eliminate the causes of these failures.

Region II of Fig. 7.4 identifies an area of nonrandom failures. The preferred definition of nonrandom failures (or pattern failures) is the second occurrence of the first "random" failure. The manufacturing contribution to the failure rate in region II is much smaller than in region I. One example of a manufacturing-caused nonrandom failure deals with shipping products which will wear in use, i.e., whose critical dimensions do not conform to the specification.

Table 7.4 denotes the causes of unreliability found in region II as a result of an inadequate manufacturing process. As can be noted, the relative contribution of problems which are the responsibility of the design process far outweigh those which are manufacturing-controlled.

TABLE 7.4 Region II Random Failures

1. Insufficient design margins	6. Predictable failure levels
2. Misapplication—overstress	7. Nonpattern failures
3. Use in wrong environment	8. Intermittent failures
4. Inadequate design margins	9. Inherent manufacturing leakage rate—sampling
5. Cause "unknown" failures	

Region III is similar to region II with respect to manufacturing-caused failures. The rationale is that the design, materials selection, and tolerance of the design for the use determine the life. From a more pragmatic viewpoint used in many reliability measurement programs (failure allocation responsibility), if manufacturing defects are present, they will show up before the end of design life is reached.

Table 7.5 again illustrates one or two categories that by themselves have a high content of manufacturing-caused failures; however, the majority of the categories is still a function of the design. This holds true whether we are talking about systems, equipment, components, or devices.

TABLE 7.5 Region III Wearout Failures

1. Scratching	6. Inadequate or improper preventive maintenance
2. Material wear	7. Assembly interference fits
3. Aging	8. Loose hardware
4. Incipient stresses	9. Misalignments
5. Limited-life components	

7.4 BENEFITS OF RELIABILITY IN MANUFACTURING

The main manufacturing-related costs of an unreliable product are operating costs, warranties, and external costs of failure. More subtle, but just as real, are the benefit of reliability to an equipment manufacturer when the devices and components used in the equipment provide continuous flow in the production area without stoppage for troubleshooting and repair. The value of these benefits is in decreased inventory and carrying costs.

The effect of a mature design being placed in the production department at start-up also creates the same benefits, because it is virtually impossible to maintain line continuity when there is a large amount of design and material change.

To be successful in satisfying the customers, there must be a design that will achieve close to the mature reliability when starting up production. It is also necessary to have a well-defined and disciplined production process. The three critical subphases of the production process are therefore:

1. Transfer of the design to the production department
2. The activities accomplished during the internal production start-up
3. The delivery of initial production to the customer

7.5 RESPONSIBILITY MATRIX

The line management is responsible for the ultimate quality and reliability of the delivered product. Middle management, manufacturing and production engineers, material handlers, as well as the industrial work force, have a contribution to make to product reliability. Figure 7.5 illustrates the responsibilities of the various organizations, functions, and levels of personnel in regard to minimizing the degradation of the reliability of the design.

7.6 RELIABILITY ORGANIZATION— PRODUCTION PHASE

Defining the proper organization of the reliability department is based on many factors, and an organization or business may have any of several workable forms. The basic form of the reliability organization, however, in the production department should be based on:

Reliability activity	Production engineer	Quality engineer	Reliability assurance engineer	Finance	Production control	Purchasing	Training
Reliability targets			Ⓧ*				
Reliability plan	X	X	Ⓧ	X		X	
Component reliability	X	X	X	X	X	Ⓧ	
Reliability screening	X		Ⓧ			X	
Reliability testing		X	Ⓧ		X	X	
Data collection		Ⓧ	X			X	
Failure analysis	Ⓧ		X			X	
Corrective action	X	Ⓧ	X			X	
Reliability training	X	X	Ⓧ	X	X	X	X
Reliability improvement	X	X	Ⓧ	X		X	
Program							

* Ⓧ = Primary responsibility for definition; X = Contributes to definition.

FIG. 7.5 Reliability program responsibility matrix—production.

1. The desired role of the reliability functions
2. The composition of the business; key interfaces should exist to the reliability organization
3. The support of the reliability organization to the line functions
4. The expectations and support of top management, which enable proper usage of the skills and expertise of the reliability organization
5. The long-range desire of the organization to provide career growth to the members of the reliability organization

The role of the reliability organization in the production department will also be, by definition, a staff support to the line organizations. Refer to the reliability functions described in the responsibility matrix (Fig. 7.5).

In order to summarize the forms of organizational structure for the reliability function, reference will be made to Fig. 7.6, which shows advantages and disadvantages of the four following alternative locations:

- Product assurance
- Centralized reliability assurance
- Decentralized reliability assurance

Organizational form	Advantages	Disadvantages
Product assurance	• Good organizational positioning	• Tend to be generalists
	• Flexible on reassignments	• External to line functions
	• Home for reliability engineers	• Tend not to be dedicated
	• Strong working relationship with design reliability group	• Seen as excessively large organization
Centralized	• Identify with peers on reliability issues	• External to line functions
	• Facilitates reliability training and upgrading	• Seen as staff advisors, not doers
	• Home base for future assignments	
	• Good interface to design reliability group	
Decentralized	• Part of production team	• Exposed to line pressures
	• Major part of quality assurance	
	• Directly involved with day-to-day problem solving	• No home base
	• Help integrate reliability throughout line organizations	• Less than optimal communications with peers

FIG. 7.6 Characteristics of reliability assurance functional organizations.

The key characteristics of each of these is summarized in the following sections.

7.6.1 Product Assurance

Product assurance is the name of an umbrella form of organization that includes reliability. It usually is headed by a director or vice president who reports to the president or general manager of the enterprise. The functions span the design and development phase as well as the production phase. The product assurance function can itself be either "centralized" or "decentralized."

7.6.2 Reliability Assurance

Centralized Organization.   A centralized reliability assurance function generally is performed by a staff organization combined with the quality assurance functions of the production department. Reporting levels are of a director or vice president of quality/reliability to the director or vice president of manufacturing operations.

Decentralized Organization.   A decentralized reliability assurance organization is generally used where the mix of products or the geographic arrangement of facilities demands a strong local reliability function.

7.6.3 Staff

A summary of the key staff functions from the responsibility matrix of Fig. 7.5 defines the following staff roles:

Develop policies and procedures

Provide consultants with special skills

Provide reliability specialists

Maintain a data center

Perform systems and components reliability functions

Provide system specialists

Provide strong and respected reliability management that can interface with the design functions (both line and reliability engineering management)

Provide ability to deal effectively with internal and external customers, agencies, or individuals

Provide ability to develop business-oriented reliability improvement programs

7.7 THE PRODUCTION RELIABILITY ASSURANCE PROGRAM

In order to keep perspective on the reliability function, we should reduce the total subject to four areas. In describing these areas we shall range from the strategic business objective of the company to the fundamental functions of the reliability program. The information and functions expected to be provided by the reliability assurance organization can be broad and technically demanding; therefore, it is important to focus on the broad issues rather than the details.

The four basic elements of a company's production reliability assurance program, which can be different from but complementary to those of the design and development department, are simply:

1. A set of strategic and tactical objectives
2. A definitive reliability program composed of the elements to be accomplished by each organizational segment
3. A measurement process which is complementary to the reliability measurement process of the organization on each side of the production department (e.g., design and field)
4. A very strong feedback process of two types: corrective action and preventive action

The distinction between corrective and preventive actions is as follows:

1. Corrective actions deal with reliability problems found during the production and field use phases of a current product.
2. Preventive actions are steps taken after combining the data from the corrective action process with input from the total upstream organization (e.g., design, corporate engineering, marketing, program management). These data are used to develop the policies and procedures to be used during the design and development of subsequent products to reduce or eliminate generic causes of product unreliability.

Figure 7.7 shows a flow diagram of the four major aspects of the reliability assurance function. Applying this approach to production develops the following actions:

- Establish operational reliability objectives which relate to strategic business issues.
- Isolate those that are controllable by the production department.
- Define the metrics (subobjectives) that are directly influenced by the production process.
- Define and implement the reliability assurance activities that will achieve these reliability metrics (whether at the system, subsystem, or component level).
- Apply the resources of the reliability organization in only one direction (i.e., take corrective action to achieve quantitative reliability objectives).
- Act as the surrogate customer during the production and field use phases to provide the analysis and data needed to make reliability improvements.
- Look, as part of the total business, at opportunities to implement new or update existing policies and procedures to yield mature product reliability (superior to field reliability of existing product) to the customer.

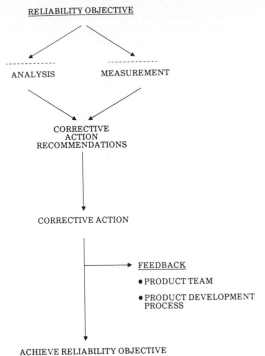

FIG. 7.7   Four basic steps of the reliability assurance program.

7.8 RELIABILITY REQUIREMENTS

In production, the reliability requirements and metrics vary as a function of:

1. The supplier and customer relationship
2. Whether a commercial or aerospace user
3. The type of product (e.g., electronic or electromechanical)
4. The consequences of failure

In all cases, the common issue to consider with regard to the production reliability requirements is that the production-controllable causes of unreliability relate to the overall reliability requirement. Since the reliability requirement applies to the product as delivered to the customer, there will *not* be a separate reliability requirement per se for the manufacturing process.

In some cases (examples are given in a later subsection), there will be departmental (e.g., design, manufacturing, field) reliability objectives, but this is done to assure accountability of the responsible department. These departmental objectives can be developed from various sources, such as historical departmental contributions, allocation from the product requirement, consideration of the relative contribution of each department, and the need to provide information about problems into independent departmental corrective action systems.

In the design of large systems, and in some cases large equipments, where businesses depend on subcontractors and suppliers of major components, product reliability requirements will be included in the procurement specifications, thereby imposing reliability duties onto suppliers. The production organization that assembles and delivers the end item to the customer can integrate the reliability efforts of the entire business team in achieving the system reliability requirements.

7.9 DEVELOPING RELIABILITY REQUIREMENTS—AN EXAMPLE

The manufacturing department uses a three-phase approach to developing and describing quantitative field reliability requirements for each new product program in order to be responsive to the evolutionary nature of product development. These three phases, which relate to the corporate planning cycle, are described below.

7.9.1 Feasibility Phase

Estimated reliability levels are developed during the feasibility phase. These are based on the demonstrated reliability of a known (similar) product. Estimated predictions are also developed from demonstrated reliability levels of products currently in the production phase.

7.9.2 Definition Phase

Preliminary reliability predictions are developed during the definition phase. These predictions, which are the commitments of the manufacturing department to satisfy the needs defined by the marketing department, define the optimum balance between field reliability (unscheduled maintenance calls),* committed resources, and schedule of placements.

7.9.3 Production Phase

The third phase, which represents the commitment of top manufacturing department management to the corporation, reflects the optimum relationship of the level of reliability to investment resources and schedule; this occurs prior to the production phase.

Numerous product and program options are evaluated over the development cycle. An explicit reliability prediction and level of confidence are developed for

---

* UM is used to identify unscheduled maintenance calls or incidents of product failure.

each option. The relationships for various options, particularly where time (schedule) and resource levels are affected, are described by reliability predictions which range from low reliability with high confidence through high reliability with low confidence.

Initial (feasibility phase) goals are established generally with high reliability at a low confidence level. As resources are defined, knowledge gained, and decisions made, the confidence increases from a low level to a high level. Meeting a high level of confidence is a criterion for transfer from the definition phase to the design phase. Implementation of the manufacturing reliability program during the definition and design phases establishes the plans and resources required to achieve the field reliability requirement. These sources are the basis for the production phase reliability program. Predictions and planned objectives for a new product would be defined as in Table 7.6.

TABLE 7.6 Typical Reliability Objectives

	Program life	
	Definition phase	Production phase
Reliability, UM/10^{6*}	52	30
Confidence, %	80	80

*In this example UM/10^6 defines unscheduled maintenance calls per million copies made by the copier machine. Similar general reliability goals should be defined by relating unscheduled maintenance to appropriate measure of operation of system under consideration.

7.9.4 Field Reliability Phase

Three characteristics are used to determine the field reliability requirements for a new product. Two of these are defined by what the customer requires of a product, whereas the third is defined by the reliability level felt realizable by each of the operating departments: engineering, manufacturing, and field operations.

7.10 FORMS OF MANUFACTURING FIELD RELIABILITY TARGETS

The standard form of manufacturing reliability requirements is to relate service calls to total operation time after installation of the product in the customer's account. The secondary form, monthly product usage, relates the product's frequency of failure to program life and provides a common base for integrating the manufacturing targets with targets of the engineering and field operations departments. Defining the targets for a series of blocks requires that the field measurement data system "normalize" all machines in a given block; that is, to assume that all machines were installed at the same time (for each block).

7.11 DEVELOPMENT OF MANUFACTURING TARGETS

Manufacturing reliability targets are developed from considerations of past performance, product comparisons, and future expectations. Product reliability characteristics are shown in Table 7.7. Values of future expectations are applied as a function of the degree of modification of the existing product required to produce the new product. All growth factors are related to calendar time, assuming past knowledge is always carried forward and that failure modes associated with manufacturing problems are continually reducible, given the proper application of resources and time.

A family of curves representing the manufacturing reliability targets for a new product is shown in Fig. 7.8.

7.12 MANUFACTURING CONTRIBUTIONS TO FIELD RELIABILITY

A standard method of partitioning manufacturing failures is to classify each failure either as workmanship or component related. The typical contribution of failure modes to these categories is

Category	At installation	Maturity
Workmanship	60%	20%
Component	40%	80%

TABLE 7.7 Product Reliability Characteristics

	Past performance	Product normalization factors		Future expectations
Block-to-block growth	Improvement during month 1 over 18-month period	Complexity factor, K_1	Comparison of new to old product as function of active part count and technology differences	λ launch: Starting point of new product reliability adjusted for product normalization factors
Within-block growth	Reliability improves 90% over the first month to the sixth month of machine life	Learning curve, K_2	Start-up effect of manufacturing plan	
Maturity reliability, λ_{mat}	Level of product reliability at month 1 of maturity	Volumetric effect, K_3	Effect of monthly copy volume on reliability	α_1: Block-to-block reliability growth
				α_2: Within-block growth
				α_3: Stable level, month 7 through end of life
				λ_{mat}: New product reliability at month 1 of maturity

TABLE 7.8 Reliability Elements for Typical Complex Airborne Electronic Equipment

Effect	Element
Random part failures	1. Prediction (part type, stress level, temperature)
	2. Normal commercial inspection practice
	3. Investigation and failure analysis program
Quality of workmanship	1. Component burn-in, screening
	2. Card conditioning
	3. Module-subsystem burn-in
	4. Validated processes and handling process
	5. Reliability and quality requirements on suppliers
Pattern part failure	1. Failure reporting, analysis, and corrective action
	2. Specification control drawings
	3. Test data analysis
Design caused	1. Derating criteria
	2. Allocation
	3. Qualification test
	4. Design review process
	5. Circuit analysis
System effect	1. System test—qualification, AGREE
	2. Failure mode and effect analysis
	3. Performance specifications
	4. Configuration control

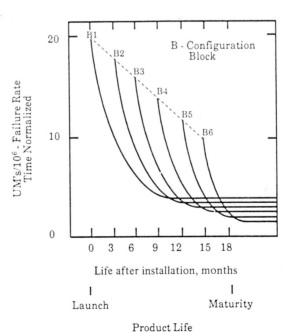

FIG. 7.8 Manufacturing reliability targets (typical). (UM = unscheduled maintenance.)

7.16

These ratios are fairly standard and, as expected, with time and effort reflect improvements in training and skill levels in the workmanship category. The component control program is a direct application of aerospace methods and results in a level of control.

7.13   RELIABILITY PROGRAM ELEMENTS— PRODUCTION DEPARTMENT

Management of product reliability through the design, production, and use phases is only possible through the timely execution of a total integrated reliability program. The reliability elements implemented during the production phase are both dependent on and interactive with the quality control and quality assurance elements. There is extensive reference material on the reliability functions and elements applicable to the design and development phases. Also, published military standards (i.e., MIL-STD-285) treat the design reliability elements. Military standards (e.g., MIL-Q-9858) and AN91 Standard Z-15 describe the quality control and quality assurance functions applicable to the production phase.

7.13.1   Reliability Assurance Program

By referring to Table 7.8, the relationship of reliability elements to causes of unreliability can be seen. This table includes the design reliability elements as well as some elements that must be included in the reliability assurance program. The table also points out that the selection (or tailoring) of the elements should be governed by:

- Causes of unreliability
- Product type
- Consequences of failure
- Resources available

7.13.2   Scope of Reliability Assurance Program

Figure 7.9 is included to

- Provide the reader with an overview of what is to be described in this section
- Display the reliability assurance elements and the general level to which they apply
- Provide a multidimensional array to point out that some elements apply at all equipment levels and others apply to equipments only

It must be emphasized also that there is a highly interactive relationship between the elements of the array, and at the same time, there must be balance among these elements if the reliability assurance program is to be considered adequate.

Product level	Rel. targets	Rel. plan	Comp. reliability	Reliability screens	Data collection	Failure analysis	Corrective action	Rel. training	Rel. testing	Rel. improvement process
System	X	X	—	X	X	—	X	X	X	X
Subsystem/module	X	—	—	X	X	—	X	—	X	X
Component	X	—	X	X	X	X	X	—	X	X
Device	X	—	X	X	X	X	X	—	X	X

Reliability assurance activity

FIG. 7.9 Scope of the reliability assurance program.

7.13.3 Reliability Prediction—Manufacturing Phase

Most of the literature on prediction processes and models does not fit the manufacturing process because of the assumptions used, e.g., a constant rate of failure and failures caused by electrical and environmental overstress. The predictive processes one must use during the production phase are generally based on empirical results. As long as we understand this in using these techniques, they are a very powerful ingredient of the program.

7.13.4 Prediction Techniques

The reliability prediction techniques that are relevant to the production phase are

- Weibull plot
- Duane plot
- Fix effectiveness model

Where these techniques are used, it is the product in production that is being evaluated, not just the manufacturing unreliability.

Weibull Plot. The Weibull plot is used most effectively to evaluate the reliability parameters of components (e.g., electromechanical) and special components (i.e., photoreceptors). Weibull analysis (either graphical or computerized) is very valuable during the developmental phase of a component.

Duane Plot. This predictive model is a very powerful measure of product reliability over time as yielded by the manufacturing process. The predominant use of the Duane model on electronic equipments is for the quantitative measurement of the effectiveness of the reliability program.

Fix Effectiveness Model. This model uses measurements from system reliability tests, and its main purpose is to predict the adequacy (effectiveness) of individual corrective actions, but most importantly it is used to determine when product reliability requirements will be achieved. Another important feature of this process is that it involves all members and functions of the reliability program. Significant benefits provided to the production department are:

- An early problem definition is input to the manufacturing process.
- The reliability management process is assured that reliability requirements were met *before* production was started.

The knowledge gained in the use of any one of these three predictive models is very powerful and should be retained. By working with the line organization, the causes of product nonreliability introduced in the production department can be identified and corrective action taken. Just as important, this information can be used in the early phases of a new program by the manufacturing team members.

7.14 COMPONENT RELIABILITY ASSURANCE

The major reason to have a comprehensive component* reliability assurance program is because the product's reliability is determined by the reliability of the weakest element, i.e., the component. The Pareto principle (the so-called 80–20 rule) states that 20 percent of all components will contribute to 80 percent of product unreliability. This is important in regard to helping plan the budget for the reliability assurance program. The major elements of the component reliability assurance program implemented during the production phase are:

1. Knowledge of design requirements via design specifications
2. Sourcing strategy and technical support to obtain vendors that will provide reliable components
3. Integration of the plans and efforts of the engineering department responsible for qualifying the component*

The relationship of the equipment manufacturer (customer) to the component supplier will have a lot to do with the success of a component reliability program. Basic principles which should be used to assure success are:

- Common business objectives
- Well-defined reliability requirements
- Well-defined responsibility for the supplier to implement the necessary actions to achieve reliability requirements
- Clear and agreed-to supplier quality assurance program
- Early involvement of suppliers to address customer needs

The activities associated with good equipment and vendor quality assurance programs involve the use of well-published methodologies in statistical quality control to achieve both a stable and acceptable process. By achieving component conformance to specifications through quality programs, reliability of the procured component is improved. As a result, the contribution of these commodities, both components and materials, will then support the efforts to yield reliable subsystems and systems.

7.15 PRODUCTION DEPARTMENT ORGANIZATIONAL INTERFACES

During the design and developmental phase and into production start-up, there are numerous program activities. These activities can be grouped into two categories: (1) program team activities, and (2) checkpoints.

---

* The term *component* as used herein is typically a motor, fan, clutch, power supply, or small electronic assembly (printed wiring board).

7.15.1 Program Team Activities

Typical program team activities are:

- Inclusion of the production engineering function as part of the product development team, both from a management and technical viewpoint
- Scheduling of management and business reviews of the reliability assurance program
- Colocation of the manufacturing and design functions to facilitate the joint definition for the design process to the manufacturing process

7.15.2 Checkpoints

Typical checkpoints are

- Preliminary and critical design reviews
- Production readiness review
- Design issue and release

During each of these checkpoints, the outputs of the reliability assurance program should be used to define problems with the hardware as well as design documentation so that corrective actions can be taken. Design review checklists should be used to take advantage of various improvements to minimize incompatibilities between

1. The design
2. The manufacturing processes
3. The hardware (e.g., systems, subsystems, components)

Historical reliability data or familiar assembly processes, components, and vendors should be used by the design team to keep previous production problems from being incorporated in the new product.

Product transfer from the design phase to production start-up is both a major business event as well as an important checkpoint. Management of this transfer can be facilitated by a review of the product's key business indicators (e.g., product cost, measured vs. predicted reliability, continuity of material and component supply). If the proper integration of the design and production teams has occurred over the product development cycle, there should be no major problem at this time.

7.16 SUBSYSTEM RELIABILITY ANALYSIS

Considerable attention is given to systems reliability activities and component and device reliability programs as described in this handbook. The key area that often creates reliability problems is that of major subsystems or modules. The problems of misapplication of components, excessive settings, very complex configurations which are not amenable to simple error-free assembly are only some examples.

The use of a well-organized data system to provide a data base for the design team can be invaluable. At the subsystem level, a failure modes and effects analysis (FMEA) should be conducted during the design phase (Table 7.9). The subsystem analysis has two purposes:

TABLE 7.9   Subsystem Analysis

Type of problem	UM/ machine	Failure mode	UM/ machine	Reported or probable failure cause	Corrective action considerations
Optics jams (includes false shutdowns)	0.093	Cable broken or sheared	0.005	Cable misaligned with idler pulley	Add requirement to idler bracket and add self-locking flange-ER-1105
			0.006	Secure pin not properly seated	
			0.012	Mounting hardware loose	
		Dirty rails or bearings	0.003	Contamination from toner	a. Felt rail wiper (ER-00138) pending approval
					b. Optics cavity contamination level reduced, ER-00119
		Wrong type rail installed	0.001	Improper assembly	
		Return spring malfunction	0.029	Improper tension	
			0.003	Improperly seated	
		Timing switch failure	0.003	Defective switch	
		Dashpot malfunction	0.034	Defective, misadjusted, loose, binding	Inspection instruction modified
		Trip panel interference	0.011	Improper installation, set screw loose	
Noise: squeaks	0.057	Lamp reflector binding	0.006	Improper assembly	Final test critical check
		Optics rail dry	0.018	Inadequate lubrication	
Slams		Part missing	0.009	Rubber tip missing from dashpot	Qualify new source after redesign
			0.018	Defective dashpot (collapsed)	
			0.006	Defective dashpot (loose sleeve)	

Note: UM = unscheduled maintenance.

1. To define what design features must be included to allow manufacturing to assemble and inspect the machine easily, thereby minimizing failures due to poor workmanship

2. To provide the manufacturing department with a base for planning the assembly, developing the manufacturing process sequence, and assigning quality control plans

7.17   RELIABILITY SCREENS

The major reason for implementing reliability screens (i.e., burn-in, run-in, etc.) is the actual or expected high incidence of "infant mortality" failures. There are also reasons why reliability screening should be an unnecessary part of the reliability program and efforts should be made by the supplier to provide a product that does not require it.

1. Unfavorable return on investment (supplier should provide reliable product)

2. Impact on production throughput (time required to perform tests, isolate failures, effect repair, etc.)

It is essential that the reliability program eliminate the need for burn-in through a dedicated effort that focuses on the cause of the problems as determined from the data generated by the program. These data should be used to change the product or process and eliminate the need for burn-in.

Table 7.10 illustrates the major elements of a 100 percent screening and burn-in program. These criteria pertain to the responsibilities and actions of the design and production departments.

Table 7.11 illustrates the use of screens at various equipment levels and some typical details. This table, however, is for illustration only since the type of burn-in reflected is a composite of experience in the aerospace and commercial industries. Also, what works well for electronics assemblies and equipments does not necessarily apply to electromechanical equivalents. In reference to device

TABLE 7.10   Elements of an Equipment, Assembly, or Component Screening Program

- Development of criteria for reduction or elimination of burn-in programs.
- Definition of a method for determining the proper form of burn-in considering the makeup of the system and costs. (If burn-in requires incremental capital equipment and resources to support the production schedule, it must be in place at production start-up.)
- Definition and implementation of a measurement analysis program that allows the shape of the "infant mortality" curve to be analyzed, which then leads to identification of the causes of the infant mortality effect.
- Creation of an ongoing data collective and failure analysis, problem definition, and correction action program to eliminate the problems that created the need for the burn-in program.

TABLE 7.11 Typical Burn-in and Screening Programs

Equipment level	Description	Conditions
System		
Extended final test	20% more time than standard hours	Room ambient
Failure-free final test	Final test hours 10% of specified MTBF requirement	Room ambient, 0 failures allowed to make accept decision
Burn-in	96 hours 1 month copy volume	Room ambient
Customer compliance test	Final test procedure simulates all customer use conditions plus 10–20% more test hours than standards	Room ambient
Subsystem or module		
Production environmental test (PET)		Random or sinusoidal vibration
Burn-in	48 hours	±125° C, rapid thermal excursion functional test specification
Component		
Printed wiring board assembly	96 hours	8-hour cycle over range of 25 to 70°C, input power applied. Measure to test specification
Power supply	8–10 hour run-in	Outputs loaded, overnight soak

(e.g., resistors, transistors, etc.) manufacturers, the equipment manufacturer procures electronic assemblies from suppliers whose responsibility is to deal with the device manufacturers.

From a production perspective, the testing portion of the reliability assurance program during the production phase should meet the following criteria. The tests should be

1. Complementary and consistent with the reliability test program accomplished during the design phase
2. Continuous during the production phase
3. Highly correlated to reliability performance of the products in the field

Various system reliability test programs useful during the production phase are summarized in Table 7.12. In order to put this material in context, the testing programs should be designed to:

1. Protect the customer from receiving products that do not meet requirements.
2. Assess conformance of the product and the manufacturing process with the quantitative reliability requirements.

TABLE 7.12   Reliability Assurance Testing—Production Phase

Type of production reliability test	Description	Purpose
Production reliability assurance test (PRAT)		Monitor performance to reliability requirements
New production (monthly)	Samples selected from samples used for measuring product quality.	Detect significant shifts in as-built reliability
Refurbished product (at startup)	Each sample exercised for one month of operation	Evaluate adequacy of refurbishment process
Bimonthly AGREE*tests	Small sample tested under same environmental conditions as qualification test	Detect significant shifts in the as-built reliability from as-designed reliability
Monthly reliability environmental tests	Type A & B tests—samples of product subjected to temperature and humidity tests	
Reliability requalification tests	Periodic (semiannual) repeat (simplified) of design qualification tests	Reassess total product reliability against reliability requirements

* Advisory Group on the Reliability of Electronic Equipment (see MIL-STD-217A).

3. Define the problem in such a way that corrective action can be taken to solve short-term problems. Also, provide opportunity for generating solutions to problems arising from the business system to allow for long-term reliability growth.

4. Provide meaningful business indicators of programs.

The suitability of a reliability testing program for refurbished products is an area that is often neglected. Many industries have businesses that are heavily dependent on recycling, upgrading, and replacing products long after the design team has moved on.

The production reliability assurance testing recommendations are also included in Table 7.12. The entries of Table 7.12 are in addition to the type of screening and burn-in program previously discussed. In order to draw a line, all screening and burn-in programs are performed on 100 percent of the products, and as a result are considered to be part of the production cost or as part of the internal failure costs as defined by cost-of-quality principles. In order for the production reliability testing program to be of value, there must be continual feedback from the customer and/or field sites. This will be discussed later on in this section. One of the reasons that this is critical is that all internal test programs are based on specifications that are believed to represent the customers' requirements; often, however, particularly in the commercial market, there is a wide dispersion of customers as well as use conditions. This information loop between what the customer sees versus how it is being evaluated in the internal reliability test program must be continually monitored.

Manufacturers of large systems or systems made up of several large subsystems, each of which is manufactured and individually subjected to reliability assurance testing at widely separated manufacturing sites, should keep in mind the following important principles:

- The customer judges the product and company by the reliability over the early months of operation.
- Conformance of the major subsystem, even to the best test specification, will not guarantee the entire system will perform as required.
- About 10 to 20 percent of all system reliability failures are a result of interactions between the subsystems.

The recommendation, therefore, is that the reliability test program for systems be built around the configuration of the systems to be installed in the customer's location. This means that companies with a large number of diversified manufacturing sites would:

1. Have each site responsible for implementing a reliability test program consistent with *customer* usage.
2. Implement ongoing reliability testing programs (on a sampling basis) but use production hardware at a central site.
3. Feed back (internally) the problem information to the respective manufacturing sites.

7.18   DATA COLLECTION, FAILURE VERIFICATION, FAILURE ANALYSIS, PROBLEM DEFINITION, CORRECTIVE ACTION

The production department has an important role to play in maintaining a strong reliability data sourcing system as illustrated in Fig. 7.10. The main reason is that the production department has both the interface responsibilities to the customer (field) and to the design department. Using Fig. 7.10 as an outline, the following will briefly describe each of the key elements.

7.18.1   Data Collection

Sources of data are shown in Fig. 7.10. It is generally easy to implement data systems to capture failure information. The reliability data collection system must also include success information. The failure and success information needs to be combined to describe reliability performance over time. In addition, there must be a tight linkage of the description of the problem symptoms to the reliability statistics. Systems (internal and external) must be implemented mainly by the line organization to obtain the replaced hardware to move to the next step in the process, failure verification.

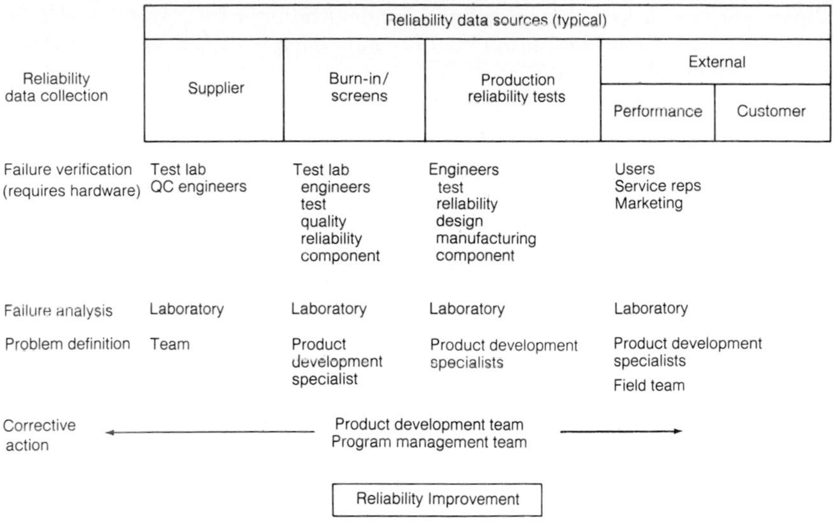

Reliability data collection	Reliability data sources (typical)				
	Supplier	Burn-in/ screens	Production reliability tests	External	
				Performance	Customer
Failure verification (requires hardware)	Test lab QC engineers	Test lab engineers test quality reliability component	Engineers test reliability design manufacturing component	Users Service reps Marketing	
Failure analysis	Laboratory	Laboratory	Laboratory	Laboratory	
Problem definition	Team	Product development specialist	Product development specialists	Product development specialists Field team	
Corrective action	←		Product development team Program management team	→	

Reliability Improvement

FIG. 7.10 Typical functions of the data collection, failure verification, failure analysis, problem definition, and corrective action system.

7.18.2 Failure Verification

There often will be a large number of system malfunctions where the diagnosis will be incorrect because the wrong component is removed, because of interactions between two or more functions or components, or for other similar reasons. To ensure against incorrect failure assumptions, there must be a system established to evaluate the component quickly against its specified requirements. This process is referred to as failure verification. Up to 30 percent of the components replaced are verified as being acceptable either because the removed component did not cause the system malfunction or because the component specification is inadequate.

7.18.3 Failure Analysis and Problem Definition

Upon completion of the failure verification step, the component should be subjected to failure analysis. In this manner, material, circuit, and application mechanisms of failure can be determined and serve as the basis of properly identifying and defining the problem (see Chap. 14).

7.18.4 Corrective Action

The short-term corrective action is primarily based on an investigation, an integral part of which is the failure analysis report. The comprehensive corrective action plan should in most cases consist of some degree of off-line testing to establish both the validity and effectiveness of the planned corrective actions.

Implementation of corrective action is handled by either a corrective action committee or a configuration control board. Cut-in dates for the corrective action are based on many factors, such as significance of the problem, material value of inventory, and availability of new parts.

7.19 FAILURE REVIEW BOARD

A very effective method of integrating the functions of a program team that is often applied where formal reliability measurement tests are being conducted is the failure review board. This reliability function has been used by manufacturers of aerospace equipment as well as of commercial products. It is an effective centralization of program functions on both internal and external reliability measurement programs. It serves as an effective link between the reliability measurement program and the problem definition functions shown in Fig. 7.10.

Failure review board meetings are conducted weekly and are chaired by a program manager who administers the work of the team. Generally representatives of the line organizations (design, manufacturing, and field) review failure information. MIL-STD-785 defines the functions of a failure review board as

Assure that an action responsibility is assigned for all reliability failures.

Input all relevant problem information to the respective departmental problem investigator and/or corrective action system.

7.19.1 Reliability Assurance Plan

The guidelines for the development of the content of the reliability assurance program as well as what should be documented in the program plan are as follows:

1. Causes of unreliability determine *what* should be included in reliability assurance program plan.
2. Market and/or customer technology requirements determine *how* much depth and hence cost of reliability assurance program.
3. Company policy.
4. Organization of company determines *who* is responsible to accomplish or to input task.
5. Phase of product development determines *which* elements are accomplished *when*.
6. A reliability assurance program plan represents *what* will be done, *when* it will be done, and *how much* cost will be incurred for the *company*. It must integrate all product reliability elements regardless of who is organizationally responsible for accomplishing the reliability element.
7. It is implicit (and should be explicit) that a cost-benefit analysis is performed to support the reliability program.
8. Cost-benefitanalysis should include the value of intangibles even though the value cannot be expressed in an economic form.
   8.1. Customer satisfaction
   8.2. Company reputation
   8.3. Consequences of failure

System	Third generation copier			Manufacturing site		Issue date						P. ___ of ___

No.	Task	Respon-sibility	Inputs	R. procedure	Schedule 7/86	8/86	9/86	10/86	11/86	12/86	Completion date	Task output/control
1	● Develop/issue system R requirements	Mktg.		R702						△	12/86	Issued
2	● Define/issue components' reliability assurance program	Rel.		R710				△			10/86	Issue program
3	● Define/issue critical parts lists	Rel	Rel. qual.	Q142					△		11/86	Issue list
4	● Define R requirements for critical parts	Design comp.		R714						△	1/87	Design standard
5	● Define system, subsystem burn-in/screening program	Rel.	Mfg. design	N/A					△		11/86	Issue plan

FIG. 7.11 Reliability assurance program summary.

7.29

An effective format for summarizing the major elements of the reliability program is shown in Figure 7.11. This format is effective for program reviews and has been effective in the aerospace business as well as in the commercial business. The plan reflects the reliability activities to be accomplished. The organization with prime responsibility (line organization) and secondary responsibilities (staff) plan start and completion dates. The outputs associated with each activity are also included. The structure of the plan should be simple, effective, and reflect the resources to be applied to enable the product to achieve the reliability requirements.

7.20 RELIABILITY TRAINING

In order to avoid the problem of assuming the reliability manager is responsible for product reliability, formal training in reliability is a very important element of a company's ability to succeed. The reliability training objectives should be established within the company for the production department and should consider that:

1. Top management, middle management, production engineers, procurement buyers, and reliability engineers have need for different forms and levels of training in reliability but should all have at least some basic understanding of the objectives and functions of reliability.
2. The production reliability training program should be more relevant to the business, customer, and product.
3. The program should be complementary to other reliability training within the company to have consistency and a common language.
4. Integration of the reliability training program with training in quality is highly desirable.
5. Strong involvement is necessary, particularly with the company's training department.
6. Instructors should be recognized as competent in their field of expertise.
7. There should be a range of reliability trainers, some of whom can train top management and others who can teach the detailed reliability activities.

There is not an extensive amount of information available on training programs which have been designed for the various functions of the production department. There are a large number of consultants, as well as the American Society for Quality Control, who have supported the development and implementation of reliability training programs for the production department.

CHAPTER 8
ACCEPTANCE TESTING

Robert W. Smiley
Aerojet Strategic Propulsion Company

8.1 PURPOSE OF ACCEPTANCE TESTING

Acceptance testing is carried out by both the producer and the customer. For the producer, the purpose is to (1) weed out defective or unsatisfactory product and (2) eliminate "infant mortality" before delivery to the customer. For the customer, acceptance testing is carried out to assure that the product meets the reliability specifications included in the purchase contract. Acceptance testing provides both the producer and the customer with considerable data which should be used:

1. In assessing the reliability of the product population
2. For subsequent product improvement programs
3. For monitoring the production process for unwanted variations

Many factors influence the scope of the acceptance test program. Among them are

Product characteristics

Consequences of failure during manufacture or after delivery

User needs

Industry norms

Economic and government climate

Cost

Schedule

Acceptance testing should be viewed as only one aspect of the total test program. It is from the totality of testing in the program, from the earliest feasibility tests through design and qualification testing, production acceptance testing to the tests performed on the delivered product, that the required reliability of the product population is attained and assured.

Acceptance testing for reliability is broadly defined to include all of the testing which ensures that the product, at different levels of assembly, will function in accordance with the design requirements. Reliability acceptance testing differs from quality assurance acceptance inspection in that reliability is a characteristic related to the time, cycles, or occurrences over which the product will perform the specified functions in the specified environment. Therefore, testing must be done over a time span, whereas the quality assurance inspection simply compares the result of an inspection with a specific specification that can be done almost instantaneously. In some cases, especially one-shot devices, nondestructive testing, such as x-ray or magnetic particle inspection, may be used for reliability testing, but those methods will not be discussed in this chapter. Design qualification testing is sometimes looked upon as acceptance testing, but since that is to accept only the design and not the product, it will not be discussed. On the other hand, production assessment testing is involved with the reliability of the product and will be treated as one aspect of the acceptance of the manufactured product.

Although this chapter primarily addresses the acceptance testing of large systems, the principles can be applied to the development and application of acceptance testing programs for small systems, subsystems, and consumer products (such as TV sets, washing machines, and automobiles). The specific techniques used are influenced by the same factors listed above that influence the scope of the acceptance test program.

The development of the acceptance test program must begin with and proceed concurrently with the development of the product itself in order to assure that:

1. Test points are available in the product
2. The test program is "qualified" during the design qualification to be sure the acceptance test program will indeed assure that the hardware meets performance requirements
3. Acceptance test equipment is designed, built, and ready for use when the hardware is ready to be delivered
4. Acceptance testing and test equipment are compatible with testing and test equipment used in development and qualification testing

8.2   KINDS OF ACCEPTANCE TESTING

Acceptance testing may be classified in many ways and intelligent test planning requires that these options be considered and an optimum choice made from among them. Seven categories into which tests may be subdivided are:

1. Variables vs. attributes testing
2. Destructive vs. nondestructive testing
3. Ambient vs. environmental testing
4. Level of tests
5. Acceptance testing vs. quality control
6. Production assessment testing
7. Screening and burn-in

8.2.1   Variables vs. Attributes Testing

In variables testing, the actual values of the parameters are measured and recorded. In attribute testing, the test equipment compares each measured value with predetermined limits, determines whether the measured value is inside the limits, and makes a go–no go decision (acceptable or not). The measured value is not displayed or recorded, and the history is limited to the number of items that passed or failed the test criteria.

Advantages of variables testing are:

1. It collects and records more data, permitting estimation of means and standard deviation of parameters.
2. It allows for trend analysis to be made of the production process (as in the classic control chart fashion) with limits to trigger corrective action.
3. It extracts the maximum product information for any given size sample.
4. The cost of variables testing may not be much greater than for attribute testing if automated test equipment is used (but cost may be greater if tests are performed on manual test equipment and data are recorded and analyzed manually).
5. Any desired confidence level can be attained with substantially smaller sample than by attribute testing.

Advantages of attribute testing are:

1. It is excellent for separating acceptable from unacceptable product.
2. Automated equipment for attribute testing is usually less expensive and less complicated than for variable testing.
3. It provides data for p (fraction defective), np (number of defectives), or c (number of defects) control charts.
4. Manual attribute testing can be performed by lower-skilled labor with a small amount of training.

8.2.2 Destructive vs. Nondestructive Testing

A simple definition of a destructive test is one that leaves the tested hardware unfit for its intended use, whereas a nondestructive test does not prevent the intended use of the hardware. The hardware subjected to a destructive test may not be delivered to the customer, but it may be used for additional testing to failure to obtain information on stress limits, failure modes, and other reliability data. The advantages of these other benefits may be sufficient to justify the use of destructive tests rather than nondestructive tests, especially if the number of items to be tested can be reduced substantially. Some items can only be tested functionally by destructive tests, such as the firing of an explosive device, an electrical fuse, or a signal flare.

The choice of destructive vs. nondestructive testing is usually based on the economic factors involved with each and the purposes served by the tests. Factors to be considered include:

1. Number of nondestructive tests to attain the same level of confidence as by a number of destructive tests.
2. Value of the hardware being tested. The more expensive the hardware the more incentive there is to use nondestructive testing.
3. Ability of nondestructive test equipment to measure economically the important function-related parameters.
4. Investment and operating costs of the different types of test equipment.

8.2.3   Ambient vs. Environmental Testing

Testing performed with the specimen at the temperature, humidity, and other conditions existing in the factory is called ambient testing. Environmental testing is performed with the specimen subjected to temperatures, humidity, shock, vibration, or other conditions not normally encountered in the factory. If the product is to be used in a home, factory, or office, ambient testing is usually sufficient to determine the acceptability of each specimen tested. When the product must function in environments considerably different than the factory, it is often necessary to perform the acceptance testing at or near these conditions. However, because of the higher cost and complexity of environmental testing, correlation studies are often performed during development to find an ambient test program, generally with tightened acceptance tolerances, which will assure that product passing these ambient tests will perform properly in the expected environments.

Sometimes for acceptance purposes ambient testing of every unit of product is combined with additional tests in environments of occasional samples. This is a more stringent test program which will detect deleterious changes in the manufacturing process which would pass through ordinary ambient acceptance testing but result in product that would fail in the use environment. The environmental portion of the test program is often called production assessment testing (see Sec. 8.2.6).

Environmental testing is usually performed in an environmental test facility, either in-plant or as a purchased service. Sometimes laboratory testing is not suitable and actual environments must be used. This occurs when:

Parts are too large

Environments are not reproducible in a laboratory

Specific combinations of environments are not reproducible in a laboratory

It may be more cost effective to use natural environments than to reproduce them in a laboratory when limited quantities of specimens are to be tested.

8.2.4   Level of Tests

Acceptance testing is performed at all levels of assembly, from piece parts through final product. The aggregate of all these tests is the acceptance test program. Power output of an automobile, for example, can be measured on a test track in the completed vehicle; on a dynamometer in the factory with the chassis and engine assembly; or on an engine test stand at either the assembly plant or the engine plant. The choice of level depends on the following factors.

Factors Favoring Higher-Level Testing

1. Customers prefer testing at the highest levels for maximum assurance that defects have not been introduced after the acceptance testing. Furthermore, they want infant mortality burn-in to be performed on the total product before delivery.
2. Testing costs are lower with higher-level testing because more attributes are tested at once in a single test.

3. Functional spare parts are most efficiently and effectively tested at their assembly level.
4. Higher-level testing weeds out adverse tolerance build-ups which would not be evident in testing of the subassemblies.

Factors Favoring Lower-Level Testing

1. Some important attributes are covered up and inaccessible for testing at high levels of assembly and must be tested in subassemblies.
2. Low-level testing weeds out defective material before it is used in assembly and contributes significantly to productivity. "Drive the failures to the lowest possible level."
3. Subassemblies produced remotely (e.g., overseas) should be thoroughly tested before shipment to preclude exorbitant turnaround time and cost.
4. Purchased material should be tested at that level by either the buyer or the seller so that payment can be facilitated for the seller's convenience but with the buyer's assurance that good material was received.

8.2.5 Acceptance Testing vs. Quality Control Testing

The purposes of acceptance testing are (1) to demonstrate that the product meets the important functional requirements or to identify nonconforming product so that it can be repaired or reworked; and (2) to acquire data to permit reliability evaluation of the population. Quality control testing, on the other hand, is performed to weed out defective material before it is used in further manufacture, or to generate data which can be used to detect unwanted manufacturing variations before the process drifts out of control. Although many tests serve both purposes, they are planned differently. The acceptance testing program is largely determined by the contract or customer terms, or by government regulations requiring demonstration of specific performance. It is jointly planned by reliability and design engineers. The quality control testing is determined by economics and is performed when the cost of the testing is less than the expected cost to replace the defective parts or subassemblies in higher assemblies that the quality control testing would weed out of the process. Manufacturing engineers work with quality engineers to determine where quality control tests should be added to the test program.

8.2.6 Production Assessment Testing

When the product is required to perform satisfactorily in environments which are markedly different and more severe than the factory environment, the factory ambient testing may not detect changing or deteriorating manufacturing processes or parts which deleteriously affect performance in these stringent environments. Production assessment tests (see Sec. 8.2.3) are environmental tests which are performed to detect such degradation. The tests

1. Are performed on periodic samples drawn from production. The sample rate is nonstatistical, typically being monthly at the beginning of a production run and stretching to quarterly if the process is very stable. Testing at too frequent intervals does not allow time to complete the test, analyze the results, determine whether corrective action is required, and to take that action before the next sample is selected.
2. May be performed at any level of assembly, but usually are done at the higher levels to save testing cost.
3. Are performed with the specimen exposed to a limited range of environments, typically shock, vibration, temperature, and humidity.
4. Are nondestructive, so the specimen can be delivered. Environmental levels on the order of 75 percent of the qualification specification levels are often selected arbitrarily as being nondestructive.

The production assessment program should be planned simultaneously with the design qualification test program, and the test equipment should be designed with the capability for both programs. This may require some compromises, because design qualification tests measure many more attributes and will require more extensive test capability than production assessment testing requires. On the other hand, production assessment testing is repetitive, and optimum test equipment design should aim for economy of test cost. Despite the compromise, a single set of test equipment usually provides substantial cost savings to the project. Furthermore, identical test equipment for both test programs provides maximum compatibility of the test results, which is advantageous (1) in detecting, diagnosing, and correcting deleterious shifts in production processes from those existing at the time of release to production; and (2) for providing correlatable data for use in reliability assessment of the population as it is being produced.

8.2.7 Screening and Burn-in

Screening and burn-in of parts and subassemblies are often performed to weed out infant mortality before use or delivery. Occasionally whole systems are burned in at the factory before delivery. This testing

1. May be performed at ambient conditions, and only powering the device or running it through simple operating cycles.
2. Is often performed with vibration to accelerate mechanical failures. The selected levels are low enough to keep from inducing failure in good hardware, typically 75 percent of the expected use environment. This is especially useful for finding workmanship errors such as poor solder joints or inadequately supported components.
3. Is also often performed with temperature cycling, especially on electronics components and assemblies. If the temperatures are increased in step fashion, this is called step-stress testing (see Fig. 8.1).

Most large suppliers of electronic components have the necessary equipment to perform screening and burn-in economically, and often have standard add-on price lists for Hi-Rel, TX, or other widely used screening and burn-in regimes.

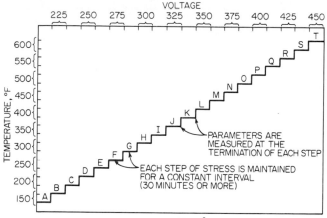

FIG. 8.1 Typical stress test regime ($°C = \frac{5}{9} (°F - 32)$).

8.3   TEST PLANNING

An acceptance test program must be planned in a thorough and timely manner if all of the necessary elements are to be available at the right time and place and in the right quantity. The planning should start very early in product development so that design of special test equipment can proceed and be ready in time for testing the first research and development (R&D) hardware. Furthermore, the preliminary test plan must be available to product designers so that the necessary test points can be designed into the product.

Reliability and/or quality assurance engineers usually plan the R&D acceptance testing by utilizing inputs from the product designers and adding the desired quality control and reliability data test points to those the designers specify for measuring performance. This arrangement assures that the data necessary for reliability evaluation are collected on the early development hardware, and permits meaningful reliability engineering input to the design effort.

8.3.1   Test Planning, Timing, and Scheduling

The importance of starting the acceptance test planning at the beginning of the project cannot be overemphasized. This planning should be done at the same time as the planning for those tests directly associated with the development effort (research, feasibility, evaluation), so that the many valuable reliability data will be collected from the acceptance tests of early hardware. Timely planning precludes losing the basic correlation data comparing performance in controlled environments (collected in the development tests) with performance in the ambient test conditions (collected from acceptance testing). These data are required to complete the production acceptance test planning process. Timely planning also requires that much other company, customer, and product information, planning, and results be available and be used. See Fig. 8.2 for a summary of the more important of these.

The entire integrated reliability test program, which includes acceptance

Company policies	Manufacturing plans	Make or buy lists
Contract or customer requirements	Integrated reliability test program plan	Spare parts lists
Drawings, specifications		Critical parts lists
Reliability prediction and assessment requirements		Field service test plans
	Classification of characteristics	Organization charts
R&D and qualification test plans, procedures, and results	Failure modes and effects analyses	Test facility and capability lists
Product cost analyses	Tolerance studies, including tolerance funneling	Plant layout
Test and inspection cost analyses		

FIG. 8.2 Prerequisites to planning an acceptance test program.

testing, should be blocked out at the beginning of the project, with best estimates made of the parameters to be tested and the environments to be used for each individual test program. All of these tentative test plans are laid out with the understanding that the planning will be revised at frequent intervals, perhaps monthly in the early stages and changing slowly to quarterly or semiannually by the time the project has reached the design-release stage. Changes in the program test plan will result from many factors and events, including:

1. Changes in hardware design concepts
2. Changes in production techniques or location
3. Changes in intended use of the hardware (which in themselves may result from changes in design)
4. Failure feedback information from testing
5. Discovery of new weaknesses and modes of failure which should be screened in the production acceptance testing

The scheduling of key events leading to the acceptance test program is also important, and the schedule should be charted on Gantt-type time-based charts against such key project dates as R&D and production design releases, acceptance of first production hardware, and qualification test spans (see Fig. 8.3).

8.3.2 Planning Documents

For each item to be tested, an *integrated reliability test plan sheet* should be prepared outlining all of the testing to be performed on that item throughout the entire project (see Fig. 8.4). Integrating the planning in this way allows comparison of the test programs, ensuring coordination and compatibility of test conditions among the different programs. For each assembly to be tested, a sheet is prepared, listing

Product assurance acceptance test program	Ground link transceiver		
	19 X 3	19 X 4	19 X 5
	J A S O N D	J F M A M J J A S O N D	J
P.A. program plan submittal	△		
Safety hazards analysis	△		
Subcontract test plan submittal	△		
Safety hazards classify data submittal	△		
Acceptance test procedure submittal		◇ △	
Equipment test procedure submittal		◇ △	
Initiate production testing		◇ △	
Prototype test series		△	
Revised test series		△	
Final test		△	
Test reports		◇ △△△△	
Test equipment design	△——△	◇◇	
Test equipment build	△———————◇—△		
Test equipment proofing		◇ △	

◇ — Accelerated schedule

FIG. 8.3 Sample page from acceptance test program schedule and status.

    All of the attributes (in words) to be tested

    The specification requirements

    The quantity to be tested in each environment

    The environmental requirements

    Document numbers for the implementing plans and procedures

These integrated plan sheets are very useful because they allow the totality of testing to be examined, ensuring that adequate reliability data will be generated. They also indicate where design of experiment techniques may be used to economic advantage.

    The integrated reliability test plan spread sheet is incomplete as a planning document from which individual test procedures can be written. It does not contain the input stimuli, parameter values and tolerances, or test equipment identification and errors. The following supporting details and numbers are required:

1. For test equipment designers to be able to design test equipment
2. For measurement engineers to compute or demonstrate the test equipment error for each measurement
3. For calibration engineers to write test equipment calibration procedures

INTEGRATED RELIABILITY TEST PROGRAM PLAN

PART NAME: GYRO PACKAGE, FLIGHT CONTROL SUBSYSTEM FLIGHT CONTROLS	PART NO 1963822-C SPECIFICATION WS 13898	TEST PLANS: PARA DOCUMENTS:	EET OD 22546 / OD 22545	PPT OD 22548 / OD 22547	VAT OD 22550 / OD 22549	PAT OD 22552 / OD 22551	FLD OD 22554 / OD 22553

Environment conditions / tolerances

Environment	Condition	Tolerance
ACCEPTANCE	LAB AMBIENT TEMP. 60–90°F, HUMIDITY >90%	TEMP. ±5%, HUMIDITY +10% −0%
RANDOM VIBR.	20–4,000 CPS, 5.1 grms	±10% G
SINUSOID. VIBR.	FREQ. SWEEP 10–50 CPS, 5 MINUTES/CPS, 2 HOURS AT MAX RESONANCE, MIL STD 167	±2% OR 1 CPS
SHOCK	50G PEAK 10 ±1.0 MSEC, 300G PEAK 2 ± 0.5 MSEC	±10%
ACCELERATION	10G, 5 SEC MIN; 3G, 60 SEC MIN	±5%
TRANS. TEMP.	−40°F, 12 HOURS; +150°F, 12 HOURS, 4 CYCLES	< 5%
FLIGHT TEMP.	120°F, 0°F; 600°F FOR 6 MINUTES	< 5%
HUMIDITY	MILE 5272, 120 HOURS MIN, 5 CYCLES	+10% −0%

Parameter quantity matrix (each group: EET, PPT, VAT, PAT, FLD)

Parameter Tested	Specification	ACC (E/P/V/A/F)	RANDOM VIBR. (E/P/V/A/F)	SINUSOID. VIBR. (E/P/V/A/F)	SHOCK (E/P/V/A/F)	ACCELERATION (E/P/V/A/F)	TRANS. TEMP. (E/P/V/A/F)	FLIGHT TEMP. (E/P/V/A/F)	HUMIDITY (E/P/V/A/F)
EXAMINATION, NONOPERATING — VISUAL EXAMINATION OF MATERIALS, DESIGN, CONSTRUCTION, DIMENSIONS, WEIGHT, COLOR AND FINISH, IDENT AND WORKMAN		4 / 4 / 1 / 1 / −							
LEAKAGE NONOPERATING	NO VISIBLE LEAKAGE	4 / 4 / 2 / 2 / 1	4 / 4 / 2 / 2 / —	4 / 4 / 2 / 2 / 1	4 / 4 / 2	4 / 4 / 2	4 / 4 / 2	4 / 4 / 2	4 / 4 / 2
POWER CONSUMPTION OPERATING	15.0 WATTS ± 2.5 WATTS AT AMB, ± 3.0 WATTS IN ENVIRONMENT	4 / 4 / 2 / 2 / 1	4 / 4 / 2 / 2 / 1	4 / 4 / 2 / 2 / 1	4 / 4 / 2	4 / 4 / 2	4 / 4 / 2	4 / 4 / 2	4 / 4 / 2
INSULATION RESISTANCE	40 MEGOHM MIN	4 / 4 / 2 / 2			4 / 4	4 / 4	4 / 4 / 2	4 / 4	4 / 4 / 2
DAMPING RATIO OPERATING	SEE SHEET 3	12 / 12 / 6 / 6 / 6	6 / 6 / 3 / 3 / 3	6 / 6 / 3 / 3 / 3	9 / 4	9 / 4	9 / 4 / 3	9 / 4 / 3	4 / 4 / 2
SENSITIVITY OPERATING	SEE SHEET 3	12 / 12 / 6 / 6 / 1	6 / 6 / 3 / 3 / 3	9 / 6 / 3 / 3 / 3	9 / 4	9 / 4 / 2	9 / 9 / 4	9 / 9 / 4	9 / 9 / 4
NULL DRIFT OPERATING	MAX INCR − 3.6 MV (0-PK) PER G² (RMS), PARA 3.24.6	6 / 6	12 / 12 / 6	12 / 12 / 6			6 / 6	6 / 6	6 / 6
ELECTRICAL INTERFERENCE	WS 13898	6 / 6	6 / 6	6 / 6	6 / 6	6 / 6	6 / 6	6 / 6	6 / 6
STARTING CURRENT	250 MA MAX	4 / 4 / 2	4 / 4 / 2			12 / 12	4 / 4		
AXIS ALIGNMENT OPERATING	OUTPUT OF EACH 2 AXIS ≤ 500 MV, AMBIENT OR ≤ 650 DURING ENVIRONMENT	12 / 12 / 2 / 1 / 6	6 / 6 / 2	12 / 12 / 2	12 / 12 / 2	12 / 12 / 2			

EET = ENGINEERING EVALUATION TEST
PPT = PREPRODUCTION TEST
VAT = VENDOR (FACTORY) ACCEPTANCE TEST
PAT = PRODUCTION ASSESSMENT TEST
FLD = FIELD TEST

EET, PPT: QUANTITY INDICATED IS TOTAL QUANTITY TESTED
VAT, PAT, FLD: QUANTITY INDICATED IS SAMPLE SIZE PER PRODUCTION LOT

PAGE 1 OF 3

FIG. 8.4 Sample integrated test-program sheet.

8.10

4. For procedure writers to write acceptance test procedures

The necessary supporting details are set forth in additional documents. The values for each parameter to be tested, including the tolerances and the input conditions with their tolerances, are conveniently set forth in a *parameters document* (Fig. 8.5), as well as in a *parameters spread sheet* (Fig. 8.6), which permits comparison of test values at different levels of assembly. There is one of these spread sheets for each subsystem and for each higher-level assembly to be tested. These two documents are used to ensure that each test is designed in coordination with all other tests of the same parameter performed in the various test programs. This is particularly important in larger organizations where the individual work of many different suborganizations is required to establish the details of the overall reliability test program. It also ensures that tolerances are properly funneled (see Sec. 8.3.3).

For each item to be tested, the following must be given:

1. A test program, as outlined in the integrated reliability test program plan sheet
2. The parameters, funneled as set forth in the parameters spread sheet
3. Test input conditions as outlined in the individual parameters document sheets

Then:

1. The test engineers can design the test equipment.
2. Product designers can provide test connection points.
3. Test planners can lay out floor space, personnel, standard test, and/or environmental equipment requirements.
4. Production schedulers can plan the necessary test spans into the production plan.

8.3.3   Funnel of Tolerances

A particular facet of test planning arises from the fact that hardware "flows" through the assembly process, from piece parts through various levels of assembly into the final product level and into customer use. Tests are performed at each of these levels, and often the same or related parameter will be tested at one or more subsequent levels. Many functional attributes, especially in electronic and hydraulic equipment, drift with time, handling, or functional-use cycling. If the acceptance-rejection limits on these attributes are set identically at successively higher levels of tests, there will be a measurable percentage of hardware with attributes just inside the limits at one level of test which will drift outside the limits in the next higher test and be rejected back to the lower level for rework. To preclude the resulting circulation of hardware, the tolerances of a single attribute are established in a funnel arrangement, with the tightest tolerance at the lowest level of assembly (Fig. 8.7). The tolerances are tighter at each lower test echelon, so that room is provided for some functional parameter drift or degradation with time, use, and transportation. A 5 percent tightening per echelon is a common goal, although it is not always possible to attain that much when the attribute is tested at more than two or three levels. Under no circumstances,

O.D. 22551   PART IV

PARAMETERS DOCUMENT SHEET

REFERENCE:
TEST POINT # 201

SHEET 1 OF 7   REV C

PART NAME		PART NUMBER	
GYRO PACKAGE, FLIGHT CONTROL		1963822-C	
SUBSYSTEM Flight Control	DATE PREPARED 7-31-63	DATE RELEASED 9-15-63	RELEASE AUTHORITY JRA 52PD-20105C
PREPARED BY D.R. Keith (System Analysis)	CHECKED BY P.P. Parish	RELIABILITY Fallon	
DESIGN CHECK & EE Broughort	PROJECT OFFICE WA Steven		CUSTOMER APPROVAL J.E. Van Eau

TEST PLAN ITEM	ATTRIBUTE	PARAMETER	TOLERANCE

1	**RANDOM VIBRATION**
	1.1 Input: 115 volts a-c rms ±0.5%, 400 ±2cps @ 70 ± 6°F
	1.2 Load: Non-reactive impedance of 15K ± 300 ohms for
	    pitch, yaw and roll outputs
	1.3 The package shall be subjected to random vibration
	    along each of its three mutually perpendicular axes
	    for one minute. The level of vibration shall be
	    5.1 grams contained within the spectral density
	    limits below:
	    1.3.1 Upper Limit		

	Frequency (cps) Envelope of Peaks (g^2/cps)
	        20 0.005
	       500 0.007
	       750 0.014
	     4000 0.025
	    1.3.2 Lower limit shall be one fourth of upper limit
	
	1.4 Measure the in-phase null degradation of the demod-
	    ulated output signal through a 0 to 4 cps filter
	    (8 db/octave), of each gyro when vibrated in each
	    axis (measure 0 to peak).
	    1.4.1 Pitch axis
	    1.4.2 Yaw axis
	    1.4.3 Roll axis
		
		1.7 mv 0-p/$(gms)^2$ max
		1.7 mv 0-p/$(gms)^2$ max
		1.7 mv 0-p/$(gms)^2$ max
		(210 mv 0-p max, per axis)	

2	**POWER CONSUMPTION**
	2.1 Input and load same as 1 above
	2.2 Measure
	    2.2.1 Running power
	    2.2.2 Power factor
	    2.2.3 Time to reach running power
	    2.2.4 With squelch circuit applied, measure running
	        power and power factor
		
		15.0 watts
		.960 ldg to .960 lag
		50 sec
		
		10.0 watts
		
		1.0 to 0.7 lag
			
			±2.5 watts
			
			max
			
			+1.8 watts
			-2.3 watts

3	**INSULATION RESISTANCE** 3.1 Apply 500 volts d-c minimum, between all isolated circuits and between circuits and case for 5 seconds minimum 3.2 Measure insulation resistance	 40 megohms	 min.
4	**UNDAMPED NATURAL FREQUENCY AND DAMPING RATIO** 4.1 Input and load per 1 above 4.2 Apply sinusoidal change of rate to produce 90° phase lag to pitch, yaw and roll rate gyros. 4.3 Measure the natural frequencies and damping ratios		

FIG. 8.5   Parameters document page.

PARAMETERS SPREAD SHEET

Volume II, Section 2
Flight Control Subsystem
Gyro Package, Flight Control 1963822-C

	PERFORMANCE REQUIREMENTS		PARAMETERS							
ITEM	CHARACTERISTIC	CONDITIONS AND REMARKS	NOMINAL	In-Flight Tolerance	Submarine Tolerance	Tender Tolerance	Depot Tolerance	Factory Tolerance	Vendor Tolerance	Units of Measurement
GENERAL CONDITIONS										
1.0	External resistive load	(all output circuits) 15,000 ± 300 ohms								
ENGINEERING REQUIREMENTS										
2.0	Input									
2.1	Voltage	Single phase	115	±3%	±2.7%	±2.4%	±2%	±1.5%	±1.0%	volts a-c rms
2.2	Frequency		400	±0.8%	±2	±2	±2	±2	±2	cps
2.3	Harmonic content			2 max	2 max	2 max	2 max	2 max	2 max	per cent
3.0	Electrical									
3.1	Power consumption									
3.1.1	Starting current			125 max	125 max	125 max	125 max	125 max	125 max	milliamps a-c rms
3.1.1.1	Transient	Time to reach running power limits		50 max	50 max	50 max	50 max	50 max	50 max	seconds
3.1.2	Running power		15.0	+4.0	±3.8	+3.6	+3.3	±2.5	±2.0	watts
3.1.3	Running power factor	Leading		0.960 min	0.960 min	0.960 min	0.960 min	0.960 min	0.960 min	
		Lagging		0.960 min	0.960 min	0.960 min	0.960 min	0.960 min	0.960 min	
3.1.4	Squelch power		10.0	+1.8 -2.3	+1.8 -2.3	+1.8 -2.3	+1.8 -2.3	+1.8 -2.3	+1.8 -2.3	watts
3.1.5	Squelch power factor	Lagging only	1.0	-0.3 max	-0.3 max	-0.3 max	-0.3 max	-0.3 max	-0.3 max	watts
3.2	Insulation resistance	The resistance of insulation between isolated circuits and between circuits and the case with a potential of 500 volts d-c applied for not less than 5 seconds		40 min	40 min	40 min	40 min	40 min	40 min	megohms
3.3	Output frequency		36	±6.0	±5.94	±5.86	±5.75	±5.5	±5.25	cps
3.4	Damping ratio	At a natural frequency of 30 cps		0.91 max 0.49 min	0.90 max 0.50 min	0.89 max 0.51 min	0.88 max 0.52 min	0.85 max 0.55 min	0.82 max 0.58 min	
		At a natural frequency of 50 cps		1.32 max 0.38 min	1.31 max 0.39 min	1.30 max 0.40 min	1.27 max 0.42 min	1.23 max 0.45 min	1.19 max 0.48 min	
		(As the natural frequency increases from 30 to 50 cps the damping ratio limits shall increase linearly between the values given.)								

FIG. 8.6 Parameter spread sheet.

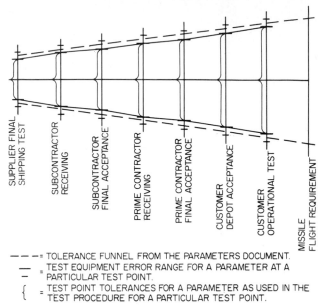

SUPPLIER FINAL SHIPPING TEST

SUBCONTRACTOR RECEIVING

SUBCONTRACTOR FINAL ACCEPTANCE

PRIME CONTRACTOR RECEIVING

PRIME CONTRACTOR FINAL ACCEPTANCE

CUSTOMER DEPOT ACCEPTANCE

CUSTOMER OPERATIONAL TEST

MISSILE FLIGHT REQUIREMENT

─ ─ ─= TOLERANCE FUNNEL FROM THE PARAMETERS DOCUMENT.
─
─ = TEST EQUIPMENT ERROR RANGE FOR A PARAMETER AT A
        PARTICULAR TEST POINT.
{ = TEST POINT TOLERANCES FOR A PARAMETER AS USED IN THE
        TEST PROCEDURE FOR A PARTICULAR TEST POINT.

FIG. 8.7   Tolerance funnel.

however, should the slope between successive layers become negative. The error inherent in all test equipment is somewhat related to this subject; it is discussed in Sec. 8.4.4.

8.3.4 Attributes

The selection of specific attributes to be tested at each test position is a complex matter, with the final selection resulting from a balance of conflicting factors. These factors include:

1. The need to demonstrate functionability under all use conditions
2. The need to demonstrate reliability
3. The cost of testing, including test and environmental equipment
4. The time to perform each test
5. Equipment and personnel available to perform tests
6. Customer requirements
7. Need to assure spare part interchangeability
8. Desire to provide optimum process and quality control
9. The required reliability of the part or system
10. The cost of, and the cost of replacement of, the part tested

When these factors have been considered and a test program selected, then a test plan should be prepared for each tested item. See Fig. 8.8 for a page from a typical test plan.

TEST PLAN SHEET   225/52   PART I-C

PART NAME				PART NUMBER	TEST POINT	REV
GYRO PACKAGE, FLIGHT CONTROL				1963822-C	201	B

Specification	Test Procedure	Test Sta.	OCD POINT Yes X No	Page 1 of 4
WS 13898	22634	No. 52X220	TEST DATA REQD. Yes X No	

Prepared By	Checked By	Date Prepared	Release Authority
Johnson	P P Parish	5-13-63	JRA 52TP-22552

QA Engineering	Design	Project Office	Customer Approval
RA Hold	PD Stroun	WG Stevenson	JA Van Ess

OCD CAT	ITEM NO.	Attribute Tested and Test Condition
	1	**RANDOM VIBRATION**
		1.1 Input: 115 volts a-c rms $\pm 0.5\%$, 400 ± 2 cps @ $70 \pm 6^\circ$F
		1.2 Load: Non-reactive impedance of 15K $\pm$ 300 ohms for pitch, yaw, and roll outputs
		1.3 The package shall be subjected to random vibration along each of its three mutually perpendicular axes for one minute. The level of vibration shall be 5.1 grams contained within the spectral density limits below:
		1.3.1 Upper limit:

Frequency (cps)	Envelope of Peaks (g^2/cps)
20	0.005
500	0.007
750	0.014
4000	0.025

OCD CAT	ITEM NO.	Attribute Tested and Test Condition
		1.3.2 Lower limit shall be one fourth of upper limit
MAJ		1.4 Measure the in-phase null degradation of the demodulated output signal through a 0 to 4 cps filter (8db/octave), of each gyro when vibrated in each axis (measure 0 to peak).
	2	**POWER CONSUMPTION**
		2.1 Input and load per 1 above
		2.2 Measure
MAJ		2.2.1 Running power and power factor; time to reach running power.
MIN		2.2.2 With squelch circuit applied, measure running power and power factor.
	3	**INSULATION RESISTANCE**
		3.1 Apply 500 volts d-c minimum between all isolated circuits and between circuits and case for 5 seconds minimum.
MAJ		3.2 Measure insulation resistance.
	4	**UNDAMPED NATURAL FREQUENCY AND DAMPING RATIO**
		4.1 Input and load per 1 above
		4.2 Apply sinusoidal change of rate to produce 90° phase lag to pitch, yaw, and roll rate gyros.
MAJ		4.3 Measure the natural frequencies and damping ratios of each of the following gyros:
		4.3.1 Pitch rate
		4.3.2 Yaw rate
		4.3.3 Roll rate
	5	**SENSITIVITY, IN PHASE AND QUADRATURE AND HARMONIC COMPONENTS**
		5.1 Input and loads per 1 above
		5.2 Apply the following rates to the pitch, yaw, and roll axes
		5.2.1 75°/second CW and CCW
		5.2.2 35°/second CW and CCW
		5.2.3 15°/second CW and CCW
		5.2.4 0°/second CW and CCW
		5.3 Measure the signal phase for each of the above rates
		5.4 Measure the in-phase voltage component (0° or 180°)
		5.5 Measure the out-of-phase voltage component (Quadrature and Harmonics).

FIG. 8.8 Test plan sheet.

8.3.5   Classification of Characteristics

Except for the simplest of parts, it is not practical to test all functional attributes of a part in every test regime, nor generally in any one regime. Consequently, it follows that any test plan represents only a sampling of the attributes. To arrive at the final list of attributes, it is essential that the importance of each attribute be classified in accordance with its importance to the final use or mission of the product. This classification can be done informally, but a better method is described in DOD-STD-2101(OS), *Classification of Characteristics,* which is used by the military. In this system, each attribute is classified as critical, major, or minor in accordance with its effect on coordination, life, interchangeability, function, and safety (the CLIFS) of the product. Attributes are classified as critical if, when the attribute is defective, there will be an adverse effect on safety of the product in use; as major if, when the attribute is defective, there is not a safety consideration, but there could be a significant degradation in performance; minor is any other attribute. The advantage to using a formalized system is that classification of attributes is standardized from project to project, as well as from item to item and from test program to test program within a project. Furthermore, many side advantages of classification can be realized throughout the whole life of a project, including:

- Automatic designation of inspection and test sampling plans
- Automatic emphasis on important attributes for such efforts as failure diagnosis, corrective action, reliability data collection and analysis, inspection, and design change control
- Automatic base for the quality incentive in incentive contracts

However, the immediate benefit to test planning is that the otherwise completely subjective process of selecting attributes is reduced to objective application of an agreed-upon set of ground rules, largely eliminating the subjectivity. Thus it could be agreed that all criticals must be tested at the highest level and as late in the production process as possible, which might require that a special test connector be added to the product so that the attribute can be tested in the final product configuration; or that all majors must be tested in-house, which could require duplicating suppliers' testing. See Fig. 8.9 for a typical inspection classification sheet; it can also be used for classifying functional attributes.

8.3.6   Selection of Input and Environmental Conditions

Concurrently with the selection of items and attributes to be tested, the input and environmental conditions associated with each test point are specified, because the designation of an attribute to be tested is meaningless unless the conditions of the test are also specified. As a general rule, input conditions should be at least as rigorous as those encountered by the part in its ordinary use. This will often dictate that two or more sets of input conditions be provided: the "high" and "low" conditions which can be encountered as the part ordinarily functions (e.g., both high and low values of input power). In some instances, as for example when a test of linearity, hysteresis, or sensitivity is required, other intermediate points will also be required to establish the performance curve. In instances where the part is operating in a derated condition, it may be satisfactory to utilize a one-point

CLASSIFICATION OF CHARACTERISTICS

PART NAME				DWG. NO. 2363057		INSPCTN. POINT 13034		REV. 00
Valve, Injector Assy.				DWG. REV.		PAGE 1	OF 3	
PREP. BY		SQA SIG.				DATE	SAMPLING	

	CLASS-IFIC-ATION	ITEM	ATTRIBUTES TO BE VERIFIED	REFERENCE	INSPECTION METHOD	TABLE	AQL PLAN
	Crit		None				
M	Maj	*101	1.117 ± .002/.000 dia. daturm - B	11D	Mics.	8A	1.5%
S		102	1.006 ± .002/.000 dia.	11D	Mics.	--	---
S		103	Datum Dia. - B - Shall be perpendicular to datum surface - M - within .001	11D	Indicator	--	---
M		*104	.330 ± .010 dim.	11D	Mics	8A	1.5%
S		105	.250 ± .005/.000 dim.	11D	Go-No Go Plug GA.	--	---
C		106	Keyway orientation	9C	Visual	--	---
S		107	1.000 ± .010 dim.	10C	Indicator	--	---
S		108	.187 dia. max.	10B	No Go Plug GA.	--	---
S		109	Datum Surface -M- shall be flat within .001.	8B	Indicator	--	---
S		110	.500 ± .010 dim.	9C & 5D	Mics.	--	---
S		111	.375 dia. max. (6) places	8A	Mics.	--	---
M		*112	.375 dia. max (6) places shall be in true position within .020 dia.	8A	Indicator 1870246	8A	1.5%
S		113	.468 ± .035 dim. (12) places	9C	Mics.	--	---
S		114	1/4 -28 UNF 3B THRD. (12) places	5D & 5A	Go-No Go THRD Plug GA	--	---

This Unit of Product is Defective if any of the Requirements Listed Herein are not Met.

Notes: A. This classification of Defect, originated by PA Inspection Engineering list those attributes which, when used in conjunction with the documents listed below, comprise the total inspection acceptance criteria:
1. Applicable Drawing and Specifications
2. Applicable PA & TS Policies & Procedures
3. Applicable Product Assurance Operating Instructions
B. Asterisk (*) before item number indicates an OCD attributes.

Remarks or Sampling Tables.

FIG. 8.9 Classification of characteristics. Although in this example, all attributes to be tested happen to be classified "major," other parts also have attributes classified "critical" or "minor."

nominal input condition, provided the allowable output tolerance is tightened by an arbitrary amount from what would be allowed if the input were both high and low. Since such a one-point approach always represents less than optimum assurance and is usually dictated by cost rather than quality or reliability considerations, it should be the exception rather than the rule.

Environmental exposure should be used wherever possible if high reliability is paramount. Ambient testing, except when the part will be used only in an ambient condition, is at best an approximation, and uncertainties always exist in such testing. Unless test cost is prohibitive or the part has been designed with a limited duty cycle in its operating environment, a general rule for the project should be established that all testing will be done in the use environment. The practical constraints of time, equipment, and money will mitigate, in many instances, against the rule, but starting with the rule makes it administratively necessary that departures from it be justified, and the end result will be that a larger percentage of the testing will be under environments.

In particular, the vibration environment should be utilized wherever possible, primarily because vibration is one of the most economical and effective quality control tools available to the test engineer. The probability of detecting

Intermittent operation

Loose and cracked parts

Inadequate mounting or protection

Poorly soldered joints

Workmanship defects

is much higher in this environment than in any other. From a practical point of view, the part may be operating during the vibration exposure, so that little additional time is added to the overall test time or cost because of it. Caution must be exercised in selecting the level of vibration, however, for with some delicate parts the exposure can be degrading if the use environment level or time limit is exceeded. However, the need for such care should not be accepted as an excuse for not using the environment. (Testing may include both random vibration to simulate the actual-use power density spectra and sinusoidal vibration to permit meaningful failure diagnosis.) Shock, temperature, and humidity extremes are other common environments easily attainable and frequently used.

Consideration should be given to the use of combined environments—more than one at a time. The interaction of environments is difficult to assess; in a complex-item test program it is almost impossible to do so mathematically, and empirical data are necessary to plan the production assessment testing properly. Not all environments can be combined simultaneously (e.g., both high and low temperature), but usually a series of three or four tests will suffice. These include high and low humidity with high and low temperature, combined in each case with shock and vibration.

8.3.7   Sampling vs. 100 Percent Testing

It is generally true that production *assessment* testing is performed on samples drawn from production, while production *acceptance* testing is performed on every unit of product produced (100 percent). However, sampling theory is as applicable to testing as it is to inspection. Acceptance testing of one-shot items, like explosive devices, can only be performed on samples. Even for electronic devices, however, if many identical items are to be produced and the homogeneity of the population is reasonably good, sampling can be utilized. MIL-STD-105, *Sampling Procedures and Tables for Inspection by Attributes,* and MIL-STD-414, *Sampling Procedures and Tables for Inspection by Variables for Percent Defec-*

tive, are the most widely used documents describing the details of acceptance sampling, and the consumer and producer risks (AQL, acceptable quality level) can be chosen to ensure that the desired reliability level of the population is maintained by the sampling acceptance test program. Reliability calculations can be based on sampled data if confidence intervals are adjusted to account for the added uncertainty. See Chaps. 18 and 19 for detailed information on sampling plans, risks involved, and confidence levels.

Sampling is more difficult with low-volume production projects, since there is no statistically valid sampling rate for very small lots. This situation is not uncommon in weapons systems or in large industrial equipment projects, where the production for a year may total from 10 to 100 units. In these situations, it is necessary to subjectively choose statistically nonvalid sampling plans. Typical rates may vary from every other unit to 1 sample from 10, with the exact rate dependent upon

1. The tolerable degree of risk
2. The probability of uniformity of successive units of product
3. The potential effect of failure of the item
4. The ability of the production process to produce sufficient hardware for both delivery and test within the time available
5. The cost of the hardware and the test

If the low-volume production is to run for several years, the sampling rate may be decreased with increasing maturity of the production process and the attendant increase in product homogeneity.

8.3.8 Test Procedures

Preparation of detailed test procedures is the last step in test planning. Performance of acceptance tests should not be left to the skill and knowledge of the test technicians, but should be carefully controlled by the audited use of detailed procedures. The more important reliability is to the success of the project, the more important detailed and controlled procedures become.

Test procedures should describe and control three separate and distinct areas of the testing:

1. *Calibration of the test equipment* in accordance with calibration procedures, comparing the test equipment to standards traceable to the National Bureau of Standards. The calibration should be performed at the interface between the test equipment and the hardware to be tested (i.e., at the test leads and commonly called end-to-end calibration), and should include not only verifying the measuring or comparing section of the test equipment but also the section which establishes the input and environmental conditions as well. See Fig. 8.10 for a page from a typical end-to-end calibration procedure. A detailed calibration procedure like the example should be prepared

- To ensure consistency of calibration methods, which ensures consistency of product test results
- To ensure that calibration results are available in consistent format so that test equipment drifting can be analyzed and recalibration intervals adjusted

---

EXAMPLE   10

Page from End-to-End Calibration Procedure

SCP-OD 26345
Revision 1
12 September 1963

3.4.2.13.1 Console   Thermal   AC   Load   Tolerance
            Load   Transfer   Ammeter   Resistance
            Resistor   Standard

                    _____ Volts   ____ Amps   ____Ohms - 230 Ohms WIR

            To obtain the value of the Load Resistance divide the voltage

            observed on the Terhmal Transfer Standard by the current observed

            on the AC Ammeter.

** 3.4.2.14 Observe the lead power factor as indicated by the Phase Angle

            Meter (Unit 1.2.1.3). Record on the Test Data Sheets.

3.4.2.14.1 Load Power   Power Factor   Tolerance
            Meter   Meter

                    _____   -.95 to † .95

            NOTE: Paragraphs 3.4.2.15 through 3.4.2.24 are to be accomplished

            only if 3.4.2.13 is out of tolerance because of inadequate trimmer

            range.

3.4.2.15   The variable resistor under the Control Panel AC Milliameter has

            a capability of trimming the Console Load Resistance approximately

            7.5%. In the event that this trimmer does not have a range that

            will permit setting the load resistance, in paragraph 3.4.2.13, to

            230 ohms nominal, proceed as follows, and then accomplish paragraphs

            3.4.2.13 and 3.4.2.14.

3.4.2.16   On the Control Panel (A6), adjust the Variable Resistor control

            under the Control Panel AC Milliammeter and labeled SET TO 500 MA,

            to the center of its range.

3.4.2.17   On the Power Amplifier (Unit 1.2.1.13), position the POWER switch

            to the 30 SEC WARMUP position.

FIG. 8.10   End-to-end calibration.

- To ensure that a traceable (auditable) trail exists from the test equipment to national measurement standards, ensuring the interchangeability and coordination of products at all levels of assembly with interfacing hardware

Section 8.4.5 further discusses test equipment calibration.

2. *Proofing the test equipment,* with the hardware to be tested, in accordance with an equipment test procedure is the next step; see Fig. 8.11 for an example of such a procedure. This operation demonstrates that the test equipment performs its intended function when coupled with tested hardware, and it is run to uncover unanticipated anomalies such as ground loops or variations in input conditions with variations in loading or in line voltages. Proofing is particularly important the first time a new test equipment design is used with a specific configuration of product hardware; it should be routine at the start-up of production testing.

3. *The test procedure* describes in detail all of the adjustments, hook-ups, and switch and button operations required to perform the test. An example is shown in Fig. 8.12. The test procedure should include detailed data sheets (if the recording is to be done manually), which will ensure that all of the desired data, both input and output, are recorded and in the units desired. See Fig. 8.13 for a representative data sheet. The sheets set forth the accept or reject limits for the test, as derived from the parameters' document tolerances, from which the test equipment error (see Sec. 8.4.4) has been subtracted. They should also include spaces for recording nontest information such as test area ambient or environmental conditions, date, precise hardware configuration tested, test operator and inspector identification, and any other administrative data which will permit reconstruction of the test in the event of subsequent question.

Configuration management controls should be established over all of these test procedures to ensure that the original release and subsequent changes are authorized by appropriate levels of management and by interested activities. When a customer desires to control the project closely, customer approval will also be required.

8.4 TEST EQUIPMENT AND FACILITIES

8.4.1 Test Equipment Defined

Test equipment includes:

1. The equipment which provides the input stimuli to the hardware being tested
2. The measurement equipment which detects the output
3. The display equipment, including meters, or comparison equipment which compares the output with known standards and displays the comparison or prints out either the direct reading or the comparison
4. The environmental equipment that provides the environment to which the hardware is exposed during the test

In the discussion following, all the above will be included in the term "test equipment."

EXAMPLE   9

Page from Equipment Test Procedure   ETP-OD 21681
                                        Page 10

9.1.4   (continued)

          RATIO BRIDGE AC POWER ON
          800 CPS POWER SUPPLY LINE-ON
          Visicorder POWER ON
          LOW FREQUENCY FUNCTION GENERATOR POWER
          COMPUTING DIGITAL INDICATOR POWER-ON

9.1.5   Position DIGITAL VOLTMETER Mode switch to AUTO.

9.1.6   Position AC-OFF switch of +35 VDC panel to AC. Observe that
          indicating light illuminates. Record results on Test Data Sheet.

9.1.6.1   Adjust +35 VDC panel voltage adjustment control for an indication
          of 35 ± 1 vdc on the panel meter.

9.1.6.2   Position CONTROL PANEL VOLTAGE SELECTION switch to +35 VDC and
          adjust Power Supply voltage adjustment control until DIGITAL
          VOLTMETER indicates +35 ± 0.5 vdc. Record results on Test
          Data Sheet.

9.1.7   Position AC-OFF switch to -30 VDC panel to AC. Observe that
          indicating light illuminates. Record results on Test Data Sheet.

9.1.7.1   Adjust -30 VDC panel voltage adjustment control for an indication
          of 30 ± 1 vdc on the panel meter.

9.1.7.2   Position CONTROL PANEL VOLTAGE SELECTION switch to -30 VDC and
          adjust Power Supply voltage adjustment control until DIGITAL
          VOLTMETER indicates -30 ± 0.3 vdc. Record results on Test
          Data Sheet.

9.1.8   Position CONTROL PANEL VOLTAGE SELECTION switch to +10 VDC.
          Verify that DIGITAL VOLTMETER indicates +10 ± 0.5 vdc. Record
          results on Test Data Sheet.

9.1.9   Position CONTROL PANEL VOLTAGE SELECTION switch to -10 VDC.
          Verify that DIGITAL VOLTMETER indicates -10 ± 0.5 vdc. Record
          results on Test Data Sheet.

9.1.10   Position VOLTAGE SELECTION switch to OFF.

9.1.11   Position DIGITAL VOLTMETER Function switch to AC.

9.1.12   Position OUTPUT-ON switch of 800 CPS POWER SUPPLY to ON.

9.1.13   Position VOLTAGE SELECTION switch to 28.75 VAC.

9.1.14   Adjust COARSE and FINE controls of 800 CPS POWER SUPPLY for an
          indication of 28.75 ± 0.25 vrms on DIGITAL VOLTMETER. Record
          results on Test Data Sheet.

FIG. 8.11 Sample page from equipment test procedure.

8.4.2 Comparative Features of Test Equipment

It is convenient to consider several features of test equipment separately. The
features are *purpose, type of control, calibration*, and *readout*.

General Purpose vs. Special Purpose.   Test equipment can usually be classified as

6.3   <u>Insulation Resistance</u>

6.3.1   Connect Cable 1607447 (ref para 3.3.2) to INSULATION TEST jack on test console and to unit under test.

6.3.2   Actuate INSULATION MODE and MEGGER ANALOG switches

OCD 106   6.3.3   Actuate REF C switch and all pin switches except C sequentially. Actuate Megger foot switch after each pin selection. Record results on Test Data Sheet as instructed in paragraph 6.3.6.

OCD 107   6.3.4   Actuate REF D switch and all pin switches except D, E, F, and G sequentially. Actuate megger footswitch after each pin selection. Record results on Test Data Sheet as instructed in paragraph 6.3.6.

OCD 108   6.3.5   Actuate REF H and all pin switches except H and T sequentially. Actuate megger foot switch after each pin selection. Record results on Test Data Sheet as instructed in paragraph 6.3.6.

6.3.6   Maximum indication on DVM shall be 1.198 volts (DVM will measure current: –1 volt = 10μA). Record results on Test Data Sheet.

6.4   <u>Natural Frequency and Damping Ratio.</u>

6.4.1   Pitch Gyro

6.4.1.1   Mount the unit in the Holding Fixture (para 3.3.1). Mount the Holding Fixture securely on the Simulation Table (para 3.2.8) with face "C" down on the table (ref. Figure 2). Connect the circuit as shown in Figure 5.

6.4.1.2   Actuate RATE, NORMAL LOAD, PITCH and GYRO ON switches. Allow 60 seconds for the gyros to spin up. Actuate COUNTER EXT. switch and adjust counter to read frequency times 10. Allow the counter 10 minutes to warm up.

6.4.1.3   Position Servo Analyzer (part of 3.2.8) E2-DC, AC switch to the DC position. Position E1-DC, AC switch to AC position. Adjust gain on the amplifier to 5. Position Frequency Range switch to A position and adjust the Test Signal Amplitude control for approximately 30. Adjust the Frequency Control for 1 cps. Position E1/E2, Damping Ratio switch to E1/E2 position. Balance E1/E2 reading for 0 db by using appropriate db attentuators and assure that the Overload lamps are extinguished. Position E1/E2 Damping Ratio switch to the Damping Ratio position and zero Damping Ratio Meter.

6.4.1.4   Position Frequency Range switch to C position and adjust the Frequency Control for 90° indicated phase shift as read on the Phase Angle Meter (assure Phase Angle Meter Sel. switch is in the A position).

OCD 109   6.4.1.5   Record the frequency as measured on counter (part of 3.2.1) as Natural Frequency on the Test Data Sheet. The Natural Frequency

FIG. 8.12 Sample page from acceptance test procedure.

either special or general purpose, depending upon whether the equipment is usable, respectively, on only one or on more than one type of test article. General-purpose test equipment should be chosen whenever possible unless some aspect of the test program dictates the use of special-purpose equipment. Among the factors which may so dictate are the following:

8.24   INTRODUCTION TO RELIABILITY

Test Data for ATP/OD 22634	Page 19
Report Number	

OCD No.	Procedure Para. No.	Function	Requirement	Actual Reading
107	6.3.4	Insulation Resistance		
		Ref D to Pin A	−1.20 volts max	_____
		Ref D to Pin B	−1.20 volts max	_____
		Ref D to Pin C	−1.20 volts max	_____
		Ref D to Pin H	−1.20 volts max	_____
		Ref D to Pin K	−1.20 volts max	_____
		Ref D to Pin S	−1.20 volts max	_____
		Ref D to Pin T	−1.20 volts max	_____
108	6.3.5	Insulation Resistance		
		Ref H to Pin A	−1.20 volts max	_____
		Ref H to Pin B	−1.20 volts max	_____
		Ref H to Pin C	−1.20 volts max	_____
		Ref H to Pin D	−1.20 volts max	_____
		Ref H to Pin E	−1.20 volts max	_____
		Ref H to Pin F	−1.20 volts max	_____
		Ref H to Pin G	−1.20 volts max	_____
		Ref H to Pin K	−1.20 volts max	_____
		Ref H to Pin S	−1.20 volts max	_____
109	6.4.1.5	Natural Frequency, Pitch Gyro	30 through 50 cps	_____ cps
	6.4.1.5	Damping Ratio, Pitch Gyro (determine requirement from 6.4.1.5 and note it)	- - - - - -	_____
110	6.4.1.6	Natural Frequency, Yaw Gyro	30 through 50 cps	_____ cps
	6.4.1.6	Damping Ratio, Yaw Gyro (determine requirement from 6.4.1.5 and note it)	- - - - - -	_____
111	6.4.1.7	Natural Frequency, Roll Gyro	30 through 50 cps	_____
	6.4.1.7	Damping Ratio, Roll Gyro (determine requirement from 6.4.1.5 and note it)	- - - - - -	_____
112	6.5.4	Pitch: CW 75°/sec In-Phase VR	+0.7287 through +0.86713 mv max	_____ mv
		CW 75°/sec Quad & Harmonic	680 mv max	_____ mv
		CW 35°/sec In-Phase VR	+0.36857 through +0.43143 mv	_____ mv

FIG. 8.13 Acceptance test procedure data sheet.

1. No general-purpose equipment is commercially available to make the test.
2. General-purpose test equipment error is sometimes too large and consumes too much of the product tolerance. This usually occurs when the product design is pushing the state of the art and permissible product variation is small.
3. General-purpose equipment setup time is too long considering the frequency with which the proposed test will be performed.
4. The general-purpose equipment utilization factor is too high to permit tying it up in a permanent setup.

5. Test time with general-purpose equipment is excessive, and the frequency of the test performance is high enough to warrant the cost of designing and building special equipment.

General-purpose equipment provides greater flexibility than special-purpose equipment, i.e., it can be used in many different test setups but at a reduced efficiency of testing because it is not the optimum design for the particular test. In large projects, the sheer number of instruments and test equipment that must be utilized make it possible to choose special test equipment whenever the economics of a particular test dictate. In small company test laboratories, however, the individual situation is not so easily settled, since quantities of standard general-purpose test equipment are usually limited.

Automated vs. Manual Control. Test equipment can be controlled or programmed either manually or automatically. As a general rule, manual control is cheaper in first cost and in maintenance, but it is usually more expensive to operate. Often with repetitive testing the additional costs of automation can be recovered if the testing continues for a year or two. Perhaps more significant than the saving in direct operation cost is the saving resulting with automation from reduction in operator error in programming (or in recording data). Furthermore, automated test equipment normally provides more repeatable results, permitting easier understanding and diagnosis of test failures or anomalies (which represents additional cost savings), more uniform testing and more consistent quality of product. Hence first consideration should be given to the use of automated test equipment, with manual equipment utilized only as the exception. Particular attention should be paid to automating the cyclic environmental equipment utilized in parts screening and for vibration testing, since there can be substantial cost savings for these.

Calibration. Calibration is defined as the comparison of the indication of a measuring or testing device with a known standard, the known standard itself being compared to more accurate standards in a series of controlled echelons up to a national standard held by the National Bureau of Standards. This process of calibration is often termed *traceability of instrumentation.*

The normal calibration method entails taking the test equipment to a calibration and standards laboratory for periodic calibration. Each meter, input device, or comparison standard is individually calibrated in the laboratory. Two other methods should be considered: *built-in calibration* and *special calibration systems.*

It is economically feasible to build into the test equipment or setup sufficient standards and circuitry to enable the operator to run through a calibration procedure at the test equipment, which significantly reduces the test equipment out-of-service time for sending meters and other components to the calibration laboratory. When this is done, transfer standards are provided so that calibration laboratory technicians bring the transfer standards to the test setup and calibrate the built-in standards.

When a complete test equipment system is used, it is also feasible to design and build a special dedicated portable system which can be hooked up to the test equipment at the test station. The portable special equipment is returned periodically to the calibration laboratory for calibration. Although this is the costliest of the three methods in first cost, savings in calibration costs can be significant.

Manual or Automated Readout.   Readout equipment can be completely manual, with the operator reading dials and meters and recording by hand on data sheets. This method has the cheapest first cost and is warranted if only a few tests will be performed. An improvement in cost and accuracy can be obtained by providing recording devices which will automatically print or plot the output readings, or the readout can be completely automated, with or without a printout of readings. In this method, the readings are compared with the acceptance limits, programmed into comparators, and only the deviating values are printed.

If manual readout is used, data sheets should be provided to minimize operator error in recording the results. If machine readout is used, it should be compatible with the data handling system and equipment being used for reliability calculations.

8.4.3 Standardization of Test Equipment

Reliability calculations are based on the assumption of homogeneity of the product population. Inherent in this assumption is that the data generated by the testing of the units of that population are also homogeneous—unbiased by test equipment aberrations. For this implication to be true, overt action must be taken by test equipment design personnel to assure that all testing of the product is indeed compatible, i.e., homogeneous. Incompatible testing will also result in unnecessary hardware circulation between seller and buyer when the product is tested at both places and between successive levels of assembly.

Incompatibility between test equipments arises mainly from two sources. One is from differences in circuitry in the test equipments, typically the interface impedances of either the stimulus section or the measuring section. By assigning the responsibility for design of all project test equipment to a single organization and instructing them to maximize compatibility of such circuitry, this source can be minimized. The second major source is the use of presumably interchangeable instruments, like power supplies, from different manufacturers without a compatibility test. This source of error can be eliminated by designers specifying a particular manufacturer, which prevents buyers from ordering supposedly equal (and less costly) alternates.

8.4.4 Test Equipment Error

All measuring equipment has an inherent error which must be considered in the design and use of test equipment. The errors are not apparent or observable.

1. Proper calibration only ensures that the device operates within its rated accuracy (percent error).
2. Instruments are not perfectly repeatable, i.e., a number of readings or calibration comparisons made in a row will vary. (This is called the precision of the instrument, and the variation or repeatability can be treated statistically.)
3. Error exists not only in the reading, sensing, or comparing portions of the test equipment but in the portions which provide the inputs and environmental conditions as well.
4. Equipment error is different than operator reading errors; digital readout devices reduce operator reading errors, but do not in themselves reduce test equipment error.

5. The amount of error can be a significant percentage of the allowable tolerance on the parameter being tested, and degradation of the functionability and reliability of the product can result from ignoring test equipment error.

For complex special test equipment, the total error should always be measured. The measurement is properly made at the interface between the test leads and the article being tested, where it will include the errors introduced by the cabling and connectors, and with the test equipment and the product both energized. Multiple readings should be obtained and averaged. In error analysis calculations, it is permissible to utilize statistical techniques (summing the errors by root sum square) to combine the individual errors. Figure 8.14 is a sample of a test equipment error analysis report, indicating one useful way of analyzing test equipment errors and combining them for an overall estimate.

Test equipment error should not be split around the limits of the parameter, but taken from the engineering tolerance limits. This is common practice with mechanical measuring devices, where the "gaugemaker's tolerance" (another name for test equipment error) is taken from the drawing tolerance. Thus, if the product limits are 90 and 100 V, and the test equipment error is 2 V, the product should be rejected if the equipment reads less than 92 V or more than 98 V. Or an oven with a 90 percent accuracy should be set to 606°F (319°C) if the test specification calls for exposure at 600°F (316°C). The product may in that case "see" 612°F (322°C), but product reliability doctrine demands rigorous testing to not less than the requirements.

8.4.5 Test Equipment Calibration

Test equipment is calibrated to ensure that test data taken over time from all sources are compatible, and to ensure that no more than the planned test equipment error occurs (test equipment error analyses are based on the assumption that the equipment is within error limits, i.e., is properly calibrated). All test and measuring equipment utilized in the project, including that used by R&D personnel, should be calibrated at specified intervals. These intervals are usually set empirically by standards engineers after thorough analysis of instrument drift rates and are based on the economics of having to recall product erroneously accepted by test equipment which has drifted out of tolerance.

Most large companies now have their own calibration laboratories. Smaller companies without standards laboratories usually utilize the services of commercial laboratories. When none of these is available in the area, arrangements sometimes can be made with the nearest Department of Defense activity, such as arsenals, shipyards, or air stations, to have the calibration performed on a time-and-material basis.

There are three kinds of calibration:

1. *Calibration of individual instruments,* such as meters, gauges, or power supplies. These are usually transported to the calibration laboratory. When the instruments are part of a complex test setup which is in continuous use, it is economical to calibrate only the scale on the instrument which is used in the setup. With large fixed instruments or with delicate moving-coil instruments, it may be necessary to carry a fixed standard from the calibration laboratory to the instrument.

Equipment Title: Flight Control Gyro Test Station 52X220			Parameter Document No 22351 Part IV		Prepared By		Approved By MSL	Page 1 of 4
Equipment Drawing No 2334536			Test Point # 201		Checked By		Customer Approval	
			Date Prepared 8-26-63					

Function			Measurement Device	Accuracy (T_1)		T_2/T_1 Ratio	Error Formula	Final Readout Limit
Title	Nominal Value	Tolerance (T_2)	Description	Full Scale	Measurement Point			
Gyro output (pk-pk) Para 1.3	968 mv	Max.	Ballantine 3165/2	±3% rdg	±28.8 mv	N/A	$(3 \cdot 10^{-2})(968) = 28.8$ mv	939 mv
In-phase null degrad Para 1.4	90 mv	Max.	Keithley 151R Eqpt panel assy	3% rdg / 3% F.S. / Total	3 mv / 7 mv / 10 mv	N/A	$(3 \cdot 10^{-2})(90) = 2.7$ mv + panel error	80 mv
Power consumption input Para 2.1 (115 vac input is adj; to desired value by setting Gertsch to .869565 and read output of Gertsch with DVM)	115 vac 400 cps	±1%						
	100 vac	+.00v, -2.00v	Gertsch Ratiotron NLS V358, NLS125E Converter	.0058% / 0.5% I.V. / 0.05% FSx / .3 (% 3rd harm. distort)	Negligible / 500 mv / 50 mv		$(5 \cdot 10^{-2})(100 \cdot 10^3) = 500$ mv / $(5 \cdot 10^{-3})(100 \cdot 10^3) = 50$ mv / $(3 \cdot 10^{-2})(100 \cdot 10^3) = 300$ mv	
				IV 300 mv / TOTAL 850 mv ERROR	585 mv	N/A	$[500^2 (50)^2 (300)^2]^{1/2} = 585$ mv	
			TOTAL RSS					
Running Power Para 3.2	15.0 watt	±2.5 W	Valtron #20.035	2% F.S.	±.40W	6.0	$(2 \cdot 10^{-2})(20W) = 0.4W$	15 ± 2.1 watt
Power factor	0.960	Max ld/lag	Valtron #20.036	- -	±0.02	N/A	Full Scale = 1, 2% = .02	±.980
Time to rch running pwr	50 sec	Max	Operator reaction	- -	±1 sec	N/A		49 sec
Squelch applied: Running power	10.0 W	±1.8W / -2.3W	Valtron #20.035	2% F.S.	±.40W	4.50 / 5.75	$(2 \cdot 10^{-2})(20W) = .4W$	11.4 watt / 8.1 watt
Power factor	.85 log	±.15	Valtron #20.036	- -	±0.02	7.5	Full Scale = 1, 2% = .02	.85 ±.13
Insulation resistance Para 3.2	500 vdc		Wiley 5P-2 P.S.			N/A	$(4 \cdot 10^2)(2 \cdot 10^{-2}) \cdot 4 \cdot 10^5 = 41.6$ megohm min	
Para 3.2	40 megohm min		Wiley 5P-2 with analog output to NLS V358 DVM	±2%	±1.7 megohm corresponds to .042%		$\frac{5 \cdot 10^5}{(41.6)10^5} = 12.0$ microoamp max = -1.2 vdc max	
(Mard in terms of	-1.0 vdc	(Max)					$-1.2[(-5 \cdot 10^{-4})] - 1$ digit = 1.198 v max = 41.7 megs	
Output Frequency Measurement (Para 4.3)	36	±5.5 cps	Micro Gee 64A Beckman 7350	No error ±0.5 cps		11/1	0.5	36 ± 5.0 cps
Damping Ratio @ 36 cps	0.70	±0.15	Micro Gee 64A Data Log 204A Servo Analyzer			N/A	Table I for T/f error vs CPS	
@ 50 cps	0.84	±0.39		5% rdg				
Sensitivity, In-Phase & Quadrature and Harmonic 5.2.1 75°/second	8V	±650 mv	Gentico C181 Gertsch ACR8	1% rdg / .007% F.S. / TOTAL	+8.0 mv / +0.7 mv / +8.7 mv	78.1		8000±641.3 mv

FIG. 8.14 Sample test equipment error analysis report sheet. This spread sheet is used to calculate the amount that a specification tolerance (columns 2 and 3) is to be tightened to compensate for test equipment errors. If more than one error contributes, as in power consumption, the errors (500, 50, 300) are combined statistically, using the root square (RSS) formula, shown in column 8.

2. *Calibration of systems* of complex test or environmental equipment is usually done in place. Sometimes specially designed transfer standards, specifically tailored to a particular test station, are used to reduce the cost of repeated calibrations in a long production run.

3. *Calibration of standards* is usually performed by comparing each standard to one of higher accuracy in the same or a higher-level laboratory. It can also be performed by cross-checking with one or more like (same level) standards in the laboratory, with only a small reduction in accuracy but at a significant savings in upward calibration costs.

The usual ratio of accuracy for each echelon of calibration for mechanical measurements is 10:1. This ratio is difficult to achieve in many other measurement areas, and it is necessary to accept a lesser accuracy ratio. Studies by the National Bureau of Standards and others have shown that a 4:1 ratio results in a reasonably acceptable compromise between the cost of the more accurate 10:1 calibration and the cost of scrapped material resulting from the inaccuracy.

Errors in calibration can be minimized by the preparation and audited use of written procedures for all calibrations. These procedures (see Fig. 8.10) should be prepared in considerable detail, including extensive use of hook-up diagrams, and they should include data sheets which require recording both the as-received and as-adjusted readings (to permit engineering analysis of drift rates). Calibration procedures prepared by instrument manufacturers are not always in enough detail to provide a satisfactory level of discipline and must often be rewritten.

8.4.6 Test Facilities

Testing can be performed, partially or totally, either in house or contracted out. The economic trade studies prerequisite to decision making are similar to those used in production make-or-buy analyses. Other factors to be considered are as follows:

Factors Favoring In-House Testing

Coordination of test schedules with manufacturing schedules is easier.

Company capital asset utilization is improved.

Work is performed by company personnel, and it is easier to control discipline, enhancing the quality of the testing.

Flexibility of test scheduling is enhanced.

Unnecessary to train ''foreign'' personnel in company techniques, standards, methods, and systems of technical and financial management.

Hardware transportation time is minimized.

May broaden the company's testing capability, thereby improving the company's competitive posture.

Hardware is continuously under direct company control.

Risk of industrial espionage is reduced.

Factors Favoring Contracting Out

Schedules for the testing may be the overriding consideration, and the in-house laboratory cannot provide the technical capability, personnel skills, or capacity in time.

Commercial laboratories may be underloaded, and the total cost may therefore be cheaper.

Some tests may have special requirements which can be provided by existing commercial laboratories without a prohibitive capital investment and when there is only a small chance that the capability will be needed again soon.

Customer or government may dictate that some or all of the testing be performed in their facilities or in an "unbiased third-party" facility.

In-house work load can be leveled by sending out peak-loading work.

Corporate management may dictate placing some test work in underutilized laboratories in sister divisions.

A business decision may be made to develop and maintain an outside test source for backup in case of fire or other unplanned in-house disaster or emergency.

In mass production situations, most companies decide to perform all levels of acceptance testing in house because the testing schedules must be integrated into the manufacturing and delivery schedules. Reliability screening and burn-in testing fit this situation too. When the production rates are very low, e.g., one per month, there is more latitude for contracting out acceptance testing. Reliability assessment, however, is usually more independent of the production process and the decision can go either way.

University and Government Laboratories.   One often-overlooked source of outside laboratory support is the capability which exists in many universities and government activities (Departments of Defense and Transportation and NASA, for example, have extensive laboratories for materials, component, and system testing, with excellent environmental equipment). The work will usually be performed only on the basis of time-and-material billing, but experience has shown that their costs are well estimated and controlled and very competitive with commercial laboratories. They offer the advantage of completely unbiased testing, and many have specialized capabilities and skills, developed to support pure research, that are uneconomical to establish or maintain in commercial laboratories. If work is placed in a university, it is advisable to ascertain the qualifications of the specific personnel to be assigned to perform each test, since universities want to utilize as much undergraduate help as possible to ensure that the work will provide maximum student training. This can be acceptable, but only if an adequate number of fully qualified supervisors closely direct the undergraduate work; a one-for-one ratio is quite good.

8.5 MANAGING THE ACCEPTANCE TESTING

Data from testing is the basis for reliability prediction, estimation, and assessment, and it therefore must be accurate and timely. Managing the test program is important to ensure the availability and accuracy (quality) of the data. The same quality control techniques that are applied to product design and product manufacture must be applied to the test program, and a meaningful audit system should be established to assess continuously the data quality and to pinpoint unacceptable practices. It is also important that the test organization meet its schedules and commitments to other organizations.

Since the principles of organization and management apply as much to managing testing as any other area at work, only the following areas that significantly affect the quality and timeliness of the testing will be discussed. The areas covered are:

Planning work loads and monitoring status

Station proofing

Recording variations

Interpreting results

Inspection coverage

Controls

Failure diagnosis

Organization

8.5.1 Planning Work Loads and Monitoring Status

Inasmuch as the work involved is actually performed by a very wide range of organizational elements, not generally reporting to a single functional manager, it is wise to perform the administrative planning in a committee comprised of scheduling representatives of the many organizations involved. Firm commitments should be required from each for such functions as:

Preparation of overall plans

Preparation of detailed procedures

Design of test and environmental equipment

Ordering and delivery of test hardware

Proofing of test stations

Start and completion of testing

Preparation of test reports

Analysis of data

Milestone charts (see Fig. 8.15 for an example) should be prepared for each part of the test program, and a monitoring group assigned specifically to the test program to monitor performance against the commitments. Management attention should be directed to the behind-schedule activities so that decisive and timely corrective action can be taken.

Considerable attention should be given in the planning and make-ready phase of the test program to the preparation and release of supporting information. Included particularly are the product drawings and specifications, for it is in these documents that the basic criteria to be tested are defined. The end goal of the

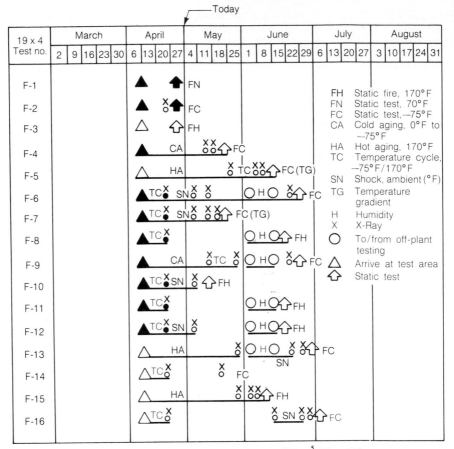

FIG. 8.15 Production assessment test schedule and status [$°C = \frac{5}{9}(°F - 32)$].

planning and monitoring function must be to ensure that *everything* that is needed to perform a scheduled test is available to the test group at the right time.

Overall responsibility for all aspects of planning, budgeting, and executing the acceptance test program should be vested in the quality assurance organization. There are two reasons:

1. Responsibility should be vested in a single organization because the test program is based on a myriad of interlocking requirements, including technical, budget, schedule, data, test equipment and test hardware, all of which must be coordinated in an optimum balance of trade-offs and compromises. Splitting the responsibility among several organizations makes optimum decision making more difficult.

2. Responsibility should be vested in quality assurance because testing should be performed by an unbiased third party having no proprietary interest in either the design or production efforts.

8.5.2 Proofing Test Equipment Stations

Sparked by more stringent demands from consumers of all types for increased assurance of the safety, reliability, and quality of delivered product, acceptance testing has grown increasingly more extensive, complex, and difficult. Increasing sophistication in product design has further exacerbated this situation. This increased sophistication more often requires special design of test equipment (rather than an aggregation of interconnected standard instruments), design and procurement of which takes almost as long as the design and procurement of the initial product. Therefore, the production acceptance test equipment is often not available during R&D and will not have been seasoned by being used to test prototype product hardware. Consequently, a special proofing program should be established as a planned part of the overall acceptance test program.

This proofing consists of a meticulous in-place test of the test equipment, mated with prototype product hardware, if available, or with deliverable hardware as a last resort, which ensures that the test equipment actually tests all of the attributes it is supposed to test in the manner and to the accuracy desired. Proofing is a series of tests which starts with a "smoke test" (power applied to the test equipment with no product hardware attached), continues through activation of each of the discrete sections of the test equipment without and then with prototype hardware, and ends with a complete run of the test in the full final configuration. During this operation, the test and calibration procedures are also proofed out, with corrections made as the proofing progresses so that a final proofed-out copy is available when equipment is released for production test purposes.

8.5.3 Recording Variations

No matter how well acceptance and assessment testing are planned, when the tests are actually performed there will be variations from the detailed requirements of the plans and procedures. These variations occur in every phase of the test program. For instance,

Hardware will be of a slightly different configuration than planned

The exact test equipment specified will not be available and a substitution will be required

Operators will make errors and push the wrong buttons or connect leads improperly

Laboratory ambient conditions will be outside the limits specified

Input environmental equipment will not be controlled accurately enough or will suddenly develop trouble and be incapable of meeting the entire spectrum specified.

The list is endless, and although managers use all of the quality control and management techniques available, the variations will occur. Therefore a procedure should be established to both control and record them. The controlling portion mainly concerns procedures to be followed to obtain permission to continue the testing or to ensure analysis of the variations. The recording of variations should be accommodated by having, in every test procedure, a section specifically earmarked for recording them. The section should be conspicuously placed at the beginning of the report, where all those who use the report will be sure to see it. Likewise, a place must be provided in the test data sheets for recording the variations. And lastly, test personnel must be instructed that recording variation is not so much an admission of error as it is the means of informing the test data users that something other than the specified test conditions occurred in order to permit meaningful analysis and use of the data.

8.5.4 Interpreting Results

Much of the value of a test lies in the manner in which the results are interpreted. If, for example, tests are run at periodic intervals on production hardware, an examination should be made of the trends in the test data, so that one of the important side benefits of acceptance testing (predicting that a process is going out of control) will be realized. Therefore, a test analysis program should be established to ensure that the data are adequately utilized. This program should specify how the data will be analyzed and for what purpose or purposes. These purposes could include:

Acceptance of product for either further processing or for delivery

Process control or trend analysis

Reliability assessment or prediction

Screening product for different customer requirements

Aging surveillance

Burn-in or stress testing

Until all of the product is removed from use, all test results for a project should be carefully filed or stored in a computer in such a manner that the results of individual tests will be available for comparative analysis. The results of early tests should be used as a base against which current or later tests are compared, and analysis personnel should look for trends or shifts in the center points in attribute histograms. Computer storage of data facilitates these analyses. Of particular value in analyzing such shifts are the repetitive tests performed on the same serial number specimens (which will reveal aging effects if the tests are separated in time) or the periodic assessment tests on samples drawn from production (which will reveal shifts in the production process). Every sudden shift in center value should be investigated until a satisfactory explanation is found.

8.5.5 Inspection Coverage

Testing is commonly performed by one of the divisions of the quality assurance or reliability departments. Since these are the departments that normally provide the checks-and-balances function for production and design, respectively, it is not unusual that no checks and balances are set up for the test work performed by quality assurance. However, the test function is an original effort of considerable complexity and is fraught with probability of error. Since accurate and complete data are of paramount importance to a reliability program, it is worthwhile usually to apply a check and balance to the testing effort. This can best be provided by setting up inspection or close auditing of the testing work by a different division of the quality assurance department.

The function should be organized to provide assurance that:

The detailed requirements for the tests are met scrupulously

The procedures are followed precisely or that any variations are properly recorded and witnessed

The test equipment is properly calibrated, proofed, and then sealed and placed under break-of-inspection control

Test specimens are of the specified configuration and properly inspected before delivery to the test area

The specimens are sealed and placed under break-of-inspection control

The data have been properly recorded (if manually recorded) or entered into the computer

Inspection coverage is particularly valuable when tests are performed in an outside laboratory, where the presence of a company inspector has a most salutary effect on the laboratory operation. In this situation, the inspector can also act as company representative and provide technical liaison to the laboratory.

Inspectors assigned to testing work should not be given authority to waive or modify any of the test requirements, since such authority tends to degrade their inspection posture. The authority to grant such waivers or to make such modifications should be reserved by the test engineer, to whom the test operator can appeal with the inspector's tag.

8.5.6 Controlling the Testing

Closely allied with the foregoing discussion of providing inspection coverage for the testing is controlling the testing. There should be a system of control over the preparation, review, signature, and release of the test documents, identical to that over drawings and product specifications. There should be an equally tight control over changes, waivers, deviations, or variations to the prescribed test. *In a high-reliability project, everything that can be defined should be.* An example of looseness is the commonly used notation "or equal" in test procedures, referring to the test equipment to be used. These words grant carte blanche authority to anyone, including the test operator, to decide what test equipment is equal to that specified. A much improved wording is "Substitutions of test equipment must be approved by the test engineer." Only in this way will there be assurance that such problems as impedance matching will be considered at the engineering level.

From the preceding, it is apparent that the controls established should ensure:

That technical testing documents are properly approved and released

That changes to them are also released and approved by the same level of personnel as the original release

That any approved variations are recorded in the test reports to permit an analysis of otherwise unexplainable data anomalies

In a customer-financed and -controlled project, a rigorous customer will want to approve all the original releases of test and calibration procedures, as well as any changes or variations to them, so that the customer's engineering force can be used as an effective check and balance.

8.5.7 Failure Verification and Diagnosis

Failure verification and failure diagnosis are two closely related but not identical functions. *Failure verification* is the first step required when a failure indication results from a test and consists of ensuring that the indicated failure has indeed resulted from something wrong with the test specimens and not from an operator error or test equipment malfunction. To permit meaningful failure verification, it is necessary to require that at any indication of test failure the operator will make no additional runs, but instead will summon a failure verification engineer (or team) to examine the test setup. It is important that the operator be instructed not to move, change, or touch any part of the setup, including connectors or cables, since too frequently such actions will destroy the evidence and make verification impossible.

Once the failure has been verified as being a product failure, the related second step in this process is *failure diagnosis,* which consists of performing necessary additional quantitative tests or examinations to pinpoint the exact cause of the failure in the product. This action is necessary not only to permit the product to be repaired or reworked but also to ensure that corrective action can be taken on the parts, design, or manufacturing process to preclude the recurrence of the failure mode.

It is important that the failure verification and failure diagnosis results be well documented and provided to data analysis groups to permit collation and trend analysis. Only from such data analysis will it be possible to identify specific pieces of poor test equipment, untrained or marginal test operators, poorly prepared test procedures, and other factors contributing to inefficient testing. This data analysis is also necessary to find the production processes or parts which are drifting out of control. Control chart techniques are useful in this effort.

It is generally not adequate to have the failure verification and diagnosis performed by the test operator or the mechanic who replaces test equipment parts or makes adjustments, since the "investigation" performed will be cut-and-dried and the data will be relatively meaningless. Although the use of engineers makes the verification and diagnosis effort expensive, the cost is warranted because the data from this effort are the beginning of the true reliability and quality control function, since they are the foundation on which corrective action is to be taken.

8.5.8 Organization for Testing

Many organizational units have either a direct or indirect interest in the planning for or the conduct of the reliability tests, including acceptance and assessment tests, and in the results. A common weakness in present industrial practice is the assignment of the different test programs to that organization which has the most immediate direct interest in that test. Thus the design-oriented tests (feasibility, evaluation) are assigned to engineering, and the production-oriented test (acceptance and assessment) to quality assurance. Such a division of responsibility almost guarantees that the test results from these many test programs will be incompatible and incapable of being combined into an overall reliability data bank. Assignment to the reliability organization, the user of the data, will help assure data usability. In addition, since testing can be viewed as a check-and-balance function, assignment to reliability will preserve the check-and-balance principle.

Test Planning Committee. Test planning is not a one-time process with an inflexible result, but must be a changeable selection of items and attributes which will vary with time and knowledge. Testing should be viewed as the principal means of uncovering weaknesses or anomalies in either the design or the production process. As a project progresses, new weaknesses, requiring different tests, will be discovered, and old weaknesses will be corrected. Therefore the planning process continues and should not be assigned as an ad hoc effort.

Since the work is continuous and since it requires close coordination of many diverse organizations, a permanent test planning committee is a convenient way to perform the planning and replanning. The committee should be chaired by a member of the reliability department, the organization most interested in assuring that test data are continuously compatible and useful for population analysis. Other members should include quality assurance, product design, test equipment design, production planning, and test laboratory personnel.

In the committee operations,

The product designer will specify those items and attributes which need to be tested to ensure functional operability of the product

Quality assurance will specify the additional items and attributes which must be tested for process and quality control

Reliability will dictate the reliability verification and data requirements

Test equipment designers will provide the limitations on testing capability

Product designers will limit the number of test points and provide the allowable duty cycles.

The committee as a whole will find the optimum balance between risk and cost.

Test and Calibration Procedure Preparation. The detailed test and calibration procedures should be prepared by test engineers, separate from the test operators. Test engineering and inspection engineering (where testing is measurement of functional attributes and inspection is visual examination and mechanical measurement) are sometimes separated, with inspection planning and procedures preparation performed by an engineering group in quality engineering and the similar function for testing by a group in reliability engineering. The test planning committee is likewise sometimes split, and two separate committees consider inspection and testing, respectively. This split permits grouping of different technical specialties under a single supervisor.

Monitoring Status. The complexity of the overall reliability test program, which includes acceptance and reliability assessment testing, and its sheer size in terms of labor-hours and dollars, makes it almost mandatory that a special budget and labor-power group and a special status-monitoring group be established. These are sometimes staff functions reporting to the product assurance manager; they may also be in an administrative department within product assurance, as shown in Fig. 8.16.

All of these concepts are represented in the suggested organization chart shown in Fig. 8.16, which is an actual organization chart of a large company which successfully manages overall reliability test programs for complex electronic equipment.

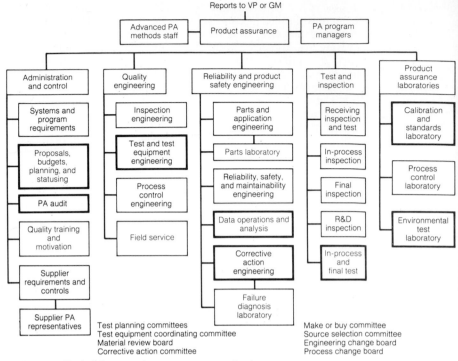

FIG. 8.16   Typical large product assurance organization.

BIBLIOGRAPHY

See Chap. 5 for an annotated list of U.S.A. military standards, specifications, and handbooks as well as standards and specifications from other organizations. Many of those deal specifically with testing programs.

MIL-T-5422, *Testing, Environmental, Aircraft Electronic Equipment.*

MIL-D-8706, *Data and Tests, Engineering: Contract Requirements for Aircraft Weapon Systems.*

MIL-STD-410, *Nondestructive Testing Personnel Qualification and Certification.*

MIL-STD-690, *Failure Rate Sampling Plans and Procedures.*

MIL-STD-750, *Test Methods for Semiconductor Devices.*

MIL-T-18303, *Test Procedures; Reproduction, Acceptance, and Life for Aircraft Electronic Equipment, Format For.*

MIL-STD-810, *Environmental Test Methods.*

MIL-STD-1235, *Single and Multiple Continuous Sampling Procedures and Tables for Inspection by Attributes.*

MIL-STD-1635 (EC), *Reliability Growth Testing.*

MIL-STD-2068 (AS), *Reliability Development Tests.*

Books

Barlow, R. E., and Frank Proschan, *Statistical Theory of Reliability and Life Testing,* New York, John Wiley & Sons, 1975.

Kapur, K. C., and L. R. Lamberson, *Reliability in Engineering Design,* New York, John Wiley & Sons, 1977.

Nelson, Wayne, *Accelerated Testing: Statistical Models, Test Plans, and Data Analysis,* Schenectady, N.Y., published by the author, 1986.

O'Connor, P. D. T., *Practical Reliability Engineering,* New York: John Wiley & Sons, 1985.

ENGINEERING FOR RELIABILITY

CHAPTER 9
ENGINEERING IN RELIABILITY

Richard Y. Moss II
Reliability Engineering Manager, Hewlett-Packard
Corporation

9.1 IMPACT OF RELIABILITY ON "QUALITY COST"

The statement "reliability must be designed in a product as well as manufactured in" is widely repeated by managers and engineers. Unfortunately, in many cases this acceptance appears to be merely lip service to a popular notion, with no real understanding of why the statement is true or how to apply it. Worse yet, the belief that high quality and high reliability are more costly is almost universally held, even by those who know better. In fact, the costs to remedy quality or reliability defects are much larger than the assessment and prevention costs, often comprising two-thirds to three-quarters of the total quality costs. (It would be more accurate to use the term *unquality cost* in such situations.) In studies of large, supposedly well-managed corporations, it has been shown that the quality costs can be in excess of 20 percent of total revenues and that most of this is the cost of all those activities which take longer (or exist at all) because the product or service was not right the first time. A list of such activities in a typical company which designs, manufactures, and markets an electronic product is shown in Table 9.1.

In a high technology industry such as electronics, it is not unusual for as many designs to be abandoned as are completed; in fact, if you were to question the designers carefully, you might find that they try to discard several designs before selecting one. That is a part of the design process and is not what we are talking about in Table 9.1. By "abandoned" designs, we mean those that were completed and then were later abandoned because they were faulty or inadequate.

When a product has a quality or reliability problem, the first step is usually to divert some new product design effort to production, to investigate the problem and, it is hoped, solve it quickly and simply. Often the "fix" involves a redesign, usually after products have been manufactured and shipped to customers. Now the redesigned product must be retested, often a lengthy and expensive process where performance, reliability, and regulatory approval tests are required. When the revised design is finally approved, the documentation must be revised and issued to manufacturing, field support and service, and customers. All of these activities are costly, consuming as much as a third of the total design budget. This figure does not include the lost opportunity cost from the delays in shipping the product in trouble or the delay in the new product schedule because part of the design effort was diverted to fixes.

Unquality costs affect manufacturing just as dramatically as design and, since many of them are designed in, deserve discussion. First, *all* troubleshooting can

TABLE 9.1 Elements of Unquality Cost

Engineering:	Faulty designs abandoned
	Redesign of faulty products
	Test of redesigns
	Documentation of redesigns
	Design effort diverted to production
Manufacturing:	Troubleshooting and fault diagnosis
	Rework and retest
	Scrap of defectives
	Obsolescence due to redesign
	Increased inventory ("safety stock")
	Increased work-in-progress
	Production change documentation
	Increased expediting
	Increased overtime
Marketing:	Higher selling costs
	Higher order processing costs
	Higher installation costs
	Higher service and support costs
	Higher warranty costs

be considered an unquality cost, since it is only necessary when the product is defective. In studies of test time of electronic assemblies, it is common for the troubleshooting time to diagnose a defective assembly to equal 10 times the test time for a good one; so if as few as 10 or 20 percent of the assemblies require troubleshooting, the overall test time can double or triple. In the case of components or small assemblies, the defective product may be scrapped without much troubleshooting, but for large or expensive assemblies rework and retest are more likely, further adding unquality costs. The possibility of rework is also raised when a design fix is introduced, and this also adds to the inventory (of parts needed for rework), to increased documentation changes, and to increased scheduling and expediting effort. In the electronics industry, it is not unusual to find that these unquality costs account for a third of the cost of goods produced, and more if severe problems occur.

Some of the unquality costs assigned to marketing are hard to assess, such as higher selling costs. Everyone will agree that it is harder to sell an unreliable product—you have to do more promotion in order to find customers who have not heard about the problems, or desperate ones who will tolerate the troubles—but this is an intangible cost that is hard to quantify.

More tangible unquality costs are such items as the higher order processing and delivery costs because shipments of troubled products are late or intermittent, and higher installation and start-up costs of those same faulty products when they finally arrive. The service and support organization must be enlarged when there are product reliability problems, with more people hired and trained, more equipment and facilities provided, and a much larger inventory of spares and replacements maintained. Finally, warranty costs increase because of the necessity to repair or replace the unreliability products at the producer's expense. It is not unusual for all of these costs to equal one-quarter to one-third of the marketing cost, not counting the higher selling costs which are difficult or impossible to estimate.

9.2 RELIABILITY ENGINEERING IN THE DESIGN PROCESS

The basic philosophy of *total quality control* (TQC) is that every activity is a process, and processes can be analyzed and improved. Viewed that way, engineering a new product is certainly a process and can therefore be broken into steps which can be measured and improved. Engineering a new product by a well-controlled system will yield a better result than by a haphazard or out-of-control process. Nearly every step in the design process involves decisions which affect the final reliability of the product. You sometimes hear objections that subjecting the design process to quality control will stifle creativity and produce reliable but costly, lower-performance products. That is not true; in fact, understanding and controlling the design process can yield two unexpected benefits:

1. Reducing the number of different alternatives the designer must investigate by imposing *design rules* and standardizing routine aspects of design and test
2. Awakening the creative urges of the designer to the concepts of engineering out failures and engineering in reliability, which may not have been design objectives before

Figure 9.1 shows the flowchart from a typical electronic product design process. It is divided into three phases, representing the different stages in the development of the new product.

9.2.1 Investigation Phase

In the investigation phase, the feasibility of the proposed product must be demonstrated, and the detailed development schedule and budget prepared. The emphasis is usually on selecting a technology that will permit the product idea to be realized, and mistakes in this phase can be very costly to the following phases, since they can lead to blind alleys and dead ends.

Setting Reliability Objectives.   Performance and cost objectives for a new product are required before the development of that product proceeds, so why not objectives for its quality and reliability? Extending this logic one step further, why not set some objectives for the quality of the process by which the product is engineered, as well as for the product itself? It is traditional to set unquality or unreliability goals for the product, such as a maximum failure rate or minimum mean time between failures (MTBF), but why not set goals for the percentage improvement in failure rate or MTBF compared to the existing products? Setting a goal of improving the failure rate of a product family only 21 percent per year will result in a 2:1 improvement every 3 years, and 10:1 in a decade. The impact of that rate of change on quality costs would be dramatic indeed. One implication of setting this type of goal is that if a new product is scheduled to take 3 years to develop, the reliability goals for it had better be twice as good as today's products. Another implication is that the whole process of developing a new product must change, and that just tacking on more tests at the end is not enough.

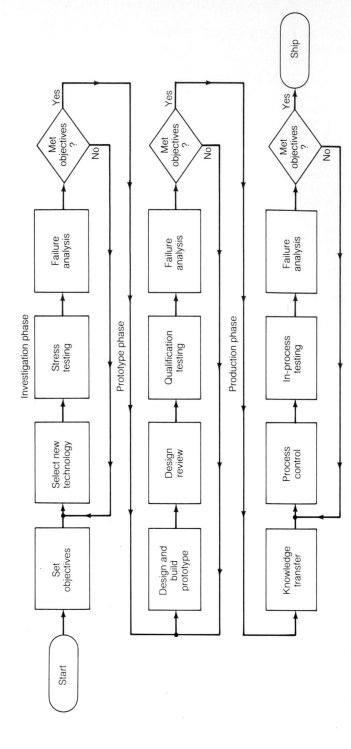

FIG. 9.1 Product development process flow diagram.

Technology Selection. Because most engineers are eager to try the latest technology and because the marketplace often richly rewards those who are first with improved performance, we seldom can afford to wait until a new technology or process is perfected before using it. In the electronics industry, many new products are available on the market before they achieve mature reliability, because waiting for perfection is asking to be left behind. For example, a rule of thumb in the semiconductor industry is that the learning curve for a new process technology is 5 years, but nobody is willing to wait that long to start using it.

Some factors are more important than others in technology selection, particularly those which involve stress and complexity. For example, in the computer field the biggest changes in reliability have been the result of dramatic changes in power dissipation and size of a bit of memory. Computers of the 1950s, which consumed thousands of watts, filled a good-sized room, and failed every few hours, can now be put on a silicon chip smaller than your little fingernail, dissipate a fraction of a watt, and have an MTBF of more than 1 millions hours. This leads us to some rules of technology selection:

1. Choose designs and materials that generate the least stress and that are tolerant of the stresses imposed by their environment.
2. Minimize the *multiplicity* of your designs, that is, the number of individual parts or components.
3. When evaluating a new component or technology, analyze the internal construction as well as the performance.
4. Stress-test samples to failure, and then redesign to reduce or eliminate their worst failure mechanisms.

The designs that generate the least stress are usually low power, so it is not surprising that the low power component families are also the ones with the best reliability records. One glaring exception to this generalization is where low power electronics are used in moist environments. In such cases, the higher power components dry themselves out by the heat generated and are less subject to the failure mechanisms associated with moisture. Minimizing the parts count is always a good idea, because each part has a failure rate, and since the failures generally are independent, the rates simply add up. Analyzing the internal construction of a new component or technology requires special facilities and expertise, but this often allows us to discover quickly a new part weakness that would take thousands of hours of accelerated life testing to predict. Despite the powerful insight which can be gained, construction analysis is not a substitute for stress testing, since none of us is clever enough to anticipate every possible failure merely by inspection.

Stress Testing. Failures are caused by failure mechanisms, which are built in and then activated by stresses. Studying the basic stresses and the failure mechanisms they activate is fundamental to the design of effective and revealing reliability tests. The correct design approach is, of course, to find and eliminate the fundamental failure causes. This means that the most successful stress tests are ones that result in failures. Successful tests are also ones that are tailored to look for particular failure mechanisms efficiently, by selectively accelerating them. Table 9.2 shows the relationship between the basic stresses, the most common failure mechanisms, and some component types which are prone to these

TABLE 9.2 Stresses and Related Failure Mechanisms

Stress type	Stress	Failure mechanism	Component types
Electrical	Voltage	Dielectric breakdown	Capacitors, semiconductor oxide layers
		Avalanche breakdown	Semiconductor junctions
	Current	Electromigration	Thin metal films
		Fusing	Conductors
Thermal	Heat	Chemical reaction	Batteries, electrolytic capacitors
		Intermetallic growth and alloying	Interconnections
		Ionic migration	Semiconductors
Chemical	Water	Dendrite growth	Metals (silver)
		Corrosion	Thin metal films
		Ionic migration	Semiconductors, insulators
Mechanical	Temperature cycles	Differential expansion	Polymer encapsulated devices
		Condensation	Cavity packages, high voltage circuits
	Shock and vibration	Fatigue	Interconnections, wire bonds, and die attach
		Conducting particles	Cavity packages

mechanisms. Despite the fact that most of these failure mechanisms have been well known for more than a decade, they continue to be the dominant ones in today's electronic products.

In electronic components, electrical stress-induced failures are major causes of unreliability, and the rate of failure is a strong function of the design. For example, dielectric breakdown in the insulating oxides of all types of semiconductor devices and avalanche breakdown of bipolar junction devices can both be minimized by reducing the applied voltage relative to the breakdown voltage, that is, by derating the device. Taking care to minimize transient peak voltages the circuit experiences is another means of designing to reduce failure rate. In the case of electrical current, the current density in thin film conductors such as the interconnection metal on integrated circuits is a critical factor in the reliability of those connections, and design rules for the maximum current density must be observed if long operating lives are expected. Fusing tends to happen because of severe short-term overloads rather than slight long-term overstress, and designing to prevent such surges will increase reliability greatly.

Heat, both internally and externally generated, is the most common enemy of reliability. One reason is that heat increases the rate of chemical reactions, and components containing chemical systems suffer accordingly. A second reason is that heat accelerates most of the other failure mechanisms listed in Table 9.2, including the electrical, chemical, and some of the mechanical ones. The famous purple plague of the early semiconductor era is an example of a gold-aluminum intermetallic compound that forms at high temperatures. Ionic migration, or mobile ion contamination as it is sometimes called, results in increased leakage currents and threshold voltage shifts in semiconductor devices because temperature increases the mobility of the contaminating ions. The best method for designing out all these failure mechanisms is to reduce operating temperatures by a combination of lower power circuitry and packaging to provide better cooling.

Chemical contamination causes many failures, and the most common contaminant is water. Because it is such a good solvent, water usually contains other elements and compounds in dissolved (ionized) form, particularly sodium and chlorine. Water containing impurities will conduct an electrical current, and this leads to leakage paths across insulators and the transport of metal atoms to form lacy short circuits called *dendrites*, from the Greek word for "tree." Silver is particularly prone to dendrite growth, and its use on integrated circuit lead frames as a plating material is to be avoided. Corrosion of thin metal films is prevalent where moisture, a corrosive ion such as chlorine or phosphorus, and bias voltage are all present simultaneously, such as in plastic-encapsulated integrated circuits operating in moist environments. Several design precautions can be taken against this serious failure mechanism, including using better passivation systems (silicon nitride rather than silicon dioxide), more corrosion-resistant metals (tungsten rather than aluminum), and hermetic packages rather than plastic for components containing thin film resistor networks or linear integrated circuitry.

Many engineers are unaware of the threat of mechanically induced failures in electronic products, yet the physical effects of temperature cycles on electronic components are severe. The problem is that every different material has a different coefficient of thermal expansion, so when dissimilar materials in contact with one another are heated, stress failures result. Cracking of silicon dice, shearing of these dice from their mounting surfaces, and fatigue failure of the interconnecting wire bonds due to repeated flexing are examples. Such examples are major causes of failure in components which dissipate enough power to heat up more than 50°C, and designing out these failures is done by reducing magnitude of temperature rise as well as selecting the best package.

Temperature cycles cause failures in another way, one that is perhaps more properly covered under the "Chemical" heading in Table 9.2 rather than "Mechanical." The failure mode is moisture condensation, which happens when temperature is lowered below the dew point, at which water condenses on the product. High voltage circuits tend to break down, and corrosion, dendrite growth, and ionic leakage currents occur. The design solutions are to encapsulate or seal the critical circuits to prevent moisture from reaching the sensitive areas, or to incorporate just enough heat generation in the circuit itself that condensation cannot occur.

Mechanical stresses such as shock and vibration can lead to failures of interconnections, ranging from failure due to resonance of a structure at a specific frequency to breakage because the wire lead is supporting a large mass which moves. The solution is to securely mount any component with a mass of more than a few grams using adhesive or a clamp, rather than depending on the wire leads to restrain it. Shock and vibration often cause extraneous conducting particles to break loose inside a component, usually debris such as solder or weld spatter. The best solution is to eliminate such contamination by better process control or a cleaning step, since it is impossible to predict when such debris will loosen and begin causing failures.

Failure Analysis. Every failure has a cause and is a symptom of a failure mechanism waiting to be discovered. The tools of a failure analysis are both statistical and physical; used together, they are a potent means for detecting the often unique fingerprint of the underlying mechanism.

On the statistical side, there are two types of information to be gathered: (1) the rate (and cost) of the failures; and (2) the behavior of the failure rate with stress and time. Determining the magnitude of the failure rate is necessary to focus the failure analysis effort on the most serious problems, since there are always more failures to analyze than time to do it. The behavior of the failure rate with stress and time is valuable information also, but is not always gathered. The reason is that failure rates can decrease, remain constant, or increase with time as shown by the three regions of the *bathtub* curve (Fig. 9.2). Most failure mechanisms exhibit only *one* of those types of behavior, so knowing which behavior the problem exhibits helps the failure analyst find the failure mechanism.

The decreasing failure rate, or "infant mortality," region of the curve is where most production and warranty failures occur, so minimizing the failure rate in this region is important to reducing quality costs. Initially defective components and those damaged by the manufacturing processes fall in this group, with the corresponding failure mechanisms including dielectric or avalanche breakdown of defective components, failure of components which have been misassembled or cracked when subjected to the mechanical stress of a temperature cycle, and leakage or corrosion failures of parts initially contaminated by water.

Constant failure rate, which implies that the failures are occurring randomly in time, cannot be traced to any particular failure mechanism, unless you consider random accidental destruction to be a failure mechanism. (In the human mortality curve, from which these regions get their names, the constant failure rate portion does occur in those age brackets where accidental death is the dominant mode.)

The increasing failure rate, or "wearout," region is caused by failure mechanisms which slowly change the component in an irreversible way, such as corrosion, dendrite growth, electromigration, fatigue, or chemical reaction. Ionic contamination can also behave this way, if the moisture must first diffuse through a permeable barrier rather than being present on the sensitive surface from the start.

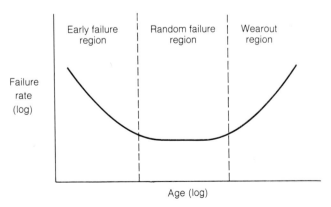

FIG. 9.2   Reliability bathtub curve.

Physical failure analysis is a sufficiently complex subject that two chapters have been devoted to it (Chaps. 13 and 14). Rather than try to summarize all the techniques discussed there, let us consider the uses of failure analysis results. First, knowing the identity of the dominant failure mechanisms allows the designer to try to minimize or eliminate them. In most cases, the failed component was designed by one organization, manufactured by a second, and designed into a piece of equipment by still another. Thus the failure analysis results are valuable objective evidence to determine which organization has the primary responsibility for eliminating the problem, avoiding much unproductive finger pointing.

9.2.2 Development Phase

In the development phase, the first complete prototypes of the new product are designed, built, and tested. Documentation such as material lists, drawings, and test procedures is created. It is not unusual to repeat this phase more than once, perhaps building the first prototype in the laboratory to test the design and the second generation in manufacturing as a test of the documentation. The design should not exit this phase until a prototype has met *all* the design goals set for it (see Fig. 9.1), although it may not be possible to demonstrate the reliability goal because of the small number of prototypes available to test.

Design Rules. It used to be thought that in order to encourage creativity, engineers had to be given total freedom from rules and restrictions. Now we realize that this much freedom results in a huge increase in the number of decisions that must be made, and hence makes the job of design much longer and more complex. The startling speed with which a new product containing digital logic can be designed and debugged is due primarily to the existence of standardized integrated circuit logic blocks, so that the design is done at the logic level rather than the circuit level. Thus, well thought out design rules contribute to engineering productivity and have an important impact on reliability as well. Design rules can prevent the use of obsolete or unreliable technologies and can specify the maximum stress allowed to be applied to the recommended technologies to minimize their failure rates. An example of derating guidelines for the most popular capacitor families is shown in Table 9.3.

There are two different sets of derating guidelines, labeled "Nominal Conditions" and "Worst Case." The nominal conditions are what designers like to call *design center,* that is, room temperature, nominal power input, and output load, etc. The worst-case conditions will vary with the product and its intended use, but generally reflect the maximum and minimum ambient temperatures at which the product is expected to work correctly, plus the worst-case combination of input power, load, and operating duty factor. Under worst-case conditions the product may have degraded performance compared to nominal conditions, but it should not suffer permanent degradation or fail.

As can be seen by examining Fig. 9.3, voltage and temperature are stresses which reduce capacitor reliability. Reducing the applied voltage relative to the manufacturer's ratings reduces the failure rate dramatically, because the failure rate due to the dielectric breakdown mechanism increases exponentially as voltage increases linearly, and the exponent ranges between 2.5 and 6, depending on the dielectric material. The temperature of a capacitor is usually determined by the ambient temperature, and the relationship of this stress to failure rate is also shown in Fig. 9.3. In some applications, alternating current ripple or direct current

TABLE 9.3   Capacitor Derating Guidelines

Type	R.E.P. code	Recommended practice	Do not exceed
Mica, fixed	CMF	To 50% rated* voltage less 2%/ ° C above 45° C	75% rated* voltage, 55° C ambient
Ceramic: Fixed Variable	CCF CCV	To 50% rated* voltage less 2%/ ° C above 55° C	70% rated* voltage, 80° C ambient
Plastic film, fixed (mylar, polycarbonate, polypropylene)	CPF	To 40% rated* voltage less 2%/ ° C above 55° C	60% rated* voltage, 70° C ambient
Plastic film, variable	CPV		
Electrolytic (aluminum)	CAO	20–60% rated* voltage less 2%/ ° C above 55° C to 1-V reverse bias	80% rated* voltage, 65° C ambient, 1.5-V reverse bias
Electrolytic, (solid tantalum)	CST	To 50% rated* voltage less 1%/° C above 55° C to 5% reverse bias	75% rated* voltage, 80° C ambient, 10% rev. bias @ 25° C, 5% rev. bias @ 70° C
Air, variable glass, variable plastic, variable	CAV CPV	To 50% rated* voltage less 0.5%/ ° C above 55° C	70% rated* voltage, 80° C ambient

 * 25° C rating.

leakage through the capacitor causes an internal temperature rise, and in those cases that internal temperature rise must be added to the ambient temperature to arrive at the internal temperature of the component. The effect of temperature on each technology is different, since different failure mechanisms dominate; for example, in a typical mica capacitor, an important effect of temperature is to increase the rate of silver migration from one plate to another, leading eventually to internal short circuits. In the aluminum electrolytic, increasing vapor pressure of the liquid electrolyte leads to leakage and catastrophic failure; in the case of the plastic film dielectrics, softening and expansion of the dielectric material lead to failure if dielectric breakdown does not happen first. Finally, you will note that reverse polarity must be avoided in all types of polarized electrolytic capacitors, since it causes damage to the dielectric, leading to short circuits and explosive destruction due to overheating.

Packaging.   Packaging is much more difficult than is generally recognized, since it is often completely original rather than being just an original arrangement of standard components, and it involves poorly understood areas of design such as heat transfer, electrical shielding, mechanical strength, and environmental containment. Packaging mixes elements of function and style, so individuals with no technical knowledge may be involved in specifying or approving it. Finally, fabricating the package may well involve expensive tooling or new manufacturing processes, so there are strong temptations to cut corners or use makeshift methods to save money.

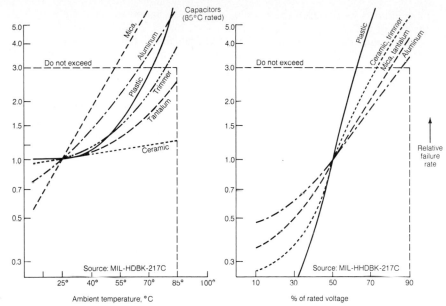

FIG. 9.3 Capacitor stress vs. failure rate.

The functions of packaging which most strongly affect reliability are structural support, heat transfer, and acting as a barrier to chemical contaminants and electrical signals. In the case of structural support, the old rule of thumb that "if it does not move, it will not fail" is not strictly true, since the motion may be microscopic as a result of expansion and contraction with temperature changes, yet that motion may be sufficient to cause fatigue or fretting failures. Designing out that type of failure may be as simple as avoiding using plastic encapsulants, which have high coefficients of thermal expansion, in intimate contact with delicate, low-expansion-coefficient materials such as fine metal wires. Another factor to remember is that the total motion is the coefficient of expansion multiplied by the length of the material being heated, so long narrow structures will create more problems than round or square ones.

Heat transfer is an important function of packaging, and one which has a direct impact on reliability. Heat is transferred from a warm object to a cooler environment by a combination of three means: conduction, radiation, and convection. (Evaporative cooling is unusual in electronic equipment and will not be discussed.) Conductive cooling requires contacts of materials with high thermal conductivity to be effective. Some materials, such as metals and certain ceramics, have much higher thermal conductivities than other package materials such as plastics. This means that a component or product encased in plastic must have some other means of cooling or it will run very hot, and consequently be unreliable.

Radiation is the means by which heat escapes from a hot body to a cooler surrounding space without any contact, so radiation is effective even in a vacuum. The problem is that radiation depends upon the temperature difference between the hot object and the cooler space, so in products where we are trying to design for a maximum temperature rise of only 5 to 10°C, radiation accounts for less than 1 percent of the cooling.

Convection, the third means of cooling, depends upon the presence of a gaseous or liquid medium in which convection currents can circulate. Convection is the primary means by which heat in most electronic products is dissipated into the surrounding air, but it depends upon the density and velocity of the air, the area and orientation (horizontal versus vertical) of the hot body, and the temperature difference. The thermal conductivity of the hot surface also is a factor, so the best cooling is from a large, vertical metal plate with a high velocity airflow over it. Forced-air convection is much more effective than natural convection, providing that fan failure, a dirty air intake filter, or other disruption of the airflow does not occur. It is a good idea to design some sort of airflow or temperature monitor into a product which could be damaged by cooling air failure.

The barrier functions of a package should not be overlooked. The reliability of the product may well depend upon the package keeping contaminants out of the circuitry. At the component level, we generally expect sealed packages, impervious to liquids and small particulate contaminants. In this context, it is important to remember that encapsulating plastics are not hermetic, that is, water will diffuse into and completely through the common packaging plastics in time, generally within 1 or 2 weeks. This means that circuits packaged in plastic are not moistureproof unless other means of moisture control such as internal heat generation or nitride passivation are used. At the equipment level, the barrier functions are more electrical, shielding to stop radiated electrical signals and to surround the product's circuitry with protection against large foreign objects. The trend toward plastic cabinetry means that the shielding function is not being performed, unless there is an electrically conducting layer added to the plastic.

Design Review. This is the moment when the new design can be measured, compared to the previously set goals, and improved. One effective design review technique is the *peer review,* at which several experienced and qualified engineers meet with each design engineer and review his or her part of the design. This is not a management review, although managers are permitted if technically qualified to contribute. The important items to have at the design review for hardware are a sample of the product or assembly being reviewed, diagrams and material lists, a part-by-part stress analysis listing the measured voltages, currents, and temperatures of each component compared to the maximum ratings, and test data on the product or assembly. The reviewers are looking for overstress, misapplications, lack of margin between requirement and specification, or potential problems which would cause a minor failure to escalate into a serious one. Failures which can start fires or cause shock hazard are such potential problems, and they can nearly always be designed out. A failure rate estimate should be made at this point, and the estimated value compared to the budget or target for the product being reviewed. Designers should expect that the outcome of the design review is a list of changes or at least tests to perform, for without changes the product has not been improved by the review. Because changes are an expected outcome, the earlier these reviews are held, the less the schedule disruption and cost.

Qualification Testing. Qualification tests are of two different types: margin tests and life tests. *Margin tests* are concerned with assuring that the threshold of failure—the combination of conditions at which the product just begins to malfunction—is outside the range of specified conditions for the product's use. Because these tests are generally performed on a small number of products, the test conditions may be more severe than the specified conditions to allow for the possibility that the unit being tested is a better-than-average unit. Another way to assure that the product has adequate margin is to vary more than one stress at a time, as is shown in Fig. 9.4. In case A, one stress at a time is varied as tests 1 to 4 are conducted. In case B, both stresses are varied at once, and the resulting circle of performance is much larger. This is also the point at which compliance with safety and regulatory requirements, such as electromagnetic interference or shock hazard, is tested. Finally, it should be remembered that margin tests are not reliability tests; the duration is too short and the sample size too small to draw a valid inference about the reliability of the new product.

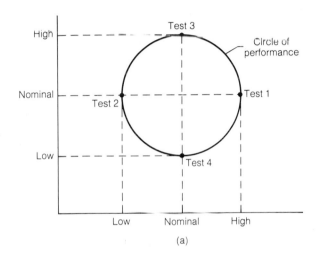

(a)

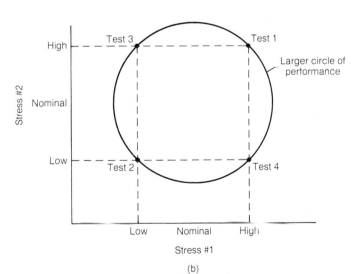

(b)

FIG. 9.4 Single stress vs. multiple stress testing.

Life tests differ from margin tests primarily in the number of units tested and the duration of the test. Life tests are intended to find patterns of failure which occur too infrequently to be detected by engineering tests on one or two prototypes. Tailoring the tests to specific failure mechanisms may not work very well because the product is a mixture of different components, each of which has different dominant failure mechanisms. Thus to limit life tests to only one or two stresses is foolish, and the probability of overlooking a serious problem is high. Broadband, multistress tests designed to uncover a wide assortment of ills are the best choices.

One test philosophy based on this concept has been named *STRIFE,* a name based partly on a contraction of the phase STRess-lIFE, and partly as a play on words. The basic concept of STRIFE testing is that simultaneous multiple, cyclical stresses are much more powerful than single, constant stresses at discovering failure modes in products with a mix of technologies and large number of parts. A typical STRIFE test used in the electronic hardware industry requires as many prototype products as can be afforded, up to 25 or more in the later stages of development, for 1 to 6 months. Temperature cycling below 0 and above $+50°C$ at a rate of change of 1 to $5°C/min$, power on-off cycling during the temperature cycling, and continuous running of diagnostic tests is required, and an accurate log of every failure or malfunction must be kept. Every failure is analyzed, and an attempt is made to design it out. The purpose of the test is to find patterns of failure, so a test with no failures is *not* successful, while one where failures have occurred, design changes have been implemented, and no repetition of the failures has occurred is considered successful. It is not unusual to cut the failure rate of a new product in half by employing this approach, while also lowering the infant mortality region slope to the more nearly flat failure rate of the mature product.

Failure Analysis.   Just as it is the role of stress testing and failure analysis at the component level to help the device designer and process engineer improve a component, so failures during equipment-level testing will help uncover product-level failure modes as well as additional component-level modes. The problem is that a component analysis may not suffice; the failure cause may be a combination of factors at the system level and may require the combined talents of a failure analyst, design engineer, and production engineer to understand and solve it. Happily, not all problems are this complex; often the failure is a single component, overstressed or damaged in a straightforward way. The most important point is that product reliability improvement only occurs if something is changed, and a failure ignored or explained away will return at the most damaging moment. Of course, not all failures can be neatly analyzed and eliminated, but experience has shown that 50 to 75 percent can be, particularly those failures that occur more than once.

9.2.3 Transfer to Production

Knowledge Transfer.   Contrary to popular opinion, the design is not finished when the transfer from development to production takes place—quite to the contrary. The problem with this situation is that changes at this stage of product development are very costly and tend not to be evaluated with the same thoroughness as the original alternatives. Finding a fix fast is the order of the day, and preferably finding a fix that does not require extensive retooling or scrap. Still, there are some legitimate reasons why changes in the design occur after release to production, such as the discovery of phenomena that occur infrequently so that they are not discovered until a large number of products are manufactured. Another reason is that with the pressure to develop new products in a shorter time, concurrent design is often practiced. This means that a new manufacturing process is developed simultaneously with a new product using that process, rather than sequentially. This is a risky approach, but one that is gaining popularity because it saves time.

The formal instrument of transfer of knowledge about the design is the documentation: drawings, material lists, assembly and test procedures, tooling, samples, and the like. This is usually not enough, and one method that has been successful is to transfer one or more of the design engineers into manufacturing at the same time, to become the production engineers for the first 6 months to 1 year. This used to be met with howls of protest about being shunted out of the technology mainstream and so forth, but if the assignment is offered to an engineer with management aspirations, it can be sold as part of a broader experience and is valuable as such. In places where this technique is used and where graduates of this program are seen to advance in management as a result, the benefits have been substantial. In at least one company, the engineers who took this path and then returned to design were quickly seen to be more effective, and the company began a program of bringing new design engineering hires into the company with a 6-month assignment in production engineering first. Obviously this program will not work for everyone; some designers simply are not going to be happy or successful in production, and some are going to decide to remain in production permanently.

Process Control.   As was mentioned before, when manufacturing problems arise, there is a tendency to look for a quick fix. One type of solution that is wrongly perceived as quick and easy is adjusting the manufacturing process to minimize the problems, rather than changing the design. Perhaps that is because the process documentation, if it even exists, is internal and not shipped to the customer along with the product. More likely, it is because the process is under the jurisdiction of production and consequently does not require design engineering's approval to change it. The problem is that the situation can deteriorate to the point where there is a customized process for each product, nothing is standard, and the process is out of control most of the time. When design rules and process parameters are both being varied at the same time, the situation quickly becomes too complex to understand or control, and quality suffers.

The correct solution is to optimize the process, get it under control, and keep it that way. Then the designs can be modified so that they fit the standard process, producing stable and predictable yields day after day. (There is a belief among semiconductor processors and users that a high level of rejects is a signal that there may also be an accompanying high level of hidden defects, leading to reliability problems. In other words, astute customers do not want even the good parts from a bad lot or process.)

The traditional method of judging the quality of a manufacturing process has been to look at the product yield, so that is another reason why process parameters are often varied rather than the design. This is possible using sophisticated statistical techniques, but it is not a very good way to run a process. For one thing, the end result may be a higher yield, but of a product with varying characteristics, such as in heat-treated metals. A much better approach is to run specially designed test parts through the process concurrently with the product to be sold and to design those test parts to be sensitive to the basic process parameters. In summary, the product yield is important, but the process should be controlled by the yield of test devices, not product.

In-Process Testing.   There are two purposes to in-process testing: to eliminate defects before so much value is added to the product that the cost of diagnosis and remedy is too high, and to adjust or calibrate the performance in some way. Calibration is usually done at nominal conditions so that the performance is most accurate in the same environment as the average customer, but performance verification is much more effective if done with varying stress, so as to assure margin between specified and actual performance over a range of conditions. Design margin is a powerful tool for increasing reliability, as has been previously discussed, and margin testing on large numbers of products in manufacturing is much more revealing than the same testing done only to a few prototypes.

Testing to eliminate defects is also a way to improve product reliability. The common method is to employ some sort of burn in of the completed product, or at least of major subassemblies. Unfortunately, experiments have shown that burn in of complex products or assemblies is a waste of time, at least if by the term *burn in* you mean aging a product with infant mortality behavior so that the failure rate after the burn-in period is significantly lower than it was at the start. One reason for this is that the typical product or assembly contains a mixture of components, and hence a mixture of failure mechanisms. One stress, or even a combination of stresses, is not going to accelerate all those mechanisms equally, or by a large enough factor to do much aging. In a study of warranty failure rates of electronic products in the field, it has been found that the average failure rate declines to about one-half its initial value in the first 12 months of use (Fig. 9.5). Thus, to age the product enough to cut the failure rate in half would require operating it a year in simulated field conditions, clearly an impractical suggestion. Even increasing temperature to the maximum that could be tolerated by the product without risk of damage would only increase the aging rate by 4:1, so that the burn-in period would still be 3 months.

There are usually reliability improvements that can be achieved by production stress testing. As was mentioned in the section on life testing, simultaneous multiple, cyclical stresses are far more effective at discovering failures in complex assemblies than single, constant stresses. Experiments have shown that a heat run

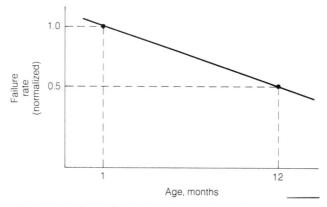

FIG. 9.5 Early failure behavior of electronic products.

constructed along the same guidelines as STRIFE tests, but shorter in duration, can have two benefits:

1. Products which have been exposed to a cyclical, multiple-stress heat run exhibit field reliability which is 20 to 30 percent improved, rather than the 1 to 5 percent seen in equipment level burn in at constant stress.

2. Such a heat run, even if only 24 to 48 h duration rather than the 1 to 6 months recommended for STRIFE, will cause multiple failures which must be repaired in the factory. With multiple failures available for statistical and physical analysis, the dominant failure modes will quickly become apparent, and with the failures occurring in quantity right on the premises, designers can be quickly convinced that the problems are real and a design change is needed.

It is not unusual for the *Pareto principle* to be reaffirmed by such a program; that is, for 80 percent of the failures of a new design to be caused by 20 percent of the failure modes, usually two or three. Design changes to minimize or eliminate these have immediate effects on product reliability. One final remark: once the dominant problems have been solved and the failure rate reduced, it may not be economically justifiable to continue this testing on a 100 percent basis. A continuous sampling plan might work very well to assure the outgoing quality level, thus freeing heat-run facilities for the next new product.

9.3 CONCLUSIONS

As was stated at the outset, design is a step-by-step process. Reliability is designed in or not designed in at each step of the process. The important rules are to reduce stress, increase margin, and decrease parts count. Reliability improvements of 2 or 3:1 are achievable in each of these areas without any dramatic breakthroughs in technology, and if all three are improved by that ratio, an overall improvement of 10:1 is well within reach.

BIBLIOGRAPHY

Crosby, Philip B., *Quality Is Free,* McGraw-Hill, 1979.

Scott, Allan W., *Cooling of Electronic Equipment,* John Wiley & Sons, 1974.

Ott, Ellis R., *Process Quality Control,* McGraw-Hill, 1975.

Lipson, Charles, and Sheth, N. J., *Statistical Design and Analysis of Engineering Experiments,* McGraw-Hill, 1973.

Bailey, R. A., and Gilbert, R. A., "Strife Testing," *Quality,* November 1982, pp. 53–55.

Moss, R. Y., "Modeling Variable Hazard Rate Life Data," *Proceedings, 28th Electronic Components Conference,* April 1978, IEEE 78CH1349-0, pp. 16–22.

Carter, A. D. S., *Mechanical Reliability,* John Wiley & Sons, 1976.

CHAPTER 10
SPECIFICATION OF RELIABILITY REQUIREMENTS

Burton S. Liebesman, Ph.D.
Bell Communications Research

A good reliability design starts with an understanding of the needs of the customer. These needs must then be translated into well-defined requirements that designers and manufacturers can understand. Requirements should be broad in coverage and should include standards for the manufacturer's quality and reliability management system in addition to the numerical values of specific reliability parameters. Specifying only the values of the reliability parameters will not guarantee a reliable product. In fact, just specifying these values will often lead to unsatisfactory results because the means of attaining desired reliability are based on a sound management system. In this chapter, we take a broad perspective and describe the specification of the reliability program requirements.

Specification of reliability requirements should occur early in the product's life. The earlier the customer considers the detailed needs, the better the outcome. Visibility of the requirements is an important aspect of success in providing high reliability. All persons associated with bringing the product to market should understand the importance of reliability and should include reliability considerations in their work. There is a close relationship between quality and reliability requirements. We will assume that quality requirements are being satisfied. However, during the discussion of contractual provisions, a section on the factory and initial service period includes quality assurance provisions.

The material in this chapter is aimed at complex, "high-tech" products. For less complex products it is not necessary to apply all of the provisions. In fact, the provisions applied to a product should be economically tailored to the needs of that product.

10.1 FOUR PHASES OF THE PRODUCT LIFE CYCLE

This chapter is organized in the following way. First, we will describe phase 1, the conceptual and advanced planning phase, of a product life cycle. Since this is mostly a planning phase, we will discuss management procedures for the most part. Early technical requirements are developed during this phase, whereas the more comprehensive technical development is done in phase 2, the design and development phase. The bulk of the material in this chapter will cover phase 2 actions, including (1) environmental analyses, (2) failure definition, (3) reliability prediction, (4) reliability growth management, (5) human factors, and (6) reliability demonstration. The discussion will end with a description of contractual requirements. These need to be specified unambiguously so that supplier-customer relationships will be completely open and an atmosphere of cooperative effort will prevail. Phases 3 (production) and 4 (deployment) will not be discussed in depth. However, the contractual provisions described at the end of this chapter provide the framework for viable production and deployment reliability specifications during these phases. Proper insertion of provisions into a contract will assure accurate measurement of reliability and provide feedback to correct reliability problems.

In the final portion of the chapter, we will discuss the major references available to planners. In this chapter, we can only briefly describe the various elements of the specification process. The references will provide in-depth help with the many important details.

Before we leave the abstract discussion of reliability specification, we should remind ourselves that the aim of a reliability effort is user satisfaction during the life of the product. Keep this in mind so that the reliability tools do not become ends in themselves. Also keep in mind that many people have responsibilities which are part of the overall reliability effort. For example, when the designer selects parts, reliability requirements should be included in the selection criteria. If this is not the case, the selection may result in very costly changes during manufacture. It is important to consider reliability specifications early in the design process and monitor them throughout the life of the product. The main purpose of this chapter is to discuss specification of reliability requirements. However, maintainability and human factors are closely tied to reliability. We shall bring them into our discussion as the need arises.

Consider the product life cycle as divided into four phases: (1) conceptual and advanced planning, (2) design and development, (3) production, and (4) deployment. Figure 10.1 depicts the elements of the reliability specification process. The road map described in that figure will be followed in the next sections of this chapter.

Phase 1 is the conceptual and advanced planning phase. During this period, the user identifies the technical and reliability needs. The reliability planning program is set up and the reliability requirements are defined in the form of a specification that can be provided to an outside supplier or to the company's advanced design organization when the product is developed internally. The most common initial contact with outside suppliers is through the "request for proposal" (RFP).[*] This

---

[*] Other terms commonly used are "request for information" (RFI) and "request for quote" (RFQ).

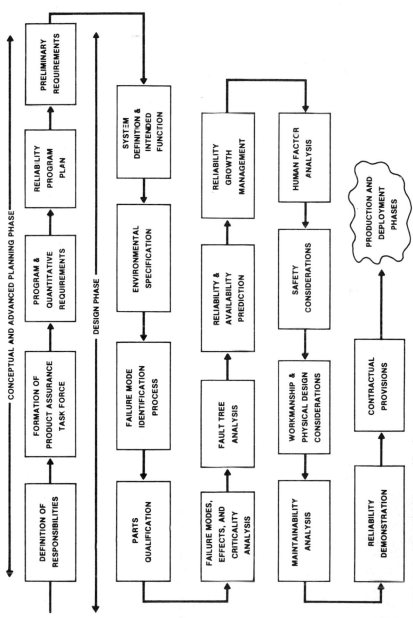

FIG. 10.1 The reliability specification process.

10.3

document should specify the technical characteristics of the product and provide information on the reliability program to be followed. It should also include the reliability objectives for the product.

Phase 2 in the product life cycle is the design and development phase. During this period, the reliability requirements are formalized. Reliability growth procedures should be started at this time. The reliability needs of the product will often change or evolve as the development proceeds. When the product is produced by an outside supplier, contractual reliability requirements are developed at the end of this phase. These requirements are based on the specification developed during phase 1 and a growth in understanding of the reliability needs during phase 2.

Phase 3 is the production phase. During this period the project administrators are directly concerned with reliability considerations in the manufacturing process. This implies a need to measure reliability parameters (often during a "pilot" production phase) and to report problems to the engineering and manufacturing organization; this often leads to a partial redesign of the system. Hence, personnel involved with phases 1 and 2 should be available for consultation during phase 3. Otherwise, the characteristics of the product may be changed and the design intent may not be met.

Phase 4 is deployment. The product enters the marketplace during this phase. Does it meet the user's needs? Does it provide the reliability required? Means of measuring the product effectiveness should be provided. Data collection mechanisms and feedback paths need to be developed. These will often lead back to people involved in the development during earlier phases. New reliability requirements may be determined or old ones modified. A well-defined reliability program is invaluable in controlling changes during this phase.

Specification of reliability requirements includes specifying the reliability management program. For almost all products, the specification work occurs during the first two phases. This includes off-the-shelf purchases for which a purchase contract is developed. The development of the contract is considered part of phase 2. We will concentrate on phases 1 and 2 in this chapter. During these phases, the framework of the full reliability program needs to be developed into an effective operation.

10.2 CONCEPTUAL AND ADVANCED PLANNING PHASE

Refer to the first row of Fig. 10.1 during the discussion of this phase.

10.2.1 Definition of Responsibilities

The primary reason for clearly defining the responsibilities of each part of the reliability program is efficient utilization of reliability resources. The first step is to define the major responsibilities of the various organizations involved in the program and to include their inputs in the development of the reliability program plan. Of course, the specific organizational structure will dictate actual assignment of these responsibilities. We can, however, define the basic responsibilities and indicate their part in the total program. Also, we will point out the specific skills needed to create a strong reliability program. The organization should find people with these skills early in the development process.

Two types of organizations are generally used in industry. These are a quality-assurance-based organization and an engineering-based organization [9]. Figures 10.2 and 10.3 depict the management chain of command in each of these organizational structures. The quality-assurance-based organization has the advantage of placing the closely related quality and reliability functions in the same organization. This is shown in Fig. 10.2, where the reliability engineering function is under the quality assurance (QA) manager. On the other hand, the engineering-based organization has the advantage when there is a considerable amount of innovation required. In Fig. 10.3, the reliability engineering function is under the engineering manager. In both types of organizations, it is critical that the manager responsible for the reliability function have a direct line to the product manager, independent of the other functions.

10.2.2 Product Assurance Task Force

A special task force should be created with the responsibility of overall program control. In this chapter, we will call this task force the "product assurance task force." Other authors may use different names such as the "design review committee." The chairperson of the task force should be from the product management organization. Generally the chairperson will utilize existing organizations to do the various jobs. The task force responsibilities can be classified as either program requirements or quantitative requirements.

10.2.3 Program Requirements

Three major reliability requirements are reliability engineering, reliability accounting, and reliability program interfaces. If the product has a large software component, a fourth major requirement is software management.

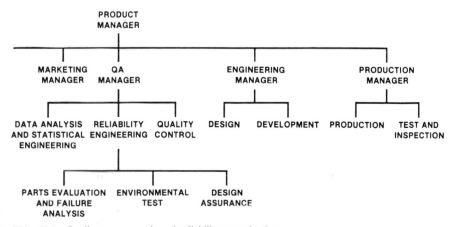

FIG. 10.2 Quality assurance–based reliability organization.

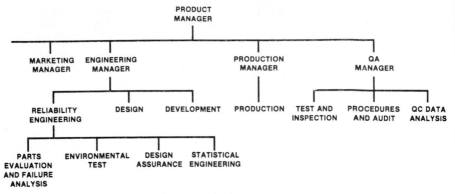

FIG. 10.3 Engineering-based reliability organization.

Reliability engineering's main task is to identify reliability problems early in the conceptual phase and thus avoid large costs in later phases. For example, if the product is to be used in a hot environment, components should be tested at high temperatures.

Reliability accounting provides the information necessary to monitor the reliability program efficiently. The results of the program should be clearly understood by all organizations involved in the process, and this should be done in a cost-effective manner. The reliability accounting group should provide design and engineering information from past experience to aid in developing the initial design.

Next, there is the matter of reliability program interfaces between internal organizations and with subcontractors. Clearly, this is the responsibility of all organizations involved. However, a special responsibility belongs to the product assurance task force. Finally, there is the matter of software management. A software life-cycle plan needs to be developed. This plan should provide the tools for testing, error measurement, change control,* and documentation.

10.2.4 Quantitative Requirements

Strong statistical skills are needed because of the quantitative requirements of the reliability program. Organizations charged with these responsibilities must have people available who are familiar with mathematical modeling, statistical hypothesis testing, and reliability analysis. Also, these organizations must include people who are familiar with the electronic and mechanical components of the system and the system software. Ideally, personnel with strong engineering, statistical, operations research, and software engineering skills have the best background for developing the quantitative requirements.

In developing requirements for the system and for critical vendor items, great care should be taken to include appropriate quantitative reliability and maintainability requirements and effective verification tests. To be effective, these

---

* Also commonly called "configuration management."

TABLE 10.1 Elements of a Reliability Program Plan

1. Definition of preliminary reliability requirements	9. Reliability demonstration methods
2. Description of program tasks	10. Identification of key personnel
3. Procedures to evaluate status and control of each task	11. Description of the organizational structure
4. Description of task interrelationships and interfaces with other systems	12. Source of reliability design guidelines or design review checklists
5. Schedules for the tasks	13. Description of the reliability contribution to the total design
6. Identification of known reliability problems, their impact and suggested solutions	14. Inputs from operational and support personnel
7. Action item register	15. Software life-cycle plan
8. Designation of milestones	

requirements must exceed minimum acceptable values as established by the development plan, and the tests should include sufficient data to make the test results significant.

10.2.5 The Reliability Program Plan

The reliability program plan is the starting point for the reliability specification process. The goals are to achieve a disciplined approach to life-cycle reliability management and to optimize the use of reliability-related resources. The plan starts with a definition of preliminary reliability requirements and then describes the tasks needed to meet these requirements. This includes functional responsibilities and interrelationships between tasks, and it describes procedures to evaluate the status of the tasks.

Other major functions of the plan are to: (1) set up a mechanism for identification and correction of problems (an action register), (2) set up means for updating and disseminating documentation, (3) identify key personnel, and (4) provide design checklists. If subcontractors are to be used, the plan should define procedures to monitor them. This should include procedures for evaluating their reliability program and a definition of contractual provisions which will monitor reliability requirements. Finally, the plan should specify a schedule with milestones and responsibilities. Table 10.1 summarizes the elements of a reliability program plan as described in task 101 of Military Standard 785B, *Reliability Program for Systems and Equipment* [1].

10.3 THE DESIGN PHASE

Refer to rows 2 through 5 of Fig. 10.1 during the discussion of this phase.

TABLE 10.2   Design Factors for a Reliability Specification

1. System definition and intended functions	6. Safety
2. Environmental specifications	7. Workmanship and physical design characteristics
3. Failure definition	8. Maintainability requirements
4. Reliability and availability requirements	9. Reliability demonstration methods
5. Reliability growth requirements	10. Software control requirements

10.3.1   Stating the Reliability Requirements

The best time to define the parameters of the reliability requirements is during the design phase. This is because reliability problems encountered during the production and deployment phases will cost much more than problems identified during the design phase. In addition, definition of parameters during the conceptual and advanced planning phase will often be premature. However, preliminary requirements should be published during that phase and redefined during the design phase.

The key inputs to the definition of reliability parameters are as follows: (1) the intended environment(s) of the product and (2) the intended use(s) of the product by the customer. These must be clearly understood before the designer can proceed. Models of the various environments should be developed so that the product assurance task force can analyze and conduct tests of the product in surroundings similar to its intended use. The task force should visit a future site and talk to the principal users. If the product will replace an existing system, that system's operation must be clearly understood. There are many methods of testing, such as computer simulation, scale models, and physical simulators which can be used to help understand environmental and usage problems.

It is very important to define the meaning of failure during phase 2. For example, a software failure in a telephone switching system could be defined as an action that causes a restart of a program module or one that causes a reinitialization of the whole system. In the former case, a connected customer may not be affected, while in the latter case, the customer will lose the connection. However, a restart of the program module will cause the maintenance crew to take some action and will be an expense to the telephone company. The definition of a failure must take into account the effects on service, maintenance, and sparing levels.

Table 10.2 contains a list of those factors needed to be considered during the design phase. The means for providing these inputs will be described in later sections.

10.3.2   System Definition and Intended Functions

The first factor in Fig. 10.1 that must be dealt with is a description of the system and a definition of its intended functions. This is obviously necessary, because the purpose of the system is to satisfy a customer's needs. The system definition should include its reliability requirements. For example, a telephone switching system will have functional requirements such as call processing, service and traffic performance, switching, signaling, transmission, and synchronization. It

TABLE 10.3 Local Switching System Reliability and Availability Objectives

Objective	Value
Cutoff calls (CC)	Probability of cutting off a stable call: $\leq$ .000125
Ineffective machine attempt (IMA)	Probability of an IMA: $\leq$ .0005
Line failure rates (LFR)	LFR $\leq$ 15,000 FITS (failures in 10^9 hours)
Line downtime (LD)	LD $\leq$ 28 min/year
Total system downtime (TSD)	TSD $\leq$ 3 min/year
Trunk downtime (TD)	TD $\leq$ 28 min/year

will have physical requirements such as power, control of electromagnetic interference, installation, and environmental requirements. It will also have operational requirements such as maintenance, testing, and administration. All these and other requirements need to be defined at the outset of the design and development process.

In particular, the switching system in our example will have reliability and availability objectives that it must meet. Table 10.3 summarizes some of the objectives as defined by Ref. 2 for local switching systems. The more complex objectives have been left out of the table.

10.3.3 Environmental Specifications

The first step in the development of environmental requirements is to identify the conditions under which the product will be stored, handled, transported, and used. The environmental analysis should include identification of possible misuse, modes of maintenance, and special operating conditions. A product may be used in many different environments. For example, an aircraft that flies from the southern part of the United States to the northern part of Canada will undergo severe climatic changes. As a second example, consider telephone equipment which could be placed in either the controlled environment of a central office or in an open field where extremely high or low temperatures may affect its operation. In both examples the design must consider the different environments to which the product will be exposed. Table 10.4 lists a number of environment factors which must be considered. A good reference document for the testing of environmental requirements is Military Standard 810 [3]. It contains a description of testing methods for a large variety of environmental conditions.

Today, one of the major environmental problems for electronic systems is electrostatic discharge (ESD). Even small voltages can destroy electronic componentssuch as large-scale integration (LSI) devices. Work is in progress to identify ESD problems and find solutions. Reference 4 provides guidelines for establishing an ESD program.

10.3.4 Parts Qualification

An early task during the design phase is the development of a qualified parts program if one does not already exist. The use of standard parts will help reduce the uncertainties that lead to design problems. Task 207 of MilitaryStandard 785B

SECTION I - GENERAL APPLICATION PARTS

SUBSECTION A - MECHANICAL

INDEX NO	DESCRIPTION	DOCUMENT NO	PART NUMBER	REMARKS
A0001B	ADPTR, AL AL, .250 FEM PIPE THD TO .250 MALE FLD	2A156	2A156-4-4 82742-12	
0002	ADPTR, TUBE TO HOSE, LP NOSE, PART OF AN6270, 1/2 TUBE SIZE	MIL-A-38728	MS27404-8D	CRITICAL PART, LONG LEAD TIME
0009	O-RING, BOSS SEALING, 160F 0.924 ID, BOSS	MIL-G-5510	MS28778-10	

SECTION I - GENERAL APPLICATION PARTS

SUBSECTION B - ELECTRICAL AND ELECTRONIC

INDEX NO	DESCRIPTION	DOCUMENT NO	PART NUMBER	REMARKS
0006	CAP, TA, SLD, 22-330 μF, 6-100 Vdc, CSR-13	MIL-C-39003/1	M39003/01-****	FAILURE RATE LEVEL S, QPL AVAIL, CRITICAL PART, REVERSE VOLTAGE
0007A	CAP, TA SLD, 0.47-18 μF 6-75 Vdc, CSR-09	MIL-C-39003/2	M39003/02-****	FAILURE RATE LEVEL S, QPL AVAIL
A0010	CAP, TA, FOIL, 4-500 μF 15-150 Vdc	92A643	92A643-1-2 130J46-3 439X-72J20	CRITICAL PART, HIGH COST AND LONG LEAD TIME

(a)

10.10

SECTION II - LIMITED APPLICATION PARTS

SUBSECTION A - MECHANICAL

INDEX NO	DESCRIPTION	DOCUMENT NO	PART NUMBER	REMARKS
B0101	BEARING, BALL END, PRCN, SELF-ALIGN, .250 BORE	XYZM140	XYZM140-48CR	USE RESTRICTED TO XYZ CO ONLY
B0102	NUT, BLIND RIVET, .025-.081 GRIP RANGE, .112-40 THD, DRES	NAS1330	NAS1330C04K81	USE RESTRICTED TO XYZ CO ONLY
J0197	CONNECTOR, HOSE QUICK DISCONNECT FEMALE, ANTI "G" SUIT	MS27755	MS27755-1	USE RESTRICTED TO JONES CO ONLY SEE SAME INDEX NUMBER SECTION I FOR STANDARD PART

SECTION II - LIMITED APPLICATION PARTS

SUBSECTION B - ELECTRICAL AND ELECTRONIC

INDEX NO	DESCRIPTION	DOCUMENT NO	PART NUMBER	REMARKS
A0200	CAP, CHIP, CER, XRC-10	123-091	123-091-***	FOR PRODUCTION USE MIL-C-55881, SEE SECTION I
A0201	CAP, TA, SLD, UBG9A	123-093	123-093-1	FOR PRODUCTION USE MIL-C-39006/17, SEE SECTION I
B0203	CAP, METZD, PLSTC, FILM, .001-10 μF, 100 Vdc,$\pm$1%, CHR01	MIL-C-39022/9	M39022/09G**	USE RESTRICTED TO XYZ CO ONLY
B0209	MCKT, OP AMP COMP		LM111	USE RESTRICTED TO XYZ CO ONLY; CRITICAL APPLICATION PART; FOR PRODUCTION USE M38510/10304BXX

(a)

FIG. 10.4 Sample format for a qualified parts list.

TABLE 10.4 Environmental Factors

Natural environments	Induced environments
1. Temperature range	1. Transportation vibration and shock
2. Temperature cycling and/or shock	2. In situ vibration and shock
3. Pressure	3. Nuclear effects
4. Humidity	4. Electromagnetic effects
5. Moisture (rain, snow, ice, condensation)	5. Chemical effects
6. Wind	6. Biological effects
7. Sand/dust	
8. Salt	
9. Solar radiation	
10. Plant growths	

[1] and QR-800-S, *Reliability Program for Systems Equipment Development* [5], outlines a parts control program. Military Standard 965 [6] should be used in conjunction with task 207. Reference 9 is also a good source of information about parts qualification procedures.

A parts control program generally consists of the following elements: (1) a parts control board, (2) a qualified parts list, and (3) procedures for approval of nonstandard parts. The parts control board is responsible for elements (2) and (3). Figure 10.4 is an example of a qualified parts list from Military Standard 965 [6].

The use of qualified parts helps in the early development of a system because the characteristics of these parts are well known. Some environmental testing must be done, however, to verify the survival capabilities of the parts; and if qualified parts are not available, new parts may have to be designed and tested. Weaknesses in the area of parts qualification need to be identified early in the design phase or they may result in costly changes during later phases.

There are four methods defined in task 223-MI of Ref. 5 of analyzing the survival capabilities of a component. The first is comparative analysis (i.e., the part has been subjected to stringent military specifications). The second method is similarity (i.e., the part is similar to an acceptable part). The third method is inference (i.e., the part works in a higher-level assembly). The final method is by direct testing of the part. The last method is preferred.

The end result of the component qualification process should be an understanding of the failure characteristics of the component. These characteristics can then be used in the analysis of higher-level failures.

10.3.5 The Failure Mode Identification Process

The failure mode analysis process generally falls into one of three categories: (1) bottom-up, (2) top-down, or (3) a combination of bottom-up and top-down.

In the bottom-up method, we first analyze the failure modes of the basic components. Then failure modes of successively more complex units are analyzed until the complete system is analyzed.

The top-down method is the opposite process. In this method, we determine the system failure modes and then decompose each into its subassembly and component failure modes. Often, both methods are used simultaneously in practical situations. We will start by describing the bottom-up method.

10.3.6 Bottom-up Analysis: Failure Modes, Effects, and Criticality Analysis

The basic reference for failure modes, effects, and criticality analysis (FMECA) is Military Standard 1629A [7]. FMECA is the basic tool of the bottom-up method. The analysis should be initiated as early in the design phase as possible and iterated throughout the life cycle of the product. In this way, feedback to the designer can be achieved and reliability improvement will be the end result.

There are five tasks (101 through 105) defined in Military Standard 1629A, *Procedures for Performing a Failure Mode, Effects and Criticality Analysis* [7]. The first of these is task 101 (failure mode and effects analysis). In this task, the system under study is defined. Block diagrams of the system architecture are very useful in this definition. Figure 10.5 is an example of a block diagram for a typical local switching system call-processing path. This diagram can be used as a starting point in the analysis of the ability of the local switching system to meet the reliability objectives shown in Table 10.3. A call essentially passes through seven system modules as it isbeing processed. Failure of any of the modules would cause a failure during call setup or the cutoff of an existing call. The main objective of task 101 is to identify all predictable failure modes and their causes. These should be listed on an FMEA worksheet. See Fig. 10.6 for an example showing two failure modes in a local switching system. The first results in failure of a memory card and the processor memory, whereas the second results in failure of a line card.

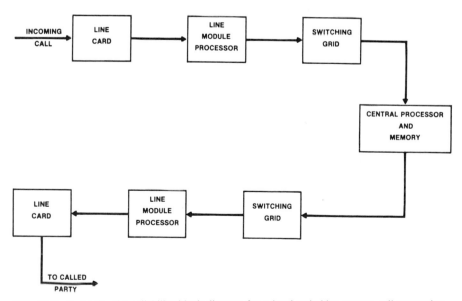

FIG. 10.5 Example of a reliability block diagram for a local switching system call-processing path.

FAILURE MODE AND EFFECTS ANALYSIS

SYSTEM LOCAL SWITCHING SYSTEM
INDENTURE LEVEL _____
REFERENCE DRAWING _____
MISSION _____

DATE _____
SHEET _____ OF _____
COMPILED BY _____
APPROVED BY _____

IDENTIFICATION NUMBER	ITEM/FUNCTIONAL IDENTIFICATION (NOMENCLATURE)	FUNCTION	FAILURE MODES AND CAUSES	MISSION PHASE/ OPERATIONAL MODE	FAILURE EFFECTS			FAILURE DETECTION METHOD	COMPENSATING PROVISIONS	SEVERITY CLASS	REMARKS
					LOCAL EFFECTS	NEXT HIGHER LEVEL	END EFFECTS				
1.	MEMORY CARD FAILURE	STORE DATA	MEMORY FAILURE	OPERATIONAL	MEMORY CARD FAILURE	MEMORY DOWN	TOTAL SYSTEM DOWN	TEST	SPARE MEMORY	MINOR	TOTAL SYSTEM DOWN
2.	CODEC ON LINE CARD	CODE & DECODE	LINE CARD FAILURE	OPERATIONAL	LINE CARD FAILURE	NONE	1 LINE DOWN	TEST	NONE	MINOR	1 LINE DOWN
3.	· · ·										

FIG. 10.6 Example of a failure mode and effects analysis (FMEA) worksheet.

TABLE 10.5 Failure Mode Categories

Level	Type	Probability of occurrence during operating time interval
A	Frequent	Greater than .20
B	Reasonably probable	.10 to .20
C	Occasional	.01 to .10
D	Remote	.001 to .01
E	Extremely unlikely	Less than .001

Failure modes are generally grouped by the blocks in the block diagram. For example, all failure modes affecting a line card would be grouped together and an aggregate failure rate for a line card would be determined. If the block diagram has differentlevels of aggregation, the analysis will continue until the full system level is reached. The effect of each failure mode is also listed. There are three classifications of effects which help in the aggregation process: (1) local, (2) next higher level, and (3) end effect. Other pieces of information listed on the worksheet are the failure detection method, compensating provisions, severity classification,* and remarks.

Once the failure modes have been listed, each can be classified according to its severity and end effects; then, the designers must decide on the acceptance of the failure mode or a redesign of the equipment. Designers can use information such as diagnostic features of the system to help make these decisions.

Task 102 of Military Standard 1629A [7] (criticality analysis) is done in conjunction with task 101. Failure modes are categorized as level A through E depending on the probability of occurrence. Table 10.5 summarizes the definitions of levels A through E. Military Handbook 217D [8] and Bell Communications Research Reliability Procedures [14] are two sources of methods for calculating probabilityof occurrence. See Sec. 10.3.9 for a description of failure rate predictions used in estimating the probability of occurrence.

Figure 10.7 is an example of criticality analysis for the failure modes introduced in Fig. 10.6. The calculations result in the classification of failure mode 1 as remote (level D) and failure mode 2 as reasonably probable (level B). Designers should be alerted to the potential problems of failure mode 2 and an effort should be made to reduce the probability of its occurrence.

Task 103 of Ref. 7 describes FMECA-maintainability analysis. It includes subtasks covering failure predictability, failure detection means, and basic maintenance actions. The purpose of this task is to provide early criteria for maintenance planning. The worksheet provided in task 103 of Ref. 7 is similar to Fig. 10.7 with the addition of acolumn for "basic maintenance actions."

Tasks 104 and 105 of Ref. 7 provide "damage modes and effects analysis" and guidance for the development of an FMECA plan. The worksheet provided in task 104 is similar to Fig. 10.7 with the addition of columns for local, next higher, and end damage effects.

Examples of the use of FMECA techniques can be found in Refs. 10 to 12. In Ref. 11, the FMECA analysis led to the conclusion that the worst failure mode for the wind turbine being analyzed is overspeed. This failure mode will often lead to the turbine throwing its blade. Twodesign changes were made, which consisted of the addition of a redundant switch to control the brake and a redesign of the disk brake to ensure stoppage at full power.

---

* A measure of the worst potential consequences resulting from a design error or item failure as defined in Ref. 7, p. 9.

CRITICALITY ANALYSIS

SYSTEM ____ LOCAL SWITCHING SYSTEM ____
INDENTURE LEVEL ____
REFERENCE DRAWING ____
MISSION ____

DATE ____
SHEET ____ OF ____
COMPILED BY ____
APPROVED BY ____

IDENTIFICATION NUMBER	ITEM/FUNCTIONAL IDENTIFICATION (NOMENCLATURE)	FUNCTION	FAILURE MODES AND CAUSES	MISSION PHASE/ OPERATIONAL MODE	SEVERITY CLASS	FAILURE PROBABILITY FAILURE RATE DATA SOURCE	FAILURE EFFECT PROBABILITY (β)	FAILURE MODE RATIO (α)	FAILURE RATE (λ_p)	OPERATING TIME (t)	FAILURE MODE CRIT * $C_m = \beta\alpha\lambda_p t$	ITEM CRIT * $C_r = \Sigma (C_m)$	REMARKS
1.	MEMORY CARD FAILURE	STORE DATA	MEMORY FAILURE	OPERATIONAL	TOTAL SYSTEM DOWN	X	.001	.01	10^{-6}	1.5×10^5	1.5×10^{-3}	1.5×10^{-3}	LEVEL D
2.	CODEC ON LINE CARD	CODE & DECODE	LINE CARD FAILURE	OPERATIONAL	1 LINE DOWN	X	1.0	1.0	10^{-6}	1.5×10^5	.15	.15	LEVEL B
3. ...													

*CRITICALITY = PROBABILITY OF OCCURRENCE

FIG. 10.7 Example of a criticality analysis worksheet.

10.16

10.3.7 Safety Margins

Reliability safety margin tests are defined in Ref. 5. The concept described in that reference is the "test-to-failure concept." This consists of developing an experiment or experiments to determine the environments most important to the functioning of the product and to find the one(s) most severe. Stresses are applied to the product while it is in the various environments until the system or component fails. Safety margins can then be determined from the data.

The end result of the FMECA and safety analyses is often a derating of the components. Derating is the design procedure by which the electrical, thermal, mechanical, and other stresses are reduced. These stresses may be reduced to as little as 50 percent of the normal operating stress on the component. See O'Connor [9] for a table of device derating guidelines. Also, see Appendix B of Ref. 5 for a detailed description of the Army Missile Command derating procedures. Figure 10.8 illustrates the parameters used to define the derating factor. The overstress condition defined in that figure occurs when the actual stress value (worst case) exceeds the maximum accepted stress value. The stress conditions shown in the figure do not include the overstress case. The figure does define the derating factor as the ratio of maximum stress value divided by the rated stress value.

10.3.8 Top-Down Analysis: Fault-Tree Analysis

Fault-tree analysis (FTA) is the basic top-down procedure. In FTA, the analyst considers the failure modes of the overall product and works down to the identification of the causes. Detailed logic diagrams are used to describe the functionality of the product. See Ref. 9 for common symbols used in FTA.

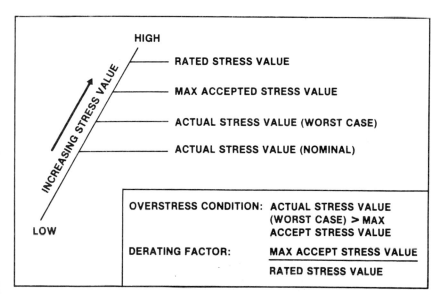

FIG. 10.8 Example of a derating factor determination.

One popular technique used in FTA is the "cause-and-effect" diagram [13]. Figure 10.9 is an example of a cause-and-effect diagram used in the analysis of a cutoff call in a telephone switching system. "Cause and effect diagrams are drawn starting from effects and moving toward causes" [35]. In Fig. 10.9, there are six major causes of a cutoff call. One of these, line card failure, is due to either a codec failure or a component failure.

Note that there will be a separate cause-and-effect diagram for each of the reliability and availability objectives listed in Table 10.3. Each objective will have associated with it a different set of failure modes. It is through a combination of

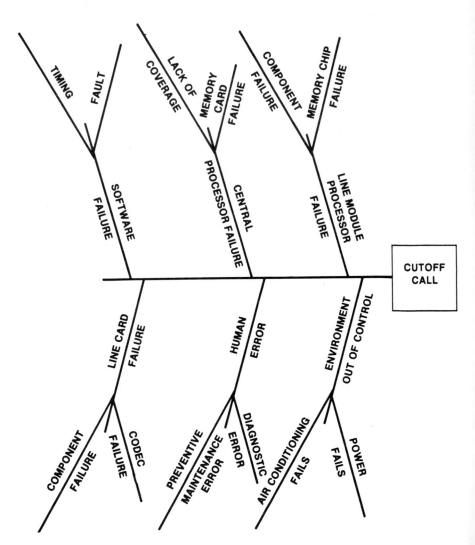

FIG. 10.9 Cause-and-effect diagram for cutoff calls.

failure mode analysis and fault-tree analysis that the complete reliability and availability model of the switching system was developed.

10.3.9   Failure Rate Predictions

Task 102 of Military Standard 1629A [7] includes the calculation of failure rates in the probability of occurrence calculations. The most common unit of failure rate measurement is FITS, or failures in a billion hours. The process starts with component failure rate predictions. Two basic references for the component predictions are Military Standard 217E, *Reliability Prediction for Electronic Systems* [8] and the Bell Communications Research Reliability Prediction Procedures (RPP) [14]. A manufacturer should also consider developing its own component failure rate data base. Other methods of component failure prediction are described in Ref. 17.

The next step is estimation of unit (i.e., circuit pack) failure rate predictions. Three methods of unit prediction are defined in the RPP: (1) the parts-count method, (2) the laboratory method, and (3) the field-tracking method.

The parts-count method in the RPP is based on the Military Standard 217E [8] procedures. This is a method in which the components on a unit are assumed to be in series and the failure rate of a unit is calculated as the sum of the individual component failure rates. The component failure rates are estimated as the product of the following terms: (1) an environmental factor, (2) a quality factor, (3) a stress factor, and (4) a temperature factor. Unit failure rates are calculated as the sum of the associated component failure rates. Figure 10.10 is an example from Ref. 14 of a unit consisting of 10 bipolar integrated circuits (ICs), 5 MOS ICs, 5 transistors, 5 capacitors, and 1 light-emitting diode (LED). The environmental factor (π_E) is 1.5, the stress (π_S) and temperature (π_T) factors are 1.0 for all components, and the quality factor (π_Q) is as shown. The resultant failure rate prediction is 2333 FITS (failures in 10^9 hours). A similar analysis was done on the memory card of the local switching system with a resultant estimate of 1000 FITS. This translates into a $\lambda_p = 10^{-6}$ as shown in Fig. 10.7.

The laboratory method of prediction is based on a properly designed reliability test program. The program must take into account the infant mortality period. This method also uses the parts-count method results as an input. The estimate is a weighted average of the parts count prediction and the laboratory prediction. Reference 36 describes the analysis which led to the development of the Bayesian estimator used in this method.

The third method, field tracking, is based on the statistical analysis of in-service failures obtained in properly designed tracking studies. The field-tracking studies should be run according to the criteria described in Ref. 16. Key elements in these procedures are inventory control and traceability. Failed units are returned to a repair location where the failure modes are analyzed. A proper field-tracking system must identify the removed components and be capable of determining the failure mechanisms. In many instances field-tracking studies are very expensive and other means should be used to estimate failure rates.

Figure 10.11 illustrates a typical field-tracking study. In this figure, a hardware replacement results in a trouble ticket being sent to reliability analysis, while the hardware is returned to repair. The failure modes are determined at repair, with the test results being sent to reliability analysis. Other information may also be used to characterize system troubles. The characterization results in improvements to the product, and the field reliability study can be used to track these improvements.

				DATE 1-1-84	PAGE 1 OF 1	
	WORKSHEET			UNIT EXAMPLE 1	MANUFACTURER XYZ INC.	
DEVICE TYPE	PART NUMBER	CIRCUIT REF SYMBOL	QTY (N_I)	FAILURE RATE (λ_{G_i})	QUALITY FACTOR (π_{Q_i})	TOTAL DEVICE FAILURE RATE* $(N_I\, \lambda_{G_i}\, \pi_{Q_i}\, \pi_{S_i}\, \pi_{T_i})$
IC BIPOLAR NON HERM 30 GATES	A65BC	U1-10	10	10	4	400
IC MOS NON HERM 200 GATES	A73X4	U11-15	5	40	4	800
TRANSISTOR SI PNP PLASTIC ≤ 6W	T16AB	Q1-5	5	25	2	250
CAPACITOR	C25BV	C1-5	5	8	1.5	60
CERAMIC LED, PLASTIC	L25X4	CR-1	1	15	3	45
SUBTOTAL						1555

TOTAL = (λ_{SS}) = $\pi_E\, \Sigma\, N_i\, \lambda_{G_i}\, \pi_{Q_i}$ = (1.5) (1555) = 2333   * $\pi_{S_i} = \pi_{T_i} = 1$

FIG. 10.10   Example of a unit failure rate calculation.

10.3.10 System Reliability and Availability Predictions

Once the failure modes are identified, the unit failure rates are estimated, and the fault-tree analysis is completed, the system reliability and availability parameters may be estimated. Reliability and availability models are discussed in most texts on reliability. References 9 and 17 give the basic general background and provide ample references to more advanced concepts.

There are two types of reliability models: (1) basic models and (2) complex models. Two assumptions are generally part of these models: (1) software is perfectly reliable, and (2) the human element is perfectly reliable. The basic model considers the units to be part of a simple series structure. The reliability of a

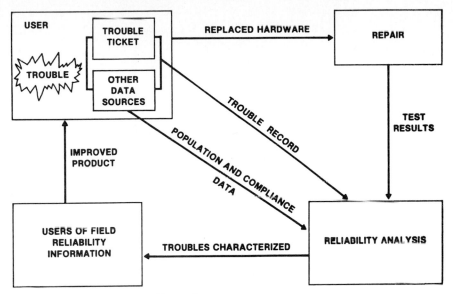

FIG. 10.11 A typical field-tracking study.

simple series structure is just the product of the individual unit reliabilities. The basic model is used to estimate maintenance demand and logistic support and forms an input to the complex model.

The complex model considers the design in terms of the ability to complete the mission successfully. In a complex model, the system is considered to be a complex series-parallel system. Calculations are made using the results of the basic models as inputs. For example, the formula for calculating the reliability of the system shown in Fig. 10.12 is:

$$P_S = (2P_C - P_{C2})P_A + [2P_BP_C - (P_BP_C)^2] (1 - P_A)$$

where P_A, P_B, and P_C are the reliabilities of units A, B, and C.

Availability is defined as the probability that the system is operating at any random time. The term "operating" in this definition means "has not failed." Techniques for calculating availability fall into two main categories: (1) analytical, and (2) Monte Carlo. The most commonly used analytical technique is the Markov chain method described in Ref. 18. Monte Carlo simulation methods are used when the system architecture is too complex for the analytical methods. They are also very useful in analysis of systems that have multiple missions. Commercial programs can be used to do the system reliability and availability prediction functions [19, 20].

The reliability and availability estimates should be updated during the design phase as an understanding of the system improves or as the design changes. The procedure should continue throughout the life of the product, with the goal being improvement in the reliability and availability of the system. A formal program to accomplish this is called *reliability growth management*.

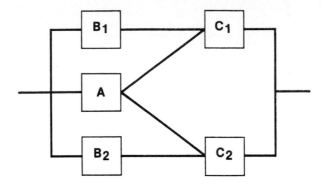

$$P_{B1} = P_{B_2} = P_B$$

$$P_{C1} = P_{C_2} = P_C$$

$$P_S = (2P_C - P_C^2)\,P_A + \left[2P_B P_C - (P_B P_C)^2\right](1 - P_A)$$

FIG. 10.12 Example of a system reliability calculation.

10.3.11 Reliability Growth Management

Reliability growth is a positive improvement in a reliability parameter(s) over a period of time due to changes in product design and the manufacturing process [21]. Two common procedures used in reliability growth are screening components and burn-in procedures [33].

A number of growth models have been proposed. The most common of these is this Duane method. O'Connor [9] provides background on this method. Equation (9.4) of that text gives an equation for the cumulative mean time between failures (MTBF) during a growth program. The cumulative MTBF (Θ_c) can be calculated by dividing total time by total failures. An equation for Θ_c in terms of parameters Θ_0, T, T_0, and α is:

$$\Theta_c = \Theta_0 \left(\frac{T}{T_0}\right)^{\alpha} \qquad (10.1)$$

where Θ_0 is the cumulative MTBF at the start of the monitoring period T_0, T is the time at which the calculations are made; and α is a measure of the rate of MTBF growth. The equation can be used in two ways: (1) to estimate the rate of growth, or (2) to estimate the time required to attain a particular MTBF at a given or estimated rate of growth. See Refs. 9 and 21 for the details of the Duane model and other models of reliability growth.

An example of the use of the equation for Θ_c is given in Ref. 9; the instantaneous (MTBF, Θ_i) is shown to be related to Θ_c by the equation

$$\Theta_i = \frac{\Theta_c}{1 - \alpha} \tag{10.2}$$

In the example we want Θ_i to be not less than 500 hours. For an $\alpha = 0.5$, $\Theta_c = (1 - \alpha) \Theta_i = 250$. During the first qualification test we observe 11 failures in 600 hours. How much testing should be planned, assuming $\alpha = 0.5$?

Solving Eq. (10.1) for T and substituting in the above values, we get Θ_0 equal to $600/11 = 54.4$ hours, and

$$T = T_0 \left(\frac{\Theta_c}{\Theta_0} \right)^{1/\alpha} = (600) \left(\frac{250}{54.4} \right)^{1/0.5} = 12{,}674 \text{ hours} \tag{10.3}$$

Hence, we must run the test for another 12,674 hours to get the desired MTBF.

10.3.12 Human Factors Analysis

Human factors analysis covers a wide range of issues which can be generally categorized into three subjects: (1) safety considerations, (2) workmanship and physical design characteristics, and (3) maintainability considerations. Safety considerations are important during the entire life cycle of the product. All safety problems should be considered critical and should be dealt with as soon as they are discovered. Workmanship problems affect the production phase and may lead to long-term reliability problems. Maintainability affects the deployment phase. Chapter 5 of Refs. 22 and 23 provide discussions of human factors analysis and give references to other publications on the subject.

10.3.13 Safety Considerations

The starting point for the development of a safety program is the FMECA described previously. Failure modes that affect safety of personnel and equipment are critical and should be dealt with by providing adequate safety margins which will reduce the likelihood of failure. O'Connor [9] describes load-strength analysis and safety margins. Also, considerations should be given to controlling the environment such that the probability of high loads on the system are extremely small.

10.3.14 Workmanship and Physical Design Characteristics

The second human factors area that must be considered in the specification of reliability requirements is that of workmanship and physical design of the parts of the system. Can the part be manufactured consistently to requirements, or will there be difficulty in meeting the workmanship and physical design characteristics? This is especially important in the electronics industry in relation to the manufacture of circuit packs. A sample of characteristics which affect the long-term reliability of circuit packs are insulation resistance, conductivity, solderability, and high-voltage breakdown. Most major manufacturers have developed workmanship standards which are used on the production line to prevent defects from showing up as long-term reliability problems. The product assurance task force should review these standards and their application to the product to ensure that there will not be long-term reliability problems due to workmanship or physical design. Reference 24 is an example of published physical design characteristics.

10.3.15 Maintainability

Maintainability analysis is the keystone of the human factors program. New complex electronic equipments are being designed with built-in test capabilities. Menu-driven programs provide on-line analyses aimed at fast evaluation, quick correction of problems, and return to full service. The trend is to require designers to include good test elements in their hardware and software designs.

Maintainability analysis encompasses more than the testing of the system. It also includes the associated logistics and the economics of maintenance. Systems may be maintained by on-site personnel, by remote analysts, or by roving maintenance crews. The availability of the system is a function of the speed of maintenance actions. Hence, various scenarios should be investigated to determine the trade-off between availability and cost of maintenance. For example, clusters of small local telephone switching systems can be maintained from a central location. Each site would be unmanned. Hence, maintenance analyses must include travel time to the site.

A preventive maintenance program is an essential part of the overall maintenance of a system. Planners must consider this during the system design. In complex electronic systems, time must be allocated to run diagnostics. Often this is done in the background during system idle periods. For example, the coverage problem occurs when a single fault in a duplex environment causes system failure when a working component is available. This event may occur when diagnostics are not run frequently enough to detect the failure and take action to make the duplex component the active component.

A comprehensive maintenance analysis and demonstration program is described in Refs. 25 to 27.

10.3.16 Reliability Demonstration

The product assurance task force must be concerned with the demonstration of reliability characteristics during all phases of the product life cycle. The most effective means of monitoring reliability is to include it in the periodic design reviews. Reference 5 lists items that are considered essential in review of complex systems. These include reliability growth management, sneak circuit analysis, electronic parts or circuit tolerance analysis, and reliability monitoring of critical items. The most effective means of demonstrating the reliability of a system is to run reliability tests on prototype equipment. Military Standard 781C [28] defines qualification procedures assuming an exponential distribution of failure occurrence.

The major objective of the tests is to support the long-range goal of user satisfaction with a minimum expenditure of resources during development and production. The test design is a compromise between accuracy of the estimates,

TABLE 10.6   Elements of a Good Software Reliability Program

1. Software life cycle plan	6. Configuration management
2. Management commitment and organization	7. Problem reporting and corrective action
	8. Data collection, analysis and use
3. Development support environment	9. Customer engineering and operations
4. Documentation	
5. Verification and validation procedures	

the high cost of development, and the achievement of customer satisfaction. Two numbers are generally specified for each test, the design target and the minimum acceptable value. A test of hypothesis is set up with associated error risks and sample sizes. Reference 28 defines the procedures and gives acceptance criteria for the various tests. Unfortunately, these procedures do not take into account distributions other than exponential. Reference 15 defines one method of reliability demonstration for systems that experience nonexponential failures. In that reference, infant mortality failures are modeled using the Weibull distribution.

The most comprehensive demonstration tests are those run during the actual operation of a prototype system. For example, Ref. 29 describes the field demonstration of gas-insulated substations. Also, the reliability demonstration test described in Ref. 15 has been used on working telephone switching systems during installation audits. The reliability demonstration test is designed to assure that the system has matured (in reference to design) and aged sufficiently to be put into operation without further reliability testing.

10.3.17   The Software Quality and Reliability Program

It is very common today for products to be software-controlled. In this section we discuss software requirements aimed at producing highly reliable products. Many of the previously stated functions also apply to software. There are, however, good software practices which if applied consistently should result in a highly reliable product. Reference 34 describes the elements of a good software program. These elements are listed in Table 10.6.

The first three elements, software life cycle plan, management commitment and organization, and development support environment are fundamental to the quality program. This is especially true of the support environment, which should include modern facilities, equipment, methodology, training, and tools. The remaining six elements build on the foundation of the first three.

Software should be developed in pace with the hardware. Often, because it is difficult to measure software quality and reliability, a product is deployed with bugs in the software. This may lead to constantly changing software and never-ending system problems. In these cases, good software control practices are essential.

10.4   CONTRACTUAL PROVISIONS

This section contains a list of generic quality and reliability provisions that should be considered for inclusion in contracts. The result of including these provisions will be coverage of the more important areas of potential quality and reliability problems and the development of mechanisms to achieve resolution of those problems should they arise. Insertion of these provisions into contracts will lead to identification and elimination of reliability problems during phases 3 and 4.*

The generic provisions listed below fall into four main categories covering: (1) the factory and initial service, (2) reliability concerns, (3) software quality, and (4) customer support. Provisions used for an individual product should be tailored for that product's needs. The customer and supplier must weigh the cost of these provisions against the quality and reliability benefits they will provide. The development of a contract between supplier and customer may take place at any time during the design phase. The earlier it takes place, the greater the customer's influence. The timing will affect the provisions that the agreement will contain.

10.4.1 Factory and Initial Service

These provisions are intended to assure that products are designed and constructed to meet the customer's needs. They consist of the following: (1) facility review and quality program analysis,† (2) quality surveillance program, (3) submission of supplier inspection data, and (4) submission of first production samples. These provisions are considered to be quality-related and the question arises, "Why are they included in a chapter on specification of reliability requirements?" They are necessary because of the strong relationship between quality and reliability. Poor control of quality functions will lead to long-term reliability problems. Suppliers should provide evidence during the facility review that they have the capability to build a good quality and highly reliable product under the assumed demand requirements.

The contract should specify that the factories where components, modules, and the complete system are manufactured will be open to a quality program analysis (QPA) either by the customer or customer's agent. A QPA is an in-depth study of the supplier's manufacturing procedures. The purposes, objectives, and procedures of a QPA are described in Bell Communications Research Technical Reference TR-TSY-000039 [30]. Provision should be made for periodic QPAs as factory processes and product designs change. Table 10.7 lists the 14 elements of a quality program analysis.

Provision should be made for the customer or agent to perform routine quality surveillance to assure that the customer's standards are being met. This should consist of sampling and inspection of product, monitoring the supplier's quality control program, and review of the supplier's inspection data. Written sampling procedures should be part of any agreement.

The supplier should provide inspection data to the customer which will enable an accurate estimate of product quality to be made. Typically this means final inspection data on rejected as well as accepted lots. These should be used by the customer to reduce the frequency of inspection. First production samples should be provided to the customer or agent prior to the first shipment of the product.

---

* It should be noted that some of these provisions may be included in a request for proposal during the early part of the design phase. The goal of such an inclusion will be to assure that an adequate reliability plan is developed by the manufacturer.

† These should be included in the contract for two reasons: (1) These activities may not have been accomplished prior to contract signing, and (2) new facilities to be used at a later date should be subject to both activities.

TABLE 10.7 Elements of a Quality Program Analysis

1. Management commitment and organization	8. Control of nonconforming material
2. Documentation of the quality system	9. Storage, handling, and packaging
3. Control of design changes	10. Corrective action program
4. Control of procured material	11. Product identification
5. Manufacturing controls (in process)	12. Periodic product qualification
6. Completed item inspection	13. Quality information
7. Equipment calibration and maintenance	14. Collection and analysis of field performance data

The customer should test these samples to assure conformity to requirements. Documents should be provided prior to the signing of an agreement. These are listed in Table 10.8.

10.4.2 Reliability Provisions

The supplier should provide a description of the entire reliability program. This should include the reliability program plan (see Table 10.1) and a description of the design factors (see Table 10.2) considered during the development process. In addition, the supplier should provide documentation which summarizes the status of the reliability program at the time the contract is signed. Table 10.9 contains a list of documentation that should be provided to report the status of the reliability program.

10.4.3 Software Provisions

All stages of the software life cycle should be considered. In addition, software should be defined to include associated firmware. The three major elements of the software program are: (1) the software life-cycle plan, (2) the software manufacturing process, and (3) software error measurement.

A total software life-cycle plan should include a description of each phase, the organizational responsibilities, the interfaces, and release procedures. The impor-

TABLE 10.8 Quality Documentation

1. The quality manual	7. Incoming inspection procedures
2. Written quality control procedures	8. Vendor qualification and requalification procedures
3. Inspection and acceptance sampling procedures	9. Procedures for disposition of nonconforming material
4. Typical factory quality data	10. Calibration procedures
5. Data analysis procedures	11. Physical design and workmanship procedures
6. Final auditing procedures	

TABLE 10.9 Reliability Documentation

1. Reliability program plan	8. Safety analysis
2. Reliability design factors	9. Maintainability analysis
3. Design review history	10. System or subsystem reliability demonstration results
4. Supplier environmental test data	
5. Device burn-in procedures and data	11. Supplier field tracking procedures and data
6. Reliability and availability predictions (including failure mode and fault-tree analyses)	12. Human factors analysis
	13. Configuration and change control procedures
7. Reliability growth analysis	

tant mechanics of support should be described. These include source code control data bases, tools, methodologies, programming languages, and other support elements.

The supplier should describe the processes of manufacture, installation, maintenance, and support of the software. This should include a description of load creation, verification, validation of software in place, and methods of handling software changes and enhancements. The supplier should also describe procedures for predicting and measuring software reliability rates. Information should be provided on field error rates, history of software fixes, new release rates, and verification results. Table 10.10 summarizes the software documentation that should be provided prior to signing an agreement.

If a supplier was not subjected to a software quality program analysis (SQPA) prior to the signing of the contract, provisions should be included to have the supplier undergo the formal analysis. The elements of a SQPA were discussed previously in Sec. 10.3.17 and are listed in Table 10.6. Reference 34 should be used as a guide during the SQPA.

10.4.4 Customer Support Provisions

The supplier should provide support both before shipment and after service turn-on. Table 10.11 provides a list of the general procedures which should be provided as part of the product support services. Documentation describing each of the provisions should also be provided to the customer.

The quality and reliability review group should have representatives from the customer and the supplier. Its purpose is to resolve quality and reliability problems in the factory and the field. The group is also responsible for resolving

TABLE 10.10 Software Documentation

1. Specifications	6. Release statistics
2. Program listings	7. Verification test results
3. User manuals	8. Estimates of software reliability paramenters
4. Programming practices	
5. Error statistics	

TABLE 10.11 Customer Support Provisions

1. Engineering, ordering and additions	6. Repair procedures
2. Field tracking support	7. Technical specification
3. Establishment of a quality and reliability review group	8. Emergency assistance
	9. Installation assistance
4. Complaint assistance	10. Change control procedures
5. Warranty procedures	

long-term reliability problems and for providing for equitable compensation for unresolvable problems. An important tool for the group is an action register for tracking problems.

10.5 SUMMARY

The purpose of this chapter was to provide a comprehensive view of the reliability specification process. This process must consider the four phases of the life cycle in order to be effective. The keystone of the process is the reliability program plan. This plan should be developed during the conceptual and advanced planning phase and tracked throughout the life of the product.

The majority of the reliability specification effort is spent in the design phase. The main tools for use in understanding the reliability characteristics of a product are the FMECA and fault-tree analysis. The results of these analyses are used in the reliability prediction model to predict the system reliability and availability. A key factor in the model is the maintainability characteristics of the system. Finally, reliability growth management and reliability demonstration techniques are important tools used to improve the reliability of the product prior to deployment.

Each of the factors mentioned in this chapter plays a role in the proper specification of reliability requirements. In addition, it is important that certain provisions be clearly spelled out in contractual agreements so that the supplier and the customer will coordinate their efforts to bring high-reliability products into the marketplace.

10.6 BIBLIOGRAPHY

There are many good sources of material on specification of reliability requirements. The primary ones come from the military and other governmental agencies. Military Standard 721C [32] provides definitions of reliability and maintainability terms. The reader should also look at Military Standard 785B, *Reliability Programs for Systems and Equipment* [1]. This document discusses the tasks needed to develop a viable reliability program. In using this document, the planner should understand the needs of the product and tailor the requirement tasks to those needs.

Two extensions of Military Standard 785B are (1) QR-800-S, developed by the Army Missile Command, and (2) the Department of the Army Pamphlet 702-4 [5, 22]. The latter gives a good synopsis of other useful military publications.

Two textbooks give a good overview of the reliability specification process [9, 31]. The book by O'Connor [9] is especially recommended.

Military Standard 1629A [7] provides procedures for performing FMECA, which is an excellent tool for understanding the reliability characteristics of a system. Associated with this standard are three other major standards which should be part of the designer's bookshelf. Military Standard 756B [17] defines reliability modeling and prediction procedures. Military Handbook 217D [8] provides procedures and data for prediction of failure rates for standard components. The Bell Communications Research *Reliability Prediction Procedures* [14] contains similar information for telecommunications products. There are other publications which are of use to the designer for specific purposes. These are listed in the references.

Acknowledgments

Many people provided inspiration and inputs to this chapter. I would like to thank my colleagues at Bell Communications Research Quality Assurance Technology Center, the ANSI Z-1 Committee on Quality Assurance, and The U.S. Army Combat Surveillance and Target Acquisition Laboratory of Fort Monmouth, New Jersey.

REFERENCES*

1. Military Standard 785B, *Reliability Program for Systems and Equipment*, Department of Defense, September 15, 1980.
2.. Bell Communications Research TR-TSY-000064, LATA Switching System Generic Requirements, Issue 2, July 1987.
3. Military Standard 810, *Environmental Test Methods and Engineering Guidelines*, United States Air Force, Wright-Patterson AFB, Ohio, July 19, 1983.
4. Donald L. Denton, "Guidelines for Establishing an Effective ESD program," *Solid State Technology*, January 1984, pp. 127–131.
5. QR-800-S, *Reliability Program for Systems and Equipment Development*, U.S. Army Missile Command, September 29, 1982. For a copy send your request to the attention of DRSMI-QRA, Redstone Arsenal, Alabama 35898.
6. Military Standard 965, *Parts Control Program*, Department of Defense, Washington, D.C., February 16, 1981.
7. Military Standard 1629A, *Procedures for Performing a Failure Mode, Effects and Criticality Analysis*, Department of Defense, Washington, D.C., November 24, 1980.
8. Military Handbook 217B, *Reliability Prediction for Electronic Systems*, Department of Defense, Washington, D.C., October 1986.
9. Patrick D. T. O'Connor, *Practical Reliability Engineering*, John Wiley and Sons, New York, 1981.
10. M. S. Gover, "Evaluation of the Worth of Equipment Reliability in the Design of

* U.S. Military Standards and handbooks may be ordered from Commanding Officer, Naval Publications and Forms Center, 5801 Tabor Avenue, Philadelphia, PA 19120. Information concerning the ordering of Bell Communications Research publications can be obtained from Bell Communications Research Information Operations Center, 60 New England Avenue, Piscataway, NJ 08854, (201) 699-5800 (ask for latest issue of the documents).

Generating Stations," *Proceedings of the 1982 Engineering Conference for the Electric Power Industry,* American Society for Quality Control, 1982, pp. 31–39.

11. W. E. Klein, "Experience with Modified Aerospace Reliability and Quality Assurance Method for Wind Turbines," *Proceedings of the 1982 Engineering Conference for the Electric Power Industry,* American Society for Quality Control, 1982, pp. 232–237.

12. F. J. Brinkmiller and F. C. Coogan, "RAMS Prediction—Electrical Products and Systems," *Proceedings of the 1983 Engineering Conference on RAM for the Electric Power Industry,* Institute of Electrical and Electronic Engineers, New York, 1983, pp. 334–338.

13. Dr. K. Ishikawa, *Guide to Quality Control,* Asian Productivity Organization, Minato-kw, Tokyo, 107, Japan, 1976.

14. Bell Communications Research TR-TSY-000332, Reliability Prediction Procedure for Electronic Equipment, 1986.

15. G. G. Brush, B. S. Liebesman, and W. J. Martin, "Assuring Age Requirements of a Complex System," *Proceedings of the 1983 Reliability and Maintainability Symposium,* 1983, pp. 71–77.

16. Bell System Publication 10400, *Field Tracking Study Handbook,* American Telegraph and Telephone Co., Basking Ridge, N.J., 1979.

17. Military Standard 756B, *Reliability Modeling and Prediction,* Department of Defense, Washington, D.C., November 18, 1981.

18. F. S. Hillier and G. J. Lieberman, *Introduction to Operations Research,* 3d ed., Holdin-Day, San Francisco, 1980.

19. "Predictor/Results Product Summary," Management Sciences Incorporated, Albuquerque, N.M. 87110, May 1983.

20. "217 Predict: The Computer Program to Reduce Prediction Costs," Syscon Corporation, Sunnyvale, CA, 1983.

21. Military Handbook 189, *Reliability Growth Management,* Department of Defense, Washington, D.C., February 13, 1981.

22. ERADCOM Pamphlet 702-4, "Product Assurance Reliability, Maintainability, and Human Factors," Department of the Army, Electronic Research and Development Command, Adelphi, MD 20783, October 6, 1980.

23. Gerald G. Silverman, "Human Performance Reliability in Man-Machine Systems," *Quantity Progress,* September 1979, pp. 22–25.

24. Bell Communications Research TR-TSY-000078, *Physical Design Requirements for Bell Operating Company Equipment,* Issue 1, 1986.

25. Military Standard 470A, *Maintainability Program for System and Equipment,* January 3, 1983.

26. Military Standard 471A, *Maintainability Demonstration,* Department of Defense, Washington, D.C., December 8, 1978.

27. Military Handbook 472, *Maintainability Prediction,* Commander Naval Air Systems Command, Department of the Navy, Washington, D.C., May 24, 1966.

28. Military Standard 781C, *Reliability Design Qualification and Production Acceptance Tests: Exponential Distribution,* Department of Defense, Washington, D.C., October 21, 1977.

29. J. J. Dodds, "Design and Reliability Improvement of Gas-Insulated Substations through Ten Years of Operating Experience," *Proceedings of the 1982 Engineering Conference for the Electrical Power Industry,* American Society for Quality Control, 1982, pp. 248–255.

30. Bell Communications Research Technical Reference, TR-TSY-000039, *Quality Program Analyses,* Issue 2, 1985.

31. C. O. Smith, *Introduction to Reliability in Design,* McGraw-Hill, N.Y., 1976.

32. Military Standard 721C, *Definition of Terms for Reliability and Maintainability,* Department of Defense, June 12, 1981.

33. F. Jensen and N. E. Petersen, *Burn-in,* J. Wiley and Sons, New York, 1982.

34. Bell Communications Research TA-TSY-000179, *Software Quality Program Analysis,* Issue 1, 1985.

35. E. Kindlarski, "Ishikawa Diagrams for Problem Solving," *Quality Progress,* vol. 17, no. 12, December 1984, pp. 26–30.

36. G. G. Brush, J. D. Healy, and B. S. Liebesman, "A Bayes Procedure for Combining Black Box Estimates and Laboratory Tests," *Proceedings of the 1984 Reliability and Maintainability Symposium,* 1984, pp. 242–246.

CHAPTER 11
PRODUCT RELIABILITY PROGRAM

Charles R. Miller
Program Manager, Evaluation Engineering,
Carrier Corporation

11.1 INTRODUCTION

The concept of a product reliability program, as discussed in this chapter, may be applied to fully specified designs, commercially marketed products, or consumer products. For the purposes of this discussion, a "commercial" product is one that is designed for end use in the private sector, with a significant degree of design specification and control by the end user. This definition, although somewhat broad, should serve to distinguish commercial products from mass market "consumer" products which receive little, if any, direct design specification input from their ultimate users, as well as from fully specified military and aerospace designs. A commercial design may be for an actual end product, such as a piece of industrial equipment, or it may define a sublevel assembly or component intended for application in a higher-level design of an original equipment manufacturer (OEM). Commercial product designs tend to fall somewhere between the fully specified design and a consumer product in terms of cost targets, design process control, customer review authority, and other related characteristics.

The above description is not intended to imply that any of these product categories have inherently better product quality for their intended task. It does mean, however, that the design process for each is different and that the techniques for achieving a highly reliable product differ in each type of program. These are due primarily to differences in the nature and degree of risk involved in the event of a product failure. For example, the potentially catastrophic result of a system failure on a tactical aircraft system dictates that the system in question be designed for failure-free performance over a critical, fixed-length mission duration, with system cost being only a secondary consideration. On the other hand, an industrial control system may need to be designed within rigorous cost constraints in order to have a reasonable chance for success in the marketplace. In this case, the product quality and reliability must be maximized within the limits of such cost targets. Nonetheless, a serious product defect in the commercial design could have disastrous consequences for both the end user and the manufacturer.

For the commercial product designer who is already caught in the cost vs. reliability "squeeze," the challenge is intensified by present industry trends, which include rapidly escalating customer expectations of product quality and the growth of competitive strategies based on reliability and quality issues. This situation is compelling an increasing number of companies to introduce more structured design reliability programs and to invest in the resources required to support such programs. The first challenge that any such company typically faces is trying to identify the most viable and cost-effective reliability organization and reliability methodologies which can be applied to its products. The balance of this chapter is devoted to a review of the factors which bear on the commercial manufacturer's decisions about such programs, as well as a description of some of the techniques which have proved to be effective.

An overview of a well-balanced design reliability program for commercial products is shown in Fig. 11.1, as it relates to typical concurrent activities in Engineering, Manufacturing, and Marketing. Although the details of such a program will vary from product to product, the five primary areas of reliability activity need to be addressed in an effective manner. As can be seen from Fig. 11.1, these primary activities include:

1. *Specifying* the product properly
2. *Designing*-in the product reliability
3. *Qualifying* the design (i.e., verifying that it meets the accepted specification)
4. *Maintaining* the inherent design reliability throughout the manufacturing phase
5. *Auditing* the results in the field and adjusting the process as needed

Each of these areas will be further discussed below.

11.2 DEFINITIONS

Manufacturing check sample: A small sample of a new product, which is built by the designated manufacturing site using released engineering drawings to verify that the product can be manufactured to the drawings. This limited build may or may not use final production tooling and processes.

Manufacturing pilot run: A limited-quantity run of a new product, using final processes and tooling intended for full production. Pilot run samples should be fully representative of production units.

Qualification plan: A comprehensive plan detailing how a new product design or revision will be evaluated and/or tested to verify conformance to the product specification.

Qualification review panel: An independent, cross-functional body of program evaluators whose primary tasks are to approve the product qualification plans, monitor the execution of those plans, evaluate project risks, and determine the readiness of the product for production release.

11.3 BUSINESS FACTORS TO CONSIDER IN PLANNING THE RELIABILITY PROGRAM

Before exploring the practical steps to be taken in constructing the product reliability program, it is appropriate to consider some of the business-related forces which may bear on the problem. While these factors will vary in different

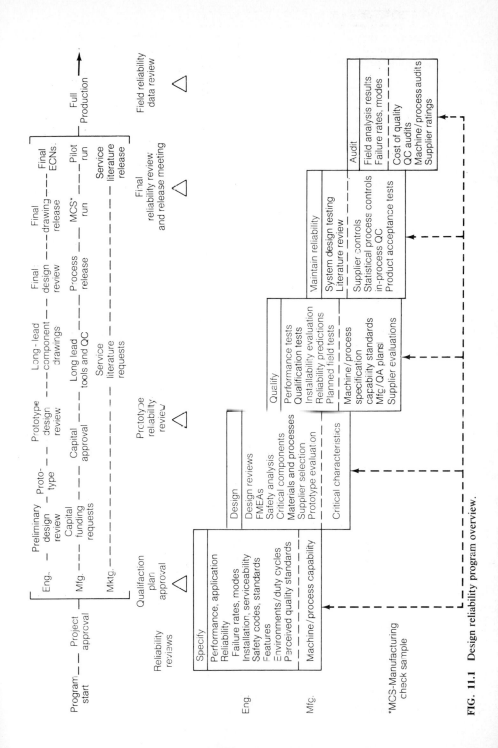

FIG. 11.1 Design reliability program overview.

11.3

businesses, they all figure into the equation. They form the environment in which the reliability manager must function—and any new product development program must take them into account.

11.3.1 The Reliability "Balance Point"

As with any type of design, the ultimate specification for a commercial product design is to meet the customer's needs. Assuming a competitive marketplace, this means that a successful commercial design must include consideration of several factors, including:

- Cost (nonrecurring and recurring)
- Reliability and safety
- Functionality
- Producibility
- Serviceability
- Program schedule

These attributes must be optimized in the proper combination and even traded off against one another in order to achieve a successful design. With the exception of product safety, no one attribute can normally dominate the design at the expense of all others. Most typically, a conflict may arise between project costs and reliability, or between project schedule and reliability. For example, a manufacturer's market strategy requirements may dictate that a new product introduction be priced below a certain maximum in order to capture an increased market share. Product success will depend heavily on careful control of both nonrecurring development costs and production costs. The reliability group's task in this case is to devise a design reliability program which maximizes the effectiveness of engineering and testing resources which are probably under a predetermined limitation. In another case, the manufacturer of a seasonal type of product must have its development program completed by a fixed date or the entire selling season could be lost. Although cost is still an important factor in this case, the reliability group must place its heaviest emphasis on "front loading" the design reliability program to achieve the highest possible confidence in the new design very early in the project schedule. This is absolutely necessary to allow recovery time in the event that design defects are found. In the most common case, *both* cost and schedule pressures are significant factors and the reliability manager's most important planning task is to devise a program which will give the customer the most reliable product without burdening the project with excessive cost factors and/or unrealistically long and complex testing schedules. Finding this reliability "balance point" is critical to product success. Even the best design reliability program will not do much for the OEM manufacturer if the customer selects a competitor's product because of price or availability. The customer may truly want increased functions and features in a new product, but how much more is he or she willing to pay for it? And how much potential reliability degradation is the customer willing to tolerate to get the increased complexity? All these factors must be weighed on each program to determine the optimum reliability program requirements.

11.3.2 Management Philosophy and Commitment

This element is key to the success of any newly implemented reliability program. General management may well believe in the overall virtue of "product quality," but frequently lacks an awareness of what this translates into in terms of specific design and production tasks. As a practical matter, one of the first tasks of the new reliability engineer or manager is to raise that level of management awareness. Unfortunately, it is a characteristic of American industry that many company managers are not engineers by training, and therefore, they usually have neither the time nor the technical background to try and understand all the details of various reliability tasks. It is the reliability manager's job to relate the value of the various elements of the program to management in terms of the potential return on the investment that management is being asked to make.

This approach acomplishes at least two important things. First, it helps convince management that the reliability department is as cost conscious and as concerned about return on investment as anyone else in the company. This helps establish future credibility. Conversely, if the reliability group becomes perceived (justifiably or not) as a bunch of zealots who are always obstructing critical projects, credibility with management is lost, the real influence of the reliability manager is diminished, and the company may soon resort to using compromises and expedients to get products "out the door." The second benefit of successfully raising management's awareness is that the very process of justifying the reliability engineering investment forces the company to examine the true life-cycle costs of putting a product into the marketplace—including all of the "hidden" costs of poor product quality. This is a healthy exercise for all concerned and the resources required for a creditable reliability program can normally be justified on this financial basis alone, without having to resort to ideology.

Figure 11.2 shows quite vividly how the corrective action costs on a project accelerate dramatically as the project progresses. Although this is only a conceptual curve, it is based on industry data from both military and commercial programs and its logic should be self-evident. A defect identified in early design analysis activities is far less costly to correct on paper than is the design flaw that

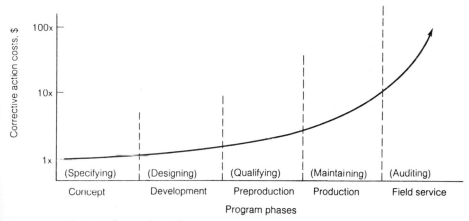

FIG. 11.2   The cost of corrective action.

evades detection until the investments in prototype hardware and test resources have been made. In addition, for defects which are not identified until the production and field service phase, the costs of correction can easily be orders of magnitude higher because of warranty charges, service labor costs, and obsolete inventory costs. In most cases, such costs come right off the bottom line of a company's profit and loss sheet.

The ultimate measure of management's commitment to reliability is found, of course, in how the company deals with the hard decisions about whether to put a deficient design into production in order to hold schedules. If the reliability manager has done a good job of management awareness training and does a proper risk visualization for that product (see Sec. 11.4.2, "Risk Management"), the right decisions will hopefully be made.

11.3.3 Technical Business Factors

In addition to understanding basic elements of the business environment, such as customer needs and management commitment, a successful commercial product reliability program must be developed around other technical factors in the particular business' marketplace. These factors may include, among others:

- *Warranty practices:* Certain standardized warranty periods and conditions may dominate a particular industry or product category. These warranty practices and their associated budgets should be a factor in establishing such product reliability requirements as design life, fault tolerance, and expected duty cycles. Conversely, advances in reliability technology can frequently yield competitive advantages, as high-reliability manufacturers become able to offer *extended* warranty protection.

- *Product cost:* In many industries, the sensitivity of the product to increases in selling cost cannot be overstated. In the so-called low-end products (whose sales margins are small), increases in selling cost can literally take a product out of competition. Development programs for such products must be carefully controlled and reliability programs tailored to meet the need. In "high-end," deluxe products, larger sales margins may permit a more intensive reliability program investment, and the product quality requirements may, in fact, demand such efforts. Such programs can often provide opportunities to prove out a new reliability engineering method which could be more broadly applied on subsequent projects.

- *Share-of-market and competitive characteristics:* In some industries, competitive strategies are based on gaining a larger share of the established market. Also, in more and more industries, such as automotive and electronics, product quality and reliability have become major competitive issues. In both of these situations, demonstrated high reliability can offer a major competitive advantage in terms of product differentiation.

- *Schedule considerations:* These factors can have an enormous impact on any development program. Obviously, certain businesses must complete new product development projects in time to satisfy seasonal market demands, e.g., automobiles, heating and air conditioning equipment, farm machinery, and construction equipment. In such programs, design defects detected late in the project could force redesigns and other corrective actions whose attendant production lead times and delays could take the company out of the marketplace for a particular product for a year or more. Such situations demand good project planning and front loading of evaluation activities in order to develop high confidence in meeting the schedules.

- *Product volume:* High-volume products present some unique challenges because the risks of financial exposure in the event of field failures are much greater. On the other hand, although field support of a low-volume product may be potentially less costly, the unit cost and complexity may severely limit the number of samples available for preproduction testing and product verification. Here again, program emphasis on front-end design evaluation makes great sense.

- *Product safety and agency approvals:* Product safety should be the paramount concern of any new product design or evaluation program. However, like reliability attributes, safety attributes need to be designed into the product through specific activities and disciplines. Hence, the model design evaluation program described below includes some key safety review and hazard analysis activities which are intended to complement and support the efforts of the design engineers. No reputable company intentionally designs products with potential safety problems, but some of the more subtle safety issues are often only detected through a rigorous third-party evaluation. Beyond the obvious ethical issues, the costs of correcting safety problems which actually reach the customer can be devastating, particularly in product recall situations or where government regulatory agencies become involved. The cost avoidance potential alone can usually provide ample justification for expenditures of resources on thorough safety analysis activities during design.

- In many industries, various U.S. and overseas product approval and standards-writing agencies (e.g., Underwriters Laboratories, American Gas Association, ASME, DOT, etc.) are required to accept a new design before it goes to market. These government and industry organizations are often involved in both standards writing and testing. They are also usually very busy, and such delays can be a real source of frustration to a project manager who is trying to meet a scheduled product introduction date. For a new product program with thorough and *well-documented* reliability and safety evaluation activities, there exists a real opportunity to reduce the overall time required to obtain such approvals, because the approval organization will frequently accept the company's analysis and test results as part of their own evaluation.

- *The after-market:* Once a new commercial product has gone into production, it must be properly supported with reliable parts and field service. Indeed, in some industries, the service business represents a significant proportion of overall corporate income. Although this may sometimes be viewed as an indictment of product quality, the fact is that manufacturers owe their customers high-quality support throughout the life of the product. Many of the elements which make up a high-quality parts and service program must be factored into the design process early in the program. The overall service and maintenance strategy must be accommodated by the unit design—with all necessary accesses and test points identified. Any special diagnostic tools must be designed concurrently with the product and service instructions prepared and reviewed. Designated service parts drawings need to be prepared and suppliers qualified. It should be self-evident that the reliability of service parts must equal or exceed that of original equipment parts. All these issues need to be factored into the design evaluation program.

11.4 ORGANIZATIONAL FACTORS TO CONSIDER IN PLANNING THE RELIABILITY PROGRAM

Although every manufacturer's organization will differ in some respects, certain basic functions are required to develop a product, put it into production, and support it in the field. Without attempting to advocate a particular organizational structure, it is safe to assert that most businesses are organized around the following activities:

- A *marketing and sales* organization to determine market opportunities, define product needs, and sell the product. Marketing is typically the internal group representing the final customer. This group may also include a strategic product planning function to look at the longer-range product strategies the company should undertake. For new products, the marketing group should define the product specifications.

- An *engineering* department to execute the design and development work required to bring the new product specification to fruition, including designing, drafting, development testing, and prototype construction. The engineering department's "products" are drawings and specifications.

- A *reliability* department (which may or may not report into the engineering organization). The role of the reliability group can vary widely in different organizations. Reliability engineers are customarily concerned with maximizing the reliability and quality of their products. This effort usually divides into two areas, providing technical reliability engineering support to the development engineers and performing a more objective evaluation or audit of the design. For the purpose of this discussion, we will call the former "reliability engineering," and the latter "evaluation engineering." Both are critical activities, but it is important to remember that while reliability engineering is deeply involved in helping to design the product, evaluation engineering is best organized as an independent and objective function, separate from the design group.

- A *manufacturing* department to receive the released engineering drawings and specs and to build a product in total conformance to the specifications

- A *quality assurance* group to support and audit the manufacturing department in order to ensure that the as-built version of the product meets 100 percent of the specification requirements

- A *field service* department to provide direct after-market customer support in the field

In addition, there are typically several other key organizations involved in fielding a new product. These may include purchasing, transportation, distribution, and other departments. The point to keep in mind here, however, is that regardless of the administrative reporting relationships, a successful product introduction depends heavily on these groups taking proper "ownership" of their part of the process. In effect, a series of "customer" relationships must be established within the organization, as depicted in Fig. 11.3.

Although marketing personnel do represent the ultimate customer, Marketing must also see itself as a "supplier" of complete and accurate product specifications to Engineering. Engineers, in turn, must see Manufacturing as one of their customers, ensuring that released drawings and specifications are equally complete and accurate—and are delivered in a timely manner. Manufacturing is, of course, a customer to Engineering and a supplier to the field—and, of course, to the ultimate paying customer. This is a very simplistic description. There are a multitude of other internal customer-supplier relationships involved in product development, but the concept is important in the structuring of an effective reliability evaluation program.

11.4.1 Resource Allocation

Many of the organizational issues encountered in product development programs relate to the optimum allocation of finite resources. The tendency, of course, is for each group to highly value its own area of activity, sometimes at the expense of other groups. Although it is ultimately the project manager's job to decide between such competing interests and properly allocate resources, two general guidelines should be kept in mind as regards product reliability. First, be sure that all foreseeable product qualification costs are factored into the project budget. These may include such costs as:

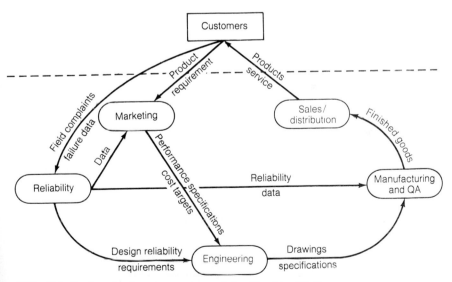

FIG. 11.3 Product development customer relationships (simplified).

- Engineering and/or technician labor-hours
- Analysis of existing field failure data
- Laboratory charges
- Design qualification test samples
- Reliability design reviews
- Field visits to customers
- Trips to meetings with approval agencies
- Supplier visits
- Capital funds for special test equipment

The second point is to frontload the evaluation activities wherever possible. Referring back to Fig. 11.2, the most effective use of limited reliability resources occurs early in the development cycle. If trade-offs must be made between labor-hours available for early design analysis and labor-hours for preproduction testing, the early design analysis work should not be shortchanged. Although *some* level of preproduction testing is usually required (and *always* advisable), there is a common inclination to put all of the reliability resources into such activity simply because it is more traditional and more visible. This temptation should be avoided.

11.4.2 Risk Management

One of the project manager's chief responsibilities is the management of risk factors associated with the program. These may be technical risks, marketing risks, manufacturing risks, schedule risks, or reliability risks, but they all are generally reducible to financial risk to the company. Evaluation of these risk factors then becomes one of the primary functions of the evaluation engineering group, and the most successful approaches to this task will use this objective, third-party approach to risk visualization.

It is recommended that the project evaluation engineer be designated as the keeper of the project risk statement, a model for which is shown in Fig. 11.4. This document should be updated and maintained throughout the product development program as various risk factors are identified and addressed. The risk items themselves should be collected from Engineering, Marketing, Reliability, Manufacturing, Service, Quality Assurance (QA), or any other pertinent source. All risk items should be expressed in terms of potential cost exposure, where

_____ Project Risk Statement Date _____		
Risk description	Current status	Risk evaluation
(Itemized list defining the nature of each risk and its consequences to the project)	(Actions taken or planned, resources required to address the risk, etc.)	(High, medium, or low probability, with the associated costs or other exposure that may result)

FIG. 11.4 Sample risk evaluation format.

possible. The consolidated risk statement should be reviewed at each design review and qualification review meeting, with action items assigned as needed to address the risk areas.

11.5 QUALIFICATION STATUS REVIEW

In order to administer the reliability evaluation program, it is necessary to create some sort of formalized qualification status review panel. Experience has shown that the most effective of these qualification review panels are cross-functional in nature, with active participation by Engineering, Marketing, Manufacturing, QA, Service, Purchasing, and other appropriate groups, all acting under the chairmanship of Reliability or Evaluation Engineering. The members of such a panel should be relatively high-level individuals with the collective authority to make hard decisions concerning product qualification status. The review panel has three fundamental tasks:

- Formulate and/or approve the reliability qualification plan for the new product in question
- Monitor the progress of the project against the approved qualification plan
- Determine the product's readiness for production release at the end of the development program

The review panel may meet as often as deemed necessary, but at least three formal reviews are recommended, as described in Fig. 11.1.

11.6 A MODEL RELIABILITY EVALUATION PROGRAM

The model program described in the balance of this chapter is a composite, generalized example that is based on experiences from a variety of businesses and industries. Certain of the activities discussed may not be applicable to some products; other industries may require an even more intensive program. Whatever the case, the five primary activity areas—specifying, designing, qualifying, maintaining, and auditing—will still apply. The checklists, summaries, and diagrams used herein are only intended as examples and should be modified as needed for use on a particular project.

11.6.1 Specifying the Design

Some of the most powerful influences on the reliability of the future product are determined at the very start of the project—as the product specification is being prepared. The marketing group is responsible for compiling a product specification which is really a description of the type of product needed to meet the market need. Frequently, problems arise early in the program because this description is not as specific as it needs to be. A typical product specification should address (in some detail) the areas described in Table 11.1. What is particularly important

TABLE 11.1 Product Specification Items

Design requirements:
  Performance capabilities and characteristics
  Physical and dimensional requirements
  Manufacturing requirements
  Power requirements
  Control specifications
  Appearance specifications
  Installation and start-up requirements
  Service and maintenance requirements
  Environmental specifications
  Accessory requirements and standard features
  Packaging, transportation, and storage specs
  Use and foreseeable misuse
  Field training requirements
Intended functions:
  Operating life and duty cycles
  Environmental requirements
  Safety requirements and required agency approvals
Reliability requirements and goals
  Allowable failure rates and modes
  Field failure traceability requirements

---

about this product specification is that it be *accepted* and agreed to by all functional groups. It is essentially the contract between Marketing and Engineering as to what characteristics should be designed into the product.

Reliability Requirements. The design reliability requirements should be accurately stated in terms of measurable results, not just high-sounding generalities. Maximum allowable failure rates for given operating modes should be stated. These numbers should be based on actual marketing needs, with practical consideration of such factors as planned warranty reserves and field service assets. Prior product experience should also be considered, with a careful review of historical failure rates and modes carried out to help determine the goals. Finally, competitive unit reliability should be factored in when such data are available. The resulting reliability specification should be a challenging, yet achievable objective within the time, technology, and resources available to the project.

Environmental and Operating Profiles. The total environmental envelope should be specified, within which the product will be required to operate throughout all phases of its production and service life. An example of a format for specifying these environmental factors is given in Fig. 11.5.

In like manner, the product specification should describe the worst-case operating profile and duty cycles for the new product during all phases of service life. These data should be organized in terms of both hours and cycles for all expected operational modes.

	X3 Controller Project Life-Cycle Environmental Requirements			
Condition (worst case)	Production/test	Shipping/storage	Installation	Field operation
1. Temperature				
High	40° C	85° C	70° C	70° C
Low	10° C	−40° C	0° C	0° C
2. Rel. humidity	90%	90%	100%	100%
3. Vibration	(Usually refers to separately developed vibration profiles)			
4. Shock	10 g	20 g	15 g	15 g
5. Rain	N/A	N/A	N/A	N/A
6. Dust	N/A	N/A	N/A	N/A
7. Salt fog	Must survive exposure per ASTM B117-73.			
8. Other ( )				

FIG. 11.5 Environmental requirements (sample format)

Product Qualification Plan. During the specification phase of the project, the preliminary product qualification plan should be drafted and submitted to the qualification review panel for acceptance. The review panel should also ensure that project management has provided the time and resources in the project schedule and budget to allow the planned activities to be carried out. For example, if 6 weeks of accelerated life testing will be required prior to production release, the official project schedule must provide for at least 6 weeks between the time that pilot or preproduction samples of the product are available and when final release is scheduled. Likewise, capital equipment and materials needed to produce the test units must also be available at least 6 weeks before full production is scheduled.

The qualification plan itself should be a broad-based document which covers more than simply the reliability test program. All areas of project activity that could impact final product reliability and/or potential warranty exposure should be addressed. As a minimum, the areas of concern are those shown in Table 11.2.

Documented plans for all the above activities should be submitted to the review panel for approval at the appropriate stage of the project. One effective

TABLE 11.2 Qualification Plan Contents

- Marketing plan
- Product development plan
- Critical components or materials list
- Manufacturing plan
- Quality assurance plan
- Value engineering activities
- Reliability plan
- Agency certification plan
- Supplier selection plan
- Field service plan

method for organizing the various major activities requiring review is the sample checklist shown in Fig. 11.6. Note that this checklist (which is introduced at the initial qualification planning meeting) serves multiple purposes. First, it outlines the major review items for the project, in a phased manner. Second, it provides procedural references for properly executing the various tasks. Third, it allows a predetermination as to whether or not the particular procedure applies to the project in question. This list could be expanded to indicate schedule dates for each activity as well. However, this information is normally picked up in the master project schedule or on more detailed checklists used as agenda guides for the particular reviews in question.

11.6.2. Designing-in Product Reliability

Once the product specifications are approved and accepted by all participating groups, the design process may begin. It is during this phase of the project that *both* functionality and reliability are designed into the product. Just as specific efforts and resources are required to design in a particular function or feature, so there are specific tasks and resources which must be applied to the project in order to achieve the stated reliability attributes. This fact cannot be overstated. Failure to apply the proper resources to design reliability tasks results in products with "reliability by chance," and no amount of testing at the end of the design program can increase the *inherent* reliability of the design. The activities described in the remainder of this section are some of the *basic* disciplines used to achieve this inherent reliability.

Design Reviews.   This fundamental activity is one of the most effective techniques for improving designs at various stages of development. The keys to design review success are: (1) having the reviews at the right time, (2) having the right participants, (3) sticking to the published agenda, and (4) assigning specific action items to *named* individuals with assigned due dates. The design review schedule should be tied to the drawing release schedule. As a minimum, there should be at least three major design reviews. An initial review should be held at the very start of the program, using design concepts, major component lists, and product sketches. This allows other activities such as Reliability, Manufacturing, Quality Assurance, and Service to critique the approach while it is still early enough to influence the design. Another formal review should be held when prototype units are available in order to support critical drawing releases on long-lead-time materials and parts. A final design review should be held upon completion of development testing to support the final drawing releases. The reliability reviews at each stage could be held separately or concurrent with the regular review.

Failure Modes and Effects Analysis.   Particular emphasis should be placed on conducting thorough and *timely* failure mode and effects analyses (FMEAs) ("timely" means early enough in the program to have a meaningful and cost-effective influence on the design). Both functional (top-down) FMEAs and part-level (bottom-up) FMEAs may be effectively used on the new product design. If project resources and time permit, both methods are recommended, with the functional FMEA being normally performed first, in the design concept phase, and the part-level FMEA being done when preliminary drawings and parts lists are generated. In any case, FMEA activity should be given first priority for the use of any design reliability resources available to the project. If resources are

PROJECT R & Q PROCEDURE CHECKLIST

Product _____

Project No. _____

Project Team: Engrg _____ Mfg _____

Mktg _____ Quality _____

Others _____ Service _____

REF. PROCED.	APPLIES YES	NO	COMMENTS

QUALIFICATION PLANNING MEETING
1. Product Spec. Approved
2. Qualification Plan
   a. Source selection
   b. Product / component / materials quality
   c. Reliability evaluation
   d. Planned field trial
   e. Preliminary quality plans
      1. Incoming inspection
      2. Acceptance tests
      3. Quality audit inspection and test
   f. Process capability study

LEADING TO INITIAL DESIGN REVIEW
1. Critical Components / Materials / Processes List
2. Risk Summary
3. Design for Manufacturability
4. Supplier Surveys

LEADING TO PROTOTYPE RELIABILITY REVIEW
1. Failure Modes and Effects Analysis
2. Safety Analysis
3. Critical Characteristics
4. Installability, Serviceability Review
5. Failure Reporting and Analysis
6. Critical Item Traceability
7. MCS and Pilot Run Plan

LEADING TO FINAL RELIABILITY REVIEW
1. Purchased Part Specs. and Drawings Finalized
2. First Article Inspection and Evaluation
3. Pilot Run Report

LEADING TO FIRST-YEAR REVIEW
1. Failure Trend Analysis
2. Product Field Evaluation
3. Cost of Quality

FIG. 11.6 Reliability procedures checklist.

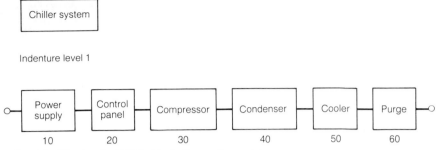

FIG. 11.7   First-level reliability block diagram (example).

limited, the part-level FMEA is recommended as the minimum analysis. There are several other techniques, such as fault-tree analysis, which can be used to perform the FMEA function. However, the tabular formats are straightforward, easy to use, and meaningful to the reader and are therefore recommended.

Functional FMEAs.   The functional FMEA is a technique of "events analysis" and is most valuable if performed early in the design program, typically in the early prototype development stage when the principal functional areas of the product design are being defined. The functional FMEA can be viewed as a top-down analysis of the system or product in which the functional requirements are first totally defined, then an assessment is made of any combination of potential events or conditions which might impair or prevent that function. The results of such an analysis will indicate critical design considerations, potential weaknesses in the design approach, and critical testing requirements for use later in the qualification test phase. Prior to conducting this evaluation, a set of reliability block diagrams must be prepared. Normally, the reliability block diagram for the product is first prepared for each indenture level. Figures 11.7 and 11.8 illustrate two levels of a typical system. As can be seen in the figures, a complete listing of inputs and outputs is made for the particular level diagram.

Using the reliability block diagrams so developed, the analyst next prepares a function and output listing which defines (in tabular form) the nature of all operating functions, their specified outputs and limits, and their relative priority. An example of such a table is seen in Fig. 11.9.

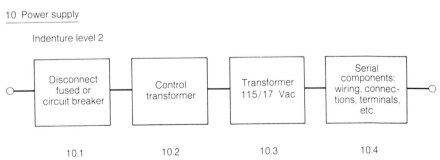

FIG. 11.8   Second-level reliability block diagram (example).

10.1   Disconnect
   1. Manual device to connect, disconnect, and isolate load devices and their controls from the power system.
   2. Prevents overload damage (limits current).

FIG. 11.9   Function and output list (example)

The next step in the analysis is to prepare a functional FMEA worksheet to evaluate the various potential failure modes for all the functions and outputs identified. An example of the worksheet is given in Fig. 11.10. Note that each potential failure mode has been assigned a level of *severity* and both probability of *occurrence* and *detection* ratings. This is normally a subjective evaluation by the system analyst that is based on all the information discussed above and the analyst's knowledge of the system. These values can be multiplied together to obtain a *risk priority number* (RPN) that conveniently identifies those failure modes which should have the greatest attention regarding corrective action.

Component FMEAs.   The component FMEA is normally undertaken at the stage of development where a preliminary layout and parts list are available (preproduction model). The analysis is done for the purpose of evaluating each potential failure mode of the individual components and relating those failure modes to their system-level effects. Hence, this analysis uses a bottom-up approach, as contrasted with the top-down approach of the functional FMEA. Once the preliminary FMEA is completed, it should be used as a baseline, with revisions incorporated as necessary to reflect changes in components that are made prior to production.

A recommended format for performing component FMEAs on electromechanical designs is given in Fig. 11.11.

Results.   The results of the component FMEA constitute an important part of the qualification record for the product. Output reports should indicate the distribution of failure severities at the system level and should clearly identify any components designated as "critical" because of either a high failure probability or severe effects of a failure mode.

When properly done, the FMEA will not only provide valuable insight into the designed-in reliability of the product, but will also have the effect of identifying and eliminating many potential failure modes which would have otherwise remained in the product as it entered production.

Safety Analysis and Review.   With product safety being of paramount concern, it makes sense to dedicate some specific resources to the safety analysis activity. This effort should consist of a formal safety and hazard analysis and at least one participative safety designreview. The hazard analysis is an extension of the tabular FMEA, wherein each component part is specifically evaluated for potential safety-related failures or hazardous conditions. A sample format for such an analysis is given in Fig. 11.12.

The formal safety review should be conducted at the prototype stage, with an updateat the time of final drawing release. This review should be led by the safety engineer and/or reliability engineer and include participation by Engineering, Marketing, Manufacturing, Quality Assurance, Field Service, and Shipping and Transportation. A detailed safety checklist should be developed for the pro-

PROGRAM:

Indenture Level 1 ______
Indenture Level 2 ______
Indenture Level 3 ______
Indenture Level 4 ______
Drawing ______

FUNCTIONAL
FAILURE MODE AND EFFECT ANALYSIS
WORKSHEET FORMAT

Project Number ________
Submitted ________
Approved ________
Date ________

Output Specification Functional Description	Failure Mode		Possible Causes	Symptom Detectability	Effect of Failure		Existing Compensating Provision	Occurrence	Severity	Detection	Risk Priority Number (RPN)	Remarks and Recommendations
	SN	Description			Local Effect	End Effect						

FIG. 11.10 Functional FMEA worksheet format.

COMPONENT FAILURE MODE AND EFFECTS ANALYSIS

Page ____ of ____

Unit Name/Subsystem Name ____________ Outside Suppliers Affected ____________ Engineer ____________
Primary Design Responsibility ____________ MCS Date ____________ Program Manager/ Dept. Manager ____________
Other Activities Involved ____________ First Production Date ____________ FMEA Date: (ORIG) ________(REV) ________

					Existing Conditions						Resulting					
Part Name Part Number	Part Function	Potential Failure Mode	Potential Effect(s) of Failure	Potential Cause(s) of Failure	Current Controls	Occurrence	Severity	Detection	(RPN) Risk Priority Number	Recommended Action(s) and Status	Action(s) Taken	Occurrence	Severity	Detection	(RPN) Risk Priority Number	Responsible Activity

FIG. 11.11 Sample format for component FMEA on electromechanical designs.

HAZARD ANALYSIS

Product: _PL 99 Blower System_
Product Function: _Air Recirculation_
Type of Review: _____

Prepared by: _G. Smith_
Date: _6/15/86_
Page _1_ of _3_

Item	Component	Hazard Mode	When Can It Occur ?	Environmental Factors	Annual Probability of Occurrence	Comments
1	Sheet metal components	Mechanical—Cut hands and fingers	Fabrication, assembly, installation	Careless handling	Medium	Review procedures for assembly and servicing. Tooling maintenance program in place.
2	Control box	Electrical—Shock	In service or at installation	Factory miswire or ungrounded, improperly maintained connection	Low	Quality control plans include electrical tests and hi-pot. Service guide emphasizes proper procedures.
3	Blower/fan	Mechanical—Cut fingers	In service	Missing/failed protective grille	Low	Grille construction meets U.L. requirements. Special tests conducted.

FIG. 11.12 Sample format for hazard analysis (with example).

11.20

duct with consideration given to establishing hazard criteria and potential hazardous modes under all foreseeable conditions of use *and misuse*, through the product life. Particular attention should be given to less obvious areas, such as user instructions, installation and service guides, and packaging. Any applicable government and industry standards should be included in the review of the mechanical, electrical, chemical, and structural safety issues.

Component and Materials Selection.   During the early design phase, activities related to the selection and qualification of components and materials should begin in earnest. First priority should be given to the early identification of *critical* components, materials, and processes that require special qualification efforts. "Critical" items can be designated as such for any of the following reasons:

- Safety related
- New component or material process
- Existing component in a new application
- New component supplier
- High cost

Whatever the reason for designating the item as critical, the important thing is to get it identified early, along with an effective item or supplier qualification plan. A sample format for such critical item identification is given in Fig. 11.13.

Careful evaluation of critical components, materials, and processes is a major contributor to product reliability. As the project progresses, *all* components, materials, and processes should be qualified for use in the intended application, either through prior usage or testing and analysis. Monitoring and supporting this effort should be a key assignment for the reliability or evaluation engineer.

11.6.3   Qualifying the Design

Following the design phase, the next area of activity involves qualification of the design, as embodied in the released drawings. This includes *verification* of performance and reliability requirements defined in the original product specification. If the design phase reliability evaluation activities have been properly and thoroughly conducted (i.e., design reviews, FMEAs, critical item analysis, etc.), a high confidence in the overall design reliability should have already been achieved. Thus, the actual performance and reliability testing activities undergo a change of emphasis in this kind of program. Whereas in a more traditional reliability program the testing phase received the majority of attention (and resources), the front-loaded design reliability program uses testing more selectively (and effectively) as a final verification of those system performance and reliability attributes that are difficult to evaluate through design analysis alone. Typical test activities are discussed below.

Component and Materials Qualification.   As the components and materials are selected during design, a qualification plan should be prepared and executed wherein all previously unqualified items are evaluated and tested under worst-case application conditions. The component-level test activities should be given highest priority when testing time or resources are limited for two clear reasons. First, we do not usually control the design process for a purchased part and must therefore rely more heavily on testing. Second, where our own test resources

CRITICAL COMPONENTS/MATERIALS AND PROCESS LIST														

Product Description _____

Project Number _____ Date _____

Prepared By _____ Page ___ of _____

PART DESCRIPTION	PART NUMBER	SUPPLIER PART NUMBER	SUPPLIER NAME	USED NOW	IS PART QUALIFICATION REQUIRED		RELIABILITY TEST REQUIRED		SUPPLIER QUALIFICATION REQUIRED		SUPPLIER VERIFICATION REQUIRED		PROCESS QUALIFICATION	CRITICAL COMPONENT LIST CATEGORIES						COMMENTS
					Yes	No	Yes	No	Yes	No	Yes	No		Safety Related	Single Source	New Supplier	New Part to Manufacturing Division	New Application	High Replacement or Service Labor Cost	

1 ● New Process—Action Required
2 ● Existing Process that will undergo significant change—Action Required
3 ● Existing Process that has been causing problems—Action Required
4 ● No Action

FIG. 11.13   Sample format for critical item identification.

must be expended, it is normally less expensive to set up a component-level test than to test that same part, to the same degree, in a full-system test.

Component and materials qualification should encompass qualification of the *supplier*, as well as the item itself. This is best done using a "supplier certification team" approach as part of a formal source selection board. Such a board, usually chaired by Purchasing, organizes the selection and qualification of suppliers for items in the new design. When supplier visits or audits are needed, they are best done by an interdisciplinary team composed of Purchasing, Engineering, Quality Assurance, Manufacturing, Reliability, and others, as needed. Standardized supplier survey forms should be developed. A list of key characteristics which should be covered in such a survey can be found in Table 11.3.

System and Process Qualification.   Once component-level qualification tasks are established, attention is turned to planning and executing the system- or product-level qualification tests. Presuming that a thorough design evaluation has already taken place (as discussed in Sec. 11.6.2), it should be possible to economize

TABLE 11.3 Supplier Survey Areas

- General
  - Product lines
  - Sales volumes
  - Physical plant condition
  - Employment
- Procedures and standards in use
- Supplier control program
- Drawing and change control
- Inspection and test equipment
- Material review procedures
- Receiving inspection
- In-process quality control and factory test procedures
- Packaging and shipping procedures
- Housekeeping and material handling
- Product audit procedures
- Process control methods

considerably in this area. System performance tests on the fully assembled product should be run, of course.

It is also recommended that worst-case environmental tests be conducted on a small number of product samples where feasible. However, long-term reliability-life tests on a statistically significant sample of units can be an expensive undertaking. For those projects where front-end design analysis has provided a high confidence in the inherent design reliability, it normally is possible to tailor a less costly test program wherein a small number of test units are subjected to a fixed-time test with cycle, temperature, and stress accelerations. These tests should be set up to provide the equivalent of the design-life operating cycles in all modes while being operated (in real time) for at least the first year's equivalent operating hours. The sample size should be determined on a case basis for each product. This kind of stress test is effective for stimulating the most likely latent design failures in a relatively short time. The testing should be conducted on production units or preproduction units which are representative of the production process. When test time is short, this kind of stress testing can usually yield enough data in a relatively short (e.g., 2000 hours) test period to allow preliminary product release decisions to be made at low risk. The testing can then be continued to whatever point is appropriate to further reduce the risks.

It is during this period that final process qualification in the manufacturing plant should also be conducted. For this reason any design qualification testing should be conducted on units built to the qualified process and tooling. Otherwise, later changes to the processes could invalidate the test results.

Field Trials and Witness Tests.   It is during this phase that any planned field trials will also be taking place. Field trials of new products can serve a variety of purposes—performance comparison, customer acceptability, installability, serviceability, etc.—but whatever the purpose, the key to a successful field trial is a thorough and well-documented plan. The purpose of the field trial should be limited in scope and clearly defined. Details of the field trial should be defined by a field trial committee with representation from the marketing, engineering, reliability and field service departments. Some of the areas the field trial plan should address are:

- Specific product identification
- Planned shipping and start-up dates
- Placement of products
- Proposed pricing and ownership of products
- Warranty procedures
- Required literature and training support
- Special product processing information
- Data recording and analysis plans
- Service and replacement parts procedures
- Actions to be taken in the event of failure

Government and industry approval agencies will frequently require witness tests on new designs. These tests should also be factored into the overall qualification test plans. In many cases, qualification test planners can save the project a lot of time by structuring their in-house tests to meet the requirements of such approval agencies. Then, with well-documented test plans and reports, the manufacturer can often convince the approval agency to accept the test results in lieu of running separate, time-consuming, independent tests.

Documentation.   Whatever test program is planned and conducted, it is highly important that thorough and accurate records be kept of all test plans, procedures, and results. This provides the important capability of tracing back the possible cause of later performance problems which might arise. It also allows more rapid evaluation of later changes in components, materials, processes, or applications. Finally, in the unfortunate event of a later product safety problem, such information can prove to be a vitally important record in potential liability cases.

Throughout the qualification test phase, all test failures and problems should be logged on a standardized failure report form. This form should allow for a description of the failure or problem which occurred, the conditions under which it took place, the party responsible for evaluation, the results of the failure analysis, and the recommended closure action. The effectiveness of the corrective action should also be analyzed, where possible. Figure 11.14 shows a sample of such a form. The reliability or evaluation engineer should maintain a log of such reports and ensure that the reports are properly acted upon and closed out. All such reports should be serialized and their closure status reviewed at the final qualification review meeting.

11.6.4   Maintaining Reliability during Production

Up to this point in the project plan, the focus of reliability activity has been to drive the design of high inherent reliability into the product. At this stage, however, the emphasis shifts to maintaining the inherent design reliability throughout the rigors of the manufacturing process. In most organizations, this translates into action on the part of the quality assurance or quality control

FAILURE REPORT

Note: * Denotes information to be
        entered by originator.
        Send all copies to Evaluation
        Engineering.

No. _____
* Originator _____
* Date _____
Update by _____
Date _____

* FAILED PART NAME, PART NO. AND PRODUCT INVOLVED:

_____

* FAILED ON MODEL _____ * SERIAL No. _____

* FAILURE DESCRIPTION (Include description of failed part; type of test; circumstance of failure; environmental conditions such as temperature, voltage, moisture, operating condition, load, etc; including date of failure.)

_____
_____
_____
_____
_____
_____

* CAUSE (Include any specific information which may indicate cause of failure.)

_____
_____
_____
_____

ACTION
    CAUSE ANALYSIS: RESPONSIBILITY: _____ REPORT DUE: _____
    DETAILS :

_____
_____
_____

CORRECTIVE ACTION: RESPONSIBILITY: _____ REPORT DUE: _____
    DETAILS INCLUDING SOLUTION : _____ EFFECTIVE DATE : _____

_____
_____
_____

QUALIFICATION ACTION : RESPONSIBILITY: _____ COMPLETE DATE: _____
    DETAILS:

_____

ACTION ACCEPTABLE: _____ DATE: _____

FIG. 11.14   Sample failure report format.

11.25

(QA/QC) groups associated with manufacturing. This change in emphasis is reflected in Fig. 11.1, at the beginning of the chapter, where the activities *below* the dashed line (i.e., manufacturing/QA) in each of the primary activity boxes begin to dominate the engineering activities.

The chief role of any QA/QC organization is to ensure manufacturing conformance to design specifications. By comparison, the focus of reliability and evaluation engineering activities, as the project moves into the production phase, should be to ensure that the requisite manufacturing and QA systems are in place (and qualified) to protect the design reliability and ensure conformance to requirements. Prior to production, some of the key tasks include:

- *Machine or process specification and capabilities studies:* The qualification review group should ensure that all applicable machine and process specifications for existing process equipment are transmitted from Manufacturing to Engineering at the beginning of the program. This allows the designers to take advantage of existing capital equipment to the fullest extent, saves capital cost, decreases the number of new machines and processes requiring capability analysis, and contributes to product quality. Where new processes *are* required, Engineering should transmit those process requirements to Manufacturing as early as possible to allow for thorough process qualification.

- *Manufacturing and QA plans:* The qualification review group should require a written manufacturing plan for the specific product in question. This should be prepared by the manufacturing facility at an early enough time to allow all manufacturing processes to be fully qualified prior to production release. The plan should also address all statistical process control (SPC) requirements. The same requirement for a written plan should be applied to the QA activity. The plan should include such key items as supplier qualification and supplier controls, in-process QA activities, product acceptance tests, preconditioning requirements, component and material quality requirements, test and inspection criteria, and procedures related to nonconforming material.

- *Manufacturing check samples (MCS) and pilot runs:* Adequate time and resources must be planned and *preserved* in the program to allow for these key activities. Once final drawings have been released, the manufacturing facility should conduct an MCS run to verify their capability to build the new product to the final drawings and specifications. This run should be limited in number and the products need not be built to final tooling (which may not yet be in place). Adequate time to evaluate the results of the MCS should be maintained in the schedule. MCS units may be usable for some preliminary product qualification testing, but they are normally *not* fully representative of the production process.

- When all final processes, tooling, and QA procedures are in place, the manufacturing facility should run a formal pilot production of a limited number of units. The results of the pilot run should then be carefully evaluated, the pilot units inspected, and a formal pilot run report submitted to the qualification review panel. Again, adequate time must be maintained in the schedule between pilot run and full production release to permit evaluation of the results *and* corrective action when required.

- The pilot run units can normally be considered representative of final production, and as the first such units produced, they are the likely source of reliability qualification test units. This means that if the planned reliability testing requires, for example, 8 weeks to a decision point, there must be at least 8 weeks preserved in the schedule between pilot run and production. The word "preserved" is important because there is a natural tendency on many projects, as delays occur, to allow the scheduled activities to be compressed toward a planned production date. Maintaining scheduled start-up dates is obviously important from the viewpoint of plant loading and scheduling; however, if high confidence in product reliability is to be achieved, the temptation to "whittle away" at the preproduction qualification schedule must be avoided.

The foregoing is not intended to be a comprehensive description of QA activities. It is only intended to highlight those issues which are most critical to product qualification planning and should, therefore, be closely monitored by the qualification review panel.

11.6.5 Auditing Reliability

As with any effective control scheme, the commercial product reliability program needs a feedback loop to help monitor results. As a product enters field service, it is critically important to have established some means of measuring reliability and quality performance so that the causes of *un*reliability can be addressed. These causes may exist in any of the preceding activities—specification, design, qualification, or maintenance of reliability.

The most effective methods for gathering such data will vary with the type of product and other factors, such as warranty policies, sales and distribution methods, inventory practices, and installation and start-up practices, to name a few. Some methods of data collection which have proved effective for various products include:

- *Warranty data screening:* If the company has a documentation system to support warranty claims, the claim forms can be designed to provide key elements of information on time to failure, failure conditions, failed part identification, etc. It may even be possible to make payment of the claim contingent on receipt of such data. In any case, the availability of personal computer data base systems makes it relatively easy to collect and use such information for field reliability estimates, trend analysis, etc.

- *Return of critical failed parts:* The above data base can be further enhanced by identifying certain critical parts for return to the manufacturer in the event of failure. There is, of course, some added cost to such an operation, but the value of information gained from such failed item analysis normally exceeds the cost of carrying the program.

- *Field audits:* In some cases it may be feasible to focus on certain products or problems by sending an active audit team into the field to evaluate the product as received by the customer. For large expensive products, this may be accomplished by sending a multidiscipline team (engineering, reliability, QC, service) to "witness" the delivery and installation of a product at the customer's site. For smaller less costly products, some companies fund the audit activity to actually "buy" products from normal distribution and bring them back to the lab for detailed evaluation.

Whatever the methods used for gathering performance data, there should always be an attempt to assess the costs associated with product failures and problems. As this "cost of quality" data base grows, it provides a useful gauge for comparing the costs of corrective action at various stages of the program. These true life-cycle cost data will inevitably provide the best justification for management support of the front-end design reliability activities that are required in order to achieve the levels of product quality demanded in the world marketplace.

CHAPTER 12
DESIGN FOR PEOPLE*

Kenneth P. LaSala
Air Force Systems Command

Arthur I. Siegel
Applied Psychological Services, Inc.

12.1 INTRODUCTION

Integrating the design effort for hardware, software, and humans is the only way to deliver a truly successful product. This chapter describes the process of managing, analyzing, designing, and evaluating person-machine systems. The chapter focuses on integrating the hardware, software, and humans at the system level. Often the product development approach has been to design the system and then to "human engineer" the details. Unfortunately, this usually results in the perpetuation of seemingly obvious deficiencies. The anticipated declines in the numbers and proficiencies of personnel [1] and the estimated 30 to 50 percent human-induced failures [2, 3] support the need for intensive system-level considerations of personnel in the design phase.

It is worthwhile to establish some basic concepts to use in designing for people. Traditionally, there has been a distinct separation—a gulf—between the disciplines that are used for designing hardware and those that are used in human factors engineering design. In reality, effective system design recognizes that the "person in the loop" cannot be separated from other system functional elements. Figure 12.1 shows the interactions of human performance with hardware functional performance in overall system readiness. This figure also shows where the system and job design processes affect the design for operation, maintenance, and, eventually, readiness. The system design process consists of designing human tasks, "jobs," that apply to system operation and maintenance. This involves configuring the tasks in terms of physical and intellective actions and then assessing the likelihood of errors or correct performance. The error data then are used to synthesize estimates of system performance and reliability. A well-organized design process evaluates the impact of the human performance and reliability and determines the need for design improvement. The resultant system performance and reliability are major contributors to the readiness of the system in the field. Later sections of this chapter provide more detail on the analysis, design, and testing actions that are included in this process.

Since much of system or process design and testing depends on quantitative measures, it is necessary to have a set of measures that includes the human component as well as the hardware and software. Table 12.1 identifies a set of measures that are analogs to conventional reliability, maintainability, and availability measures. Like their hardware reliability counterparts, the measures are

---

* This chapter is dedicated to the memory of Dr. Arthur I. Siegel, whose untimely passing in March 1985 was an irreplaceable loss to the human factors and reliability communities.

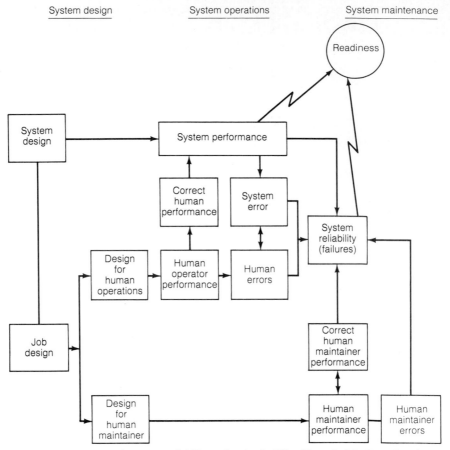

System design   System operations   System maintenance

.**FIG. 12.1** The human in system reliability and maintainability (*From Ref. 1. Reproduced with permission.*)

oriented toward the designer. They are used to evaluate specific hardware, software, and human combinations by means of the analysis techniques in this chapter.

12.2 OVERVIEW OF THE DESIGN-FOR-PEOPLE PROCESS

The most critical aspect of designing for people is the integration of human, hardware, and software-oriented engineering tasks. The key to this integration is the restructuring of human reliability and human engineering disciplines into a "top-down" design process. The fundamental areas in this process are management, analysis, design, and test. Table 12.2 lists several important disciplines that

TABLE 12.1 Reliability, Maintainability, and Availability Measures for Human-Machine Systems

1. $\text{Human reliability} = 1 - \dfrac{\text{No. of failures}}{\text{No. of attempts}}$

6. $\text{Equipment MTBF} = \dfrac{\Sigma \text{ times between failures}}{\text{No. of failures}}$

$= \dfrac{\text{Mission time} - \text{downtime}}{\text{No. of failures}}$

2. $\text{Human availability} = 1 - \dfrac{\text{Time lost or unmanned hours}}{\text{Total mission manhours}}$

7. $\text{Equipment MTTR} = \dfrac{\text{Total repair times for all missions}}{\text{No. of repairs}}$

3. $\text{Human MTTR*} = \dfrac{\text{Total time of second-try successes}}{\text{No. of second-try successes}}$

8. $\text{System reliability} = 1 - \dfrac{\text{No. of equipment failures} + \text{No. of people failures}}{\text{Iterations} \times \text{No. of equip} + \text{Human attempts}}$

4. $\text{Equipment reliability†} = 1 - \dfrac{\text{Total no. of failures during mission}}{\text{No. of missions} \times \text{No. of equipments}}$

9. $\text{System availability} = 1 - \dfrac{\text{System downtime}}{\text{Mission time}}$

5. $\text{Equipment availability} = \dfrac{\text{Equipment uptime}}{\text{Total mission time} \times \text{No. of equipments}}$

10. $\text{System MTTR} = \dfrac{\Sigma \text{ time for repairs} + \Sigma \text{ time for second-try successes}}{\text{No. of repairs} + \text{No. of second-try successes}}$

---

* "Second-try successes" are the successful remedial actions taken to correct a human error.
† Failures are those events that cause a loss in desired capability. With redundancy or on-line repair, component failure may not mean loss of capability.
Source: Adapted from *Human Reliability Prediction System User's Manual,* Naval Sea Systems Command, Washington, D.C., December 1977.

should be used to guide the development of person-machine systems. Associated with each discipline are applicable actions in the management, analysis, design, and test areas. Management actions are those contractual and control actions that steer the system development. They provide the administrative circumstances that permit the proper application of human-machine reliability disciplines. The analysis actions are mathematically oriented activities that either provide the preliminary definition of the product or support the physical design process. The design actions are those that apply to the physical realization of the system. They involve the selection of hardware and software and its arrangement in the system configuration. Test actions are simply the validation of the design through appropriate experiments or exercises.

Readers should examine their system development activities to ensure that they have devoted proper attention to each discipline by incorporating the appropriate detailed Table 12.2 actions into their programs. The applicability of each action depends on the development phase. Table 12.3 shows the recommended applicability of design- and test-related actions in the system development process.

TABLE 12.2   Design-for-People Discipline Matrix

Discipline	How managed?	Analysis approach	Design techniques	Test techniques
		Acquisition fundamentals		
1. Firm contract requirement for personnel and performance and relia-bility	1. Required by planning documents	1. Derive quantitative requirements from system analysis and trade-offs	N/A	N/A
	2. Included in solicitation material (Spec/SOW) with management approval	2. Examine program schedule, funding, technical performance, and R&M requirements to *tailor* management and technical tasks		
2. Financial incentives for reducing personnel workload and human error	1. Special contract provisions for design excellence, e.g., incentives program	1. Soliciation review	N/A	N/A
	2. Management strategies reflect 1 above	2. Perform cost-benefit analyses to define rewards to contractor for design excellence		
	3. Note—warranty probably not enforceable	3. Quantify amount of workload and error reduction		
	4. Specifically reported during design and program reviews	4. Define specific workload factors to be reduced		

Two important principles are apparent from the tables. The first is the principle of defining the product and its environment early in the conceptual phase. Many costly delays and redesigns can be avoided if product use and environment are defined specifically. The second principle is one of the early establishment of "midlevel" design strategies, described below, that address both human and hardware elements of the product.

The customary approach to designing human-machine systems is to define the hardware and software and then, as an adjunct, tailor those to the human being. Unfortunately, many operation and maintenance efficiencies are omitted by this process because the use of the human is severely constrained by the already established hardware and software configurations. A more effective approach

TABLE 12.2 Design-for-People Discipline Matrix (*Continued*)

Discipline	How managed?	Analysis approach	Design techniques	Test techniques
		Acquisition fundamentals (*Continued*)		
3. Firm human engineering source selection criteria	1. Solicitation requires contractor's proposal to provide: (a) quantitative personnel performance and reliability characteristics; (b) specific human engineering design techniques to be employed; (c) management and engineering tasks: (d) identification and solution of specific human engineering problems likely to be encountered.	1. Source selection ratings give significant impact to human engineering	N/A	N/A
	2. Request for proposal (RFP) reviewed by management	2. Selection panel has in-house design evaluation techniques, e.g., simulations		
		3. Proposal evaluation requires rating of contractor response to RFP requirements shown at left		
4. Closed-loop failure reporting and corrective action during development and deployment	1. Contract requirement	1. Analysis of frequency and criticality of deficiencies	N/A	N/A
	2. Design reviews			
	3. Tracking and reporting of deficiencies in deployed systems			
		Design fundamentals		
1. Define human functions, environments, performance levels, required reliability by mission, and mission segment	1. Contract requirement for task performance	1. Allocation of human-machine functions	N/A	N/A
	2. Preliminary design review	2. Mission simulations		

TABLE 12.2 Design-for-People Discipline Matrix (*Continued*)

Discipline	How managed?	Analysis approach	Design techniques	Test techniques
		Design fundamentals (*Continued*)		
	3. Critical design review	3. Mission profile definition	N/A	N/A
	4. Preproduction design review for close-out of corrective actions	4. Trade-off analyses		
2. System physical configuration and required human functions similar to other systems performing same basic task	Same as above	1. Task analysis	1. Physical-auditory and visual displays, controls, consoles and panels, environment, work station	1. Laboratory experiments and/or factory verification
		2. Operator sequence diagrams	2. Task content-allowable and necessary time, action stimulus, requested action, feedback, potential error consequence, skill level	2. Simulations
		3. Time-line analysis 4. Time and motion analysis		3. Field observations
3. Automation to assist operator and maintainer task performance	Same as above	1. Operator-maintainer load analysis	1. Test point selection for maintenance	Same as above
		2. Elemental task analysis	2. Decision aids	
		3. Time and motion analysis	3. Equipment portitioning to facilitate BITE	
		4. Decision analysis	4. Detection aids	
		5. Task criticality analysis	5. Response aids	
		6. Analysis of task complexity	6. Task construction to facilitate automation	
			7. Ease of use	

TABLE 12.2 Design-for-People Discipline Matrix (*Continued*)

Discipline	How managed?	Analysis approach	Design techniques	Test techniques
		Design fundamentals (*Continued*)		
4. Technical documentation for 9th-grade level of comprehension	1. Same as above	1. Operator documentation requirements analysis	1. Format guidelines	1. Comprehension tests
	2. Special documentation review	2. Maintenance technician documentation requirements analysis	2. Content and comprehension guidelines	2. Operator reliability testing
		3. Task analysis	3. Mode of presentation	3. Maintainability testing
		4. Readability and comprehensibility analysis		
5. Simplify tasks for operation and maintenance	Same as above	1. Task analysis	1. Guidelines for data displaycues	1. Laboratory and/or factory tests
		2. Operator sequence diagrams	2. Guidelines for type and complexity of decisions	2. Operator reliability test
		3. Time-line analysis	3. Guidelines for control type and manipulation	3. Maintainability tests
		4. Time and motion studies	4. Guidelines for response type and time	4. Field observations
		5. Operator load analysis		
6. Label and code instructions for operation and maintenance clearly and consistently	Same as above	1. Elemental task analysis	1. Guidelines for effective labeling	Same as above
		2. Operator and maintainer load analysis	2. Guidelines for effective coding	

TABLE 12.2 Design-for-People Discipline Matrix (*Continued*)

Discipline	How managed?	Analysis approach	Design techniques	Test techniques
		Design fundamentals (*Continued*)		
7. Design for degradation due to physical and psychological stress factors	Design reviews	1. Task analysis, especially with regard to potential errors	1. Use existing data on degradation due to various physical stress factors, e.g., temperature, vibration, humidity, noise	1. Performance measurements under controlled stress levels
		2. Simulations for human-machine reliability	2. Use existing data on degradation due to psychological stress factors, e.g., fatigue, complexity, time, skill	2. Human performance in R&M testing
				3. Field observations
8. Design for human performance variability and limitations	Design reviews	Same as above	1. Use existing human performance variability and limitation data on displays, controls, decision making	Same as above
			2. Use anthropomorphic data to design for ease of setup and use of equipment and accessibility	
9. Simplify wiring diagrams, diagnostic procedures, and computer coding instructions	Same as 5 above	Same as 5 above	Same as 5 above	Same as 5 above
10. Need for embedded training capability	1. Solicitation, proposal, contract	1. Task analysis	1. Building block concept	Same as above

TABLE 12.2 Design-for-People Discipline Matrix (*Continued*)

Discipline	How managed?	Analysis approach	Design techniques	Test techniques
	Design fundamentals (*Continued*)			
	2. Design review	2. Analysis of required practice or repetition	2. Ease of reprogramming	
	3. Training conference	3. Analysis of required levels of difficulty	3. Multiuse, multiusers' training considerations	
		4. Techniques determination	4. Performance feedback considerations	
	Test fundamentals			
1. Use actual user test subjects	1. Solicitation or contract requirement 2. Test planning reviews	N/A	N/A	N/A
2. Use controlled test methods	1. Contract required	Evaluation of environmental, mission variables	Simulator design principles	1. Models
	2. Test planning reviews			2. Simulations
				3. Controlled experiments

Note: N/A = not applicable.

employs the concept of midlevel design strategies. These strategies are the guidelines that constrain the detailed design of the product. They have their greatest effect when applied in the conceptual phase, in which the product functions are being allocated to person and machine. Some examples of midlevel strategies are (1) guidelines for operator task formulation, (2) guidance for simplifying operation and maintenance, (3) guidance for defining the type and location of automated test equipment. The design fundamentals in Table 12.2 are areas in which a product developer should establish specific midlevel design strategies to guide the development of the product. It is emphasized that while the areas covered by such strategies should remain those identified in the table, the strategies themselves will vary considerably from product to product.

TABLE 12.3 Application of Disciplines in the Development Phases

Action	Phase			
	Conceptual	Validation	Full-scale development	Production and deployment
Mission profile definition	G	G	G	N/A
Trade-off analysis	G	G	G	C
Function allocation	G	G	S	N/A
Midlevel strategy formulation	S	G	G	C
Detailed task analysis	S	S	G	C
Design fundamentals	S	G	G	C
Mockups and models	S	G	G	N/A
Simulations	S	G	G	C
Closed-loop reporting	S	G	G	G
Developmental testing	S	G	G	C
Acceptance testing	S	S	G	C

Notes: G = generally applicable; S = selectively applicable; C = applicable to changes only; N/A = not applicable.

12.3 SPECIFICATION OF DESIGN FOR PEOPLE

This section presents some of the fundamentals of specification and structuring contracts to ensure proper product design for people. The subjects that are covered include some overall program management considerations, the setting of requirements, considerations of various types of incentives, criteria for source selection, and the various design and management reviews to evaluate program implementation. The information in this section applies to contracts or agreements made between prime contractors and subcontractors or even between different company divisions or groups. The term "contract" is used loosely.

There are several general management considerations that need to be addressed before one can consider the specific mechanics of contracting. These are the activities that must take place in product development, planning, and program implementation. It is convenient to divide the product development cycle into four phases: (1) initiation, or program start; (2) concept demonstration, or preliminary verification of the system concept; (3) full-scale development (or development of production prototypes); and (4) production or implementation. Figure 12.2 provides a checklist of person-machine issues that must be addressed in each of these phases.

The success of a product development all too often depends on how well requirements are described. In the case of person-machine systems, user requirements can be specified in a variety of ways; for example, the user can specify only the human performance characteristics for the system or the user can specify only the machine characteristics. Since the combination of person and machine is important from the point of view of both instantaneous and sustained performance, the user should concentrate on defining the performance and reliability requirements for the total system. For example, the user should define the acceptable function results and the acceptable levels of human error and hardware malfunction for a given environment and a given duration.

CONCEPTUAL PHASE

1. Have the basic product requirements been reviewed to determine if any labor-power, training, and human factors engineering constraints or opportunities were used to justify the need for the product?
2. Have data been supplied that show labor-power requirements against the projected supply or availability of personnel? Do they identify specific skill levels, occupational specialties, and numbers?
3. Have any labor-power, training, or human engineering problems been identified in earlier similar products?
4. Have the constraints on training been identified? These include lead time for training development, the points at which training decisions are needed, and the evaluation of training development procedures.
5. Have labor-power, training, and human engineering been addressed adequately in any contracting procedures?

VALIDATION

1. Do product requirements address labor-power levels and options for the number of personnel, skill levels, and occupational specialties?
2. Have tests been planned that evaluate the impact of the options for personnel on product operation, reliability, and maintainability?
3. Are there any alternatives that reduce the numbers of personnel or skill levels without sacrificing proper and sustained product use?
4. Have human-machine evaluation criteria or incentives been established for required contracts and source selections?

FULL-SCALE DEVELOPMENT

1. Have specific minimum and maximum personnel constraints been established?
2. Has an assessment of the estimates for needed personnel, skills, or occupational specialties been conducted? Have the associated trade-offs been determined?
3. Have human-machine trade-off criteria been used in the development of operation and maintenance work spaces, environmental control concepts, and design concepts for controls, displays, and software?
4. Does developmental and acceptance testing address the adequacy of training, workload, human engineering, and safety? Is there a mechanism for providing the results of this testing to the product designers?
5. Have specific training programs and training equipment been identified?

PRODUCTION

1. Will training programs and equipment be available when the product becomes available to the user?

FIG. 12.2 Program implementation issues. (*Adapted from GAO Staff Study FRCD-82-5, "Guidelines for Assessing Whether Human Factors Were Considered in the Weapon Systems Acquisition Process," 8 December 1981.*)

It is necessary to translate the user requirements into a set of parameters that can be used in the design and verification process. Table 12.1 provides a set of parameters that can be used to specify total human-machine reliability and maintainability and that can be used directly by system designers. Notice that these parameters can be used to specify the system as a whole, which is preferable, or as parts that are constrained by conceptually similar design parameters.

Above all, it is essential that the required human-machine system characteristics be included in the product specification, contract statement of work, or other equivalent documents so that the producing activity is made aware of the fact that both human and machine must be considered in a cohesive manner. By including this information in the solicitation material, the procuring activity assures that any bid or proposal will have taken into account the role of the human in the product.

12.3.1 Incentives

There are a variety of positive and negative incentives that one can use to motivate consideration of the human. The most easily controlled are those that reward the designer for how effectively the actions in Table 12.2 have been employed to develop the person-machine design. In such cases, the designer would receive a positive reward for employing human engineering design disciplines effectively to discover and eliminate deficiences, such as excessive operator task complexity. The evaluation for such a reward would be developed over a series of design reviews that monitor the evolution of the product. A reward system that could be used when working with a contractor would be given in stages; for example, 30 percent of a financial incentive could be awarded when the developer established a set of valid midlevel operator job aid strategies. Another 45 percent could be awarded when a detailed design review showed proper implementation of the strategies and other human engineering principles such as those in MIL-H-46855 and MIL-STD-1472 in the prototype product. The remainder could be awarded upon successful completion of prototype testing.

Negative incentives also may be considered. These may include penalty fees or loss of options. The use of negative incentives is not encouraged because the developer frequently adds to the bid to allow for the possibility of a penalty.

12.3.2 Warranties

Warranties based on field performance are difficult to enforce because of the large number of factors that can influence the behavior of humans. They may be useful in in-plant situations where both the personnel and environments can be controlled. In some respects, a warranty can be construed as a negative incentive because it is possible that product cost may be increased unnecessarily by a developer who seeks to cover risks.

12.3.3 Source Selection

In contracting situations, the source selection process serves as a motivator for the proper design for people. A comprehensively written solicitation defines the user's requirements for quantitative system and personnel performance and reliability; the specific human engineering design techniques, e.g., those in Table 12.2, to be employed (where the specification of such is deemed essential); the management and engineering tasks; and both the identification and solution, if known, of specific human engineering problems likely to be encountered. The inclusion of this last item greatly enhances the likelihood of a successful product development.

In order for the source selection process to be effective, the selection panel also must be prepared to address the human-machine system in a comprehensive manner. This involves having a scoring procedure that gives human engineering a significance equal to that given to cost, schedule, and performance. It also involves providing the selection panel with suitable person-machine analytical tools with which to develop objective technical evaluations of the competing designs. The person-machine simulation techniques described later in this chapter are well suited to the evaluation process. If these techniques are selected, the customer should be sure that the solicitation material requires the producer to submit properly formatted data with the proposal so that the simulations can be run with a minimum of data translation.

The simulation techniques are useful for evaluating how well the proposed designs conform to the customer's quantitative requirements. As indicated above, this is only one facet of the process. There should be evaluation criteria that address the implementation of human engineering disciplines. A comparison of a contractor's proposed human engineering problem with the recommended applications in Table 12.3 may be used for this. The reader should observe that using the table by itself will only address whether a task is performed or not. The use of the table does not determine the adequacy of the performance itself. A more detailed technical evaluation by human engineering experts and reliability engineers is required to accomplish that.

After a contract is awarded, its provisions are monitored through a series of design reviews that are supported by a closed-loop reporting system. Each design review addresses progressively more detailed aspects of the person-machine interaction. At the initial conceptual and preliminary design reviews, midlevel strategies are evaluated against user requirements. Later design reviews focus on the detailed implementation of the strategies and, ultimately, on very specific human engineering issues. Blanchard and Lowry, Van Cott and Kinkade, and Woodson and Conover [4, 5, 6] provide detailed human factors design checklists and criteria that apply to both system operation and maintenance. MIL-STD-1472 also contains tables that can be converted to design review checklists. Deficiencies that are identified in the design review become action items in the closed-loop reporting system. These items are followed by management until satisfactory remedial actions are completed.

12.4 SYSTEM DESIGN

12.4.1 Missions and Allocations

One of the traditionally weak areas in the development of person-machine systems has been defining the mission. "Mission," as it is used in this chapter, is a description of the objectives and environments of the person-machine system. The important elements of this description are:

1. Environmental conditions, e.g., weather, temperature, vibration, and spatial orientation
2. Duration in terms of time or distance
3. Mission objectives in terms of the completion of definable tasks
4. Requirements for the application of strategies, tactics, or specific methodologies
5. Major decisions that must be made and the consequences of those decisions
6. Required communications with other humans or machines
7. Required human and machine performance levels
8. Required levels of human and machine reliability

Each of these elements should be described with respect to the extremes that could be encountered and the norms that are expected. If the development effort is still in the conceptual phase, mission descriptions should be developed based on the different operational and maintenance philosophies that are candidates for the final product. If several missions are contemplated for a product, the corresponding mission descriptions should be developed and assigned an estimate of the percentage of time that the mission will be required. In many situations, it is useful to display the mission as a flowchart with items 1 through 8 above clearly identified.

The mission data provide the foundation for the allocation or apportionment of functions to persons and machine. A qualitative allocation can be conducted based on Table 12.4. In general, the human is better at handling a variety of different information-processing tasks, adapting to new tasks or environments, devising new procedures, and resolving unexpected contingencies. The greatest limitations of the human are the rate of data processing and the amount of immediate memory retention [7].

A quantitative allocation of person-machine functions can be conducted by using a methodology based on conventional optimization techniques. This allows mission, fiscal, and personnel constraints to be addressed. The objective is to maximize the reliability R of the human-machine system over all its missions subject to cost, mission, and personnel constraints. This reliability is written as:

$$R = \sum_i p_i r_i \qquad (12.1)$$

The p_i's represent the probabilities that certain missions will be required and the r_i's are the mission reliabilities, which include the human element.

Mission reliability and operational readiness are the best parameters for optimization because of their wide applicability. Availability may be used in place of readiness depending on the planned use of the system. Bazovski [8] provides an excellent discussion of distinguishing between readiness and availability. The parameters provide a direct link between high-level effectiveness requirements and component-oriented design parameters such as mean time between failure (MTBF) and mean time to repair (MTTR) and both the numbers and skill levels of personnel.

TABLE 12.4 Human-Machine Capabilities

Human superiority	Machine superiority
1. Originality (ability to arrive at new, different problem solutions)	1. Precise, repetitive operations
2. Reprogramming rapidly (as in acquiring new procedures)	2. Reacting with minimum lag (in microseconds, not milliseconds)
3. Recognizing certain types of impending failures quickly (by sensing changes in mechanical and acoustic vibrations)	3. Storing and recalling large amounts of data
4. Detecting signals (as radar scope returns) in high-noise environments	4. Being sensitive to stimuli (machines sense energy in bands beyond human's sensitivity spectrum)
5. Performing and operating though task-overloaded	5. Monitoring functions (even under stress conditions)
6. Providing a logical description of events (to amplify, clarify, negate other data)	6. Exerting large amounts of force
7. Reasoning inductively (in diagnosing a general condition from specific symptoms)	7. Reasoning deductively (in identifying a specific item as belonging to a larger class)
8. Handling unexpected occurrences (as in evaluating alternate risks and selecting the optimal alternate or corrective action)	
9. Utilizing equipment beyond its limits as necessary (i.e., advantageously using equipment factors of safety)	

Source: From NASA SP-6506, *An Introduction to the Assurance of Human Performance in Space Systems*, NASA, 1968.

To apply mission reliability and operational readiness to the allocation, first identify the critical functions and their reliability and maintainability design parameters and use these to write expressions for the mission reliability and operational readiness of the human-machine system. The mission reliability plays a dual role since it provides an input to the expression for R and the mission reliability constraint. One mission reliability constraint equation and one operational readiness equation must be written for each mission.

The next step in the process is the construction of cost and personnel constraints. The most convenient way is to specify the numbers of people at discrete skill levels. This method also allows a clear definition of the effect of the human component on the system. For example, crew size and skill level can be related to MTTR.

Acquisition cost, support cost, and life-cycle cost all can be used as constraints. The reader should observe that optimizing against one of these does not optimize against the others, so consideration is required in selecting the proper cost parameter. After the constraints are defined, the optimization problem is as follows:

Maximize

$$R = \sum_i p_i r_i \tag{12.2}$$

Subject to

Mission reliability	$= R_m \geq \underline{R}_m$	
Operational readiness	$= P_{or} \geq \underline{\lambda}_m$	
Personnel	$= \bar{N} \leq N$	
Cost	$= \bar{C} \leq C$	

The constraints are written as vectors to represent the fact that there will be one set of constraint equations for each mission. Standard methods are used to complete the optimization of R in terms of design-oriented parameters such as the MTBF and MTTR of specific functions and the skill levels. In most cases, dynamic programming (Fig. 12.3) is likely to be the most common method of solution since the analysis of even comparatively simple systems becomes mathematically complicated.

12.4.2   Analysis Techniques

The design of equipments and systems for optimum utility, safety, and effectiveness is dependent on the quality of the analysis which provides a basis for defining and testing the ultimate product. This section emphasizes practical analytical techniques which can be applied early in the design stage to provide the information required for assuring that the ultimate design will meet the appropriate design-for-people standards. The techniques include computer simulation, intellective load analysis (analysis to define how much mental work the system design places on the operator or maintainer), perceptual-motor load analysis (analysis to define how much of a manipulative burden the system design places on the operator or maintainer), THERP (technique for human error rate prediction), decision analysis (analysis in terms of the error which may be induced because of faulty operator decisions), and empirical (deterministic) models.

Computer Simulation.   Computer simulation represents a technique through which the acts and behaviors of ultimate users of a product may be simulated before the product is manufactured. The result is an early determination of areas of operator error, maintainer error, time to perform various activities when employing the product, areas of required training emphasis, and areas of required redesign.

One set of reasons for the emphasis on (i.e., advantages of) behaviorally oriented computer simulation models lies in their ability to allow simulations of systems and conditions which if examined or tested directly, would (1) be dangerous when performed through physical simulation means, (2) be costly in terms of commitment of large numbers of people or of large quantities of equipment, and (3) require long periods of time (i.e., years) to set up and accomplish. Computer simulation also allows for highlighting potential problems in an actual system which has not yet been implemented, or in a situation or set of conditions which has never occurred in an implemented system.

Computer simulation of a human-machine system is also attractive when the actual system "is so fully occupied that experimentation with changes in equipment, personnel policy, or resource assignment rules may be impractical, expensive, dangerous, or unlawful" [10].

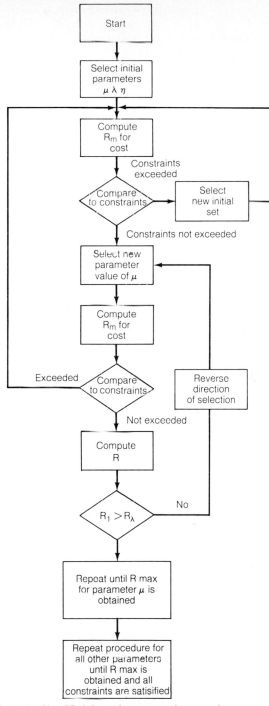

FIG. 12.3 Simplified dynamic programming procedures.
(*From Ref. 20.*)

Finally, such models provide a useful methodological tool to those behavioral and engineering scientists who are involved with: (1) the quantification and enhancement of human capability and its inherent variability, and (2) the related issue of the contributions of human unreliability to total system unreliability.

Computer models have been variously defined. However, the essence of most definitions seems to be that computer simulation attempts to mimic or to represent some aspect of real life and that the simulation process is embodied in the form of a high-speed digital computer program.

A behavioral simulation model attempts to represent logically and mathematically the psychological response, and to some extent the physiological response of the equipment user. It is concerned with human-equipment, human-human, and human-environment relationships and with their interactions, i.e., with the performance of individuals and groups under varying conditions due to the environment, training, operational doctrine, characteristics of equipment, and individual capabilities. The representation should be of such a nature that it allows manipulation of extrinsic (situational, environmental, equipment) or intrinsic (individual and group) events to determine the direct or indirect effects on some output measure.

In this context, behavioral computer simulation provides the system developer with answers to a wide variety of questions while a system is early in the development phase. Questions can be answered such as:

- What are the quantitative personnel requirements?
- What are the qualitative personnel requirements? Where, during system utilization, are operators most overloaded? Underloaded?
- How will cross training improve system effectiveness?
- Are the system operators able to complete all of their required tasks within the time allotted?
- Where in the task sequence are operators or teams likely to fail most often? Least often?
- In which state(s) of the system is the "human" subsystem (and/or components thereof) least reliable and why?
- How will task restructuring or task reallocation affect system effectiveness?
- How much will performance degrade when the system operators are fatigued? Stressed?
- How will various environmental factors (e.g., heat, light, terrain) affect total human-machine system performance?
- To what extent will system effectiveness improve or degrade if more or less proficient operators are assigned?
- How do group factors such as morale and cohesion affect system performance?

Computer simulation provides the quantitative data so that analysis can develop answers to such questions while systems are in the exploratory as well as the advanced development stages. The answers to such questions can then be used for system redesign or modification early in the developmental cycle. Once the design changes have been decided on, a series of computer simulations can be completed representing the modified system to predict the extent of the improve-

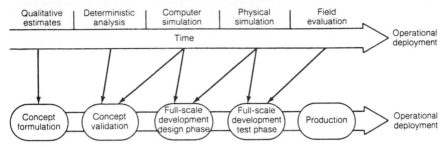

FIG. 12.4 Relationship between various system evaluative techniques and the system development continuum.

ment to be brought about by the modified system. Computers are involved in the process because of the extremely large number of calculations required for any one analysis.

The upper part of Fig. 12.4 places simulation models into perspective relative to the fidelity of various available evaluative and predictive techniques. As one proceeds along the time continuum, the techniques approach reality to a greater extent. Hence, the fidelity of computer simulation falls between deterministic analysis and physical simulation. The lower portion of Fig. 12.4 portrays the system development cycle from the conceptual phase to the product phase and operational deployment. Different evaluative and predictive techniques become appropriate during various stages in the equipment system development cycle. Qualitative estimates of system effectiveness are viewed as sufficient during the basic research phase of system development. As the system design matures and hardens in the exploratory and the advanced development stages, deterministic and computer simulation evaluative techniques play a dominant role. In the more advanced stages of development, physical simulation and various field operational and technical evaluations play a more important role.

There are a number of available simulation models. Perhaps the oldest and most substantiated are the Siegel-Wolf models [10] which apply to equipments involving as many as 20 operators. Use of these models involves:

1. Developing the required input data. These data include for each subtask in the task such items as: success probability, next subtask to be attempted after subtask success or failure, subtask(s) which must be completed before the current subtask may be attempted, and time and standard deviation of subtask completion.
2. Running the model with the above derived input data.

The output from the models provides detailed quantitative information about: the probability of completing each subtask successfully, the probability of completing the task within a given time allowance, the time required for task completion, required operator competency, areas of high time stress, and the effects of operator "stress tolerance" on task performance. Such information is then employed as a basis for equipment redesign. The redesign may then be modeled to determine its probable effectiveness.

Other available behavioral simulation models include a model developed by Boeing [12] to evaluate the physical workload on air-crew members; the human operator simulator (HOS), which was developed by the Navy to provide the user with detailed task analytic information [13]; a logistic model built to predict aircraft maintenance time [14]; PROCRU, another Air Force simulation model [15] which simulates flight crew procedures during approach to landing and incorporates both procedure- and rule-based behaviors; and a model of Kleinman et al [16] which conceptualizes an operator control model including cognitive and perceptual considerations.

Intellective Load Analysis.   Often a system or equipment will overload the user from the intellective point of view, i.e., the intellective demands on the operator or maintainer may be excessive. The operating sequences may be difficult to remember, the number of decisions which the user must make may be too great or the decisions may be too difficult, an excessive number of evaluations may be required, and the like. The intellective load analysis provides a method for identifying areas of intellective overload on the user before an equipment is produced.

The intellective load analytic technique first dissects each subtask involved in equipment employment into the major type of intellective activity involved in subtask completion. Then, the amount and the difficulty of the major type of intellective ability involved in completion of each subtask are estimated. Integration of the amount and the difficulty values within each subtask provides the information necessary for deciding whether and where intellective overload may exist.

One convenient scheme for classifying intellective activity by type was developed by Guilford [17]. As a result of an extensive factor analytic research program, he postulated the model shown in Fig. 12.5 as representative of the structure of human intellect. According to this model, intellective activity may be classified in terms of its content, operations, and products. Content referred to the type of information processed and contained four categories: figural (F), symbolic (S), semantic (M), and behavioral (B). Operations referred to the type of intellective activity involved and were classified as: cognition (C), memory (M), convergent production (N), evaluation (E), and divergent production (D). Products referred to output from the interaction of the operation with the content. Products were classified as: units (U), classes (C), relations (R), systems (S), transformations (T), and implications (I). Thus, the cell marked with an X in Fig. 12.5 may be considered to represent memory of semantic relations. Although the total model contains 120 cells, the use of considerably fewer is possible in most design situations.

Once the areas and types of elevated intellective requirements are identified and understood through such a model, design or other changes can be introduced to decrease the intellective requirements for the associated subtasks. Changes may include reallocation of functions (i.e., increased automation), reorganization of the subtask sequence so as to distribute the intellective load on the user more evenly over the total task sequence (i.e., plotting the load against subtask and then reorganizing the subtask sequence so that the graph contains no peaks), assignment of additional personnel to subtasks with high intellective requirements, and provision of job aids which decrease the intellective burden.

To complete the intellective load analysis for a given equipment, the following steps are completed:

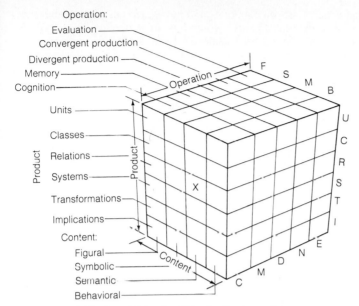

FIG. 12.5 Guilford model for classifying intellective activity.

1. Develop a sequential list of the subtasks in the total task.
2. Determine whether each subtask is intellective or perceptual-motor (see next section) in nature.
3. For each subtask identified as primarily intellective in nature, identify the type of intellective activity most required for subtask completion.
4. Assign a value from 1 to 7 to represent the amount of intellective activity involved in subtask completion. Assignment of a value of 1 would indicate "little" or "none," whereas assignment of a value of 7 would indicate "very much."
5. For each subtask, assign a value from 1 to 7 to indicate the difficulty of the involved intellective activity. A value of 1 would indicate that the intellective activity is "extremely easy," whereas a value of 7 would indicate that the activity is "extremely difficult."
6. Multiply the assigned amount and the difficulty value for each subtask. The result will be a set of intellective load values which range from 1 to 49. Subtasks containing values greater than 25 should be examined for ways to reduce the intellective load required for subtask performance.

The technique has been found useful in aircraft design, task design, radar system design, and computer interface analysis.

Perceptual-Motor Analysis. The perceptual-motor analysis, as opposed to the intellective load analysis, analyzes the load placed on the system user from the point of view of the manipulative or the manual dexterity requirements of a task.

Included within the concept are such motions and movements as: finger dexterity, eye-hand coordination, gross adjustments, two-limb coordination, and rate control.

The level of perceptual-motor load imposed by a system design may be assessed in a manner which parallels the one employed for the intellective load analyses. The steps are as follows:

1. For each subtask in the total task, decide whether the subtask is primarily intellective or perceptual-motor in nature. Continue the analysis on only those subtasks which are primarily perceptual-motor in nature.
2. For each subtask which is primarily perceptual-motor in nature, assign a value between 1 ("very little") and 7 ("very much") to represent the amount of the perceptual-motor activity involved in completing the subtask. Perform the same operation to represent the difficulty of the perceptual-motor activity.
3. Multiply the assigned "amount" and "difficulty" values by subtask to yield a perceptual-motor load value for each subtask. Subtasks containing a perceptual-motor load value greater than 25 should be examined for ways to decrease the perceptual-motor load on the operator.

Figure 12.6 presents the results of application of the intellective and the perceptual-motor load analytic techniques to a target engagement task in a proposed Army tank. The analysis was performed during the concept-development phase and served as a basis for proving and improving the design. There are five subtasks for the gunner and ten for the commander. In this case, it is evident that the design did not overload the gunner either from the intellective or the perceptual-motor points of view and confidence is increased that the design-for-people goal was not under violation for this position. However, for the commander, there was evidence of intellective overload and the need for reconsideration of the design of several of the subtasks.

Technique for Human Error Rate Prediction (THERP).   THERP [18] has evolved over the years and seems to possess special interest to those involved in human error analysis and prediction. The Nuclear Regulatory Commission has given considerable emphasis to this technique in probabilistic risk assessment.

THERP is largely deterministic in nature and, for that reason, has been criticized as not reflecting the complexity involved in human behavior. Moreover, the technique has been criticized as depending on subjective estimates and, accordingly, as susceptible to analyst influence. Validation data about the technique remain unreported in the general literature—if such data are extant.

The basic tool in applying the THERP methodology is a tree diagram representation of the actions taken by an equipment operator or maintainer to complete a task. Probability values are assigned to each successive subtask success or subtask failure branch in the tree. These probability values are compounded in accordance with the usual probability compounding principles to yield an estimate of human error probability for the task under analysis. The procedure is analogous to reliability determination for an equipment on the basis of the reliability of the components.

One of the more interesting aspects of the technique is its incorporation of "performance shaping factors." These are operator and environmental variables which influence the assigned probabilities at various points in the activity sequence. However, the treatment of these variables in the THERP technique also allows for the introduction of further analyst bias.

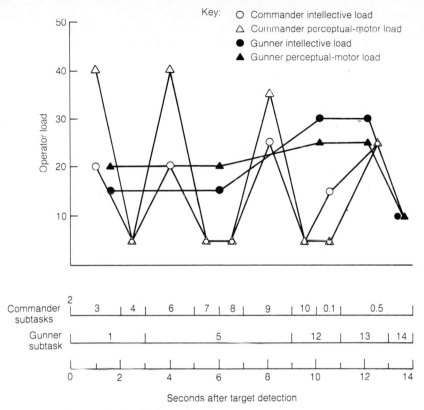

FIG. 12.6 Operator load analysis results.

Decision Analysis. As suggested in the discussion of the intellective load analytic technique, the mental or cognitive aspects of task performance represent primary issues during early equipment design. In accordance with this emphasis, there has been some focus on early analysis of the decisions which the equipment imposes on the user. Of course, the attempt is to maintain the decision difficulty or complexity within the ability of the anticipated users.

One technique for evaluating the decision-making reliability within a system was described as early as 1967 by Siegel and Miehle [19]. This technique, like THERP, is based on traditional probability compounding techniques for establishing the decision-making error probability. Within the logic of the method, the various decisions involved in completing a task are isolated and the probability of completing these decisions is calculated. Low decision-making success probability values are considered to be indicative of the need for equipment redesign, task reorganization, job aids, training emphasis, or personnel paralleling.

Empirical Models. There is a set of empirical models that addresses the relationship between the maintenance technician and maintenance times or labor power [20]. These were developed from sonar repair information. One of the more interesting aspects of the empirical methods is a concept called maintenance power, which is the product of maintenance technician experience and time put on a repair. The relationship between mean time to repair (X) and maintenance power (Y) is: $Y = 16.87X + 4.67$.

The empirical models also include a log-normal prediction model for estimating various maintainability parameters, a repair time distribution prediction method, and a method for predicting the number of required maintainers. The computations for each of these are quite direct. The methods are drawn from Navy data, and hence their specific applicability to nonmilitary data may or may not hold. However, equations specific to automobiles, computer systems, etc., and to other populations of technicians could easily be developed.

12.5 SYSTEM ENGINEERING PRINCIPLES

In this section, some of the more important human-machine engineering principles are discussed. These address qualitative rather than quantitative aspects of the design. In particular, the principles of similarity, automation, simplification, stress factors, and both human variability and limitations are discussed.

12.5.1 Similarity

To the commercial consumer, ease of operation and maintenance through product similarity long have been factors in customer satisfaction. For example, there is a minimum amount of retraining required to drive different automobiles or operate different television sets. However, as automobiles become more sophisticated, there is a considerable amount of retraining required to maintain them. The situation of the 1960s, when a mechanic could repair almost any car using the same methods, no longer exists.

Generally, product similarity degrades from product generation to generation in areas of rapidly changing technology. Lack of similarity or standardization has the following problems associated with it:

1. Extensive retraining frequently is required for those who must operate or maintain the new systems.
2. Spare part support systems become more complicated because many more parts must be carried and the likelihood of parts being out of stock increases.
3. Human-induced failure or errors increase because of lack of familiarity with the equipment.

For both the military and commercial producer, the training problem is costly. The commercial producer suffers increased costs of retraining service personnel. Customers must either retrain themselves or attend producer courses at increased cost in both time and money to themselves. The military user suffers a proliferation of personnel types and training courses and experiences delays in bringing new systems and their attendant personnel to full readiness.

The increased use of similarity or standardization has both benefits and liabilities. Training and supply costs are reduced. In the design process, lessons learned from field experience can be translated more easily into design improve-

TABLE 12.5 Similarity Costs and Benefits

Similarity design action	Relative cost	Relative benefit	Payback period
Extend use of "lessons learned"	Medium	High	Short range
Develop modular systems	High	Very high	Long range
Promote standard interface designs	Low	Moderate	Short range

Source: Adapted from Ref. 3.

ments. On the other hand, strict similarity requirements limit the innovation that can enter the design process, although this impact can be reduced from an operator standpoint if standards are set for the operator-machine interface. Table 12.5 summarizes the costs and impact of several aspects of similarity in design.

For the designer, the use of similarity to its greatest advantage consists of the proper use of analytic tools and design disciplines. Table 12.2 identifies the specific analytic tools and design factors that are used to incorporate similarity.

12.5.2 Automation

Proper design for people does not mean uncritical automation of human functions. "Automated" systems have often produced designs which contain person-machine problems. The important design-for-people consideration is what to automate or, stated alternatively, what functions to assign to the equipment and what functions to assign to the human.

The artificial intelligence concept is related to functional allocation. Artificial intelligence attempts to unburden the system's users by making decisions for them. Theoretically, such systems will incorporate learning by experience as well as the use of the full range of the available knowledge in their solution of problems. Since one of the areas of human weakness is decision-making ability under complex conditions in which the objective probabilities are uncertain, artificial intelligence would seem to introduce another area for automation. However, where the concept has been employed, it has been found that the artificial intelligence output is best considered as suggestive, with the final decision remaining with the equipment user. Part of the problem rests in the indication that the "subjective" probabilities assigned by experienced decision makers do not always closely match the "objective" probabilities of the real world. In such cases, the decisions reached by the artificial intelligence system are not satisfactory to the experienced decision maker. Accordingly, such systems represent decision-support or decision-aiding systems rather than decision-making systems.

Decision-aiding or decision-supporting systems (i.e., systems which transform information for the user or which suggest, but not prescribe, a course of action) have been found to be accepted by users and to improve the decision-making quality over that of the unaided decision maker [21, 22]. Siegel and Madden [21] reported that for a tactical military situation there was an increase in decision validity by a factor of 5 when unaided decisions were compared with aided decisions. Madden and Siegel [22] found a 25 percent increase in the remaining value of a task force after a "hostile strike" when a decision aid was employed in planning the defense against the strike.

Siegel and Madden [21] presented a flowchart of the steps which should be followed in the development of such aids. Their procedural sequence is presented as Fig. 12.7. The figure is read from the bottom to top with the considerations involved in each stage entering from the left of each box and the results of each stage exiting to the right. The number(s) above each box in Fig. 12.7 represent criteria which may be applied after each developmental stage. These criteria are defined in Table 12.6. The rounded boxes associated with each rectangular, stage box represent descriptors which may be applied as the criteria as the successive stages are met. Accordingly, an aid may be successively called "suitable," "testable," "reasonable," "valid," "effective," and "useful."

It is important to know not only that an aid does or does not possess certain attributes but also which of its characteristics contribute to the attributes. For example, an aid might be useful because it synthesizes information from a diversity of areas to give a planner new insights; might significantly reduce the amount of time and labor required to make a decision; or might allow for a more careful analysis of a broad range of alternatives. Therefore, in attempting to examine the usefulness of any aid, it is necessary to specify the factors out of which its utility may have been derived and how they may interact.

Linked with the attributes of an aid is a concern for the goals of the aid, their relative importance, and how closely they were achieved. The goals are objective expressions of what the aid should be, or do, or facilitate. Therefore, an examination of how clearly the goals of an aid were achieved and their relative importance should lead to a better understanding of what contributes to the usefulness of an aid and possibly to the knowledge of how to improve it.

12.5.3 Simplification

The design principle of simplification is another one of those seemingly obvious but frequently neglected principles. In the commercial world, it is in the category of "user friendliness." The objective of this principle is to reduce the complexity of tasks so that user skill levels and training requirements are reduced.

The analytical techniques and design factors are summarized in Table 12.2. As may be seen from the table, there are a number of facets to the simplicity. The first is the principle of operator and maintainer task simplification. The second is that of the proper use of coding to ease the burden on the user. Third is the writing of supporting documentation to the lowest practical level of comprehension, e.g., the ninth grade for most high-technology products. The last is the simplification of wiring diagrams, diagnostic procedures, and computer coding instructions. These last two probably are among the most frequently violated fundamentals in both the military and commercial sectors.

The overall impact of task simplification is shown in Fig. 12.8. Generally, the approach taken to simplify tasks consists of the following as a minimum [5, 23]:

1. Minimize the number of operational options.
2. Minimize the number of processing parameters.
3. Minimize the number of controls and displays.
4. Reduce the complexity of the steps required for operation.
5. Provide clear operating and maintenance instructions.
6. Provide cues or job aids to guide performance.

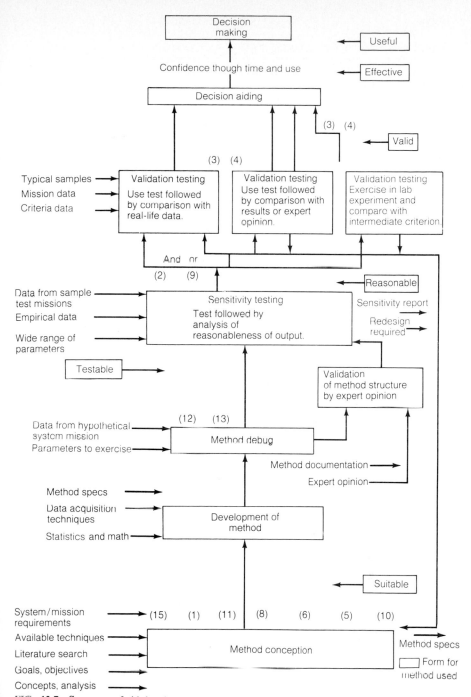

FIG. 12.7 Sequence of aid development.

12.27

TABLE 12.6 Criteria for Evaluating the Utility of a Decision Aid

Criterion	Definition
1. Internal consistency	Extent to which the constructs of the aid are marked by coherence and similarity of treatment
2. Indifference to trivial aggregation	Potential of the aid to avoid major changes in output when input grouping or conditions undergo insignificant fluctuations
3. Correct prediction in the extreme (predictive or empirical validity)	Extent of agreement (correctness of predictions) between the aid and actual performance at very high or low levels of conditions
4. Corrective prediction in mid range (predictive or empirical validity)	Like above for middle-range values of conditions
5. Construct validity	Theoretic adequacy of the aid's constructs
6. Content (variable parameter) validity (fidelity)	Extent to which the aid's variables or parameters match real-life conditions
7. Realism or "face validity"	Extent to which selected content matches each attribute included
8. Richness of output	Number and type of output variables and forms of presentation
9. Ease of use	Extent to which an analyst can readily prepare data for, apply, and extract understandable results from the aid
10. Cost of development	Value of effort to conceive, develop, test, document, and support
11. Transportability or generality	Extent of applicability to different systems, missions, and configurations
12. Cost of use	Value of all effort involving use of aid including data gathering, input, data processing, and analysis of results
13. Internal validity	Extent to which outputs are repeatable when inputs are unchanged
14. Event or time series validity	Extent to which aid predicts event and event patterns

7. Clearly label all controls, outer cases, and removable parts.

As stated above, propercoding is one means of simplifying tasks. Table 12.7 provides general guidance on the various approaches to control coding. There is one caution that must be mentioned: oversimplification can be detrimental because the human may lose interest in the task. The designer should match the task to the capabilities andcharacteristics of the performer [3].

12.5.4 Stress Factors

The "stress" which equipment or system operation and maintenance place on the user or the maintainer is considered to be a feature limiting successful and effective equipment employment. Accordingly, proper design for people includes

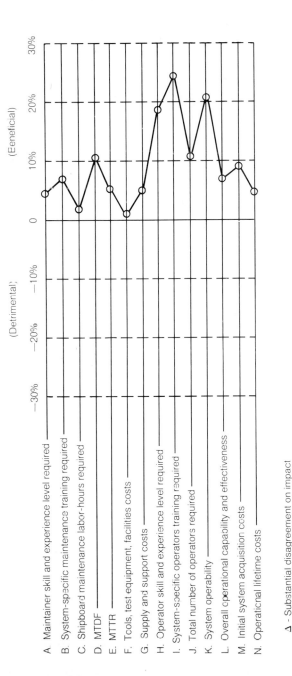

A. Maintainer skill and experience level required
B. System-specific maintenance training required
C. Shipboard maintenance labor-hours required
D. MTDF
E. MTTR
F. Tools, test equipment, facilities costs
G. Supply and support costs
H. Operator skill and experience level required
I. System-specific operators training required
J. Total number of operators required
K. System operability
L. Overall operational capability and effectiveness
M. Initial system acquisition costs
N. Operational lifetime costs

Δ - Substantial disagreement on impact

FIG. 12.8 **Judged impact of simplifying equipment operators' tasks.** (*From Ref. 23.*)

TABLE 12.7 Advantages and Disadvantages of Different Types of Coding

	Type of coding					
	Loca-tion	Shape	Size	Mode of opera-tion	Label-ing	Color
Advantages						
Improves visual identification	X	X	X		X	X
Improves nonvisual identification (tactual and kinesthetic)	X	X	X	X		
Helps standardization	X	X	X	X	X	X
Aids identification under low levels of illumination and colored lighting	X	X	X	X	(When transilluminated)	(When transilluminated)
May aid in identifying control position (settings)		X		X	X	
Requires little (if any) training; is not subject to forgetting					X	
Disadvantages						
May require extra space	X	X	X	X	X	
Affects manipulation of the control (ease of use)	X	X	X	X		
Limited in number of available coding categories	X	X	X	X		X
May be less effective if operator wears gloves		X	X	X		
Controls must be viewed (i.e., must be within visual areas and with adequate illumination present)					X	X

Source: From MIL-STD-1472B.

minimization of the stress which equipment use places on the equipment user or maintainer. Some of the earlier sections of this chapter dealt with the minimization of the intellective and the perceptual-motor loads on the equipment user. This translates, in one sense, to stress minimization through reduction in demands or workload. This type of stress minimization was also discussed above. However, other sources of stress exist. These may be classified into physical and workplace stressors, psychological stressors, and task stressors.

Physical and Workplace Stressors.   Physical and workplace stressors include any environmental factor of a physical nature which may degrade performance reliability. Examples are light level, sound level, temperature of the workplace, ventilation, accessibility of materials and items to be manipulated, design which fits the anthropometric characteristics of the anticipated user, design for adequate communications, adherence of the design to accepted human engineering principles, design for safety, adequacy of manuals and job aids, and adequacy of the job training program.

Although it may seem obvious that designs containing inadequacies are unacceptable, equipments containing such features continue to proliferate. The cost of such features in terms of productivity is difficult to estimate, but conceivably such costs could run astronomically high.

The "fix" in most of these cases is more thorough consideration of what the operator or maintainer will need to do with the equipment or system under consideration and design to support these activities. Often such design fixes, while inexpensive, are not trivial—as in the case in which incorporation of a $10 lighting fixture can help to avoid a costly industrial accident.

Task Stressors.   Task stressors involve those issues which have to do with the design of the task(s) which the operator or maintainer must perform. The matter of proper allocation of functions to the equipment or to the human in the system was discussed above. However, once a task has been allocated to the human, its execution can be compromised by inadequate design. The sequence of task performance must be analyzed into the steps of the sequence of activity and then each step must be analyzed in terms of the operator actions involved, the potential for error, the effects of the error on outputs, and the indication which will be given to the operator or maintainer when the step has been successfully completed. Each of these items will provide confirmation of design adequacy or indicate areas for needed task redesign. Industrial engineering and human factors engineering texts provide task design guidance.

However, there are other techniques which may be considered for avoiding operator or maintainer error. These include:

Task paralleling: Task paralleling involves the assignment of a task to two persons who perform the task at the same time. Paralleling assures increased effectiveness, since two persons are performing the task. That is, every step of the task has two people looking at it. Paralleling raises the reliability of task execution. Task paralleling might be considered for highly critical tasks (i.e., tasks which if performed incorrectly will induce considerable error in system function).

Task sharing: Task sharing involves the performance of some, but not all, of the steps of a task by two workers. Sharing requires coordination between the two workers, since they work at the same time on the task. Tasks which are very tedious and repetitive are likely to be performed more effectively when they are shared.

Task rotation: In tedious and repetitive tasks, performance declines as a direct function of time on the task. Although not a design item, this performance degradation can be anticipated by the equipment designer who can also recommend frequent rotation of individuals on the task or frequent rest periods.

Psychological Stressors.   The design of tasks and equipment for maximum performance reliability also depends on limiting a number of psychological stress factors. For example, the design may impose time stress on the operator or maintainer; that is, the operator or maintainer may have more to do in the time allowed than can be completed in that time. This type of situation will impose time stress; the workers will take risks and performance reliability may degrade.

Alternatively, the worker may skip certain subtasks to gain time. This course of action may also serve to induce a less than adequate end result.

If additional time cannot be allowed for task completion, time stress may be reduced or relieved by task sharing or task paralleling. It may also be reduced by task restructuring or possibly through the provision of job aids.

Other Stresses.   Other stresses may also be correlated with less than adequate design. The stress associated with working under conditions which strain the equipment user's sensory-motor or intellective capacities has already been mentioned. Similarly, the information processing load placed on the operator can yield stress. It has been suggested [24] that there is a limit on the amount of information which an equipment user can process per unit time. Overloading the operator with too much information or with more information to process than can be processed in the time allowed will lead to a failure to consider all provided information and to operator error. A related problem is that experienced users may emphasize different information than inexperienced users. This suggests that a design optimized for the inexperienced user will not be satisfactory for the more experienced user.

12.5.5   Individual Variability and Limitations

Individual differences can manifest themselves on equipment user performance in innumerable ways. There are differences between individuals in sex, aging process effects, skills and abilities, health and illness effects, handicaps and physical limitations, and susceptibility to environmental and task factors.

Regardless of the selected variable, there are mean and standard deviation changes both within and across levels of the variable. For example, Fig. 12.9 shows the changes in the mean and the standard deviation of the intelligence quotient, as measured by the Wechsler Adult Intelligence Scale, as a function of age.

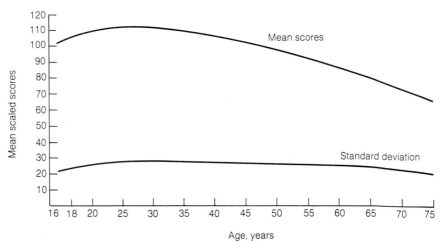

FIG. 12.9   Changes with age in full-scale scores of the Wechsler Adult Intelligence Scale, ages 16 to 75.

The point is that neither measured intelligence nor its standard deviation remain constant over time. This might suggest that depending on the user age bracket, different equipment designs are appropriate.

Other skills and abilities also vary among individuals. Table 12.8 presents the mean time to complete a simple assembly task included in the Crawford Small Parts Dexterity Test. There is about a 10 percent difference between the fastest (technical students) and the slowest (unselected applicants) on even this simple task.

Similarly, for physical characteristics, considerable differences exist. Figure 12.10 [25] presents the height of seven military populations and serves to point out both the within-group and across-group variation in height.

12.6 TESTING

Testing generally falls into two categories: developmental testing and acceptance testing. In the design-for-people process, the developmental testing serves the purpose of investigating the effectiveness of design concepts, especially the midlevel strategies. This type of testing is a "no penalties" testing that is conducted by the producer. The other type of testing, acceptance testing, is conducted by the customer for the purpose of identifying specific deficiencies with respect to design requirements. It is emphasized, however, that adequate person-machine performance is achieved through proper design, not through testing.

12.6.1 Choice of Test Subjects

There are two important testing fundamentals that should be applied in the process of designing for people: using actual user test subjects and using controlled test methods.

The user test subjects are required to assure that the design is tested using realistic human capabilities. One of the greatest shortcomings in product development efforts is the use of producer personnel who actually are more familiar with the product than either designer or user believe. This tends to provide optimistic test results and frequent complications as the product approaches

TABLE 12.8 Mean Time and Standard Deviation of Time (in Minutes and Seconds) of Various Male Subpopulations to Complete Crawford Small Parts Assembly Test, Part I

	Mean	Standard deviation
Unselected applicants	5' 20"	1' 19"
Factory applicants	5' 18"	1' 32"
War vets., Univ. of Puerto Rico	5' 01"	0' 59"
Technical students	4' 51"	0' 50"
Academic students	5' 19"	1' 08"

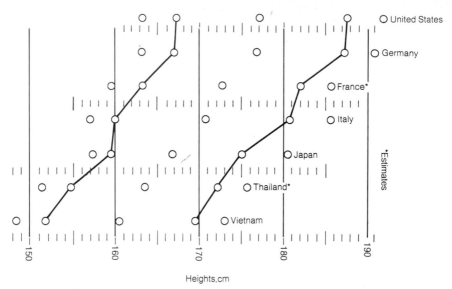

FIG. 12.10 Statures of seven military populations.

actual use. Candidate test subjects should be screened carefully to eliminate those with excessive knowledge of the design. The use of user-supplied personnel frequently is the least complicated solution to this problem. Figure 12.11 provides a convenient form for gathering subject data.

12.6.2 Developmental Acceptance

The controlled test methods are required to ensure the validity of conclusions drawn from the tests. As one might suspect, human performance at any given time is determined by a large number of both obvious and subtle factors that can provide significant variations in performance. Properly constructed tests compensate for or analyze these factors. All aspects of the testing should be fully standardized. This includes but is not limited to the following: timing of the various sections of the tests, the instructions given to the test subjects, the methods for dealing with questions that may arise, the recording of anomalies, and the nature and method of collecting data [27].

The most important factor in developmental testing is its timing. It should be performed sufficiently early so that its results can be incorporated into the design. Full-scale mockups and scale models are useful tools for exploring system concepts early in development. These can be used to evaluate the adequacy of size, shape, arrangement, and panel layout. They also can be used to determine issues such as communication distance, access, and critical human performance characteristics. Functional mockups, including computerized simulators, are important when humans serve critical functions in a system. Dynamic simulations are very effective design tools for complicated person-machine systems. They serve a secondary purpose as trainers.

Generally, person-machine acceptance testing serves four purposes:

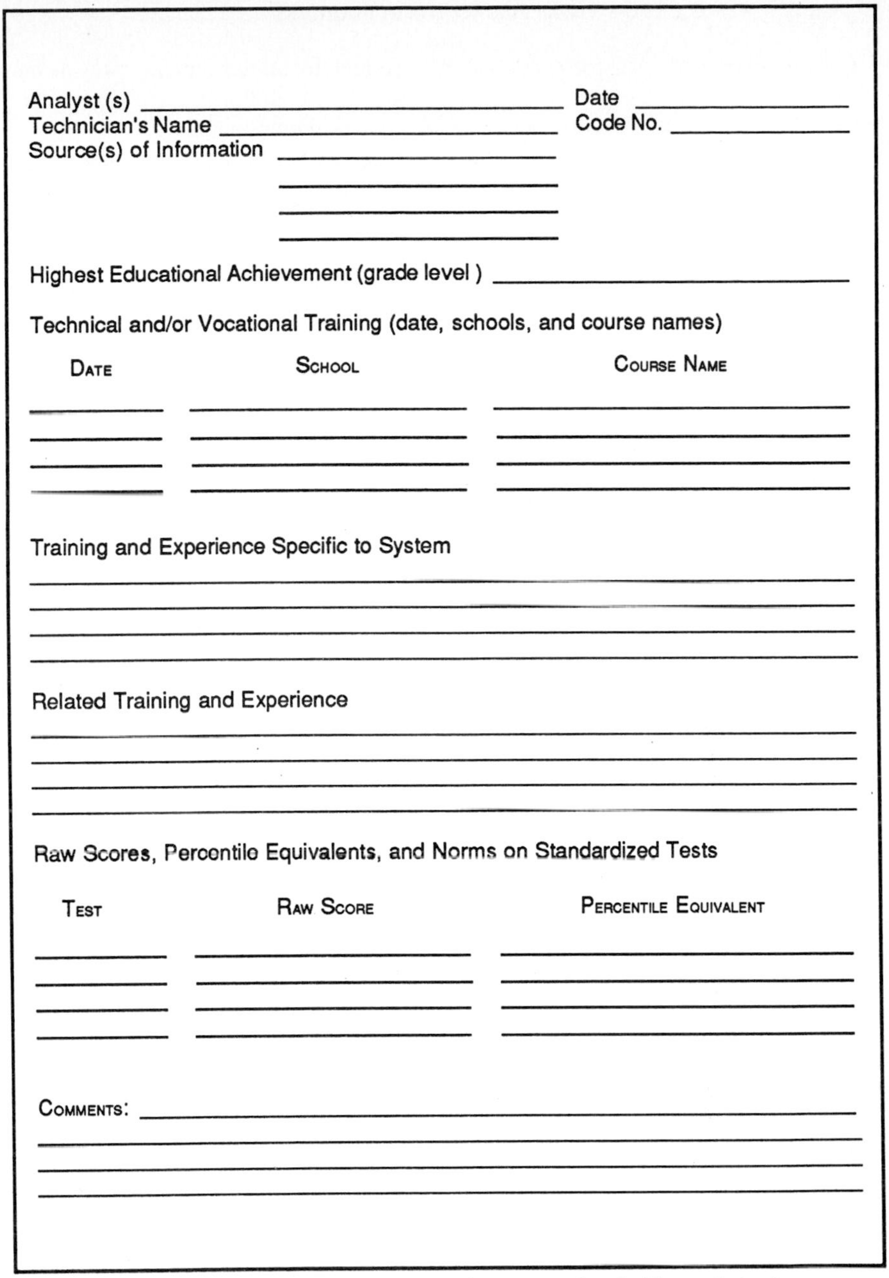

Analyst (s) _____ Date _____

Technician's Name _____ Code No. _____

Source(s) of Information _____

_____

_____

_____

Highest Educational Achievement (grade level) _____

Technical and/or Vocational Training (date, schools, and course names)

DATE	SCHOOL	COURSE NAME
_____	_____	_____
_____	_____	_____
_____	_____	_____
_____	_____	_____

Training and Experience Specific to System

_____

_____

_____

_____

Related Training and Experience

_____

_____

_____

_____

Raw Scores, Percentile Equivalents, and Norms on Standardized Tests

TEST	RAW SCORE	PERCENTILE EQUIVALENT
_____	_____	_____
_____	_____	_____
_____	_____	_____
_____	_____	_____

COMMENTS: _____

_____

_____

_____

FIG. 12.11   Test subject qualifications. (*From Ref. 26. Reproduced with permission.*)

1. Demonstrate the conformance of the product to human engineering design criteria
2. Confirm compliance with specific performance requirements
3. Secure quantitative measures of person-machine performance characteristics that are functions of person-machine interaction
4. Determine whether or not undesirable characteristics have been introduced

The findings of design reviews and exploratory testing provide specific issues for evaluation in acceptance testing. Some important general elements of the test are as follows [28]:

1. A simulation (or actual conduct where possible) of the mission or work cycle
2. A representative sample of noncritical scheduled and unscheduled maintenance tasks that does not duplicate those selected for the maintainability demonstration, if one was conducted
3. Proposed job aids, new equipment training programs, training equipment, and special support equipment
4. Typical user personnel
5. The collection of task performance data in simulated or, where possible, actual environments
6. Identification of discrepancies between required and obtained performance
7. Criteria for acceptable performance of the test

When a specific test is to be conducted, the following items must be defined as follows [27]:

1. Conditions of the test
2. Specification of the resources
3. Apparatus
4. Test equipment, tools, and supplies
5. Time and duration of test
6. Test subject and evaluation personnel selection and training
7. Control of test conditions
8. Control of test materials
9. Analysis and reporting of test data

The above list establishes a basis for deriving specific test procedures.

This section has provided only a brief introduction to the types of person-machine system testing and some of the more important aspects of that testing. Although the discussion has proceeded without referring to hardware or software testing, the reader should be aware that significant financial and technical benefits frequently can be obtained by combining tests. This requires careful planning very early in product development so that tests can be integrated to the greatest advantage.

12.7 SUMMARY

This chapter provided an overview of the disciplines that are used when products are designed for the people that must operate and maintain them. It addressed the process of setting requirements and contracting to ensure that the development effort is properly structured and funded for the design-for-people process. It also provided an overview of the requirements and system analysis processes. Finally, the roles of testing as a supplement to the design process and a checkpoint for the user were discussed.

REFERENCES

1. P. A. Watson and W. Hebenstreit, "Manpower, Personnel and Training Working Group Report," IDA Record Document D-35, Institute of Defense Analysis, 1801 Beauregard Street, Alexandria, VA 22311.

2. GAO Report PSAD-81-7 of 29 June 1981.

3. Naval Research Advisory Committee Report, "Man-Machine Technology in the Navy," December 1980.

4. B. S. Blanchard and E. E. Lowery, *Maintainability Principles and Practices,* New York, McGraw-Hill, 1969.

5. H. P. Van Cott and R. G. Kinkade, *Human Engineering Guide to Equipment Design,* rev. ed., Washington, D.C., American Institutes for Research, 1972.

6. W. E. Woodson and D. W. Conover, *Human Engineering Guide for Equipment Designers,* 2d ed., Los Angeles, University of California Press, 1966.

7. NASA SP-6506, *An Introduction to the Assurance of Human Performance in Space Systems,* NASA, 1968.

8. I. Bazovski, Sr., "Weapon Systems Operational Readiness," *Proceedings, 1975 Reliability and Maintainability Symposium,* Washington, D.C., 28–30 January 1975, pp. 174–178.

9. *Human Reliability Prediction System User's Manual,* Washington, D.C., Naval Sea Systems Command, December 1977.

10. Arthur I. Siegel and J. Jay Wolf, *Man-Machine Simulation Models: Psychosocial and Performance Interaction,* New York, Wiley, 1969.

11. William W. Haythorne, "Information Systems Simulation and Modeling," in *Information System Simulation and Modeling Techniques.* First Congress on the Information System Sciences, Report ESD-TDR-63-474-7. Bedford, Mass., Electronic Systems Division, 1963.

12. D. L. Parks and W. E. Springer, "Human Factors Engineering Analytic Process Definition and Criterion Development of CAFES," Report D180-18750-1, Seattle, Boeing Aerospace Company, 1975.

13. N. E. Lane, M. J. Strieb, F. A. Glenn, and R. J. Wherry, "The Human Operator Simulator: An Overview," in J. Moraal and K. F. Kraiss, eds., *Manned Systems Design: Methods, Equipment, and Applications,* New York, Plenum Press, 1981.

14. W. B. Askren and L. M. Lintz, "Human Resources Data in System Design Trade Studies," *Human Factors,* 1975, *17,* pp. 4–12.

15. S. Baron, G. Zacharias, R. Muralidhavan, and R. Lancraft, "PROCRU: A Model for Analyzing Flight Crew Procedures in Approach to Landing," *Proceedings of the Sixteenth Annual Conference on Manual Control,* Cambridge, Mass., Massachusetts Institute of Technology, 1980.

16. D. D. Kleinman, S. Baron, and W. N. Levenson, "A Control Theoretic Approach to Manual Vehicle Systems Analysis," *IEEE Transactions on Automatic Control,* 1971, AC-16, pp. 824–832.

17. J. P. Guilford, *The Nature of Human Intelligence,* New York, McGraw-Hill, 1967.

18. A. D. Swain and H. E. Guttman, *Handbook of Human Reliability Analysis with Emphasis on Nuclear Power Plant Applications,* Albuquerque, Sandia, 1983.

19. Arthur I. Siegel and William Miehle, *Post Training Performance Criterion Development and Application: Extension of a Prior Personnel Subsystem Reliability Determination Technique,* Wayne, Pa., Applied Psychological Services, 1967.

20. NAVSEA, *Human Reliability Prediction System User's Manual,* Washington, D.C., Department of the Navy, Navy Sea Systems Command, 1977.

21. Arthur I. Siegel and Edward G. Madden, *Evaluation of Operator Decision Aids: I. The Strike Timing Aid,* Wayne, Pa., Applied Psychological Services, 1980.

22. Edward G. Madden and Arthur I. Siegel, *Evaluation of Operator Decision Aids: II. The Strike Timing Aid,* Wayne, Pa., Applied Psychological Services, 1980.

23. Naval Personnel Research and Development Center, Report TN 79-8, San Diego, June 1979.

24. Dean Chiles, "Workload, Task, and Situational Factors as Modifiers of Complex Human Performance," in E. A. Alluisi and E. G. Fleishman, *Human Performance and Productivity,* Hillsdale, N.J., Erlbaum, 1982.

25. Kenneth W. Kennedy, "International Anthropometric Variability and Its Effects on Aircraft Cockpit Design," in A. Chapanis, ed., *Ethnic Variables in Human Factors Engineering,* Baltimore, Johns Hopkins, 1975.

26. A. I. Siegel, W. R. Leahy, and J. P. Wiesen, "Applications of Human Performance Reliability Evaluation Concepts and Demonstration Guidelines," Wayne, Pa., Applied Psychological Services, 15 March 1977.

27. MIL-H-46855B.

CHAPTER 13
FAILURE ANALYSIS TECHNIQUES AND APPLICATIONS

Dwight Q. Bellinger
Product Assurance Manager,
Command Support Division,
TRW Federal Systems Group

13.1 EVOLUTION AND CONCEPTS

13.1.1 The Concept of Failure

The term "failure" is widely used and generally understood, as long as it is not used to establish liability in a court of law or to establish reliability requirements in a contract. In such cases, agreement on the definition of failure can be difficult to achieve. Generalized abstract definitions, such as "the loss of ability to perform specified functions," are seldom satisfactory when they are used for specific products or systems without further qualification. Examples of this are abundant in the fine print found on warranty certificates. Efforts to develop better failure definitions stem from the recognition that all failed items do not necessarily reach a failed state because of inherent deficiencies. Many failures are caused by conditions for which the designer or manufacturer is not or should not be responsible. Such cases may require failure analysis to show whether or not the causative condition exists outside of the established scope of responsibility of the designer or manufacturer. In most situations involving the failure of products or systems where the design is mature and the product has an established and well-understood failure pattern, failure analysis is not performed. However, in new designs and in established products where uncertainties exist, failure analysis is often essential to

1. Obtain a better understanding of the failure event and its causative factors
2. Develop remedial actions for the prevention of failure recurrence
3. Establish responsibility for the failure and for the remedial action

13.1.2 Evolution of Failure Analysis

The first failure analyses occurred in prehistoric times when humans used subjective opinions and/or empirical trial-and-error methods, unsupported by analytical engineering processes, to evaluate and correct the failures of early tools and weapons. Failure analysis is needed to avoid repetition of this approach for the correction of failures in consumer products, industrial equipment, modern tools, and weapon systems.

Too often, it is found that system failures are either not analyzed or appropriate failure analysis methods are not used. In such cases, the conclusions may reflect unsupported opinions, popular assumptions, or biased interests rather than the facts. Examples abound in every marketplace, and most of us have had personal experience with failures in purchased articles and/or services which involved differences between the buyer and seller regarding the analysis of, and responsibility for, a failure. Sometimes failures are not what they appear to be. For example, the presence of jeweler's rouge causing failure of precision bearings in a military system was initially taken as an indication of sabotage. However, failure analysis demonstrated that the system itself was manufacturing jeweler's rouge as a result of friction between the cork and rusted iron faceplates of a magnetic clutch assembly.

Over the years, failure analysis activities have been inseparable from the experimental methods used in the development and advancement of technology in all scientific fields. A large share of our knowledge has been gained from the analysis of failures of all types. More recently, the scope of failure analysis activities in support of system reliability has been broadened from the application of physical and engineering analysis to hardware failures. Today it can involve a multidisciplined task group in the study of complex systems interacting with operating and maintenance personnel, procedures, government regulation, legislation, political action, the environment, and the general public. Examples of modern systems with failures in this category are frequently reported by the media and include nuclear power plants, missile ranges, food and chemical processing plants, offshore oil wells, rail transportation systems, and automobiles.

13.1.3 Failure Analysis in the Product Life Cycle

Traditionally, failure analysis has been performed after the fact, as a postmortem investigation. The main objective has been to identify and isolate the cause of failure so that corrective or preventive action could be instituted. For most systems, failure analysis permits the assignment of the originating cause of failure to deficiencies occurring in one of three functional areas: design, production, or end-use environments. These areas closely correspond to the phases in the life cycle of all products or systems during which failure analysis may be performed.

Design-related failures typically occur when normal operational stresses exceed design strength. Production-related failures typically occur when design strength is degraded or overstressed by factors in the production process. Use-related failures typically occur when normal operating life is exceeded or abnormal operational stresses or maintenance-related stresses exceed design strength in the use environment. However, waiting until late in the product life cycle to perform after-the-fact failure analysis to identify problems in production units can be prohibitively expensive, as evidenced by recalls in the automobile industry.

Before-the-fact failure analysis, as found in stress analysis reliability predictions and/or with other failure analysis techniques such as Failure Mode Effect And Criticality Analysis performed during the design phase are fundamental to the achievement of preproduction design improvements. By initiating preventive or corrective actions to eliminate the causes of unnecessary failures early in the product life cycle, it is possible to achieve cost and schedule savings which benefit both the manufacturer and the product end user. The earlier that these potential failures are identified, analyzed, and corrected, the greater the total savings to all concerned.

Failure analysis is an essential element of virtually all major reliability engineering activities. Beginning early in the conceptual design phase with the development of functional failure and success criteria for systems and equipment, and continuing the refinement of these criteria in each detailed application of the reliability prediction and allocation processes, failure analysis pervades most reliability engineering activities. The need for failure analysis continues throughout the normal life cycle of the hardware and in some cases goes beyond its useful life. For example, systems and equipment which require safe means of disposal, such as nuclear power systems and military ordnance, require additional failure analyses to ensure safety in disposal operations at the end of life.

Table 13.1 presents the principal types of failure analysis discussed in this chapter and indicates the phases of the product life cycle in which they are applicable and may yield the most benefits. However, the table is not intended to imply that benefits from the application of any particular type of failure analysis are limited to any phase of the product life cycle. For example, although the design analysis techniques tend to produce the greatest benefits when applied during the design phase, they can be useful for system analysis and design optimization at any time throughout the system life cycle.

13.2 TECHNIQUES AND APPLICATIONS IN SUPPORT OF SYSTEMS IN DESIGN, RESEARCH, AND DEVELOPMENT

The types of failure analysis discussed in this section are most applicable and most beneficial when performed concurrently with related design, research, and

TABLE 13.1 Failure Analysis Activities in the Product Life Cycle

Type of failure analysis activity	Research, design, and development	Procurement, production, and test	End-use environment
Destructive physical analysis	A	AB	A
Physics of failure analysis	A	AB	A
Fault-tree analysis	AB	A	A
Common mode failure analysis	AB	A	A
Failure mode, effect, and criticality analysis	AB	A	A
Single-point failure analysis	AB	A	A
Sneak circuit analysis	AB	A	A
Functional failure analysis	A	A	AB
Software failure analysis	AB	A	A
Failure analysis in logistic support	A	A	AB

Note: A = applicable, B = highest potential benefits.

development activities. They provide early feedback to technical authorities and to management of the technical information needed to optimize design characteristics affecting performance, reliability, and life-cycle cost. They are applicable at all levels of indenture of any system or equipment configuration.

13.2.1 Failure Mode, Effects, and Criticality Analysis

"For want of a nail the shoe was lost, for want of a shoe the horse was lost, for want of a horse the rider was lost..."[1]—such was the unfortunate train of events which led to the loss of the battle and then, the loss of the kingdom. This seventeenth-century example of a failure mode and effect analysis (FMEA) demonstrated the traceability of the effect of a common component failure mode to the catastrophic failure of a complex system. Had this FMEA been performed in advance, the authorities would have been alerted to the potential consequences of the failure of their communications system and might have considered system design alternatives such as,

1. Improving the design reliability of the failed component
2. Improving control of reliability factors in installation and in the preventive maintenance program
3. Incorporating redundancy in the system to eliminate the single points of failure represented by the horse and the rider

FMEA is a powerful tool for design evaluation which is applicable at all levels of any system at any time. It is an inherent element of the reliability modeling and prediction processes which are also concerned with the effects of all failures. Table 13.2 provides an example of a simplified generic FMEA performed by the author for the purpose of identifying and summarizing the common failure modes of the basic components used in standard American-made automobile braking systems and their effects.[2] The table shows that the effects of all failures are not equal in terms of their potential impact on safety and/or system performance. Therefore, to support the allocation of resources to each system design optimization alternative, the designer needs information to rank the relative importance or criticality of each potential failure. To satisfy this need, the criticality analysis function was added to the FMEA process, thereby creating failure mode, effects, and criticality analysis (FMECA).

FMECA applies a disciplined methodology to postulate and analyze the effects of single independent failure events representing each of the "modes," or ways, in which the individual elements of a system may fail. It is important to be aware of the restrictions imposed by this single-failure-oriented methodology on the analysis of systems susceptible to multiple failures which can cause catastrophic effects. In such cases, it is necessary to consider the use of other techniques, such as fault-tree analysis, to supplement FMECA with the additional multiple-failure analysis data needed for risk assessment purposes.

The methodology involved in performing FMECA is thorough and straightforward. It is also flexible because the FMECA process is open to so many alternatives regarding the scope and depth of the analysis. Therefore, to be timely

TABLE 13.2 Disc and Drum Brake System Failure Mode and Effect Analysis

Component	Failure mode	Failure effect	
		Reduction or loss of system effectiveness	Reduction or loss of directional stability
Lining	Contaminated by hydraulic fluid leaks or grease or other materials	●	●
	Lining wear	●	●
	Insufficient contact with drum or primary and secondary are reversed	●	●
	Excessive clearance to drum or glazed lining	●	
Brake shoe	Loose mounting or worn anchor pin hole		●
	Improperly installed or adjusted	●	●
	Not properly releasing	●	●
	Bent, distorted	●	
Pedal linkage	Binding	●	
	Frozen	●	
	Out of adjustment	●	
Drums	Internal diameter oversize or mismatched	●	●
	Eccentric, bell-mouthed, barrel-shaped	●	●
	Crazed or turn marks	●	
	Scored, ridges	●	●
	Heat spots	●	●
	Hard spots, checks	●	●
	Cracks	●	●
	Polished	●	
	Thin, weakened	●	
Hydraulic lines and hoses	Leaks	●	●
	Low fluid	●	●
	Air in lines	●	●
	Obstructed lines	●	●
Master cylinder	Faulty adjustment of push rod	●	
	Faulty check valve	●	
	Leaking primary cup	●	
	Plugged filler cap vent and air in system	●	
	Blocked compensatory port or piston not returning to stop	●	
	Broken stop plate lock	●	

TABLE 13.2 Disc and Drum Brake System Failure Mode and Effect Analysis (*Continued*)

Component	Failure mode	Failure effect	
		Reduction or loss of system effectiveness	Reduction or loss of directional stability
Wheel cylinder	Pistons binding	●	●
	Leaks	●	●
	Contaminated brake fluid	●	
	Push rod pin improperly located	●	●
	Loose mounting		●
Parking brake cable	Frozen or adjusted too tight	●	●
Anchor	Loose, bent		●
Backing plate assembly	Loose		●
	Mechanical interference with shoe movement	●	●
	Wrong type plate		●
Return springs	Weak or loose	●	●
	Reversed, wrong type		●
Disc rotor	Excessive runout	●	●
	Crazed		
	Scored ridges	●	●
	Heat spots	●	●
	Hard spots, checks	●	●
	Cracks		●
	Polished	●	●
	Worn thin	●	
	Loose on hub	●	●
Disc pads	Grease or brake fluid contamination	●	●
	Excessive wear and excessive clearance (pad-to-disc)	●	
	Hang up in caliper housing	●	●
Caliper assembly	Sticky or frozen pistons	●	●
	Loose mounting	●	●
	Improper alignment	●	●

TABLE 13.2 Disc and Drum Brake System Failure Mode and Effect Analysis (*Continued*)

Component	Failure mode	Failure effect	
		Reduction or loss of system effectiveness	Reduction or loss of directional stability
Self-adjuster	Insufficient adjustment action	●	●
	Excessive adjustment action	●	●
Brake fluid	Water contamination	●	●
	Air contamination	●	●
	Low fluid	●	
	Other contamination	●	●
Pressure-reducing (proportioning) valve	Blocked or restricted	●	●
	Open	●	●
Power assist unit	Vacuum leaks or blocked vacuum lines	●	
	Fluid leaks	●	
	Linkage frozen or binding	●	

and cost effective, FMECAs should be tailored to fit the unique characteristics of the individual product or system. In addition, the tailoring and allocation of resources to FMECA should be adequate to serve the needs of the principal users of FMECA, such as the design, reliability, maintainability, system safety, logistics, human factors, quality, training, and test planning organizations. This often fails to occur. "Probably the greatest criticism of FMECA has been its limited use in improving designs. The chief causes for this have been untimeliness and the isolated performance of FMECA...."[3] One effective answer to this problem has been to train designers to perform FMECA concurrently with the design development.[4]

A consensus on a standard methodology for performing FMECA does not exist. However, two standards have been issued, one originated by the U.S. government[3] and another by the Society of Automotive Engineers.[5] In addition, standardized procedures and methodologies have been developed to reduce FMEA cost and labor-hours by facilitating the computer automation of clerical functions, basic calculations, and data base manipulations.[6] Effective automation of the complex design and failure effect analysis functions of FMECA can be accomplished through future advances in the development of artificial intelligence (AI) technology and its integration with computer-aided design (CAD) systems. There are available today a number of computer programs for the automation of FMEA. In 1982, a sample survey was performed for Rome Air Development

Center by the Hughes Aircraft Company to evaluate representative programs available through computer time share service organizations for performing circuit-level FMEA and clerical support functions for higher-level FMEAs.[7]

The survey found only one clerical FMEA support program within the commercial market and commented that it "should yield some labor savings if used on an in-house machine..." but "the cost and training requirements appear to be significant limitations."[7] It also identified and recommended one digital and one analog circuit analysis program capable of supporting FMEA at the piece part level but reported that "The circuitry analysis programs do not appear to result in a substantial time savings for FMEA due to the effort required to define the circuit to the computer...."[8] Nevertheless, the complexity of analysis involved in the study of failure modes in complex microcircuits, such as microprocessors, requires the use of automated tools for analysis at this level. One example used in Ref. 8 to illustrate this point showed that any one of 65,536 failure conditions (2^{16}) could result from the failure of any microcircuit connected to the 16-bit-wide bus structure of a representative computer system, and each possible condition would require analysis with respect to the state of the microprocessor cycle, the software, and the state of the external circuitry. Today's microcircuits permit the encapsulation of major functional elements of what formerly were large electronic systems in a single component part. Each microcircuit is only one element of an infinitely more complex hardware and software system. Electronic technology growth and part complexity continue to outpace the growth of the reliability failure mode data base needed to support part-level FMECA. However, the significant growth in part reliability which has accompanied the advancement of technology is at least a partially offsetting factor. Lower failure rates reduce the importance of part-level FMECA and limit the potential for practical part-level reliability improvements which might be initiated by FMECA.

To be cost effective, all FMEA activities require careful planning and close technical supervision because of their high labor content. A good FMEA plan is an essential prerequisite for performing FMECA. Figure 13.1 illustrates the types of input information to be integrated and the major related program functions which are important to the development of the FMECA plan as discussed below.

System and Program Requirements Baselines.   All system and program requirements which relate to FMECA must be analyzed and thoroughly understood. For Department of Defense (DOD) programs, FMECA may be contractually required to a specified level of indenture in accordance with MIL-STD-1629A, *Procedures for Performing a Failure Mode, Effects, and Criticality Analysis,* and documented per applicable contractual data item description documentation with the times of deliveries specified. Detailed evaluations are required for the interpretation and integration of system functional and physical design requirements, including quantitative reliability, maintainability, and availability requirements with related mission profiles and time lines. The results are of major importance to the establishment of the failure/success criteria, criticality criteria, and the operational scenarios to be used in performing FMECA. The interface control specifications and the system environmental specification will also be important in establishing a definition of the system and its operational environment, and to characterize and establish the boundaries and definitions for the system hardware and functions to be used in evaluating failure effects. Contract and program requirements will invoke schedules and identify activities which will be closely interrelated with the FMECA activity. These include design reviews, reliability reports, and operational and maintenance planning. The requirements may also

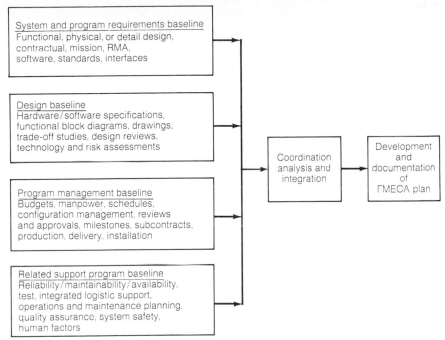

FIG. 13.1 Information flow for development of the FMECA plan.

call for the planning and implementation of related programs of importance to the FMECA effort such as test and evaluation, quality assurance, integrated logistic support (ILS), system safety, configuration and data management, and subcontract design and development. Each of these can be an important source of inputs to FMECA, as well as to users of FMECA.

Design Baseline. The data obtained from design constitute the technical baseline for FMECA. These data include system and equipment design specifications and drawings containing all physical and functional requirements, schematics, functional block diagrams, trade-off studies of design alternatives, parts and materials specifications, technical standards, technology assessment reports, design and development plans and schedules, environmental controls, thermal analyses, operational concepts, time lines and functional requirements for operational use, and failure/success baseline criteria. Collectively, these technical data should be sufficient to provide all the essentials needed for planning and performing FMECA as a design support function.

Program Management Baseline. The budget, schedule, and allocated resources for FMECA must be negotiated with cognizant management authorities. The FMECA plan will be the primary vehicle used in this negotiation. Other functions to be incorporated in the plan include support from the configuration and data management function to maintain and update the technical baseline, management reviews, the integration of FMECA into the corrective action system, FMECA coordination and support functions relating to other organizations, and the preparation and distribution of FMECA reports. In addition, when design subcontracts are involved, the plan should provide for invoking FMECA requirements on the subcontractor, since FMECA can be most effectively performed and utilized by the original design agency.

Related Support Program Baselines.   Figure 13.1 shows other program functions which can be important contributors to the planning and execution of FMECA, as well as major users of FMECA reports as discussed below.

Reliability-Maintainability-Availability (RMA).   The FMECA activity is an integral element of the RMA program, from which it obtains most of the basic assumptions and conditions used in the analysis. Other basic information obtained from the RMA program includes failure rates, failure modes, the reliability models and block diagrams, failure definitions, and failure criteria. FMECA supports the reliability and availability model development and validation and establishes the foundation for identifying reliability-critical items.

Test Program.   The test program can support FMECA verification through test reports documenting the demonstrated effects of failure modes which occur during test and evaluation activities. In addition, testing may reveal degradation modes and effects not anticipated in the FMECA. Reviews of test plans, procedures, and reports are also helpful in planning FMECA because they provide details relating to subjects such as the detailed definition of failure indications or symptoms, the investigation of failure causes, and the further development of failure criteria.

Integrated Logistic Support (ILS) Program.   The ILS program can support FMECA planning and implementation through its development and definition of the life-cycle support system which mitigates the effects of system failures in the operational environment. ILS analyses can provide important guidelines for FMECA through the identification of functionally significant items and the preparation of functional failure analysis reports. ILS corrective maintenance procedure development focuses on the analysis of failure modes and effects. In addition, level of repair analyses can assist in providing information for use in selecting the lowest indenture levels for FMECA. The data collected and used for the ILS analyses also can be very helpful in supporting FMECA.

Operation and Maintenance Planning.   This activity develops operational procedures and guidelines to assist the user in operating the system in the field. Information of importance to FMECA includes operational performance criteria, identification of failure symptoms, and procedures governing the actions of field personnel responding to operability indicators and instrumentation which provide failure enunciation. These plans and procedures contain information and criteria needed in FMECA for defining system failures, degradation thresholds, and for identifying the compensating provisions to be established to control and/or minimize the effects of failures.

Quality Assurance (QA).   Important information available from quality plans and inspection or test procedures should include the classification of criticality for the physical design and functional performance characteristics of all elements of the system, ranging from the level of parts and materials to the level of the total system. These classifications should be fully consistent with the criticality or severity ratings used in FMECA. Inspection and test data from the QA program identify failure and defect modes and can serve as valuable inputs to FMECA planning and implementation.

System Safety.   This is another function which is deeply involved in classifying the severity of failure modes and effects in the performance of various types of system safety analyses. Examples of these analyses include fault-tree analysis, preliminary hazard analysis, operations and support hazard analysis, common mode failure analysis, and FMECA. Inputs from these analyses, and the identification of safety-critical items, are of major importance to the planning and implementation of FMECA and the criticality criteria development.

Human Factors. This discipline is concerned with the human-machine interface relationships, the design of operator consoles, operator control systems, and all maintenance interfaces. Of particular importance to FMECA in the assessment of failure effects are the actions of operators in response to failure modes, and the design of the failure-enunciating hardware and systems upon which operators and maintenance personnel are dependent (see Chap. 12). Inputs to FMECA from this area will be most important to the analysis of failure effects when the effects are dependent upon proper operator response to a failure mode. The development of the FMECA plan should be based upon an integrated analysis of all of the inputs discussed above. The plan should be tailored to fit the program and the system characteristics, the established purposes and objectives, and the available resources.

With respect to resources, it should not be anticipated that all of the time or all of the money required will be available to perform part-level "bottom-up" FMEAs on all components in all systems for all environments and all conditions of use as was needed for equipment used in manned space programs. This is not to say that this level of FMECA is not essential, when properly focused, for system elements which are identified as critical by means of a top-down FMECA or by other failure analysis methods.

With respect to objectives, it is important to recognize that it is not always feasible or practical to correct, eliminate, or control completely all of the failures in a system with the potential for catastrophic effects. It has not been possible for any system to achieve total freedom from failure risks, nor to create a totally risk-free operational environment, despite the presence of fail-safe devices and other techniques which are used to provide multiple levels of protection against the effects of critical failures. A major objective of FMECA, and other failure analysis techniques, is to provide a scientific basis for guiding the allocation of resources in a manner which will reduce the probabilities of catastrophic failures to the lowest possible level. Other objectives or purposes necessary to consider in planning FMECA relate to the timely serving of the needs of the ultimate users of FMECA, such as design, system safety, maintenance planning, and quality engineering.

System design characteristics and utilization are important in determining the detailed approach of FMECA. For example, the failure rates and criticality of failure effects are not likely to be the same for different conditions of use for the same failure in the same system. The characteristics of the functional elements which make up the system, such as their unique identification, the definitions of their assigned functions, their duty cycles and operating modes, and the system of failure detection and enunciation, are important to the structuring of the FMECA worksheet and the preparation and encoding of the inputs.

The FMECA plan should be based on the analysis and integration of the information baselines illustrated in Fig. 13.1 and discussed above, and should include the following basic information:

- Budget, labor, organization, and schedules
- Definition of the FMECA purpose and objectives

- Preliminary system baseline configuration
- Functional mission profile and time lines
- Coordination and interface requirements
- Definition of applicable program requirements
- Definition of FMECA tasks and preliminary approaches
- Preliminary definition of system failure
- Preliminary rationale for determining level of analysis
- Preliminary FMECA worksheet

Planning information relating to major factors required for the implementation of FMECA, such as the detailed system design and the scope and depth of the analysis tasks, is incomplete or not available during the early design phase. Although the scope and depth of FMECA for every failure should be proportional to the significance of the failure, it is only through performing FMECA that this information becomes available. Therefore, we are initially dependent on engineering judgment and experience in scoping FMECA. However, as FMECA is implemented it can provide its own self-direction as failure criticalities are established and system sensitivities are identified.

The FMECA process which implements the plan is summarized in Fig. 13.2 and discussed below. Conducted in parallel with the design development, the implementation of the FMECA process extends and refines the original FMECA plan because it successively addresses more detailed aspects of the system design and the factors associated with its operational use. The development and implementation of the FMECA process, as illustrated in Fig. 13.2, involves a continual iteration, repetition, and interaction of tasks. This occurs partly because the FMECA process operates at many levels with a changing evolving design. It is also the result of the many decisions required for the selection of the alternatives which govern the implementation of the FMECA plan, as discussed below.

Refine and Update System Baseline Configuration.   This is a continuing task because of the evolution and changes occurring in the design and requirements baselines from the initial design through the final design. The baseline may be represented in the FMECA plan as a series of block diagrams showing the physical and/or functional elements of the system for all levels to be analyzed and may be cross-referenced to the technical documentation. The baseline need not be identical to, but must be consistent with, the reliability block diagram and correlated with the system drawings and specifications applicable to all levels of analysis. Arrangements should be made to ensure that the required configuration control data are received to update the baseline to reflect all significant changes. For purposes of simplifying entries on the FMECA worksheet, the nomenclature and part numbers for each system indenture level to be analyzed can be coded for ease of reference.

Define and Update Assumptions and Conditions.   FMECA will document and utilize a formal statement of the assumptions and conditions which apply. Generally these will include all those applicable to the reliability model but also will include additional assumptions relating to the effects of failure, as necessary

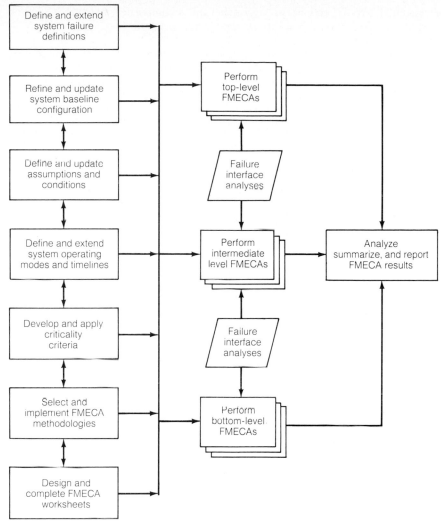

FIG. 13.2 Implementation of the FMECA process.

to support the criticality criteria. Examples include the following:

- Specification of applicable ambient environmental conditions
- Exclusions of failures arising from operator or maintenance error, acts of God, or enemy action
- Exclusion of multiple failure conditions
- Assumed interface conditions, such as the inability of external equipment failures to propagate across the interface
- Assumed failure effect conditions

- Assumed proper actions of operators and maintenance personnel, and/or successful operation of fail-safe systems when required

Refine and Extend System Failure Definitions and Criteria.   The functional and/or physical system failure definitions and criteria from the design, system requirements, and reliability baselines of the FMECA plan will require extensive interpretation and some augmentation or changes as the design definition evolves and the FMECA proceeds to lower levels. For example, the analysis will identify failure modes which result in partial degradation of one or more system functional capabilities. Since all such conditions are not usually addressed in all system specifications, coordination with design and/or user agencies is usually required to obtain approval for recommended classifications of these conditions.

Functional failure definitions which are the most universal for general application require extensive interpretation during FMECA to relate them to all of the applicable hardware failure modes and/or any software fault capable of causing the system to enter a failed state.

System failure definitions and criteria also must be extended during FMECA to establish their traceability to the definitions of failure modes and effects at lower levels. Decisions on the definitions and terminology to be used for failures at the system level and at lower levels will be assisted by the investigation and analysis of the methods to be used for indicating or detecting failures. These will include operator observations of system performance, panel lights and indicators, display consoles, and instrumentation such as built-in test equipment.

Define and Extend System Operating Modes and Time Lines for Analysis.   To be complete, FMECA should consider all of the failure modes and effects relating to each system operating mode and the related operational requirements for all applicable duty cycles, or time lines, for each element of the system. The reason for this is that the criticality of a failure mode's effect can be dependent upon: (1) the time of its occurrence, (2) the prevailing operating mode of the system at the time of the failure, and (3) the unique set of operational requirements against which the effect of the failure must be assessed.

However, when the resources available to perform FMECA are severely limited, the FMECA task may be simplified by limiting the documentation and the analysis to those operating modes and time lines during which failure effects can be most significant. It may also be decided to base the analysis upon the most adverse environmental operating conditions. Through the use of such "worst-case" approaches, the potential benefits and cost effectiveness of FMECA may be significantly enhanced. The time lines, functional operating modes, and requirements necessary to support top-level system functions must be defined for each level of the system analysis.

Develop and Apply Criticality Criteria.   Criteria must be developed for evaluating and classifying the severity of failure effects, based upon the most adverse potential impacts of each failure mode. These criteria must be integrated with a system for rating the criticality of each failure mode, based upon a combination of its severity level and probability of occurrence. Although the top-level system specifications and the reliability model are important references in performing this task, the development of the criticality rating system requires creativity and an operations research approach. The criteria and rating system to be used in any FMECA must be unique to the system and should reflect all important concerns of the system users and the system designer. Typical concerns include the relative impact of failures upon safety; system performance; mission success; cost of all types, including property damage or possible legal liabilities; impacts on interfacing systems; loss of system availability; and impact on system life-cycle costs.The probabilities of occurrence of the failure mode and the failure effect may be different and may be expressed in quantitative and/or qualitative terms. However, the criticality rating should be based upon the combination of the failure mode probability, the failure effect probability, and the severity of the effect.

Representative definitions of severity levels for failure effects, as classified and defined in MIL-STD-1629A, can be used as a guide and tailored for use in FMECA for any system, as follows:

- *Category I—Catastrophic:*   A failure which may cause death or weapon system loss (i.e., aircraft, tank, missile, ship, etc.)
- *Category II—Critical:*   A failure which may cause severe injury, major property damage, or major system damage which will result in mission loss
- *Category III—Marginal:*   A failure which may cause minor injury, minor property damage, or minor system damage which will result in delay or loss of availability or mission degradation
- *Category IV—Minor:*   A failure not serious enough to cause injury, property damage, or system damage, but which will result in unscheduled maintenance or repair

When such qualitative ratings are combined with quantitative probabilities, they can serve to indicate the criticality of each failure mode. Although the combinations of qualitative and quantitative values are effective for use in limited comparative analysis of different failure events within a severity level, they lack the common scale necessary to establish meaningful relative values across all severity levels. Therefore, if relative values are required for comparison purposes, the development of a universal rating system is required.

Select and Implement FMECA Methodologies.   A discussion of the many alternatives to be considered and the steps to be taken in performing FMECA is presented in the sections which follow.

System Approach.   A major decision involves the choice between implementing a top-down analysis, which begins at the level of system failure modes and continues successively through lower levels, or to begin at the part level with a bottom-up analysis and proceed up to the system level. A combination of both approaches can also be used. When resources are limited, a combination of the top-down approach with a bottom-up approach focused only on selected high-risk or high-failure-rate areas may be the best choice.

Analysis Method.   Another major decision involves the selection and implementation of either a functionally oriented analysis, a hardware-oriented analysis, or a combination of both. The functional method identifies the functions of system elements as outputs which are listed and their failure modes and effects analyzed at the functional level independently of the detailed hardware design. This method is very useful in the early design phase and helps to expedite any top-down analysis. Nevertheless, the analysis ultimately should be related directly to the detailed design of the system and confirmed. The hardware method requires adequate design definition as a prerequisite and is, therefore, limited to applications in later phases of the system design. However, this method can be applied effectively at any indenture level. For a complex system, a combination of both methods may be found to be the only effective method of completing FMECA on schedule and within budget. System design characteristics also influence the selection of the analysis method. For example, if a specific area of a system is known to involve design risks and/or new technology, it is a candidate for a hardware-oriented bottom-up approach which would include investigation of the generic failure modes and failure rates and effects associated with the new technology and/or areas of design risks.

Software Analysis. Another major alternative, if the system is to include software, involves decisions on how to address the hardware-software interfaces in FMECA for hardware failures and for software faults. One approach is to rely on the software error analysis techniques, as discussed in Sec. 13.2.5. Another approach is to perform hardware-software FMECA as described in Ref. 9. Still another approach is to insert faults in an operating system and thereby obtain the data required by the FMECA from actual testing in lieu of a paper study. For systems equipped with automatic performance monitoring, fault detection, and correction functions, FMECA plans must select an approach to address the analysis of these hardware-software interfaces.

Probability Measurement. Another decision involves the selection and use of quantitative and/or qualitative probability estimates for failure modes and effects. Qualitative values are often necessary for use with failure effects, for example, when the effects are dependent upon human factors such as operator interpretations of failure symptoms and correct operator actions to mitigate failure effects. Quantitative values are available or can be estimated for most component failure rates. However, limited data are available to support proportionate allocation of a failure rate to each of its constituent failure modes. Sources used by industry include in-house test data, MIL-HDBK-217,[10] and various reliability references published by the Reliability Analysis Center, Griffiss AFB, Rome, New York. Some of the material which can be found in these references includes failure rates, modes, and mechanisms for hundreds of nonelectrical devices;[11] failure rates, modes, and mechanisms for memory and large-scale integration (LSI) devices;[12] failure rates, modes, and mechanisms for discrete semiconductor devices used in airborne and space applications;[13] and small-scale integration (SSI) and medium-scale integration (MSI) digital device reliability data.[14] The IEEE has published a comprehensive source of component and equipment failure mode and failure rate information applicable to nuclear power plants.[15] In addition, the U.S. Army has published data on the rates of occurrence of predominant failure modes in high-usage components.[16]

Failure Modes. Decisions are required also on the number and types of failure modes to be addressed at each level of analysis. For example, the FMECA might require analysis of the common failure modes of "short," "open," and "out-of-tolerance" conditions at the part level and identify high-level functional failure modes for analysis at the level of equipment or subsystems. Other typical failure modes to be considered include intermittent operation, failure to operate, loss of output, degraded output or performance, premature operation, and loss of power. The FMECA workload will increase significantly as the number and types of failure modes are increased. Therefore, the decisions on the selection of failure modes are critical to maintaining both the adequacy and the cost effectiveness of the analysis.

Multiple Failures and Failure Detection. Another significant decision concerns the practices to be followed relative to the analysis of failure detection and potential multiple failure situations. Conventional FMECA methodology excludes multiple failures because the number of possible failure combinations makes this totally impractical. It does not, however, exclude the "dependent failures" which may occur as an effect of a single failure. In cases where a failure has no detectable effect, it may create a condition which could permit a second failure (not otherwise significant from a criticality standpoint) to cause the system to experience a failure effect which is critical in nature. Therefore, the extension of FMECA to address these cases also requires it to address the detectability of failures by built-in features and by operational personnel.

Other multiple failure situations should also be considered on a selective basis in performing FMECA. These should always include any common mode failures which are identified in the analyses, as discussed in Sec. 13.2.2, and system failure combinations involving fail-safe devices. Failure detectability and accurate enunciation of failures to system operational and maintenance personnel are important factors in limiting the effects and duration of failure modes. Although the analysis of failure detection systems is a normal function of FMECA, the analysis of all failure modes to ensure their detectability is a complex and difficult process. Decisions are required to govern the direction and scope of effort allocated to this area as the analysis proceeds.

Worksheet Design and Completion. The design and completion of the FMECA worksheets presents another set of alternatives for decisions. The worksheets are used to document the results at each step of the analysis and should be tailored to fit the needs of each analysis level. An example of a well-known worksheet is illustrated in Fig. 13.3. Other types of worksheets and records may be used at the circuit level, such as the matrix FMEA developed by G. L. Barbour,[17] and the outputs of commercially available circuit analysis programs.

A discussion of the content of FMEA worksheets is provided below. It addresses alternatives for the elements shown in the format of Fig. 13.3.

- *Headings:* Establish appropriate headings as necessary for correlation between the system baselines discussed in Sec. 13.2.1 and the material on the worksheets.

- *Identification number:* This column serves as an index to code and cross-reference the information to other material, such as tables or figures in the FMECA report. It may be based on work unit codes constructed per MIL-STD-780, designations used in the reliability prediction report, or any set of codes desired by the user. In cases where the item functional identification or nomenclature is considered sufficient in itself, this column should be eliminated (e.g., in a top-down FMECA).

- *Item functional identification (nomenclature):* This is used to identify the specific system element, part, or function under consideration; for example, a part reference designation, a schematic diagram symbol, an assembly drawing number, or a functional description.

- *Function:* A brief description of the item's function(s) is entered here when appropriate. Loss or degradation of the identified function(s) as a result of the failure mode will be the subject of the analysis.

- *Failure modes and causes:* Separate lines are used to describe each of the failure modes selected and identified for analysis and their possible cause(s), in-

System _____
Indenture Level _____
Reference Drawing _____
Mission _____

Date _____
Sheet _____ of _____
Compiled by _____
Approved by _____

Identification number	Item/functional identification (nomenclature)	Function	Failure modes and causes	Mission phase/operational mode	Failure effects			Failure detection method	Compensating provisions	Severity class	Remarks
					Local effects	Next higher level	End effects				

FIG. 13.3 Example of FMEA worksheet (*From Ref. 3*).

cluding those latent in interfacing elements of the system (e.g., at high or lower indenture levels). Specific guidance related to this information should be established a priori in the FMECA plan to avoid proliferation of the effort. For example, the entry of failure causes may be restricted to failures whose effects are catastrophic or critical, or limited to selected elements of the system in order to narrow the scope of the analysis and the volume of data to be entered on the form.

- *Mission phase/operational mode:*   This column identifies the particular mission time phase and related system operational modes which apply and give significance to the failure mode being analyzed. The operating time is important to the development of the failure probability. The operational mode is important to the identification of the specific system elements required to function during the specified time period. For systems and equipment having fixed operating conditions and a single operating mode, this column is not required since a uniform condition applies to all elements in the analysis. For the sake of simplicity, a worst-case approach may be selected.

- *Failure effects:*   This set of columns is used to document the significant effects each failure mode will have upon the system. As the effects of a failure mode propagate within a system, the effect of a failure occurring at one level of indenture may be considered as a "failure mode" in the FMECA performed at another level of indenture. The column "end effects" is intended to reflect the failure effect at the highest traceable level. When appropriate, include additional columns and/or use the remarks section to document the significant effects, such as severe hazards or injury to personnel, and cross-reference related FMECA worksheets when applicable.

- *Failure detection method:*   This information is important primarily to ensure that significant failures are detectable and to initiate remedial action as necessary when they are not. Such actions can be essential to limit the duration and potential impact of significant nondetectable failure modes whose presence could introduce hazards to personnel and performance and/or in combination with another failure mode seriously compromise the integrity of the system. Information on the analysis of detection methods in FMECAs is also useful to support the design of built-in test features and to maintenance planning, maintainability analysis, and availability analysis activities.

- *Compensating provisions:*   This column is used to identify operational procedures, system capabilities, or design features which eliminate or reduce the effects of significant failure modes. The presence of adequate compensating provisions, such as design redundancy or the capability to utilize alternative system operating modes, may be sufficient to eliminate the need for remedial action.

- *Severity classification:*   For each failure effect that is significant to the analysis and documented in the "failure effects" column, a corresponding severity classification should be established and entered in this column. Severity codes I through IV, as previously discussed, are typically considered for these entries. Decisions made in the FMECA plan on the scope and levels of failure effect analysis to be performed will influence the selection of severity classification alternatives. For example, analyses performed at high levels of indenture may address failure effects in all severity classifications, whereas lower-level analyses may be limited to failure effects in severity classifications I and II, due to limitations in available resources.

- *Remarks:* This column may be used for any purpose desired. It is often used to present the rationale supporting management disposition actions recommended as a result of analyses of each category I and II failure mode.

Perform Criticality Analysis. The information contained on the FMEA worksheets provides the raw data needed to identify and select those system elements and failure modes which are the leading candidates for further investigation. Criticality analysis uses probability values in combination with severity-level classifications to establish a numerical basis for comparing the relative significance of the failures and the failure modes of individual system elements. These numerical values are probabilistic criticality ratings which may be used in comparative analyses and trade-off studies for the optimization of systems and equipment. Thresholds, based on probability and the relative significance of failure effects, should be established in the FMECA plan to establish limits and guidelines for the criticality analysis efforts.

Criticality ratings for individual failure modes and individual system elements are based upon the combination of three distinct probability values as follows: (1) the probability that the element will fail λp during a specified time interval t; (2) the probability α (failure mode ratio) that the failure will involve a specified failure mode, given that the element fails; and (3) the probability β that a specified failure effect will occur, given that the failure mode occurs. Therefore, the criticality rating for a specific failure mode C_m is a probability number which can be calculated from the following formula:

$$C_m = \lambda p t \; \alpha \; \beta$$

In those cases where only one failure mode is applicable and where the same failure effect always occurs, the failure mode criticality number will simply be equal to the probability of failure $\lambda p t$ and a single severity classification will apply. In cases where several failure modes can occur and result in a variety of different failure effects, more than one criticality number and severity classification can apply. In such cases, to establish priorities for remedial action for individual system elements, an element criticality rating C_e for each item within each severity classification is calculated from the summation of its failure mode criticality numbers, using the following formula:

$$C_e = \sum_{n=1}^{j} (\lambda p t \alpha \beta) n \qquad n = 1, 2, 3, \ldots, j$$

where C_e = element criticality number
   n = element failure modes in the severity classification
   j = last failure mode of the element in the severity classification

Criticality ratings as described above are based on probability. They are usually developed only for selected system elements having failure modes with significant effects and/or high probabilities of occurrence. The criticality analysis worksheet illustrated in Fig. 13.4 was developed by the author for use in calculating C_m and C_e for each system element selected from the FMEA worksheet. This format requires less clerical effort than those which duplicate the entries made on the FMEA worksheet. It also permits the documentation of analyses for all severity classes applicable to a single line entry. The results are sometimes presented in the FMECA report by plotting the equipment identification numbers on charts scaled to represent C_e or C_m on the ordinate and each severity class on the abscissa. A simpler alternative, which may be more useful, is to present ranked listings of the C_m or C_e values in tables for each severity class. Further analysis of system elements, using the probability values represented by their criticality ratings, can be performed in trade-off studies and presented in the FMECA report to establish priorities for the allocation of resources to the optimization of system safety, performance, reliability, and other areas. For example, the numerical summation of C_e ratings across severity classes I through III might be the approach used to optimize resource allocations in corrective maintenance planning. On the other hand, separate rankings of C_e ratings within selected failure effect categories in severity classes I and II might be used in trade-off studies by system safety engineers and system designers concerned, respectively, with hazard reductions and system performance improvements. Used as relative probability indexes, these ratings provide objective user-oriented decision criteria for use in the wide range of trade-off studies involved in the total optimization of a system design. Specific goals to be achieved by their application include minimization of safety hazards to equipment, reduction of life-cycle cost, optimization of logistic planning, improved development of operational concepts, minimization of risks involving potential liabilities for manufacturers, improvement of user satisfaction in commercial systems, and improved mission success probabilities for defense and aerospace systems.

Separate criticality analysis worksheets are needed in all complex FMECAs to document significant failure effects and calculate criticality ratings using the unique criticality criteria established for the system. FMEA worksheets can be designed to contain provisions for entering the criticality ratings of system elements and failure modes or can be cross-referenced to the criticality analysis worksheets. Worksheet design can be tailored to include additional space and additional columns, as necessary, to adequately represent all important factors pertinent to the analysis. After completion of the worksheets, the results of the FMEA and criticality analyses are integrated to support the conclusions and recommendations provided in the FMECA report.

Prepare FMECA Report. The FMECA worksheets should be analyzed and summarized to extract and report the significant information for review and action by cognizant authorities. The FMECA report should address the implementation of the FMECA plan, the purposes and objectives, and specify all pertinent assumptions and conditions. It should present conclusions and recommendations which identify and rank in order of criticality the system elements with the most significant failure modes and effects. It should identify all single points of failure and common mode failures with recommendations for remedial actions. In addition, the report should list all elements with failures that are significant to the system users and manufacturers, such as reliability-critical or safety-critical items and provide recommended measures for the correction and/or control of related

Date _____
Prepared by _____
Approved by _____

Element failure probability $\lambda pt \times 10^6$	Failure mode and effect description	Failure mode ratio α	Probability of failure effect (β) by severity class				Failure mode criticality $C_m = \lambda pt\, \alpha\, \beta \times 10^6$			
			I	II	III	IV	I	II	III	IV

Element criticality rating $C_e = \displaystyle\sum_{n=1}^{j} (C_m)\, n$

FIG. 13.4 Criticality analysis worksheet

design, operational, or support factors. The FMECA report and worksheets should be distributed to all organizations requiring FMECA support, as identified in the FMECA plan.

13.2.2   Common Mode Failure Analysis (CMFA)

The widespread reliance on redundancy as a design tool to achieve reliability has been proved effective in designing failure-tolerant systems. These systems continue to operate successfully during and after the failure of a redundant element. However, the Achilles' heel of all redundant systems can be the common mode failure. Failures of this type have the capability to bridge and defeat the redundancy factor, causing system failure by simultaneously or sequentially impacting all redundant elements. Examples of common mode failures include the following:

- Simultaneous failure of redundant elements resulting from a common cause, such as a fire or electrical overload which causes the failure of all redundant system elements, or the ingestion of birds by aircraft engines which causes their simultaneous failure

- Sequential or cascading failures of redundant elements as a result of a common cause, such as contamination of a common hydraulic reservoir system resulting in the failure of all redundant pumps

- Failure of all redundant elements because of a maintenance error, improper operating procedure, or operator error

It is evident from the examples that common mode failures may originate from a broad range of causes including design, operation, maintenance, human error, and/or the environment. Therefore, the range of analytical methods applicable to the performance of CMFA can be inclusive of all technical disciplines and all types of failure analysis techniques. For example, although CMFA differs from FMECA in that the initiating failure event may be independent of, and/or external to, the specific system elements addressed in the analysis, CMFA can and should be performed in conjunction with FMECA, particularly with respect to the aspects of FMECA concerned with the identification of single points of failure. However, the characteristics of common mode failures support the use of fault-tree analysis as the one single failure analysis methodology most suitable for performing CMFA.

13.2.3   Fault-Tree Analysis (FTA)

FTA evolved in the aerospace industry during the 1960s when logic diagrams and Boolean algebra were used to represent and summarize the different events which can lead to a specific undesired event. FTA is a valuable tool for general use in multidisciplined risk assessments, single and multiple failure analyses, and in system safety analysis of failures and other events affecting safety. FTA can be used to develop qualitative and quantitative probabilistic reports, although the lack of quantitative data is usually a more difficult problem for FTA than for FMECA. FTA is most applicable at the system level during the early phases of the system life cycle, but it may be useful at any time. Early FTA reports should be

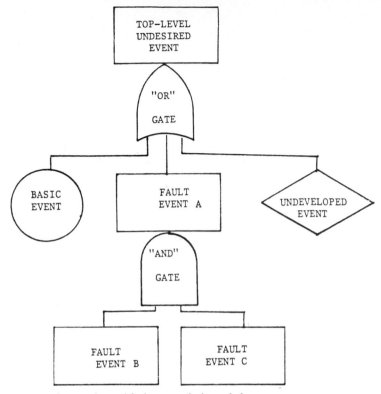

FIG. 13.5   Commonly used fault-tree analysis symbols.

updated as the system design matures and/or significant changes occur. FTA uses symbolic logic diagrams to show the cause-and-effect relationships within a system between a specified top-level undesired event and all of its possible contributing causes such as "faults" or failures of system elements. Although the contributing causes in FTAs need not be limited to system failures and may include other events such as accidents or natural disasters, this discussion is limited to FTA applications for failure analysis purposes.

Fault trees are logic diagrams developed using formalized deductive logic. Depending upon their complexity, they may be produced manually or by using automated drafting techniques with computer-aided systems. Figure 13.5 illustrates the most commonly used symbols found in fault-tree diagrams. The figure uses an OR symbol or logic gate to relate a top-level undesired event to each of three different symbols representing events which are each capable of independently causing the top-level event. The circle is used to represent a basic event or primary failure whose characteristics are such that further development in the fault tree is not considered necessary. The diamond is used to represent other events which are not developed further, either due to lack of required information or a low probability of occurrence. The AND symbol or logic gate relates the branch labeled "fault event A" to the contributing fault events labeled B and C, which must occur together to cause event A. If the fault-tree diagram were expanded

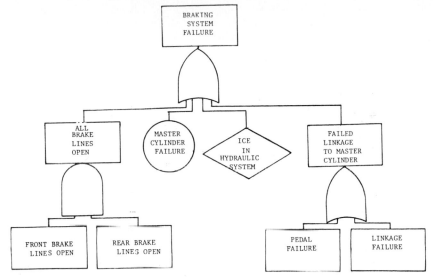

FIG. 13.6 Simplified example of a fault tree.

further, these contributing events themselves would be the end events of subordinate contributing events. The diagram—and the analytical process—begins with the identification of the undesired top-level event. This may be a system failure, as in the example, or any other type of undesired event having multiple causes requiring analysis.

Using deductive logic, the diagram is expanded using the AND gates and OR gates to connect the top-level event to all of the contributing causative events in a manner which depicts their interrelationship. These events can be considered as inputs to the top-level event, representing the states of the various constituent elements of the system.

A simplified example of a fault-tree diagram using the symbols illustrated in Fig. 13.5 is provided in Fig. 13.6. The figure shows selected types of failures which have the capability of causing the failure of hydraulic braking systems commonly used in automobiles. Beginning with the total failure of the braking system as the top-level event, the diagram models the selected system conditions that can result in total braking system failure as follows:

1. Brake lines open to front and rear wheels
2. Failure of master cylinder (a basic event not requiring further development)
3. Ice blocking the hydraulic system (an undeveloped event which is also of very low probability)
4. Mechanical failure of linkage to master cylinder involving the brake pedal or the linkage between the pedal and the master cylinder

Completed fault-tree structures can be combined with probabilistic data for the purpose of establishing the relative criticality of system failures. However, criticality analysis in FTA is more complex than in FMEA because FTA addresses multiple failure situations and other events not usually addressed in

FMECA. These additional events introduce new and complex conditional probabilities which are not represented in the total single failure rate probability baseline used in the FMECA. Therefore, FTA criticality analysis can be an important supplement to the FMECA criticality analysis (see Chap. 18).

13.2.4   Sneak Circuit Analysis (SCA)

The growth in complexity of system designs and system interface relationships which has characterized technology throughout this century has produced increased opportunities for failures of all types. Among these is an insidious type of system failure which does not involve the failure of any system element. This failure type occurs when a latent path or condition causes an undesired event to occur, and/or inhibits the proper performance of a required function, without the occurrence of a component failure. The cause of a failure of this type in an electrical or electronic system is identified as a *sneak circuit*. One example is the cross-coupling of signal wiring or electromagnetic interference which may occur as a result of improper cable routing or inadequate shielding. Another example is the capacitive coupling which may occur during the switching of high-frequency, low-level signals in a switching matrix which also carries high-level or pulse-type signals. Detection of the presence of a sneak circuit may not occur until the product is in service, since a unique set of operational conditions may be required to precipitate the failure. Sneak circuit paths can be extremely hazardous to personnel and equipment. Therefore, SCA analyses can be essential to ensure system safety for equipment and controls which are associated with high-energy systems such as nuclear power plants, petrochemical plants, and weapons systems.

The sneak circuit problem is not limited, however, to electrical and electronic equipment. It may also be present in the design of other systems, including mechanical, pneumatic, and hydraulic types. For these systems, the condition is identified as a *sneak path*. As an example, the absence of a check valve in a fuel line designed to return excess fuel from a combustion system could permit the undesired delivery of fuel to the system by a reverse flow, following shutdown of the main line. Another example is illustrated by the frequent failures of automobile gasoline engines to stop when the ignition is turned off because of the internal engine conditions which cause "dieseling." Sneak paths may also be present in the design of software systems. For example, a sneak path can permit the flow of data over an unexpected route, or the loss of data.

SCA has been used in the investigation of accidents and test failures when other techniques have failed to locate the cause. The technique was developed during the late 1960s by Boeing for NASA. It is based upon network analysis principles, using detailed design information and "as-built" drawings. The types of problems addressed include:

- *Sneak paths:*   Such as a design error which permits the flow of current over an unintended path
- *Sneak timing:*   Such as the occurrence of a circuit function at an improper time
- *Sneak label or indication:*   Such as the incorrect labeling of a switch, or misleading display (e.g., the PORV indicator light at Three Mile Island plant, TMI-2, which indicated status of power to the valve solenoid instead of valve status).[18]

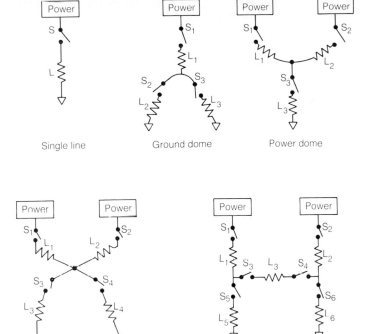

Single line   Ground dome   Power dome

Combination dome   H pattern

FIG. 13.7   Standard topological patterns used for sneak circuit analysis.

Prerequisites for performing SCA include a thorough understanding of the functional characteristics of the circuit or system and a full knowledge of its performance requirements. Basic steps in performing SCA involve the development and analysis of standard topological patterns[19] obtained from network trees representing all important features of the topography of the circuit. The network trees are developed by progressively subdividing circuits into nodes, paths, and nodal sets. Checklists are used to evaluate the standard topological patterns (e.g., Can current flow be reversed? Is logic flow correct?) to detect and identify sneak conditions which may be present. Five standard topological patterns used for SCA are illustrated in Fig. 13.7.[19]

SCA is a highly specialized activity that is both expensive and time consuming. Because it requires developing the topography of complex circuit nodes, paths, nodal sets, and the simplification of nodal sets so that all important circuit features are retained while reducing circuit complexity, the process of developing SCA analytical models can be extremely complex. For example, for digital logic circuits it requires simulation of all combinations of switching positions, transients, and timing. This involves examination of all interactions which can occur between properly functioning components in a particular circuit. As such, even with computer-aided processes, it can require extensive computer time to address all permutations and combinations involved. Because of its cost and complexity, SCA should be considered primarily for application to functions and circuits which are selected on the basis of their criticality to safety and/or performance. This selection requires a detailed engineering analysis of the potential risks inherent in the design.

SCA is not normally considered appropriate for use on proven off-the-shelf hardware, but it is applicable to new designs where critical functions and circuits are involved. The results of SCA may not only detect and identify sneak circuits but also may pinpoint documentation errors and/or reveal design sensitivities in software or hardware designs. SCA is not intended for use in the evaluation of the impact of environmental factors, operator errors, or failures which themselves initiate sneak circuit activity. These failure types are more appropriately addressed in reliability prediction analyses, FMEAs, or FTAs. Unfortunately, SCA cannot be effectively performed until late in the design phase when the design documentation is sufficiently advanced to permit the identification and analysis of all circuit paths. At this time, the results are often too late to be effective. This problem, combined with the complexity, limited application, and high cost of SCA, continues to mitigate against its widespread use.

13.2.5 Software Failure Analysis

Problems occurring in the design or operation of software, i.e., computer programs and data, are not commonly identified as failures. This term is more often used in the software world when the computer hardware fails. However, in cases where a computer program fails to perform a required function correctly during software testing, the event is usually identified as a fault, an error, or a bug (see Chap. 16). As an example, the definitions of these terms established in the recently published IEEE STD. 729-1983[20] include definitions originated by the International Organization for Standardization (ISO) and the American National Standards Institute (ANSI). ANSI defines an *error* as:

1. A discrepancy between a computed, observed, or measured value or condition and the true, specified, or theoretically correct value or condition.

2. Human action that results in software containing a fault. Examples include omission or misinterpretation of user requirements in a software specification, incorrect translation or omission of a requirement in the design specification. This is not a preferred usage.

ISO defines a *fault* as:

1. An accidental condition that causes a functional unit to fail to perform its required function.

2. The manifestation of an error(s) in software. A fault, if encountered, may cause a failure. Synonymous with "bug."

Failure is defined by ISO as:

1. The termination of the ability of a functional unit to perform its required function.

2. The inability of a system or system component to perform a required function within specified limits. A failure may be produced when a fault is encountered.

3. A departure of program operation from program requirements.

These definitions might be paraphrased and summarized as follows:

1. *Errors* are not failures but are incorrectly computed values or conditions, or are the human errors which cause faults in software.
2. *Faults* result directly from software errors or are accidental conditions which may cause systems or functional units (hardware) to fail.
3. *Failures* may be produced by faults, represent the loss of functional capability by a system element, or involve the operational departure of a program from requirements.

Why did the last definition not say "failure of the program" instead of "operational departure from program requirements"? Why is the term "failure" not used anywhere in the comprehensive DOD standard[21] governing software development? The answer lies in the common conception that software simply does not fail. This view is derived from procedural and algorithmic characteristics of software and its inherent differences from hardware.

Once developed and debugged, software can be considered analogous to a proven mathematical formula or a manufacturing work instruction which can be used over and over again without concern for error in the output produced. Unlike hardware, software does not wear out with use. Software is not subject to the laws of physics or the effects of time. The repair of faulty software is different from the repair of faulty hardware. It involves the correction of ideographic, procedural, and/or conceptual faults or errors, and results in the creation of a new software baseline configuration and a new software functional performance capability. In contrast, hardware repair involves the correction of physical deficiencies and restores the hardware to its original physical and functional baseline configuration.

When design changes are made to the software to eliminate faults and errors, they are comparable to design changes made in hardware to correct and eliminate design errors such as deficiencies discovered during development testing. In the case of a hardware redesign which has proved successful in eliminating a failure cause, or in the case of hardware departures from performance requirements which occur as a result of operator errors, the hardware design is considered free from failure and the initiating events are identified as errors rather than failures. Such precedents in the treatment of hardware failures eliminated by design changes or discounted as human error lend support to the practice of avoiding use of the term failure for software and identifying software "failures" as faults or errors. The current DOD standard[22] used for weapon system software development does not include failure or fault among its definitions, but instead uses the term "error."

Based upon these definitions and precedents, the term "software error analysis" (SEA) will be used in place of "failure analysis" to represent all types of analysis applied to the software anomalies identified as faults, errors, and/or departures of program operation from program requirements, as previously defined. SEA techniques focus upon and are tailored to meet the needs of specific elements of software development activity, and they may be related to the applicable phases of the software development life cycle, as summarized in Fig. 13.8.

Errors are introduced in software during all stages of the development process. For example, in the development of software requirements documents, user needs may not be correctly interpreted from or adequately defined in the top-level specification. Errors of this type can be the most costly because they can

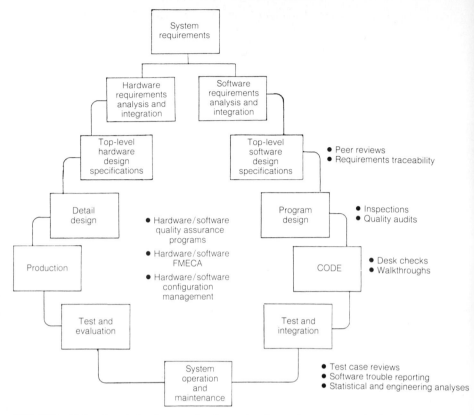

FIG. 13.8 Major steps in hardware and software development processes and related software error detection and analysis tools.

compound problems in the specification and allocation of software requirements to lower-level specifications, user interfaces, functional performance, input-output data, and hardware constraints. Additional errors occur in subsequent stages of software development during detailed design, coding, and testing. It is true that errors of all types are identified, documented, and corrected throughout the life of software systems. However, the rate and severity of errors should continue to decrease with time as the software maturity grows.

Many different types of software design errors can occur during development, such as the following:

Logic errors	Documentation errors	Requirement omission
Timing errors	Requirement misinterpretation	Data or data structure errors
Modeling errors	Missing or extra material	
Coding errors	Illegal instructions	Improper assumptions
Interface errors	Mathematical errors	
Typographical errors	Syntax errors	

As in the case of hardware failure analysis during design, testing, and in-service operation, the earlier that software problems are detected, analyzed, and corrected, the greater are the savings in terms of design and development costs, production schedules, and life-cycle costs.

During the requirements phase, SEA is the least expensive to perform and may yield the greatest benefits because of the potential for these errors to proliferate to many lower levels throughout the remainder of the software development cycle. Design review SEA techniques which involve the use of "walk-throughs," peer reviews, and requirements traceability analysis offer useful tools for error detection and analysis. Together, these analyses help to ensure the proper flowdown, allocation, accuracy, completeness, interpretation of, and compliance with top-level requirements. It is essential that these activities continue through the extension and expansion of system requirements in the detailed software development for all subtier levels of the software system configuration. Typical types of errors which can be introduced include the flowdown or development of requirements which are incorrect, incomplete, unclear or ambiguous, inapplicable, conflicting, or inconsistent.

During the preliminary design, detail design, and coding phases of software development, the opportunities for errors increase as the difficulty, detail, and scope of work expand. Since software design and development is a labor-intensive activity which involves the relatively independent efforts of many individuals, communication, standardization, and interface problems are common sources of error.

Formal inspections known as peer reviews or walk-throughs address these problems very effectively. A typical peer review involves a rigorous and detailed formal examination of software documentation by a small group of qualified technical personnel who represent peers of the author and/or skilled specialists. The types of software documentation which are subjected to peer reviews may include requirements documentation, design documentation, code listings, and test documentation.

Peer reviews should be limited to a small group of between three to six people and conducted in accordance with documented procedures. They should involve the following basic functional entities as a minimum: (1) a moderator, to lead and support the review; (2) the author, to read and respond to questions and comments on the documentation; (3) technical reviewers, to identify errors; and (4) a recorder, to prepare the error list. The moderator may also serve as the recorder. All documentation should be rigorously analyzed independently by each participant in preparation for the reviews. Checklists based upon the types of errors associated with the process are important tools for error analysis. To avoid complications and improve communication, management should be excluded from the reviews. The primary purpose of the reviews should be clearly enunciated by the moderator. It is the objective of the review to detect and document errors—not to correct them. Another purpose is to ensure that the author recognizes, understands, and accepts the detected errors for correction. The moderator holds the key to the success of the review through his or her skills in selecting the participants, defining the material and criteria to be used for analysis, guiding and arbitrating the review activities, and ensuring the proper severity classification, identification, and adequate documentation of the errors for corrective action by the author. Depending upon the results of the review, the moderator may schedule a second review of the corrected program.

Other techniques used for the analysis and correction of software errors involve the analysis of results obtained from operational testing of software at each development level. The most straightforward of these begins with the analysis of software trouble reports (STRs) and test logs. Statistical and engineering analyses are performed with these data, particularly for the purpose of reaching decisions relative to the initiation of error investigations and corrections and management decisions on the continuation or termination of software testing.

Many other diverse activities are associated with software error analysis and correction during the testing and debugging of software systems. Extensive description of these activities is beyond the scope of this chapter, and the techniques continue to grow and change with the rapid advancement of computer science. However, an indication of the scope of SEA during testing can be obtained from the very large number of analytical techniques which are used. These include the following:

Stress testing	Logic testing	Function testing
Path testing	Cause-effect graphing	Volume testing
Equivalence classes	Postfunctional analysis	Execution anaylsis
Symbolic execution	Simulation	Static analysis
Storage testing	Algorithm evaluation	Security testing
Boundary value analysis	Top-down or bottom-up testing	Modeling
Diagnostics	Timing analysis	

Examples of tools used in support of SEA include hardware such as test beds and hardware monitors, and a wide range of computer programs, such as the following:

Assemblers	Program sequencers	Data analyzers
Consistency checkers	Editors	Logic/equation sequencers
Dynamic analyzers	Compilers	Decompilers
Software monitors	Interface checkers	Flowcharters
Interrupt analyzers	Comparators	
Automated test generators		

13.3 TECHNIQUES AND APPLICATIONS IN SUPPORT OF SYSTEMS IN PROCUREMENT, PRODUCTION, AND TEST

The types of failure analysis discussed in this section are dependent on the availability of hardware, as distinct from the analyses discussed in the previous section, which require only a technical information base. Therefore, they are closely associated with all types of hardware testing and evaluation activities, including those occurring during production and in later phases of the product life cycle.

13.3.1 Destructive Physical Analysis (DPA)

DPA is the methodical dissection and inspection of unfailed parts (also see Chap. 14). It is usually performed as a receiving inspection function on small samples of parts used in high-reliability applications. The results of DPA support accept or reject decisions for the balance of parts in a manufacturer's lot. Usually these lot disposition decisions are associated with the acceptance or rejection of incoming purchased parts prior to assembly. However, unsatisfactory part performance subsequent to use in higher assemblies may initiate DPA activities. DPA uses a wide range of failure analysis tools and techniques for the inspection, dissection, and analysis of physical anomalies or process changes. It is effective for use on active devices, such as microcircuits, diodes, and transistors, and on passive devices such as capacitors and resistors of all types, connectors, crystal products, fuses, switches, thermistors, and relays. An important function of DPA is the detection of unreported changes in part design, materials, or production processes which may be significant to reliability. DPA is also an important tool for the evaluation and comparison of vendor quality systems.

A list of representative specialized laboratory equipment typically used in DPA includes:

Hermeticity test facilities	Polishing wheels
X-ray equipment	Microscopes—10× to 500×
Diamond lapping wheel	Small lathe
Abrasive lapping paper	Scanning electron microscope

The use of such equipment requires highly skilled personnel, detailed procedures, and the preparation of technical reports in close coordination with parts and materials engineering specialists. For example, a typical DPA of microcircuits might involve the following steps:

1. Inspect and photograph abnormal external conditions such as bent leads
2. Delidding
3. Depotting of organically filled parts by immersion in depotting compound
4. Visual inspection, composite photograph of internal parts, dye photograph under magnification to identify metallization pattern
5. Wire bond pull testing for lead bond strength
6. Passivating glass removal (deglassivation) by the use of buffered fluoride-containing etch using care to prevent damage to metallization
7. Dye attach integrity test
8. Substrate shear test

Details relating to each of the above steps and related standards and criteria for evaluation and reporting should be documented in procedures governing DPA for each device. They should include related standards and criteria defining characteristics such as internal workmanship standards, plating thickness, scratches, solderability, cracks, surface finish, encapsulation quality, weld quality, dielectric voids, wire bend radius, extraneous contamination particle characteristics, weld spatter, gold flaking, pinholes, burrs, and other physical dimensions and characteristics.

13.3.2 Physics of Failure Analysis

Physics of failure—sometimes called *reliability physics*—is the physical, chemical, and/or electrical analysis of failed assemblies, parts, and/or failed materials and the investigation of their failure mechanisms. Often it involves the use of laboratory equipment and specialized procedures similar to those used in DPA for the analytical dissection of failed items and/or experimental studies. It seeks to understand the detailed sequence of cause-and-effect relationships which were involved in the failure mechanism for the purpose of developing corrective or preventive measures. Physics of failure analysis can also be accomplished effectively for many types of hardware through the use of simple tools and basic inspection and test methods. A physics of failure analysis may begin with the inspection of the fractured rotor blade of a crashed helicopter and end with identification of the sharp knife used in the factory to cut the wrapping material from the blade and create the stress concentration which caused the catastrophic failure in service.

Physics of failure analysis is typically called for in situations where there is uncertainty as to the cause, e.g., during sample product acceptance testing, development growth testing, or reliability demonstration tests. In such cases, the failure analysis is essential to pinpoint the cause, e.g., design, manufacturing, vendor operations, improper operator action, or deficiencies in maintenance. This early identification is often important to product costs and schedules. For example, the early detection of a widely used defective microcircuit representing a major shift in a vendor's process could prevent costly recall and rework of many production units.

The increasingly high quality levels demanded in the components used in modern technology are driving the demand for increased physics of failure analysis and DPA in quality control and reliability. A very small degradation in component quality can result in very high product costs through low yields at final test stations, low productivity, and poor reliability. Physics of failure analysis and DPA provide strong measures of protection against these problems.

The types of equipment used in supporting physics of failure analysis testing and/or inspection of failed items are dealt with in Chap. 14.

Physics of failure analysis is not limited to advanced-technology applications. It is practiced widely in some form by electricians, plumbers, automobile repair shops, and by all organizations responsible for the prevention or correction of failures. Physics of failure analysis is not only important to designers for reliability improvement but is also important to maintain the safety and performance of the products and systems which serve us in everyday life.

13.4 TECHNIQUES AND APPLICATIONS IN SUPPORT OF SYSTEMS IN END USE ENVIRONMENTS

A goal of all failure analysis activities is to contribute to the optimization of system and equipment design reliability and other factors focused upon the end-use environment. However, the failure analysis techniques discussed in this section are more particularly oriented to the field service phase of the equipment life cycle than to other phases.

13.4.1 Functional Failure Analysis

Functional failure analysis (FFA) is the process used to identify and document the system elements, functions, and failure modes which are most important to the disciplines of maintenance and logistics planning, including reliability-centered maintenance.[23] Both FFA and FMECA focus on the failure modes and effects of system elements whose failures are significant to system safety and/or functional mission performance. Therefore, many of the inputs to FFA are the same as those used for FMECA, such as reliability and maintainability predictions and analyses, system logic diagrams, and mission profiles. However, other inputs to FFA include different material such as the results of other pertinent logistic support analysis (LSA) activities. These may include such inputs as functional block diagrams representing the system hierarchy of functionally significant items (FSIs), maintenance concepts, maintainence criticality analyses, level of repair analyses, and mission essentiality classifications.

FFA typically uses a worksheet for each selected FSI. Selections of FSIs and indenture levels are based on safety and system or mission performance considerations, using a top-down approach. FSIs may also include auxiliary system elements such as protective devices, fail-safe design features, and built-in test equipment. Worksheet entries for each element typically include FSI identification, indenture level, and application data; list of functions and outputs to higher-level FSIs; definitions of inputs to the FSI; descriptions of functional failure modes for each listed function; estimated FSI replacement cost; estimated frequency and duration of corrective maintenance; and recommendations for preventive and/or corrective maintenance investigations.

When completed, FFA worksheets are used to establish baselines and priorities for use in maintenance planning, logistic support analyses, and trade-off studies for the optimization of the support systems which sustain the reliability and availability of equipment in service.

13.4.2 Failure Analysis in Logistic Support

Failure analysis is a major element in the performance of LSA and reliability-centered maintenance (RCM) analysis. LSA is performed to identify and plan the optimum utilization of all logistic resources—maintenance, personnel, spares, facilities, test equipment, and documentation—as required to support systems during their operational life cycle. RCM analysis is an integral part of the LSA process which uses failure analysis to tailor and plan the corrective and preventive maintenance functions to support system reliability and availability most effectively.

Major failure analysis contributions to LSA include FFA and FMECA reports, failure and repair rates from reliability and maintainability predictions, and statistical reports produced by field failure reporting and analysis systems. The results of FFAs and FMECAs are used to identify the relative criticality and importance of system elements based upon the impact of failures on essential system functions and to identify the most significant failure modes. Reliability and maintainability predictions also are used to provide a quantitative basis for allocating logistic resources to system elements based upon their frequency-of-failure occurrence and downtime impact on system availability and/or the probability of mission success. In addition, fault-tree analysis and common mode failure analysis support the development of maintenance strategies to be used in allocating resources to the preventive and corrective maintenance programs for the development of RCM programs.

13.4.3   Failure Reporting and Analysis Systems

The report of a failure is the first and often the most important element of the total failure analysis process. Many different types of failure analysis activities are initiated by failure reports, including all of the types discussed in this chapter. However, failure data analysis, as discussed in this section, is the most common analytical use of failure reports.

To the system reliability budget, failure reports are like invoices to an accounting system. Failures also represent real costs. Although traditional accounting systems have not been focused on failure costs, quality cost systems and life-cycle cost analyses rigorously address the cost of failure at all levels of hardware and software throughout the product life cycle. Representative failure costs include lost system revenue due to shutdown, warranty costs, maintenance and logistic costs, the costs of discharging liabilities incurred as a consequence of failure, the cost of standby systems required to compensate for system failures, and the cost of sales lost because of an unsatisfactory product. There is growing management recognition of the fact that a well-designed and properly managed failure reporting system can result in major cost avoidance through its use in cost control and corrective action systems.

Many examples of commercial enterprises having well-developed failure reporting and analysis systems exist in the data processing industry, in transportation agencies such as airline and rail transport systems, and in large-scale industrial systems, such as nuclear- and fossil-fueled electric power generation plants. Failure reporting is now recognized as a management information system. It is of major importance to the many specialized functions and engineering disciplines involved in life-cycle management. These include the following functional areas:

Design	Operations and maintenance management
Warranty management	
Maintenance and logistics planning	Reliability and maintainability engineering
System safety engineering	
Risk management	Quality assurance
	Procurement

The design and utilization of failure reporting and analysis systems should be tailored to fit (1) the needs of management, (2) data processing and analysis requirements, (3) the characteristics of the product or system, and (4) personnel cpaabilities. Each of these design considerations should be rigorously addressed, beginning early in the design of the system and continuing through its implementation period. Any reporting and analysis system whose design does not adequately satisfy all of these design considerations will fail.

Systems may range from the minimum capability required to report, analyze, and service warranty claims for a household product to a system capable of supporting the engineering development and life-cycle management of a petrochemical plant, a power plant, or a computerized nationwide credit data system. The design of a failure reporting and analysis system for these large systems requires reporting at multiple levels of the system. Operational logs are necessary to account for time, failure modes, and status at the system and subsystem levels. An example of such a log designed by the author for use on a broad range of equipment is illustrated in Fig. 13.9. Corrective and preventive maintenance activity reports are necessary to manage, evaluate, and account for the maintenance program and identify problems. Failure reporting is needed at all lower levels of equipment to provide the details necessary to pinpoint failures by location, part type, failure mode, and other detailed information. In addition, a comprehensive data collection, processing, and analysis system is needed to provide the capability for storage, retrieval, and specialized processing, reporting, and automated analysis of data at any level of indenture for any selected equipment and/or period of time.

Even a simplified reporting and analysis system justifies extensive care in the design of all elements, beginning with the format and content of reporting forms and continuing through the design of the system for data screening, processing, storage, retrieval, analysis, and the preparation of summary reports for management use. For example, form design for reporting failures of consumer products is a multidisciplinary task. It requires analyses and trade-off studies between alternatives which involve technical, human engineering, economic, data processing, marketing, and management considerations. The failure report form must be designed to obtain accurate objective technical and product failure information on a timely basis from individuals who lack the technical capability
and the motivation to effectively provide it. It should be brief and self-explanatory and free from ambiguities, yet it must cover all failure situations and meet data processing constraints. The economics of the total system must be estimated and substantiated. Its ability to satisfy management needs and support customer satisfaction will require multidisciplinary analyses, substantiation, coordination, and demonstration in response to questions from a variety of cognizant management authorities before approvals will be granted for implementation.

Management provides resources and support for the reporting system in direct proportion to the degree to which recognized management needs are satisfied. All too often, the needs of management are not recognized and systems are designed to meet only perceived needs. During development, systems are typically focused on the needs of the design, system test, quality, and reliability organizations. During field operations, the maintenance organization typically becomes the principal producer and user of failure reports. Such approaches usually result in the design of suboptimized failure reporting and analysis systems. For example, cost data are frequently excluded from failure reporting and analysis systems.

Failure data analysis is a process which converts failure data into useful information. For large quantities of failure reports, the process requires the use of a computer which may be programmed to provide standard reports and data listing, and to respond in a "user-friendly" manner to queries from analysts at remote terminals for specialized reports and listings.

Typical products of computerized failureanalysis systems include periodic reports and listings designed and preprogrammed to meet the specific needs of various technical and management functions. As an example of failure analysis at the system level, railroads require automated data analysis of failures which affect on-time train arrivals. Output reports establish the distribution of these failures and their causes among all elements of the transportation system. Airlines are concerned with the delay rates of individual aircraft and/or aircraft types, for

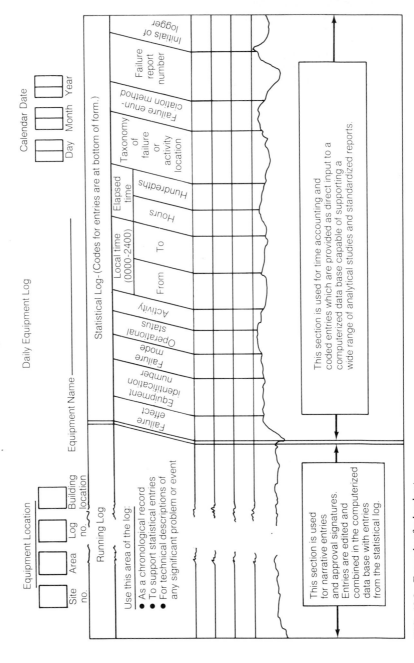

FIG. 13.9 Example of a logsheet.

example, the percentage of flights delayed for a period of more than 15 minutes beyond scheduled flight time by the failure of essential equipment. These data are used to support such decisions as changes in the types and frequencies of preventive maintenance and overhaul, the retirement of aircraft systems or equipment, the acquisition of new systems or equipment, and reliability improvement activities on existing system elements.

The scope and types of computer-aided failure analysis which may be associated with user requests are limited only by the design of the failure reporting and analysis system and the creativity of the analysts. Typically, the retrieval and manipulation of the data are focused on such areas as special investigations of specific hardware, the collection and study of data relating to particular types of system failure events, statistical studies, and trend analysis. These analyses, and also the standard output failure data analyses, are used for the ultimate purpose of developing corrective or preventive actions for failures of hardware, software, personnel, or procedures. Therefore, they are concerned with establishing cause-and-effect relationships and identifying the principal contributing factors associated with failure events. This means that they often lead to further in-depth engineering analysis. Figure 13.10 uses hypothetical data to illustrate a format developed by the author to summarize cause-and-effect data for system outages. The format also can be extended for use at lower levels of the system or in other applications. In the figure, the vertical column uses a scale of 100 percent to show the relative contribution of individual categories to the total measured value accumulated by the system. As an example, the parameter selected for this total might be the total system maintenance hours, downtime hours, or the total number of downtime failure events occurring in the period. The system element whose

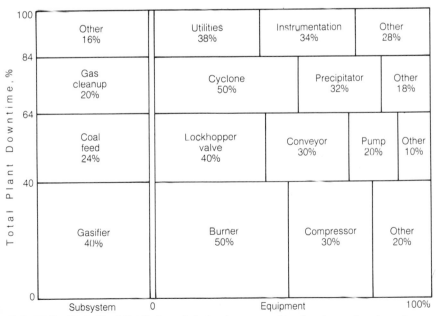

FIG. 13.10 Normalized distribution of plant outage causes by subsystem and equipment.

failures or maintenance needs caused the effect is identified in the applicable block in the vertical column. The pattern represented is typical of a Pareto distribution.

Similarly, the horizontal sections of the figure are based upon a scale of 100 percent. Each block shows the relative contribution of failure events occurring in the various equipment making up each of the system elements represented in the vertical column. Each of the blocks contains the name and/or part number of the associated subassembly with its percent contribution. The size of each block represents the relative contribution of each system element to the baseline parameter. Again, the size of the blocks also suggests a Pareto distribution.

It is important for failure analysis purposes that the design of the failure reporting and analysis system should be consistent with (1) the structure and logic of the system reliability or availability model, (2) the system specifications, and (3) the established operation, maintenance, and support concepts. This consistency also should be maintained in the criteria, formats, and reporting procedures which govern the preparation of operating logs and failure reports; the structure and organization of the data base; the design of the computer programs to be used for data analysis and retrieval; and the format of the output reports to be prepared. Finally, to support optimized life-cycle management, the reporting and analysis system design should be capable of representing the perspectives, serving the needs, and invoking the participation of both the user and the producer. For large systems these include the organizations which own, operate, and/or maintain the system, and those responsible for the design development, installation, and startup of the system.

REFERENCES

1. Herbert, George, quoted in *Poor Richard's Almanac,* Benjamin Franklin, 1758.
2. Bellinger, Dwight Q., "Component Degradation Braking Systems Performance," Report No. 11790-6001-RO-00, prepared by TRW for the National Highway Safety Bureau, U.S. Department of Transportation, December 30, 1969.
3. MIL-STD-1629A, *Procedures for Performing a Failure Mode, Effects, and Criticality Analysis,* November 24, 1980, p. iii.
4. Bellinger, D. Q., and Sims, J. R., "RMA Applications to Fossil Energy Projects," *Proceedings, 1980 Annual Reliability and Maintainability Symposium,* p. 259.
5. Aerospace Recommended Practice ARP-926, "Design Analysis Procedure for Failure Mode, Effects, and Criticality Analysis (FMECA)," Society of Automotive Engineers, September 15, 1967.
6. Davis, Richard, and Goddard, Peter L., "Automated FMEA Techniques," RADC TR 84-244, December 1984, prepared for Rome Air Development Center by the Hughes Aircraft Company.
7. Brown, M., Goddard, P. L., and Pliska, T. F., "Automated FMEA Techniques Feasibility Study," FR83-16-92, final report, dated October 1982, prepared for Rome Air Development Center by the Hughes Aircraft Company.
8. Ibid., pp. 41–42.
9. Hall, F. M., Paul, R. A., and Snow, W. E., "Hardware/Software FMECA," *Proceedings, 1983 Annual Reliability and Maintainability Symposium,* p. 320.
10. MIL-HDBK-217D, *Reliability Prediction of Electronic Equipment,* U.S. Department of Defense, January 1982.
11. Reliability Analysis Center Publication, "Non-Electronic Parts Reliability Data," 1981.

12. Reliability Analysis Center Publication, "Memory/Digital LSI Data," 1981–1982.

13. Reliability Analysis Center Publication, "Transistor/Diode Data," 1979–1980.

14. Reliability Analysis Center Publication, "Digital Evaluation and Generic Failure Analysis Data," 1980.

15. "IEEE Guide to the Collection and Presentation of Electrical, Electronic, and Sensing Component Reliability Data for Nuclear-Power Generating Stations," The Institute of Electrical and Electronic Engineers, June 30, 1977.

16. AMCP-706-196, *Engineering Design Handbook, Design for Reliability*, Department of the Army, January 1976.

17. Barbour, G. L., "Failure Modes and Effects Analysis by Matrix Methods," *Proceedings, 1977 Annual Reliability and Maintainability Symposium*.

18. *Three Mile Island, A Report to the Commissioners and to the Public*, Nuclear Regulatory Commission Special Inquiry Group, Mitchell Rogovin, Director, vol. II, part 2, p. 314.

19. Dussault, H. B., "The Evolution and Practical Applications of Failure Modes and Effects Analyses," RADC-TR-83-72, March 1983.

20. "IEEE Standard Glossary of Software Engineering Terminology," IEEE STD. 729-1983, The Institute of Electrical and Electronic Enginers, February 18, 1983.

21. DOD-STD-1679A, "Weapon System Software Development."

22. Ibid.

23. Naval Sea Systems Command MIL-P-24523, Preliminary Appendix F, "Maintenance Requirements Development Using Reliability Centered Maintenance (RCM) Methodology," November 15, 1979.

BIBLIOGRAPHY

Barlow, R. E., et al., "Introduction to Fault Tree Analysis," University of California, Berkeley, December 1973 (AD 774072).

Bazovsky, I., "Fault Trees, Block Diagrams, and Markov Graphs," *Proceedings, 1977 Annual Reliability and Maintainability Symposium*, pp. 134–141.

Bellinger, D. Q., et al., "Reliability Prediction and Demonstration for Ground Electronic Equipment," TRW, Inc., TRW Systems Group, November 1968, RADC-TR-68-280 (AD 844983).

Bjoro, E. F., et al., "Safety Analysis of an Advanced Energy System Facility," *Proceedings, 1980 Annual Reliability and Maintainability Symposium*, pp. 268–274.

Brown, J. R., and Lipow, M., "TRW Software Series: Testing for Software Reliability," TRW-SS-75-02, TRW Systems Group, System Engineering and Integration Division, January 1975.

Buratti, D. L., and Goody, S. G., "Sneak Analysis Application Guidelines," RADC-TR-82-179, June 1982.

Chin-Kuei Cho, *An Introduction to Software Quality Control*, New York, John Wiley and Sons, 1980, in conjunction with The MITRE Corporation and the George Washington University.

Collett, R. E., and Bachant, P. W., "Integration of BIT Effectiveness with FMECA," *1984, Proceedings Annual Reliability and Maintainability Symposium*, pp. 300–305.

Edwards, G. T., and Watson, I. A., "A Study of Common Mode Failures," Safety and Reliability Directorate, United Kingdom Atomic Energy Authority, July 1979.

Goldberg, F. F., Haasi, D. F., Roberts, N. H., and Vesely, W. E., *Fault Tree Handbook*, NUREG-0492, Washington, D.C., U.S. Nuclear Regulatory Commission, January 1981.

IEEE STD. 730-1981, "IEEE Standard for Software Quality Assurance Plans," The Institute of Electrical and Electronic Engineers, November 13, 1981.

IEEE STD. 829-1983, "IEEE Standard for Software Test Documentation," The Institute of Electrical and Electronic Engineers, February 18, 1983.

Naval Sea Systems Command, Product Assurance Training Center, "Failure Effects Anslysis," April 15, 1979.

Myers, G. J., *The Art of Software Testing,* New York, John Wiley and Sons, 1979.

Reifer, D. J., "Verification, Validation, and Certification: A Software Acquisition Guidebook," TRW Software Series, TRW Defense and Space Systems Group, Systems Engineering and Integration Division, September 1978.

Smith, A. M., and Watson, I. A., "Common Cause Failures—A Dilemma in Perspective," *Proceedings, 1980 Annual Reliability and Maintainability Symposium,* pp. 332–337.

CHAPTER 14
PHYSICS OF FAILURE

David Burgess
Hewlett Packard Corporation

14.1 INTRODUCTION

Failure analysis is a disciplined process for collecting data and making useful conclusions concerning a failure or a group of failures. A single analysis often leads directly to quick, effective corrective action. Savings from corrective action made possible by accurate knowledge are typically enough to pay for a failure analysis effort. Additional savings result from avoiding costly, unnecessary, nonproductive "solutions" to misunderstood problems.

When failure analysis results are examined over a period of time, trends in failure characteristics become clear. Design rules, component selection, testing, and screening choices all become more objective. This benefit of failure analysis is harder to quantify, but it is the really big payoff of failure analysis. Well-understood pitfalls can be taught to new engineers so that old failures are not repeated. Failure analysis helps management make better decisions and helps manufacturing make fewer mistakes. The difference is reflected in employee morale and in higher profits. The payoff is hard to quantify, because wrong decisions avoided by enlightening failure analysis information are never identified.

Failure analysis fosters a reliability physics approach to reliability. This is simply the belief that only one set of physical laws applies to how things work and how things fail. Failures are a rich source of information. A few failures define device limitations more clearly than hundreds of normal working devices. Where failure analysis is used effectively, failures are eagerly sought for the information they promise. At first, meaningful failures are easy to find on the production line and in the field. After "real" failures become scarce, knowledge of how stress influences failure allows simulation of years of use on prototypes before a product is introduced to market. Failure analysis on prerelease hardware allows development engineers to find and correct problems before production release, while options are still relatively inexpensive. New products are developed more quickly and cheaply, and reliability is better at introduction than was possible before. The customer sees the improvement, not the failures.

14.2 STARTING A FAILURE ANALYSIS SYSTEM

14.2.1 Sources of Failures

The success of a failure analysis laboratory depends as much on management as on technical expertise. The role of the lab must be clear. A few areas which should be addressed from the beginning are as follows:

1. *Field failures* must be analyzed; they are the very failures we are most interested in avoiding. However, field failures are discovered long after the possibility of efficient corrective action has passed. Analysis of field failures is often difficult because related information is lost. Additionally, field failures often bring with them pressure and emotion, which are distracting factors.

2. *Production line failures* are an excellent source of failure information. Failures from the product line will be a mix of good parts tested erroneously, handled inappropriately, or designed into risky applications. Causes will be both subtle and blatant. Some failures (typically less than half) will be caused by quality problems initiated by the component manufacturer. Process variations and engineering changes initiated by the manufacturer or user are often involved. Straightforward, immediate corrective action is often possible.

3. *R&D and reliability tests* are good sources of failure information because leverage is highest. Basic design changes are easier to implement. Corrective action may require a simple change rather than a major redesign. Knowledge of the physics of failure imparts a useful understanding that bare failure statistics cannot.

14.2.2   Failure Information

The following are important questions to ask and answer when setting up a failure information gathering system:

1. What information should be saved with the failure? The circumstances and history of a failure must be known. Failure analysis includes consideration of the circumstances, not just examination of the failed part.

2. What kind of reports should be written and who should receive them? Every analysis should have a short written report. Feedback to those who can apply the results and to those who gather information is essential.

3. How much time should be allowed each analysis? Quality is much more important than quantity. However, analysis should not be allowed to remain open simply because answers cannot be found. Like a news report, a clear delineation of known facts and unanswered questions helps determine the next course of action.

4. What records should be kept? Besides individual reports, a log of work in progress and cumulative summary of findings are necessary and valuable.

14.2.3   Data Collection Ideas

Production line failures are one of the most valuable sources of failed devices. The top ten causes of failure on the line can focus activity where significant results are assured. Weekly meetings among manufacturing, test, and R&D representatives can be held to correct problems defined by failure analysis. Computer-based production control systems may simplify collection of devices and information. Lacking such a system, specially marked envelopes can be a solution. One example is shown in Fig. 14.1. Part number and manufacturer, product, and

---

COMPONENT FAILURE

Part No. _____

Manufacturer _____

Model _____

S/N _____

Reference Designator _____

Source: Incoming _____

Production _____

Heat run _____

Envirn test _____

Other _____

Comments _____

_____

_____

_____

Name _____ Date _____

FIG. 14.1 Example of collection envelope for failed parts.

socket identification help determine whether the failure is dependent on board position. The remaining entries allow identification of failure sources other than production. Comments from the technician who identified the failure should be encouraged. (If appropriate, the failure should be accompanied by a tester printout.) Given a source of meaningful failures, lab operating records must be considered. Four kinds of records are required in a well-run failure analysis activity:

1. Operations log
2. Individual reports
3. Summary reports
4. Lab notes

Except for lab notes, all files can be conveniently kept with a personal computer and commercially available software. A single data base provides data for all three purposes. A working system using an HP150, Condor 3 data base, WordStar, and MailMerge is described as an example.

Operations Log.   Two kinds of data should be maintained in an operations log (see Table 14.1). First are data identifying what, who, where, and how many. This information should be recorded as the units are received, before work is begun. Second is information generated by the analysis: completion date, conclusion, methods used. These data can be added to the data base when the analysis is complete. Condor 3 is a data base program which allows data files to be created, sorted, and combined. All the information described above can be handled in a single file. Other information, such as product model numbers and/or socket identification, can be included as required.

Package Codes.   It is sometimes desirable to sort failure data by package material, outline, or size. For example, are the failure distributions for plastic devices significantly different from those for ceramic devices? Does the close spacing of surface-mount device leads introduce new problems? A four-character package code is explained in Table 14.2 and signifies material, outline, and number of leads, respectively. Figure 14.2 shows package examples.

Summary Log Report.   Data in the operations log are available in various combinations for various needs (see Table 14.3). A summary report consisting of a selected few fields serves to show backlog and progress. Simply retrieving reports with log numbers R between 84xx01 and 84xx99 lists all the jobs received in the xx month. Where Date Out is blank, the job is incomplete. Reports can easily be sorted for Date Out = 0 to list jobs in progress. Data can also be sorted by any combination of known information to retrieve a vaguely remembered report. All fields, not only the ones included in the summary log report, can be used for sorting.

Detailed Failure Analysis Report.   The detailed failure analysis report is the primary vehicle for communicating failure analysis results. Verbal reporting is inadequate even for the few privileged to hear it. Misunderstandings of the cause or its implications are inevitable as results are repeated. The only physical product of failure analysis is the written report. Reports are absolutely necessary for effective communication. The format below has proved effective. WordStar and MailMerge combined with Condor 3 generate detailed reports easily, using data in the operations log. An individual report should be produced for each and every analysis performed, even if the requester does not explicitly require one. The following order and content are suggested.

TABLE 14.1 Operations Log Data Description

Item	Variable name	Variable type/size	Note
Log number	R	NR (numeric required)/6 digits	The format suggested is yymmnn. nn is the sequential number of reports in the month.
Division	D	AN (alphanumeric)/4 characters	
Date in	DIN	mm/dd/yy	
Date out	DOUT	mm/dd/yy	
Source	S	A (alpha)/2 characters	S defines source of the failure: II = incoming inspection; SC = screen (100% conditioning); PD = production (from the line); HR = heat run (end product shake-out); RT = reliability test (evaluation); PT = prototype (prerelease design); FD = field (from shipped product)
Requester	RQ	A/15 characters	
Package code	P	AN/4 characters	See Table 14.2
Company part number	PN	AN/9 characters	Example, 1816–2346
Manufacturer	M	A/9 characters	Nine characters allows clear identification. Some abbreviations must be used and should be consistent.
Manufacturer's part number	MPN	AN/7 characters	
Country of origin	C	A/9 characters	Abbreviations should be consistent.
Quantity	Q	NR/2 digits	
Date code	DC	N/4 digits	yyww: example, 8434
Fail code	F	AN/3 characters	
Quantity 2	Q2	N/2 digits	Q2, DC2, F2, Q3, DC3, F3 allow three fail codes or three date codes to be handled in one failure analysis report.
Date code 2	DC2	N/4 digits	
Fail code 2	F2	AN/3 characters	
Quantity 3	Q3	N/2 digits	
Date code 3	DC3	N/4 digits	
Fail code 3	F3	AN/3 characters	

TABLE 14.2 Package Codes

Four-digit package code: MSnn

M indicates body material
   C = ceramic   P = plastic   G = glass   M = metal
S indicates shape or outline
   C = can, circular shape
   D = dual in-line
   A = axial
   G = pin grid array (use 99 for more than 98 pins)
   S = surface mount [includes small outline transistors (SOTs) and plastic leadless
      devices (PLLs)]
   B = leadless
   F = flatpack or single in-line
nn indicates number of leads up to 99

Boiler Plate: Identification of part number, manufacturer, source, reference, etc. This information is included in the operations log and can be automatically read from the Condor data base.

Background: Two or three sentences are usually enough to describe the problem and its significance. Special instructions affecting the analysis can be

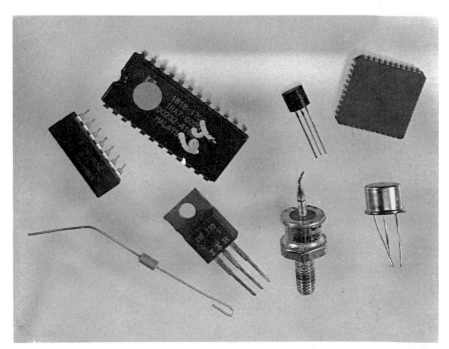

FIG. 14.2 Package examples.

TABLE 14.3 Summary Failure Analysis Log Example

| Failure Analysis Log | | | | | | 06/16/85 |
Report	Requester	Part #	Manufacturer	Date	Qty.	Fcode
850604	John Tech	20-1234	Hightek	060885	3	T01
850605	Tom Field	16-3456	Toptek	060985	2	P07
850606	RD Wiss	18-9876	Futatek	060985	1	E01
850607	Tom Field	16-3456	Toptek		6	
850608	RD Boss	20-8765	Misstek	060785	3	E02

noted here. To prevent misunderstanding, information generated by the analysis is not included in the background section

Conclusion: The conclusion, not the steps taken to obtain it, is stated here. The idea is to emphasize the conclusion, if one has been drawn. If the exact cause is still not known, that fact should be clearly stated. A few more words can be used to enumerate probabilities and eliminate impossibilities. Ultimate care should be taken to separate fact from opinion.

Recommendation: If the analyst feels that intimate knowledge of the circumstances enables him or her to suggest and/or evaluate possible corrective actions, this is the place to make those suggestions. Ideas for further study may also be appropriate. (*Note:* This report should not be counted on to initiate specific action. The report is meant to inform. Other means are required for generating specific action.)

Analysis: This section can be as brief or detailed as circumstances dictate. It should contain enough information to support the conclusion, but not necessarily every detail. It should logically show the sequence of steps performed for each device, identified by number. Reasons leading from one step to the next should be included.

Reference: Identification of previous analyses, reference documents, manufacturer's data book. This information makes subsequent analysis of the same device easier.

Photographs: For convenience in reproducing copies for distribution, photographs are numbered and grouped together at the end of the report. Annotation should be sufficient for photos to make sense without the text. (Many readers read only about a dozen lines from the background and conclusion before skipping to the pictures.) Photos should be identified by device serial number and magnifications and conditions should be given. Reports are often neglected because of lack of time or secretarial help. The forms shown in Figs. 14.3 and 14.4 illustrate how data from the operations log can be incorporated automatically into a final report. The analyst need only fill in the body of the report. The WordStar commands and headings do not appear in the printed report.

Lab Notes. What is known for each device must be known with certainty. Notes are required to separate facts and observations on many devices. Notes need not be elaborate, but they should be identified and complete.

Bound notebooks are preferred because loose pages are easy to misplace. Entries should be identified by R number, and ample space should be allowed for notations about each device. One page per device is not too much. For each

```
.PL65
.mt0
.OP
.DF B : TFILE
.RV  R,D,DIN,DOUT,S,RQ,P,PN,M,MPN,C,Q,DC,F,Q2,DC2,F2
.RV Q3,DC3,F3
.FO                                        ^B&R& PAGE #^B
^BHEWLETT-PACKARD        ^SFAILURE ANALYSIS REPORT&R &^S
Palo Alto, Building 28B                    Date: &DOUT&
Analyst: Quincy, 555-9292 ^B
Our Part Number ^S&PN&^S        Requestor &RQ&
Manufacturer: &M&               Division &D&
Mfr's Part Number : &MPN&       Source : &S&
Country of Origin : &C&

^BFor &Q& devices, failure code is ^SF&^S, date code is &DC&.
For &Q2& devices, failure code is ^SF2&^S, date code is &DC2&.
For &Q3& devices, failure code is ^SF3&^S&, date code is &DC3&.
_____^B
```

^B^SBackground ^S^B	After typing CTRL-KR and specifying FAFORM as the file to read, this form, or any form stored under file name FAFORM will appear on screen. The screen data can be changed in any way desired using all the editing power of WordStar.
^S^BConclusion ^B^S	The dot commands will not be printed. .OP means omit page numbering .PL65 means page length is 65 lines .mt means start type on first line .DF B: TFILE defines file on disc B. This must appear before the .RV
^S^BRecommendation ^B^S	.RV means read TFILE data in the order listed. Not all variables need be used. .RV means keep reading the variables which didn't fit on first line.
^S^BAnalysis ^B^S	.FO ------^B&R& PAGE #^B means print what's on that line as a footnote to every page. In this case ^B means start bold type. The R within &'s means insert the value of R from
^S^BReference ^B^S	TFILE. # will be substituted by the page number.

FIG. 14.3 Form showing WordStar dot commands and format characters. The finished report is shown in Fig. 14.4.

FAILURE ANALYSIS REPORT 870988
DATE: 8/29/88

HEWLETT-PACKARD
Palo Alto, Building 28B
Analyst: Quincy, 555-9292
Our Part Number   5260 S
Manufacturer : ABC
Mfr's Part Number : Prefect
Country of Origin: Utopianot

Requestor: John Engineer
Division 380
Source: Imaginary

For 6 devices, failure code is X03 , date code is 8745.

Background	After typing CTRL-KR and specifying FAFORM as the file to read, this form, or any form stored under file name FAFORM will appear on screen. The screen data can be changed in any way desired using all the editing power of WordStar. The dot commands will not be printed.
Conclusion	.OP means omit page numbering .PL65 means page length is 65 lines .mt means start type on first line .DF B:TFILE defines file on disc B. This must appear before the .RV
Recommendation	.RV means read TFILE data in the order listed. Not all variables need be used. .RV means keep reading the variable which didn't fit on first line.
Analysis	.FO ------^ B&R& PAGE #'B means print what's on that line as a footnote to every page. In this case 'B means start bold type. The R within &"s means insert the value of R from
Reference	TFILE. # will be substituted by the page number.

870988 Page 1

FIG. 14.4   Example of detailed report created from WordStar.

device, markings front and back are recorded, and observations during visual examinations are noted. If no extraordinary observations resulted, that is notable. Such notes will be used over and over again. For example, after delidding and seeing moisture inside one of several hermetic packages, the natural questions is

"Was this the device that didn't have the normal filet?" Unless specifically noted, there will be a tendency to assume that this must have been the one.

Note the time in and out of ovens, the temperature, opening method, etc. Attach photographs and test data. The notebook becomes part of laboratory records when it is filled.

An alternate method is to keep similar notes on 8 × 11 inch sheets of paper. The notes then are filed along with published reports. The disadvantage is that loose notes get lost.

14.2.4 Failure Mechanism Codes

Table 14.4 is a summary of failure mechanism codes which have been used for several years. These codes allow meaningful grouping of data. Finer definition of mechanisms is not necessary to make meaningful groupings. To apply the failure mechanism code, select the code that fits the fundamental cause, not a result of the cause. For example, a device whose cap fell off and subsequently failed because of metal corrosion should be assigned code P03, not M06. A device which was marked with the wrong pin 1 designation should be labeled P02, not E01, even though electrical overstress was indeed the result.

14.3 BASIC PROCEDURE AND LOGICAL PROCESS

The basic procedure for failure analysis is outlined by the 16 steps listed in Table 14.5. These steps should be kept in mind for all analyses, simple and complex. Notes on each step follow.

14.3.1 Notes on Failure Analysis Procedure

1. *Review initial information:* This critical first step should never be overlooked. Sort out information known, unknown, assumed, and claimed. Be alert for inconsistencies in the description of the failure mode. Try to understand the problem about to be analyzed. Look for answers to questions such as:
   a. *What failed?* One, some, all, just the shiny ones?
   b. *How was failure detected?* Was the component itself tested as a failure or did the assembly it was part of fail? If the component was tested, what tests does it pass, what tests does it fail? If the assembly failed, how was failure assigned to a particular component? Does the same part work in other applications?
   c. *When did failure occur?* Be careful! "Failed in production" can mean "Detected in production." Is there any evidence that the part functioned previously?
   d. *What did not fail?* Is there any other place or time where failures might have occurred, but did not?
   e. *What changes have occurred?* Manufacturing process, new lot of materials, date code of part, component manufacturer, test condition.
2. *Examine package using low-power microscope (x-ray optional):* Record

TABLE 14.4   Failure Mechanism Codes*

M Metallization (Includes polysilicon)

1 Open at oxide step	8 Migration across surface causing short
2 Open at contact	9 Interlayer short
3 Scratched open	10 Metal masking fault
4 Smeared causing short	11 Under- or overetching
5 Lifted or peeling	12 Open via (via is window for connecting layers)
6 Corroded or pitted	13 Thin-film resistor open
7 Migration causing open	14 Programmed resistor grow back
	15 Over alloy

S Surface (These codes include descriptive modes which usually are not broken down further)

1 Inversion/channel	5 Voltage sensitive
2 Foreign matter on surface	6 Temperature sensitive
3 Beta (h_{fe}), gain shift	7 Marginal electrical
4 Threshold shift	8 Noise

P Package

1 Mismarked	8 Corrosion
2 Index marked wrong	9 Solderability
3 External leads or seal damaged	10 Excess seal material
4 Hermeticity lost	11 Package broken, cracked
5 Die attach	12 Contamination on leads
6 Loose particles	13 Tar in capacitor
7 Package construction	14 Mechanism
	15 Sticky

B Bonding (Grouped by interconnect technology)

1 Wire broken	15 Bonding pad lifted
2 Wire corroded	16 Bond tail causing short
3 Wire mislocated	17 Intermetallic growth
4 Wire missing	18 Excessive bonding pressure
5 Extra wire	21 Beam broken
6 Wire to scribeline short	22 Beam lifted
7 Wire to lid short	23 Silicon/beam separation
8 Wire to pin short	24 Silicon cracked at beam
9 Wire to case short	25 Beam shorted to silicon
11 Bond misplaced	26 Beam open
12 Chip bond lifted	31 Bump open
13 Package bond lifted	32 Bump shorted
14 Bond broken at heel	33 Chip metal lifted at bump

O Oxide (Includes nitride layers)

1 Field oxide pinholes	4 Capacitor pinhole
2 Oxide fault	5 Gate pinhole short
3 Passivation defect	

TABLE 14.4   Failure Mechanism Codes* (*Continued*)

C Crystal
1 Crystal imperfection
2 Cracked chip (die)
3 Chipped chip (die)

D Diffusion
1 Spike or other diffusion anomaly
2 Isolation defect (includes isolation masking fault)
3 Masking fault

E External (Includes application excessive electrical, mechanical, or thermal stress.
This category indicates no device fault.)
1 Electrical overstress
2 Electrostatic discharge
3 Wrong application (works OK, but appli
   cation requires something else)
4 Latch up (the device may be damaged or
   not from the experience)
5 Second breakdown
6 Interconnections melted
7 Solder bridge (assembly fault, not device
   fault)

U Unknown (The distinctions among types are nebulous but may serve some purpose)
1 Cause unknown
2 Failure confirmed, cause unknown
3 Confirmed specification failure, cause
   not determined
4 DC failure, cause unknown
5 AC failure, cause unknown

N No Problem Found
0 Good device   2 Device retested OK
1 Problem not verified   3 Obsolete device (use when meets old
                                            specification but fails new requirements)

T Testing (These categories used by device manufacturer to isolate causes of shipping
parts never good)
1 Room temperature dc fail   5 Wrong ROM code
2 Room temperature ac fail   6 Wrong chip revision
3 High- or low-temp dc fail   7 Wrong device in package
4 High- or low-temp ac fail   8 Obsolete program
                                        9 Faulty test setup

* Each fail code consists of a letter to represent a general category and two digits to represent a specific defect or mechanism within that category.

all markings, note existence or absence of abnormalities. Serialize devices if more than one.

3. *Verify failure and perform clarifying electrical tests:* Test to verify that

TABLE 14.5 Basic Failure Analysis Procedure

---
 1. Review initial data
 2. Examine using low-power optics
 3. Verify and clarify failure mode
 4. Check hermeticity
 5. Perform conditioning tests
 6. Measure effect of conditioning tests
 7. *Think*
 8. Delid
 9. Examine using low- and high-power optics
10. Physical analysis
11. Draw conclusion
12. Reevaluate data for consistency with conclusion
13. Consider corrective action
14. Write and issue report
15. File data and samples
16. Follow up corrective action

---

the failure is real. Translate the product-level complaint into a more precise component-level failure mode. For example, the initial complaint may be "display won't reset." The component failure mode might be that output low level is high by 0.2 V because of high resistance, not due to an offset.

 4. *Check hermeticity (on hermetic packages):* This step is logically omitted if the source of failed devices makes hermeticity an unlikely factor.

 5. *Perform conditioning tests if applicable:* Temperature aging often "cures" failures caused by surface phenomena; failures due to cracks typically get worse; electrical overstress failures do not change.

 6. *Measure effect of step 5 on key parameters.*

 7. *THINK:* All steps to this point extract information without destroying the sample. Unless the failure is well defined, there is an excellent chance that the next step will completely eliminate all chances of understanding the failure.

 8. *Delid or decapsulate device:* Choose the most logical method based on results above.

 9. *Examine interior under low- and high-power optics.*

10. *Physical analysis:* Repeat enough electrical testing to know whether the condition whose cause is in question still exists. This step varies according to results obtained.

11. *Evaluate data, make conclusion.*

12. *Reevaluate conclusion for consistency with all the data:* Be honest, are all the critiera satisfied by conclusion?

13. *Consider and discuss or recommend corrective action.*

14. *Write and issue report:* Keep it short and to the point.

15. *File data and samples:* Include labeled lab notes to answer possible questions.

16. *Follow up corrective action:* Do this not only to assure that action is taken, but to verify the validity of analysis. Analysts learn confidence, humility, and new ideas from this step.

14.4   FACILITIES AND EQUIPMENT

14.4.1   Facilities

Failure analysis facilities should be designed for efficiency and safety. Laboratory size and shape are usually dictated by available space, but a few requirements must be addressed. For efficiency, all basic equipment must be grouped together. For safety, the equipment should be used exclusively for failure analysis, not sundry production tasks. Four hundred square feet of space is near the minimum. For discussion, consider the actual laboratory layout shown in Fig. 14.5.

Fume Hood and Sink.   The hood should be placed on a wall *away* from the door where it cannot block an exit path. The hood should not be next to a highly traveled path because brisk walking in front of the hood can cause eddy currents which pull fumes into the work area. Hood ventilation must be separate from general ventilation and should draw 100 ft^3 of air per minute average (75 minimum) across the hood opening. Appliances such as ultrasonic cleaner and hot plate are best recessed in to the counter so airflow is not blocked. Controls for appliances must be outside the hood to avoid the possibility of sparks. The sink itself should be flat-bottomed so beakers can safely stand upright. Polyethylene is the recommended material for sink construction because of its acid resistance. A typical hood is shown in Fig. 14.6. Cold water is necessary in the sink. Hot water and dry nitrogen are desirable. A small water heater can be installed under the

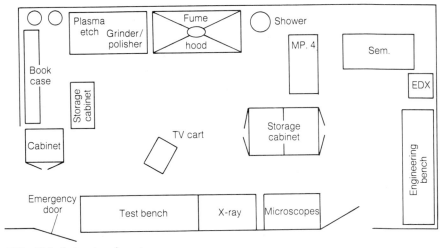

FIG. 14.5 Laboratory layout.

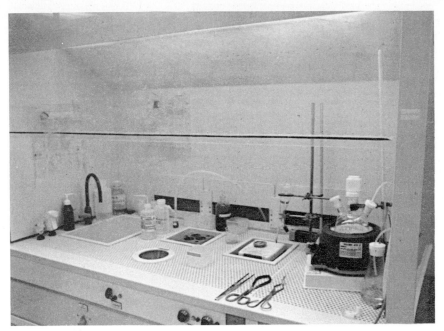

FIG. 14.6 Fume hood and sink.

sink. Aerosol cans can be used if nitrogen is not available. Check local regulations for environmentally safe disposal of acids and solvents. Generally, small volumes of acids, except hydrofluoric, can be flushed well diluted into standard drains. Hydrofluoric acid and solvents, even in small amounts, must be separately collected. Where permanent plumbing is impractical, waste can be safety collected in 1- to 2-gal containers and picked up every one or two weeks for approved disposal.

Safety Shower. A safety shower and eye wash must be located near the chemical sink. The shower should be supplied through a 1½-inch pipe to achieve 55 gal/min flow—the rate required by law in some regions. No drain is recommended; reasonable installations can handle only a few gallons per minute.

Metallurgical Facilities. Grinder/polishers are placed near water supply and drain. Quarter-inch plastic tubing is sufficient for the 0.25 gal/min required for cooling. One-inch plastic tubing is sufficient for the drain. See Fig. 14.7.

Solvent Storage and Plasma Etcher. In the layout shown the solvent cabinet and the plasma etcher are where plastic pipe feed their exhaust into the fume hood ventilation system. See Fig. 14.7.

Microscope Table. The microscope table should be located away from the metallurgical area for cleanliness, but access should be direct for efficiency. The table should be 30 inches tall or less. The table shown in Fig. 14.8 is split, and each side is separately mounted through vibration-isolating supports. The laminar flow hood keeps dust away from the optical equipment.

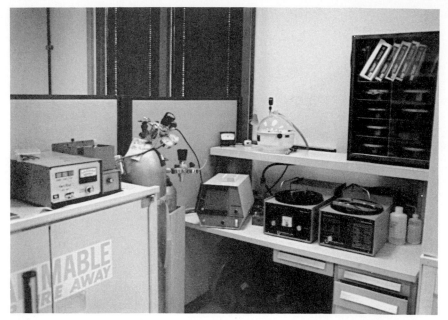

FIG. 14.7   Polishers and etcher.

Scanning Electron Microscope and Energy Dispersion X-ray.   The curtain around the scanning electron microscope (SEM) in Fig. 14.9 allows this area to be darkened independently from the rest of the lab. An area at least 10-feet square is needed for maintenance. Required cooling can be achieved with a recirculating system if plumbing is not convenient. For the setup shown, liquid nitrogen is conveniently available from a dewar located directly through the outside wall. However, the few liters of nitrogen required each week can be carried to the SEM in a thermos jug.

14.4.2   Equipment

Minimum necessary equipment includes a curve tracer, stereo microscope, and metallurgical microscopes. Highly desirable are a probe station, metallurgical polishing equipment, a plasma etcher, and an electron microscope. Failure analysis places special needs on equipment. Most equipment is designed with production needs in mind, not failure analysis needs. Thoughtful choices can save thousands of dollars by omitting unneeded features.

Stereo Microscope.   The stereo microsocpe is used continuously, so professional features are recommended. Although almost any stereo can be used, a wide field of view, an 8:1 zoom range, and good photographic capability are extremely valuable. Wild M8 and the newer Olympus SZM have unexcelled features and quality. Consider these models as a reference for comparison. The comparison

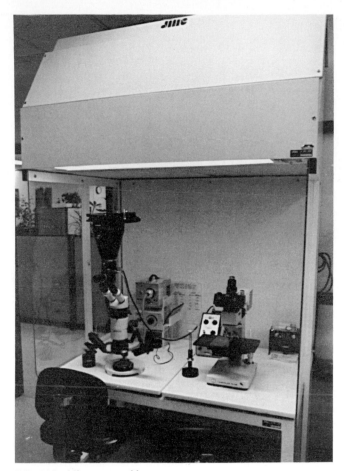

FIG. 14.8 Microscope table.

should be made on configurations including one or two incident lamps, a photographic tube, and a coaxial fiber optic ring light. Price the microscope without a photographic system. The photographic system should be selected to fit both the stereo and metallurgical microscope. Combining photographic equipment with different brands of microscopes is not a problem.

Metallurgical Microscope.   Metallurgical microscopes by Ziess, Lietz, Nikon, Olympus, and others have excellent optics. However, prices differ significantly, and typical configurations do not meet failure analysis requirements. The stage should be able to accommodate wafer-thin samples and samples 1.5 inches tall. Long-working-distance objective lenses are required above 10× to enable the analyst to see devices within a delidded package. Objective lenses cannot generally be interchanged with different brands. Long-working-distance lenses are not available from some manufacturers. (5, 10, 20, 50, 100× power is a good

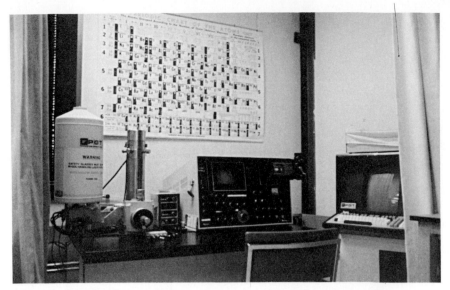

FIG. 14.9   Scanning electron microscope.

assortment.) A five-turret nose piece is required to hold five lenses. Do not remove and replace lenses routinely, as the very fine threads are easy to damage.

Microscope Camera.   An automatic photography system is worth the extra expense. Although all manufacturers have systems available, the Nikon projection lens system is excellent and is priced competitively. It allows maximum light to the film. Some procedures require photography in low-light situations. The photography system chosen should include Polaroid 4 × 5 or 3 × 4 film, and 35 mm for slides.

Macro Camera.   A third camera is required to photograph large packages and to reproduce photographs. The Polaroid MP-4 is a good choice. The standard stand, camera body, and illumination are sufficient for most work. Be sure to include the sliding head, which allows focusing without removing the film holder. Two lenses, the 135 mm and the 50 mm, will allow photographing from 0.5 to 8×. A universal camera mount will allow the camera stand to be used with conventional 35-mm cameras for reproducing slides.

Curve Tracer.   A curve tracer is important to the analyst because it provides a picture of various characteristics rather than point measurements given by digital testers. There are only a few models available. The correct choice depends on the kinds of devices to be analyzed. The Tekronix 576 is best for power devices because of its high current capability. Tekronix 577 is best for general integrated circuits, especially linear devices. The 577 is challenged by Hewlett Packard's digital curve tracer, the 4145A parametric analyzer. The 4145A has unequaled leakage current capability down to the picoampere range. It is programmable and designed for HP-IB interfacing.

   A companion to the curve tracer is a "switch box," which enables the many pins of an integrated circuit to be selected for connection to the few inputs of the curve tracer. Most labs design custom switch boxes, but a lot of time and money are often wasted before a satisfactory design is implemented.

Probe Station.   Many probe stations are available, but most were designed first for wafers, second for packaged devices. For failure analysis, the mechanism for testing packages is more important than automatic stepping or very large stages required for large wafers. Select a system with Bausch & Lomb MicroZoom optics, four or six probes, and ultrasonic cutter, easily exchangeable probe tips, and a TV monitor. Use delicate, very fine tips only if required by state-of-the-art geometries. Many companies manufacture excellent probe stations in this highly competitive field. The best choice depends on price and the subjective "feel" of the analyst. No matter which brand is selected, do not accept the TV monitor offered without considering high-resolution models better suited for *recording*. Recording can be valuable in failure analysis. High-volume commercial TV outlets can often provide the improved performance for less cost.

Metallurgical Equipment.   Only a few pieces of equipment are required for decapsulating and cross-sectioning semiconductor parts. A low-speed diamond saw, a vacuum impregnator, a variable-speed polisher/grinder are sufficient for failure analysis if proper techniques are used. Two companies, Buehler Ltd., Lake Bluff, Ill., and Leco Corp., St. Joseph, Mo., supply complete lines of metallurgical equipment. Technology Associates, Portola Valley, Calif., offers equipment, accessories, and instructions designed specifically for failure analysts. Technology Associates' nonencapsulated cross-section kit is a semiconductor industry standard.

Plasma Etcher.   A simple plasma etcher is an increasingly necessary tool for decapsulation and removal of nitride passivation. The Plasmod and close variations are the only units priced reasonably for failure analysis. Other systems are tailored for production needs. The basic etcher is available from March Instruments, Concord, Calif., and by Spi Supplies, West Chester, Pa. March Instruments also offers a gas control module which makes the necessary repeatability practical. Two gases are required, pure O_2 and CF_4.

X-ray.   The Hewlett Packard Faxitron x-ray model 43804 is a simple, manual control model well suited for failure analysis. Magnification is 1:1. The most frequent use of the x-ray is to locate metal frames within plastic packages prior to decapsulation. Causes of failure in switches and other mechanical components are identified easily with an x-ray.

Hobby Lathe and Mill.   A small lathe and mill combination is an invaluable tool for various routine decapsulation problems.

Scanning Electron Microscope, Energy Dispersion X-ray.   All microscopes produce good pictures. Selection of a microscope for failure analysis should be based on other factors. Failure analysis typically requires magnification of less than $30,000\times$. Voltage contrast and EBIC (electron-beam-induced current) are most important.

14.5 TECHNIQUES

This section will introduce commonly used techniques and the equipment needed to perform them.*

14.5.1 Examination and Testing

Low-Power Microscope Examination. Stereo microscope examination is used initially and frequently throughout the analysis. Initial examination documents package condition and labeling. Often, as in Fig. 14.10, the stereo microscope is a key factor in identification (or communication) of the cause of failure.

Pin-to-Pin Testing. Pin-to-pin testing provides information missing in standard testing to specification. Pure pin-to-pin testing involves observing the current-voltage characteristics of each input pin to ground and each pin to its adjacent neighbor. In practice, the input-output characteristics often are observed under normal powered-up conditions. Input-output characteristics always help define the failure mode. Input-outputs are either normal or abnormal. Required equipment includes a curve tracer and switch box. Figure 14.11 is a curve-tracer

---

* For additional detail, see D. Burgess and O. Trapp, *Failure Analysis and Yield Handbook,* Portola Valley, Calif., Technology Associates, 1986.

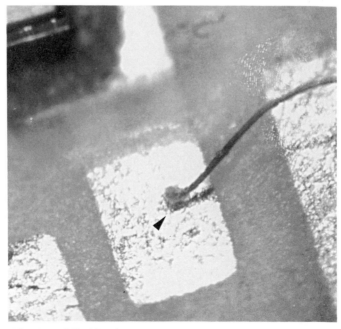

FIG. 14.10 Lifted bond.

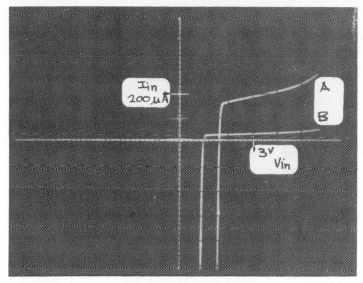

FIG. 14.11   Example of collection envelope for failed parts.

photograph of the input characteristic of a single pin observed under different input conditions. This observation alone proves that the junctions directly connected to the input pin are not damaged. Failure depends on interaction with other circuit elements.

Hermeticity Testing.   Hermeticity is a measure of the degree to which a component package isolates its internal atmosphere from the outside atmosphere. The tests are conceptually simple, yet in practice they are often wrong. A complete hermeticity test must be done in two parts, gross and fine. Fine leak test involves forcing helium or radioactive krypton through leakage paths of packages. Gross leak test is required to detect large holes in packages. Bubble testing is the most popular gross leak test. A low-boiling-point liquid is forced through package fissures, then the package is submerged into a glass tank of heated, clear liquid. Pressure buildup within packages containing liquid causes bubbles to be released. Hermeticity tests are performed on hermetic packages during failure analysis to help establish whether moisture-related failures were caused by water vapor that was sealed in or which seeped in later. The failure analyst may also be asked to determine why a device failed a hermeticity test.

Hermeticity testing is often omitted from routine failure analysis. For most cases, omission does not have a serious effect. However, the low-power visual examination should include an explicit search for cracks or corrosion. Failure analysis of hermeticity, especially fine leaks, is not always a simple task. Figure 14.12 shows the seal area of a ceramic package. Incomplete wetting of the solder resulted in a bubble test failure. Water entered the package during cleaning after wave solder and corroded aluminum wires.

Conditioning.   Various conditioning tests are possible to characterize the failure beyond initial tests. Shock, vibration, or high-temperature testing may be done to

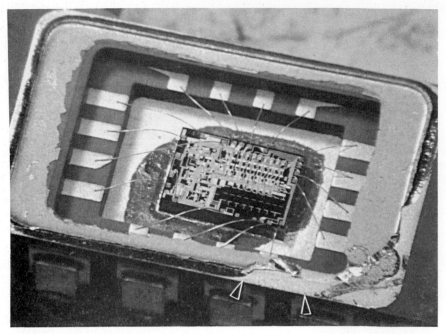

FIG. 14.12   Hermeticity failure.

re-create an intermittent failure. AC failures, leakage failures, and gain failures are often caused by charges in or on the surfaces of a semiconductor chip. Surface effects become more likely in devices which fail noncatastrophically after time under voltage. A high-temperature bake will often cure the failure by making the charges mobile. The failing condition disappears after the charges redistribute themselves. The procedure is simply to place the unopened package wrapped in aluminum foil into an oven or on a hot plate for about 16 hours at 150° C. (Two hours at 200° C can be used if materials permit.)

Figure 14.13 shows curve-tracer characteristics of a transistor before and after bake. The recovery suggests that failure was not due to physical damage. If high-temperature reverse bias causes the failing condition to recur, ionic contamination is associated with the failure. Identification of the origin of the contamination requires further analysis or experimentation.

X-ray.   X-ray is an optional failure analysis step. One time to consider including x-ray is when pin-to-pin testing locates an electrical short between adjacent leads. For the epoxy package shown in Fig. 14.14, the bonding wires were "swept" into each other as the uncured epoxy filled the cavity during molding.

14.5.1   Decapsulation

The next step is decapsulation, a process which will introduce heat, mechanical stress, and chemicals to the device. Decapsulation can destroy failure evidence or the defect itself. Now is the time to decide whether additional testing is warranted and to choose the best decapsulation method—a method compatible with procedures required after the package is opened.

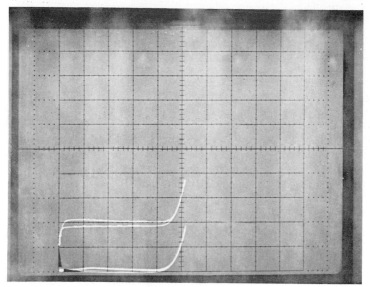

FIG. 14.13 Transistor characteristic.

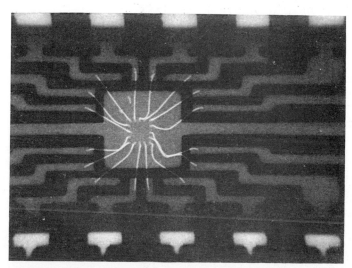

FIG. 14.14 X-ray image.

14.23

Optimum decapsulation of plastic devices or delidding of hermetic devices is quick, clean, and leaves the exposed semiconductor electrically unaltered. The external leads remain intact to allow further testing. For some packages, nearly optimum procedures exist. For others, one or more of the ideal characteristics must be compromised. A few procedures are outlined here. They can be adapted to most components. *Some of the procedures involve acids and solvents which must be used with proper equipment for safety.*

Power Device Metal Can Packages. General-purpose tools such as the diamond saw can be used. The process takes about 10 minutes. Omit machine oil or substitute water to prevent contamination. Figure 14.15 shows a T03 device opened with the diamond saw. The semiconductor within this type of package is often covered with a silicone gel to improve high-voltage performance. The silicone gel can be dissolved in a commercial solvent.

Power Stud Packages. The cap of a power stud package is welded to the base and to the terminal posts. Both connections must be cut to remove the cap. The diamond saw is first used to cut the weld in the posts. Figure 14.16 shows a hobby lathe being used to free the cap at its base. Figure 14.17 shows all the pieces from a stud package.

CERDIP Packages. Ceramic dual in-line packages (CERDIP) consist of a metal frame sandwiched between two pieces of ceramic. Each ceramic piece has an internal recess to provide space for the chip and its bonding wires. Glass surrounds the frame and bonds the ceramic pieces together. Removing the ceramic top without breaking the fragile glass which secures the leads presents a tricky problem. The nutcracker method is simplest. Crack the package using a modified clamp available at hardware and welding supply stores. This method is very quick and usually does not break electrical connections, but it always contaminates the semiconductor surface with dust. The dust can be removed by ultrasonic cleaning in mild soapy water. The opened package is fragile, but the leads can be made strong by laying a bead of epoxy along the side of the package for reinforcement. Figure 14.18 shows a modified clamp being used to open the device shown in Fig. 14.19.

FIG. 14.15 Delidded T03.

FIG. 14.16 Hobby lathe.

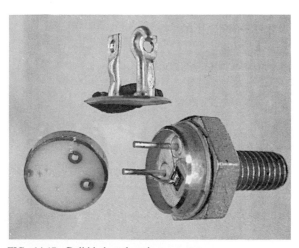

FIG. 14.17 Delidded stud package.

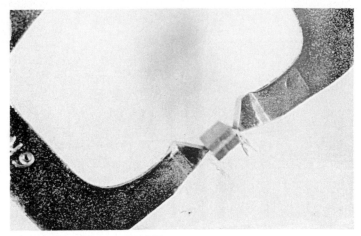

FIG. 14.18   Delidding a CERDIP.

Epoxy Dual In-line Packages (Epoxy DIPs).   Epoxy DIPs are the most common configuration in industrial electronics. Several opening methods exist. The best choice varies for different situations. Similar methods can be applied to other package types.

1. *Red fuming nitric acid dropper technique:*   This technique is an excellent, inexpensive, general-purpose method. First, a flat-bottomed cavity is milled in the epoxy above the semiconductor using a hobby mill like the one pictured in Fig. 14.20. The package is placed on an aluminum heating bar and heated to 140° C within a fume hood. The analyst, *protected by gloves and glasses,* uses a medicine dropper to fill the milled cavity with a few drops of red fuming nitric

FIG. 14.19   Delidded CERDIP.

FIG. 14.20 Hobby lathe as mill.

acid (see Fig. 14.21). (Other forms of nitric acid do not work.) The acid is allowed to work for about 30 seconds, then the package is rinsed in acetone (never water or alcohol). A few repetitions of this process are typically enough to clear the chip. After all etching is complete, the package can be thoroughly cleaned in soapy water. Figure 14.22 shows a completed device.

2. *Sulfuric acid submersion technique:* This technique is applicable to all of the packages above plus a few packages made of highly cross-linked material which does not yield to nitric acid. The submersion method is not as convenient as the nitric acid method. The hot acid is messy, tends to splatter if used improperly, and is more likely to etch aluminum bonding pads. However, it is faster if many devices must be decapsulated at once. Again, a cavity is made above the chip as already described. The analyst, *protected by gloves and glasses,* heats 50 ml of acid to boiling (339° C) in one 100-ml beaker. Another beaker of acid is positioned nearby at room temperature. Using self-locking tweezers, the device is submerged in the hot acid for about a minute. The device is cooled in room-temperature acid, then rinsed in acetone. The process is repeated as required. *Note: The device must never be moved directly from hot acid to acetone. Violent splatter will result.*

3. *Plasma etch:* Plasma etching is clean and effective on almost all encapsulants. Material removal rate is slow. (Processing time is measured in hours, not minutes.) Removing as much material as possible by machining reduces processing time, but the process remains slow. The device is prepared by grinding on the 70-micron diamond parallel to the top surface until the top of the semiconductor is clearly visible through a thin film of epoxy. Not more than 0.003 inches of epoxy should be left for the plasma to remove. Grinding is complete when the widest part of the gold ball bonds is exposed at the

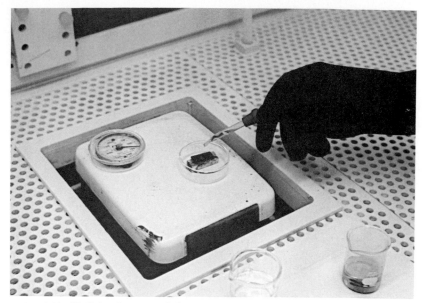

FIG. 14.21 Medicine dropper technique.

ground surface. Figure 14.23 shows a device prepared in this manner. The device is placed into the plasma etcher and etched with pure oxygen for 1 or 2 hours. The dust created is then blown away periodically until the chip is exposed.

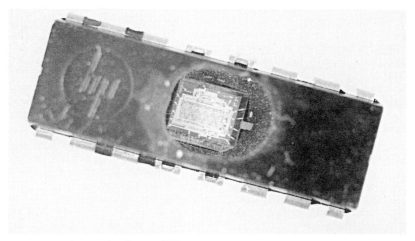

FIG. 14.22 Decapsulated epoxy DIP.

FIG. 14.23 Epoxy DIP prepared for plasma.

14.5.3 Electrical Test

Microprobe. The first step after decapsulation and optical examination is to electrically test key parameters to determine whether the failure characteristic has changed. If wires have been broken, probing is an easy way to access the disconnected leads. Probing can also extend the electrical characterization by allowing measurement or excitation of circuit nodes not available at outside pins. Ultrasonic cutters allow portions of the circuit to be isolated so that individual elements can be measured. Some of the techniques described below are gaining popularity as probing becomes more destructive on today's small geometries. Nevertheless, probing remains a valuable tool for the analyst.

Liquid Crystal Hot-Spot Detection. This old technique has been refined so that a few microwatts of power can be precisely located within a circuit. Figure 14.24 shows the exact area where dielectric breakdown has resulted in a short between gate and source of a power FET. The detection process was achieved in a few minutes. In this case, no optical defect is visible from the top of the device. Even when a blemish is visible, one-to-one correspondence between the suspected defect and the failing characteristic is invaluable to the analyst.

Liquid Crystal Voltage Contrast. One of the most difficult tasks for an analyst is understanding how a new circuit works. Voltage contrast allows the highs and lows to be mapped as a function of input conditions. Because gross voltage levels

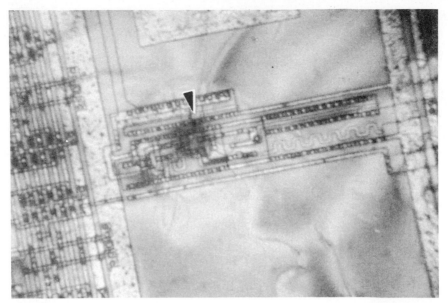

FIG. 14.24 Liquid crystal technique.

can be seen at once across the circuit, much time is saved over the alternate method of mechanical microprobing, line by line. Perhaps more important is that voltage contrast is nondestructive. This inexpensive process is not widely used, perhaps because its sensitivity is second to voltage contrast obtained on the scanning electron microscope.

Scanning Electron Microscope Examination. The scanning electron microscope (SEM) has become a versatile tool for failure analysis. Images produced by electrons resemble familiar optical images, but they are incredibly less limited in magnification and depth of field. Although spectacular images would alone be sufficient to make the SEM an incredibly valuable tool, new techniques utilizing various interactions between the electron beam and the device are developing at a dazzling pace. Noncontact voltage measurements are possible with millivolt precision, and abnormal electron-hole recombination can be detected using current induced within the semiconductor by the beam. A whole new field of electron beam testing promises to be at least as important to the analyst as the incredible photographs. Figure 14.25 shows a voltage contrast image where logic state of lines is visible in "black and white." Dark metal lines are at a positive voltage, light-colored lines are in low voltage states. The device in Fig. 14.25 is shown in the condition which produced the high leakage shown in Fig. 14.11. This information leads to the understanding that an undesired parasitic transistor caused the malfunction.

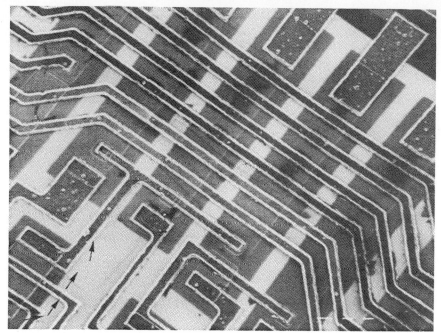

FIG. 14.25 SEM voltage contrast.

14.5.4 Material Exposure

Layer Stripping. Semiconductor construction consists of layers of glass, nitride, metal, and silicon. In order to see defects hidden within these layers, the analyst often must remove the layers one at a time until the critical layer is exposed from the top. The stripping process is sometimes simple and sometimes impossible; however, innovative analysts see few impossibilities. On the simple side, deposited glass can be etched away without removing metal or oxide grown on the silicon substrate. The metal can be removed without endangering the silicon itself. Figures 14.26 and 14.27 are photographs of an electrostatic discharge (ESD) failure. Immediately after decapsulation, no fault is visible. After removing layers sequentially and etching the silicon itself, the damage is recognizable. New processes bring new challenges, but this well-developed art remains a powerful tool.

Cross Section. A view perpendicular to the surface of a semiconductor device often provides information no other perspective can give. Perpendicular and angle cuts are incredibly easy to achieve using the nonencapsulated cross-sectioning technique. Figure 14.28 shows electrical overstress damage (EOD) involving several diffuse layers of a power semiconductor.

Surface Analysis. Several methods are available to identify materials encountered during analysis. Energy dispersive x-ray (EDX) is the most common and can be attached to an SEM as an accessory. Other methods include wavelength dispersion (WDX), which is more sensitive and accurate, but more time consum-

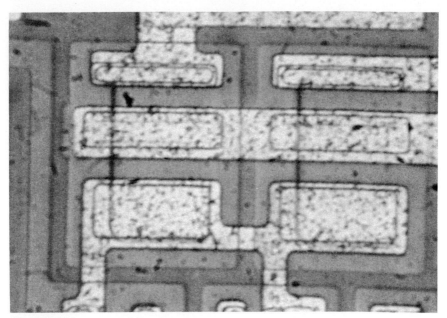

FIG. 14.26 Damaged device appears normal.

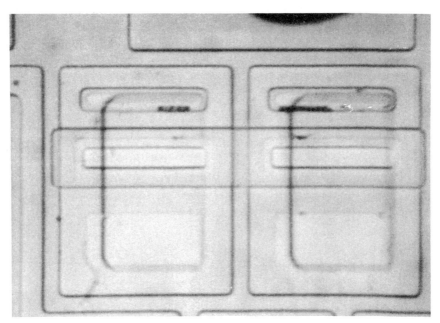

FIG. 14.27 Damaged device after staining.

FIG. 14.28   EOS failure angle lapped and stained.

ing as well. Either of these two capabilities will serve well in a basic failure analysis laboratory. Because of the cost and interpretation of results required, other analysis equipment such as Auger and ESCA are probably best left to specialty labs which contract services.

14.6   PERSONNEL TRAINING

Failure analysis requires an objective engineering perspective and broad range of interest. No new laws of physics need be learned by an analyst, who will need to become more multidisciplined than most circuit designers or process engineers, and will have to become familiar with the fascinating history of things that have gone wrong in the past. The best source of history is the annual *Proceedings of Reliability Physics Symposia* (IRPS). This international symposium, sponsored by IEEE, is the preferred forum for revealing new knowledge of failure mechanisms. Back copies are available from the IEEE Publications Department, Piscataway, N.J. Maximum benefit is gained by attending or participating in symposia. The informal gatherings occurring during these annual spring sessions are motivating as well as informational.

Another symposium dedicated to education and advancement of failure analysis techniques is the International Symposium for Testing and Failure Analysis (ISTFA). This symposium features somewhat broader, less theoretical papers than the IRPS. Mechanical reliability as well as electronic reliability is included.

CHAPTER 15
MAINTAINABILITY AND RELIABILITY

John W. Kraus
McDonnell-Douglas Corporation

When the concept of maintainability was first developed, maintainability was considered a subset of reliability. It was not until the mid-sixties that maintainability was recognized as a separate discipline and was separately established in its own organizations. Maintainability is now sometimes associated with such other disciplines as reliability, system safety, and human factors engineering under the names of effectiveness engineering, product effectiveness, and, with the addition of quality assurance, product assurance. In other companies, maintainability is included as part of integrated logistics support. The organization is frequently a reflection of a company's prime customer organization or of the type of product the company makes. In any case, however, maintainability is still closely related to reliability. Maintainability data are incorporated in the failure modes and effects analysis (FMEA), which is prepared by the reliability department. System or equipment availability depends on parameters which relate to reliability and maintainability, and failure rate data of various types provided by the reliability section are required to calculate many maintainability parameters, as will be seen later in this chapter.

Maintainability is part system integration and part design technology. The process of implementing maintainability is represented by Fig. 15.1. A system or equipment is developed to meet an operational need. Its ability to fulfill this need is in part determined by its maintainability. The first block in Fig. 15.1 represents the operational requirements which determine the requirements for reliability, availability, and maintainability (RAM). The maintenance concept is responsive to both the operational requirements and the RAM requirements. The latter requirements are frequently established by a process of trade and sensitivity studies designed to maximize system availability while minimizing cost, generally life-cycle costs.

The maintenance concept is developed to meet operational requirements and to enhance system or equipment maintainability. The maintenance concept is concerned with the levels of maintenance (operational, intermediate, and depot); the geographic locations of operational and support equipment; and such other concerns as the frequency of periodic tests and the acceptable amount of scheduled maintenance, if any. The maintenance concept will be covered in greater detail in Sec. 15.6.

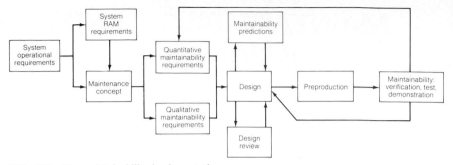

FIG. 15.1   The maintainability implementation process.

After the maintenance concept and the top-level maintainability requirements have been established, qualitative and quantitative maintainability requirements at the equipment level are developed and imposed on the designers, frequently as the maintainability section of contract end item (CEI) or other forms of product specification. During the design process, maintainability engineers keep an ongoing interface with the designers to help the latter meet the maintainability requirements. In addition to the everyday informal contacts, maintainability predictions are prepared to estimate the degree to which quantitative requirements are being met. Participation in both internal and customer design reviews is intended to discover any discrepancies in meeting either quantitative or qualitative requirements.

The final step in the maintainability implementation process is the performance of maintainability demonstration tests to measure, with actual hardware and software and under near operational conditions, the success (or lack of it) with which maintainability has been incorporated in the design. When differences exist between the requirements and the results, redesign may be required or the quantitative maintainability requirements may have to be reallocated or revised, as shown by the feedback arrows in Fig. 15.1.

Figure 15.2 lists the elements of maintainability programs typically applied to complex systems or equipments. These elements are applicable in varying degrees to meet the individual needs of system and equipment programs and to reflect the program phase to which they are to be applied. Appendix A of MIL-STD-470A, *Maintainability Program for Systems and Equipment,* the primary Department of Defense (DOD) maintainability document, contains guidelines for selecting the

Prepare maintainability program plan.
Perform maintainability analysis.
Prepare inputs to maintenance concept and maintenance plan.
Establish maintainability design criteria.
Perform design trade-offs.
Predict maintainability parameter values.
Incorporate maintainability requirements in subcontractor and vendor specs.
Integrate other items.
Participate in design reviews.
Establish data collection, analysis, and corrective action system.
Demonstrate achievement of maintainability requirements.
Prepare maintainability status reports.

FIG. 15.2   Elements of a maintainability program.

maintainability tasks according to program type and phase, and for defining a maintainability program which will meet system objectives in the most cost-effective manner. The appendix contains an application matrix which is reproduced in Fig. 15.3. MIL-STD-470A is very well written and makes an excellent reference for scoping maintainability programs, whether or not they are for DOD procurements.

15.1 DEFINITIONS OF MAINTAINABILITY AND MAINTENANCE

Maintainability is a characteristic of design which, when achieved, contributes to fast, easy maintenance at the lowest life-cycle cost. The word "maintenance" is often incorrectly interchanged with maintainability. Maintainability is design-related and must be incorporated into the design of a system or equipment during the conceptual design, definition, and full-scale development phases of the life cycle. Maintainability expenditures are heaviest during these periods and taper off to sustaining engineering once the system or equipment is in the field. Maintenance, on the other hand, is operations-related since it refers to those activities undertaken after a system is in the field to keep it operational or to restore it to operational condition after a failure has occurred. Expenditures for integrated logistics support (ILS), which includes maintenance, are primarily for planning during conceptual and definition phases and are relatively small, increasing during full-scale development as more details of the design become available, and are heaviest after the system or equipment is in use. The latter ILS expenditures are often the user's rather than the designer's or manufacturer's.

15.2 OBJECTIVES OF MAINTAINABILITY

Maintainability engineering is performed for the following reasons:

1. To influence design to achieve ease of maintenance, thus reducing maintenance time and cost
2. To estimate the downtime for maintenance which, when compared with allowable downtime, determines whether redundancy is required to provide acceptable continuity of a critical function
3. To estimate system availability by combining maintainability data with reliability data
4. To estimate the labor-hours and other resources required for performing maintenance, which are useful for determining the costs of maintenance and for maintenance planning

During the design of a system, the results of maintainability analyses are used to direct the attention of designers to equipment items which contribute to low availability, high costs, or high demands on personnel and other maintenance resources and to point out where to modify the design to improve maintainability.

Maintainability design improvements have made such familiar items as the automobile far more maintainable than it was. For example, the battery was

Task Title	Task Type	Concept	Valid	Program Phase		Operat System Dev (Mods)
				FSD	PROD	
101 Maintainability Program Plan	MGT	N/A	G(3)	G	G(3)(1)	G(1)
102 Monitor/Control of Subcontractors and Vendors	MGT	N/A	S	G	G	S
103 Program Reviews	MGT	S	G(3)	G	G	S
104 Data Collection, Analysis and Corrective Action System	ENG	N/A	S	G	G	S
201 Maintainability Modeling	ENG	S	S(4)	G	C	N/A
202 Maintainability Allocations	ACC	S	S(4)	G	C	S(4)
203 Maintainability Predictions	ACC	N/A	S(2)	G(2)	C	S(2)
204 Failure Modes and Effects Analysis (FMEA) Maintainability Information	ENG	N/A	S(2) (3) (4)	G(1) (2)	C(1) (2)	S(2)
205 Maintainability Analysis	ENG	S(3)	G(3)	G(1)	C(1)	S
206 Maintainability Design Criteria	ENG	N/A	S(3)	G	C	S
207 Preparation of Inputs to Detailed Maintenance Plan and Logistics Support Analysis (LSA)	ACC	N/A	S(2) (3)	G(2)	C(2)	S
301 Maintainability Demonstration (MD)	ACC	N/A	S(2)	G(2)	C(2)	S(2)

CODE DEFINITIONS

S – Selectively applicable   G – Generally applicable   C – Generally applicable to design changes only   N/A – Not applicable

(1) Requires considerable interpretation of intent to be cost effective.
(2) MIL-STD-470 is not the primary implementation document. Other MIL-STDs or Statement of Work requirements must be included to define or rescind the requirements. For example, MIL-STD-471 must be imposed to describe maintainability demonstration details and methods.
(3) Appropriate for those task elements suitable to definition during phase.
(4) Depends on physical complexity of the system unit being procured, its packaging and its overall maintenance policy.

FIG. 15.3 MIL-STD-470A application matrix.

moved from under the floorboards to a location under the hood to improve its accessibility for periodic inspection and servicing (scheduled maintenance), and the battery design was then improved so as to practically eliminate the need for such maintenance as frequent inspections and periodic replenishment of water. Flat tires are relatively easy to change by removal of a wheel and tire assembly and replacement with the spare wheel and tire rather than by removal of the tire from the wheel rim. This is an example of improved maintainability because the line replaceable unit (LRU) was changed from the tire to the tire and wheel assembly. Scheduled maintenance periods for oil change and chassis lubrication have been extended from 1000 miles to 10, 20, or even 30,000 miles. Many cars now incorporate test point connectors for use with the sophisticated analyzers used by most auto mechanics. These analyses reduce fault detection and isolation time and increase their accuracy. The principles used to make these maintainability improvements apply equally well to aircraft, commercial, military, and space systems.

15.3 MAINTAINABILITY PARAMETERS

The most commonly used maintainability parameters are mean time to repair (MTTR) and maintenance labor-hours per operating hour (MLH/OH). MTTR measures the *elapsed* time required to perform a maintenance operation and is used to estimate system downtime and system availability. As a design parameter, MTTR generally includes only those time elements which can be directly controlled by design, often called *active maintenance time*. These elements are (1) fault isolation, (2) removal and replacement of a failed item or repair of such an item in place, and (3) checkout to verify that the maintenance action has been successful. Other maintenance tasks which are largely controlled by maintenance managers, such as alerting the maintenance crew, obtaining tools and spare parts, and the time to reach and return from the maintenance site, are considered "administrative times" and are generally not included in MTTR, as shown in Fig. 15.4.

MLH/OH (or per flight-hour or any other normalizing period of time which is appropriate for the system under analysis) measures the number of labor-hours required to perform a maintenance operation and is used to estimate maintenance costs as well as for personnel planning. As with MTTR, early measures of maintenance labor-hours do not include administrative times. In using MLH/OH, the numberof maintenance persons required and the time each is actually engaged in maintenance are combined to determine total labor-hours.

15.3.1 Calculation of Maintainability Parameters

Mean Time to Repair.   MTTR is probably the most widely used maintainability parameter. It is the mean of a number of times to repair weighted by the probability of occurrence:

$$\text{MTTR} = \frac{\sum_{i=1}^{n} (\lambda_i \tau_i)}{\sum_{i=1}^{n} \lambda_i} \qquad (15.1)$$

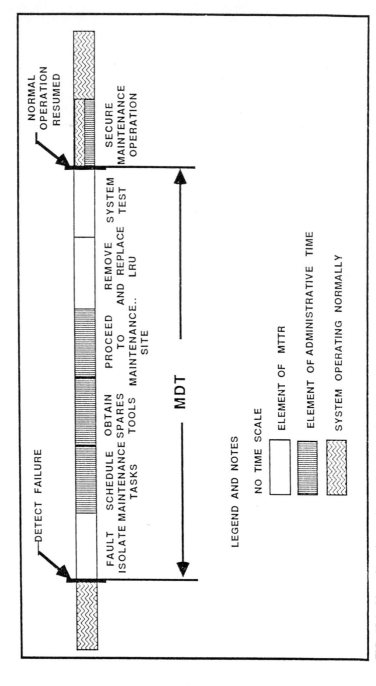

FIG. 15.4 Mean downtime and mean time to repair.

15.6

where λ_i = failure rate of the ith unit
     τ_i = time to repair of the ith unit
     n = number of units

Median Corrective Maintenance Time (MdCT).   When it is desirable to determine the time within which half of the maintenance actions can be accomplished, the median time is used. Its calculation depends on the characteristics of the system which determine the probability function to use to best describe the distribution of the times to repair. The median of a normal distribution is colocated with its mean and mode. Therefore, the median for normally distributed maintenance times equals the MTTR.

For the lognormal distribution,

$$\text{MdCT} = \text{antilog} \frac{\sum_{i=1}^{n} \lambda_i \overline{\log \tau_i)}}{\sum_{i=1}^{n} \lambda_i} \tag{15.2}$$

Maximum Corrective Maintenance Time [CMAX(%)].   The maximum time to repair for a given percentile measures the time required to perform the specified percentage of all identified potential repairs. Most commonly specified are the 90th and the 95th percentiles. Here again, the calculation depends on the probability function used to represent the distribution of maintenance times.

For the normal distribution,

$$\text{CMAX}(\%) = \overline{\tau} + k\sigma_\tau \tag{15.3}$$

where τ = mean of individual times to repair
     k = percentile factor
     σ_τ = standard deviation of distribution of times to repair

For the lognormal distribution,

$$\text{CMAX}(\%) = \text{antilog} (\log \tau + k\sigma_{\log\tau}) \tag{15.4}$$

and

$$\overline{\log \tau} = \frac{\sum_{i=1}^{n} \log \tau_i}{n} \tag{15.5}$$

$$\sigma_{\log \tau} = \left[\frac{\sum_{i=1}^{n} (\log \tau_i)^2 - (\sum_{i=1}^{n} \log \tau_i)^2/n}{n-1} \right]^{1/2} \tag{15.6}$$

The factor k in Eqs. (15.5) and (15.6) represents the percentile at which the maximum repair time is calculated. Values for the 90th and 95th percentiles are

Percentile	k
90	1.28
95	1.65

Probability of Repair within Allowable Downtime.   It is sometimes desirable to calculate the probability of performing a maintenance action within an allowable time interval $t(m)$. The ability to restore a function within an allowable time can be considered as another form of redundancy, one which is less costly for both hardware and maintenance. This parameter, $P(M)$, is defined as

$$P(M) = (1 - e^{-t(m)/\tau})\tag{15.7}$$

where $t(m)$ = the allowable downtime
       τ = the expected downtime, MTTR

The average number of failures during a mission length of time T is λT. The number of failures which cannot be repaired within the allowable time is given by

$$\lambda T e^{-t(m)/\tau}$$

where λ = failure rate in failures per hour.

The number of failures which can be repaired within their allowable downtimes is given by

$$\lambda T(1 - e^{-t(m)/\tau})$$

Probability of Fault Detection.   System availability is affected by the accuracy of fault detection. If a failure occurs which cannot be detected, availability may be reduced if the failure prevents the operation of the system when it is needed. The best source of information about detectability is the FMEA. It should identify which failure modes are detectable and give their probability of occurring. The following equation is useful for determining the probability of correctly detecting faults:

$$P_d = \frac{\sum_{i=1}^{n} \alpha_{id}\lambda_{id}}{\sum_{j=1}^{N} \lambda_j}\tag{15.8}$$

where α_{id} = portion of failure rate which is detectable for the ith item
       λ_{id} = failure rate of the detectable portion of the failure rate of the ith item
       λ_j = failure rate of the jth item
       n = all items which have some degree of fault detectability
       N = total number of items

Mean Time to Detect (MTD).   When part of a system is dormant, it frequently cannot be continuously monitored. It is customary to perform periodic tests by "powering-up" the dormant part of the system. A failure may occur at any time during dormancy, but it would not be detected until performance of the periodic test. The time between the failure and the test is its time to detect. The mean of these times for a system, MTD, affects availability. The equation for calculating the MTD is

$$\text{MTD} = \frac{T}{(1 - e^{-\lambda T})} - \frac{1}{\lambda}\tag{15.9}$$

where λ = the failure rate, in failures per hour
$\quad T$ = the test interval, in hours

At high failure rates MTD approaches T; at low failure rates MTD approaches $T/2$. MTD increases for higher values of T. Figure 15.5 shows MTD as a function of λT and is useful for rapid estimates of MTD.

15.4 GOVERNMENT SPECIFICATIONS, STANDARDS, AND HANDBOOKS FOR MAINTAINABILITY

Following is a list of specifications, standards, and handbooks for the Department of Defense and NASA. For a more complete description of specifications, standards, and handbooks, see Chap. 5.

Department of Defense.

MIL-H-46855, *Human Engineering Requirements for Military Systems, Equipment and Facilities*

MIL-STD-470, *Maintainability Program for Systems and Equipment*

MIL-STD-471, *Maintainability Verification/Demonstration/Evaluation*

MIL-STD-721, *Definition of Terms for Reliability and Maintainability*

MIL-HDBK-266(AS), *Application of Reliability-Centered Maintenance to Naval Aircraft, Weapon Systems and Support Equipment*

MIL-HDBK-472, *Maintainability Prediction*

MIL-STD-1472, *Human Engineering Design Criteria for Military Systems, Equipment and Facilities*

NASA.

NHB 5300.4 (1A), *The Reliability Program Provisions for Aeronautical and Space System Contractors*

NHB 5300.4 (1E), *Maintainability Requirements for Space Systems Programs*

15.5 MAINTAINABILITY REQUIREMENTS

Both qualitative and quantitative maintainability requirements are normally used to define the maintainability characteristics desired in a system or equipment. Qualitative requirements take the form of maintainability design ground rules and criteria. These will be discussed in detail in Sec. 15.7. They describe such requirements as accessibility, ability to detect and isolate a failure, weight limitations of replaceable units, dimensional limits to allow replaceable units to be transported from their installed location to a repair shop or for shipment to their manufacturer's facility, and design requirements to make replaceable units compatible with robots for removal and replacement in remote locations or hazardous environments. Qualitative requirements frequently include human

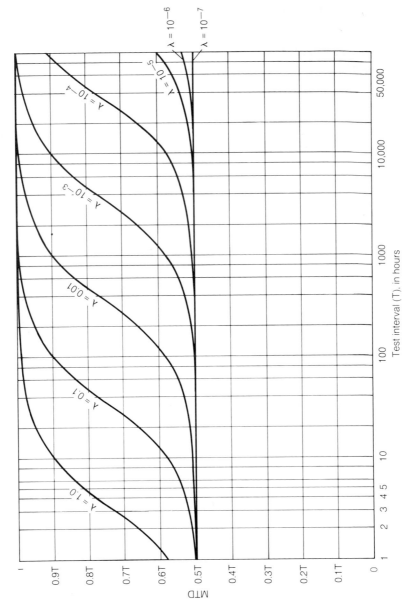

FIG. 15.5 Mean time to detect as a function of test interval and failure rate.

factors criteria such as are found in MIL-HDBK-1472, *Human Engineering Design Criteria for Military Systems, Equpment and Facilities.*

Quantitative requirements are based on desired limitations on system or equipment downtime, maintenance labor-hours, or on system or equipment availability requirements. Downtime is frequently limited to ensure that an important function is not down for so long a time as to permanently damage the system. Downtime also plays a major role in establishing availability, which is a function of uptime (a function of reliability) and downtime, as will be discussed in detail in Sec. 15.10, Maintainability and Availability Analysis.

Maintenance labor-hours have major significance in determining a system's life-cycle costs (LCC) and must frequently be limited to remain within the LCC budget. Maintenance labor-hours may also have to be limited because of limitations on personnel time such as might be found on a space station which has a fixed number of crew members whose time is needed for many tasks other than space station maintenance.

15.6 THE MAINTENANCE CONCEPT

The maintenance concept is developed early in a program to describe a maintenance process which will meet the systems operational requirements. It may be developed by either the customer or a contractor. The initial concept defines such maintenance concerns as:

1. The mission profile
2. Availability and/or reliability requirements
3. Maintenance worker constraints
4. Weight and volume limits (especially for orbiting spacecraft and their resupply vehicles)
5. Sparing policy
6. Periodic testing
7. Scheduled or preventive maintenance
8. The geographic deployment of the system
9. The levels of maintenance (operational, intermediate, and depot or manufacturer)
10. Planned types of support equipment

Later in the program when more details about the system are known, the maintenance concept is described in greater detail. This is often accomplished by functional-flow block diagrams (FFBDs) as shown in Figs. 15.6 and 15.7. The FFBDs describe the steps required to perform the system's mission. These steps include the activities of maintenance personnel and equipment required to accomplish scheduled and unscheduled maintenance.

Figure 15.6 shows the top-level FFBD for a missile system. Block 2.0, Certify Missile Subsystem, represents the steps required for a postinstallation check of the missile and its immediate support equipment. If the missile a subsystem passes the tests, part of the subsystem is continuously monitored and the balance is subject to periodic tests. (The time between periodic tests is determined on the basis of

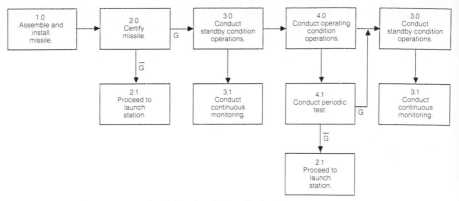

FIG. 15.6   Missile system top-level functional-flow block diagram.

reliability and availability requirements and is justified and stated in the maintenance concept.) If the missile system does not pass the postinstallation check, a
maintenance crew proceeds to the launch station, as shown in Block 2.1, with the
fault isolation test set to determine whether the missile or the missile subsystem
test equipment is at fault. During standby (Block 3.0), continuous monitoring of
launch station environmental conditions is performed, as shown in Block 3.1. If an
anomalous condition is signaled, corrective maintenance is carried out. Block 4.0
represents operating condition activities (similar to launch preparation), which
include the performance of complete tests of the missile subsystem. Since these
tests are performed on a preset schedule, they are identified as periodic tests
(Block 4.1). When a periodic test fails, Block 2.1 represents the activities of fault
isolation and corrective maintenance. These activities are represented in Fig.
15.7.

The maintenance concept also covers the various levels at which different
types of maintenance will be performed. These are generally established according to where the skills, test capabilities, and facilities required for the many
different maintenance operations are available. The using activity normally has
the least-specialized maintenance skills, test equipment, and maintenance facilities. This level of maintenance is called *organizational maintenance*. The next
level, at which more complex maintenance is undertaken, is called the *intermediate level*. Intermediate maintenance is usually performed in a facility which is
close to and centrally located in relation to several using activities. Here

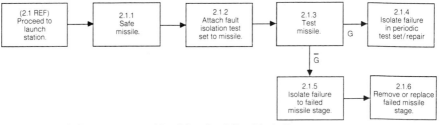

FIG. 15.7   Missile system second-level functional-flow block diagram.

maintenance is performed which is either too complex or too time consuming for the organizational level. In the case of an orbiting manned space station, any maintenance performed off line, as in a maintenance work station, is considered intermediate maintenance. When maintenance requires even more specialization than is available at the intermediate level, the maintenance is assigned to the *depot level*. This level of maintenance can be performed at a large central military or other user facility or at the item manufacturer's plant where the most specialized skills and equipment are available.

The maintenance concept frequently responds to high availability requirements by establishing that organizational-level maintenance shall consist of removal and replacement of a faulty unit. Such a unit is called a line replaceable unit (LRU) for ground-based equipment and an orbital replaceable unit (ORU) for units which will be replaced in orbit, whether part of a manned or unmanned spacecraft. Removal and replacement of a failed item is normally the fastest way to return a system or equipment to full operational capability after a failure has occurred and therefore contributes to high availability and requires less time of the first-line maintenance crew. Selection and design of LRUs are important factors in determining the speed with which they can be removed and replaced as well as in establishing what spares are needed at the organizational level and the logistics required to make them available and to use them. The effect which LRU or ORU selection has on other system factors is shown in Fig. 15.8. Table 15.1 contains an example of ORU selection ground rules and criteria applied to a space station.

15.7   MAINTAINABILITY ALLOCATION

System-level quantitative maintainability requirements are derived from system and operational requirements. They are, however, not useful as equipment design

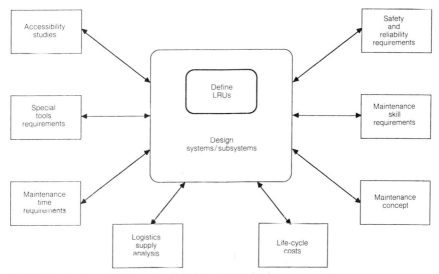

FIG. 15.8   Factors affecting and affected by LRU selection.

TABLE 15.1   ORU Selection Ground Rules and Criteria

ORU Ground Rules
1. ORUs must meet the *required* ORU selection criteria.
2. ORUs must be transportable in the Shuttle. EVA ORUs do not have to be capable of entering a module as long as they can be loaded and unloaded directly from an EVA location directly into the Shuttle cargo bay.
3. ORUs may contain ORUs. When it is technically and economically feasible to restore an equipment which has been designated an ORU by removing and replacing one of its components which has also been designated an ORU, only the lower-level ORU is considered for that maintenance action.
4. The following are *not* ORUs:
a. Standard hardware, including fasteners, latches, etc.
b. Electronic parts, including chips, transistors, connectors, etc.
c. Major truss assemblies such as keels, booms, etc.
d. Modules (habitat, laboratory, logistics)
e. The results of processes, e.g., welds, brazed joints

ORU Selection Criteria	
Required selection criteria	Desirable selection criteria
1. Removal and replacement (R&R) will not introduce unacceptable hazards.	1. The ORUC is a flight-qualified Shuttle LRU.
2. Failures of the ORU candidate (ORUC) must be detectable on orbit.	2. The ORUC has multiple space station applications.
3. Failures of the ORUC must be isolatable to the ORUC on orbit.	3. R&R of the ORUC does not require special tools.
4. If EVA, the ORUC must be capable of passing through the module hatch.	4. R&R of the ORUC does not require special crew skills.
5. Limited-life items will be designed to be ORUs.	5. R&R of an EVA ORUC can be accomplished robotically.
	6. All parts of an ORU have approximately the same reliability.

requirements because they do not address the equipment level. For this reason, it is necessarv to allocate (or apportion) the top-level requirements to the subsystem and equipment levels where the allocated value is useful as a design requirement. The principle of allocation applied to maintainability is similar to that of reliability allocation, except that two variables, failure rate and MTTR, must be balanced in a maintainability allocation as opposed to only one for reliability. This requires that an iterative process be used rather than a deterministic process as is used for reliability.

A reliability block diagram showing the relationships of the different levels of equipment and the failurerates associated with them (see Fig. 15.9) is useful for defining the system being allocated. Estimated values for the maintainability parameter to be allocated are then developed for each item at each level using the same techniques described in Sec. 15.9, Maintainability Prediction. Working from the bottom up, the top-level value of the parameter will usually not be equal to the established maintainability requirement. It then becomes necessary tointroduce a

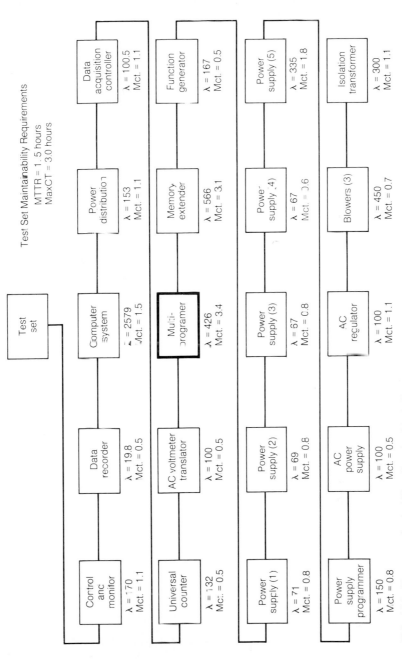

Test Set Maintainability Requirements
MTTR = 1.5 hours
MaxCT = 3.0 hours

Test set

Control and monitor
λ = 170
Mct. = 1.1

Data recorder
λ = 19.8
Mct. = 0.5

Computer system
λ = 2579
Mct. = 1.5

Power distribution
λ = 153
Mct. = 1.1

Data acquisition controller
λ = 100.5
Mct. = 1.1

Universal counter
λ = 132
Mct. = 0.5

AC voltmeter translator
λ = 100
Mct. = 0.5

Multi-programer
λ = 426
Mct. = 3.4

Memory extender
λ = 566
Mct. = 3.1

Function generator
λ = 167
Mct. = 0.5

Power supply (1)
λ = 71
Mct. = 0.8

Power supply (2)
λ = 69
Mct. = 0.8

Power supply (3)
λ = 67
Mct. = 0.8

Power supply (4)
λ = 67
Mct. = 0.6

Power supply (5)
λ = 335
Mct. = 1.8

Power supply programmer
λ = 150
Mct. = 0.8

AC power supply
λ = 100
Mct. = 0.5

AC regulator
λ = 100
Mct. = 1.1

Blowers (3)
λ = 450
Mct. = 0.7

Isolation transformer
λ = 300
Mct. = 1.1

FIG. 15.9 Test set maintainability allocation block diagram.

proportioning factor to either increase or reduce the estimated lower-level values of the maintainability parameter being allocated. The form of the equation used to accomplish this is shown in Eq. (15.10).

$$MTTR_a = \frac{\sum_{i=1}^n [\lambda_i(kMTTR_i)]}{\sum_{i=1}^n(\lambda_i)} \tag{15.10}$$

where k = the allocation factor
λ_i = the failure rate of the ith item
$MTTR_i$ = the original estimate of MTTR for the ith item
$kMTTR_i$ = the adjusted value of MTTR for the ith item

Several iterations using different values of k to obtain a top value equal to the required value are usually needed. A simple computer program or commercially available spreadsheet can be used to facilitate this process. Table 15.2 illustrates the spreadsheet approach. In Table 15.2, the initial estimates of MTTR produce a subsystem MTTR [MTTR(SS)] which exceeds the requirement of 1.50 hours. Substituting an allocation factor of 0.95 lowers the allocated value below the acceptable MTTR(SS), to 1.47. Applying the individual MTTRs thus derived would penalize the designers by reducing the time to repair each component. A third run with $k = 0.97$ produces the desired MTTR(SS). The values in column G represent the individual MTTRs which should be used.

The final allocated values, $kMTTR$, become the design requirements for the various levels of equipment. If as the design matures and more information becomes available, one or more of these values should appear to be unreasonable in relationship to the others, it may be necessary to readjust some or all of them, again by performing trial iterations until the combined failure rates and maintainability parameters result in a top-level value equal to the top-level requirement. Such disparities in equipment-level requirements frequently do not appear until the design has progressed to the point where design details show that no further improvement in maintainability can be achieved *at an acceptable cost*. At this time, a reallocation may be required, using the same iterative process that was used for the initial allocation.

15.8 *MAINTAINABILITY DESIGN CRITERIA*

Maintainability design criteria provide designers with specific guidance regarding qualitative design requirements. At the highest, or least-detailed, levels, they may be referred to as maintainability ground rules. These are usually derived from the maintenance concept. Examples are as follows:

1. The design shall preclude the need for scheduled maintenance.
2. Built-in test (BIT) and built-in test equipment (BITE) shall be the primary method of fault detection.
3. Fault isolation shall be provided to the line replaceable unit (LRU) level.

TABLE 15.2 Spreadsheet Application for Maintainability Allocation

SYSTEM	BASIC WBS #	NO OF ORUs EVA	NO OF ORUs IVA	TOTAL ORUs	FAILURE RATE EVA	FAILURE RATE IVA	ORUs x FAILURE RATE EVA	ORUs x FAILURE RATE IVA	ESTIMATED MTTR EVA	ESTIMATED MTTR IVA	ALLOCATED MTTR EVA	ALLOCATED MTTR IVA
TRUSS & STRUCTURES	16	3500	0	3500	0	0	0	0	2		1.38	0
NODES/TUNNELS	17	9	0	9	0	0	0	0				0
AIRLOCK	18	2	8	10	0	0.000005	0	0.00004	2	2	0	2
HR & T	19	1242	0	1242	8.3333E-07	0	0.001035	0	4.5		3.105	0
MECHANICAL	20	2245	3	2248	4.9383E-06	0.00002	0.01108642	0.000006	2.5	1	1.725	1
RESOURCE INTEGRATION	21	1003	338	1341	0.000001	0.000002	0.001003	0.000676	2	1	1.38	1
GN & C	22	29	29	29	0	4.2553E-05	0	0.00123404	2		1.38	0
DMS	23	219	127	346	6.3694E-05	7.6923E-05	0.01394904	0.00976923	2	1	1.38	1
C & T	24	90	389	479	2.0633E-05	2.0833E-05	0.001875	0.00810417	2	1	1.38	1
HABITABILITY/MANNED SYSTEMS	25	0	40	40	0	4.5455E-05	0	0.00181818		1		1
EVA SYSTEM	26	0	13	13	0	0.0001	0	0.0013		1		1
		8339	918	9257			0.03018251	0.02176758	2.2654	1.002	1.566	1.0018

SYSTEM	ESTIMATED CMT EVA	ESTIMATED CMT IVA	ESTIMATED CMT TOTAL	ALLOCATED CMT EVA	ALLOCATED CMT IVA	ALLOCATED CMT TOTAL	PROP FACTOR EVA	PROP FACTOR IVA	MAINT ACTIONS/YR EVA	MAINT ACTIONS/YR IVA	MAINT ACTIONS/YR TOTAL	AV CREW/ACTION EVA	AV CREW/ACTION IVA	AV CREW/ACTION BOTH	DUTY CYCLE EVA	DUTY CYCLE IVA
TRUSS & STRUCTURES	50	0	50	34.5	0	34.5	0.69	1	25	0	25	1	1	1		
NODES/TUNNELS	0	0	0	0	0	0	0.69	1	0	0	0	1	1	1		
AIRLOCK	0	0.7008	0.701	0	0.701	0.701	0.69	1	0	0.35	0.35	1	1	1		1
HR & T	40.8	0	40.8	28.152	0	28.15	0.69	1	9.07	0	9.057	1	1	1	1	1
MECHANICAL	206.4	0.2628	206.6	142.4	0.263	142.7	0.69	1	82.5	0.26	82.81	1	1	1	0.85	0.5
RESOURCE INTEGRATION	15.82	5.9218	21.74	10.913	5.922	16.83	0.69	1	7.91	5.92	13.83	1	1	1	0.9	1
GN & C	18.38	0	18.38	12.68	0	12.68	0.69	1	9.19	0	9.189	1	1	1	0.85	
DMS	85.54	29.952	115.5	59.02	29.95	88.97	0.69	1	42.8	30	72.72	1	1	1	0.35	0.35
C & T	31.21	67.443	98.65	21.533	67.44	88.98	0.69	1	15.6	67.4	83.05	1	1	1	0.95	0.95
HABITABILITY/MANNED SYSTEMS	0	7.9636	7.964	0	7.964	7.964	0.69	1	0	7.96	7.964	1	1	1		0.5
EVA SYSTEM		11.388	11.39	0	11.39	11.39	0.69	1		11.4	11.39	1	1	1		
	448.1	123.63	571.7	309.2	123.6	432.8			192	123	315.4					
Σ CMT USING PM FACTOR	515.3	142.18	657.5	355.57	142.2	497.8										
			PREVENTIVE MAINTENANCE FACTOR=						1.15							0.69

Most maintainability design criteria are more specific than ground rules and frequently are oriented to specific types of equipment such as fluid systems, electronics, instrumentation, cables and connectors, and test equipment, regardless of where these items may be situated in the system. Maintainability design criteria can also be grouped by subsystem or specific equipment. In either case, these design criteria must be tailored to the specific types of equipment found in the system and to the requirements of the system.

15.8.1 Maintainability Design Criteria Examples

1. General design features:
   a. The design shall preclude the possibility of damage to the equipment during maintenance and servicing. Guards shall be provided to protect delicate parts exposed during servicing.
   b. Minimize the need for special tools.
   c. Part reference designations shall be located next to each part and shall be legible and permanent.
   d. Keying, size, or shape shall be used to ensure that removable parts are reassembled in the correct position.
   e. Guide pins shall be provided for alignment of modules or high-density connectors.
   f. Handles shall be provided for removable units weighing over 10 pounds or whose shape makes them difficult to handle.
   g. Sharp edges, corners, or protrusions which could cause injury to personnel shall be avoided.

2. Mounting and location of units:
   a. Provide for the removal and replacement of LRUs without the removal of unfailed units.
   b. Provide for the removal and replacement of LRUs without interrupting critical functions.
   c. Provide clear access to all LRU locations. Mount units to chassis or structure rather than on other units.
   d. Mount heavy units as low as possible. Label each access for units which can be reached through it.

3. Test, checkout, calibration:
   a. Fault isolation test circuitry shall not cause failure of the circuit under test.
   b. Test points on printed circuit boards shall be located to permit in-circuit testing.
   c. Calibration and adjustment controls which are intended to have limited motion shall be provided with adequate stops to prevent damage.
   d. All adjustments shall be designed to be common in their displacement response (i.e., clockwise, right, or up to increase).

4. Cables, leads, wiring, connectors:
   a. Provide clearance around connectors for adequate viewing and hand access.
   b. Route cables to facilitate tracing, removal, and replacement.
   c. Provide service loops in cables and harnesses to facilitate installation, checkout, and maintenance.
   d. Code or label wires and cables throughout their length for easy identification.

It is frequently desirable to classify maintainability design criteria which are mandatory as *requirements* and those which are desirable but not mandatory as *goals*. The letter R may be placed in front of required criteria and the letter G in front of goals.

15.9 MAINTAINABILITY PREDICTION

Maintainability predictions are performed to estimate the degree to which the design is meeting quantitative maintainability requirements. The prediction is valuable in that it identifies specific components which need more design analysis in order to meet their requirements, or instances where the requirements can be shown to be unreasonably costly to meet.

Maintainability predictions are performed for such parameters as:

- Mean time to repair (MTTR)
- Maximum maintenance time
- Percent of faults isolatable to one replaceable item (RI)
- Percent of faults isolatable to $\geq N$ RIs
- Mean maintenance labor-hours per repair
- Mean maintenance labor-hours per operating hour
- Mean maintenance labor-hours per flight hour

Notice 1 to MIL-HDBK-472 contains extensive methods for making these predictions. Similar frequently used methods are described in the following paragraphs.

Predictions are performed from the bottom up. This means that the first analyses are performed at the lowest hardware level that has been defined or for which maintenance is planned. Maintainability time predictions are made by estimating the maintenance time required for each lowest hardware-level component and combining these estimates to reflect the number of each component, the duty cycle, and a measure of the predicted frequency with which maintenance will be required. The equation used to carry the MTTR prediction from the component level to the system level is as follows:

$$MTTR_s = \frac{\sum\limits_{i=1}^{n} (\lambda_i \, MTTR_i)}{\sum\limits_{i=1}^{n} \lambda_i} \qquad (15.11)$$

This equation is used to predict each next higher level until the system or other desired level is reached.

Predictions of maintenance labor-hours per operating hour use the following equation to perform the prediction at progressively higher equipment levels:

$$MLH/OH = t \frac{\left[\sum\limits_{i=1}^{n} (\lambda_i \, MTTR_i \, N_i) + \sum\limits_{j=1}^{n} (f_j \, MPMT_j \, N_j) \right]}{T} \qquad (15.12)$$

A third parameter of value is maintenance labor-hours per year. This parameter gives significant help to logistics support in planning their personnel requirements, including skill levels and training. An equation which addresses this parameter is as follows:

$$\text{MLH/YR} = 8760 \sum_{i=0}^{n} \left(\frac{N_i \, d_i \, \text{MTTR}_i}{\text{MTBF}_i} \right) + \sum_{j=0}^{m} (N_j f_j \bar{M} p_j) \qquad (15.13)$$

where d_i = the duty cycle of the ith element, expressed as the fraction of the calendar time the ith element is active

f_j = the number of times per year preventive maintenance is required for the jth element

λ_i = the failure rate of the ith element

MPMT_j = the mean time required to perform preventive maintenance on the jth element, in hours

MTBF_i = the mean time between failures for the ith element in hours

MTTR_i = the mean time to repair the ith element, in hours

n = the number of elements under consideration

N_i = the average number of maintenance crew members required to repair the ith element

N_j = the average number of maintenance crew members required to perform maintenance on the jth element

t = the calendar time period under consideration

T = the operating time of the system or equipment during calendar time t

The above equations apply to Eqs. (15.11), (15.12), and (15.13).

It is necessary to estimate individual times to perform a maintainability prediction. MIL-HDBK-472, Notice 1, contains times for elemental maintenance actions. These time elements are associated with specific hardware, pictures of which are provided for guidance in transferring the data to similar hardware.

15.9.1 Estimating Maintenance Times

A maintainability engineer with an industrial engineering background which includes time measurement can estimate maintenance times with satisfactory accuracy using MIL-HDBK-472 or similar time element data. Examples of time elements prepared by this approach are shown in Tables 15.3 through 15.5. Table 15.3 provides maintenance element times for the most common maintenance tasks performed at the LRU level. Table 15.4 provides factors used to modify Table 15.3 task times to compensate for different degrees of access. Table 15.5 converts general descriptions of test length, first described in minutes, to hours. The data in these tables have been validated by comparing predictions based on them with subsequent maintainability demonstration test data for the systems predicted. These time estimates can be applied to any maintenance operations performed under normal maintenance environments but would have to be modified for such abnormal environments as extreme heat or cold, zero g, or heavy rain. An example of a maintainability prediction based on the above method for a test set follows. The maintainability allocation block diagram for the test set is shown in Fig. 15.9 with the multiprogrammer selected for a detailed maintenance time analysis highlighted.

TABLE 15.3 Maintenance Elemental Time Standards

	Hours, first	Hours, each additional
Fasteners:		
1. Bolts or screws in threaded holes (hand tools)		
a. Install	0.0167	0.0083
b. Remove	0.0125	0.0042
2. Quick acting—requires less than one rotation (hand tools)		
a. Install	0.0125	0.0042
b. Remove	0.0111	0.0028
3. Quick acting—requires less than one rotation (no tool)		
a. Install	0.0042	0.0042
b. Remove	0.0028	0.0028
4. Door, hinged, with ¼ turn latch and T handle		
a. Close	0.0042	0.0042
b. Open	0.0028	0.0028
Connectors:		
1. Three-prong power plug		
a. Plug in	0.0056	
b. Unplug	0.0028	
2. Bayonet-type connector		
a. Connect	0.0056	
b. Disconnect	0.0028	
3. Screw-shell-type connector		
a. Connect	0.0125	
b. Disconnect	0.0097	
4. Circuit card or card connector		
a. Connect	0.0069	
b. Disconnect	0.0042	
5. Rectangular connector with two screws		
a. Connect	0.0223	0.0139
b. Disconnect	0.0181	0.0098
6. Terminal connection (screw/loop terminal)		
a. Connect	0.0167	0.0083
b. Disconnect	0.0125	0.0056
7. Terminal connection (screw/spade terminal)		
a. Connect	0.0125	0.0028
b. Disconnect	0.0125	0.0028
8. Banana jack connection		
a. Connect one only	0.0028	N/A
b. Connect 2 or more (pair jack/socket)	0.0056	0.0042
c. Disconnect	0.0028	0.0028

TABLE 15.3 Maintenance Elemental Time Standards *(Continued)*

	Hours, first	Hours, each additional
Operator/technician miscellaneous activities:		
1. Walk, per 10-ft increment	0.0028	
2. Handle part/assembly (one operator/technician)		
a. Replace on tracks in rack/position		
Light/small (<3#; to 64 in³)	0.0056	
Medium (3# < weight < 25#; to 1 ft³)	0.0069	
Heavy/large (25# < weight < 85#; size and shape manageable by one person)	0.0167	
b. Remove from tracks in rack		
Light/small	0.0028	
Medium	0.0035	
Heavy/large	0.0083	
c. Aside or obtain part		
Light/small	0.0028	
Medium	0.0035	
Heavy/large	0.0083	

The multiprogrammer resides in a standard drawer in an equipment rack. It consists of fifteen I/O cards in a mainframe, each of which has one of the following functions:

Quantity	Description
1	Digital-to-analog voltage converter
2	Timer/pacer
10	Isolated digital input card
1	Interrupt card
1	4K memory card

The test set self-test will isolate to a failed I/O card or to the mainframe. In the event of a mainframe failure, it will be removed and replaced with a spare mainframe, using the cards removed from the failed mainframe. In the event that an I/O card fails, only the failed card will be removed and replaced. All I/O cards have the same configuration and thus require only one type of spare for all 15 cards. (Both the mainframe and the I/O cards are LRUs.)

The first step in predicting the maintainability of the multiprogrammer is to analyze the design to identify the precise sequence of steps required to perform the necessary maintenance. Figure 15.10 lists, in order, the elemental activities (EAs) required for the removal and replacement of an I/O card. Ten EAs are identified in the second column. The third column is used to record data important to elemental activity time determination such as the quantity of fasteners, special

TABLE 15.4 Accessibility Factors

Elemental Task Time Factors	

Access - This pertains to those physical conditions which must be overcome in gaining admission to the element of interest.

	Factor
1. Good - Open, unrestricted movement; no obstacles to work motion	1.00
2. Fair - Slightly restricted work motions such as working with special tools or with tools applied at awkward angles or where there are slight physical obstructions	1.05–1.20
3. Poor - Restricted work motions, cramped movements, and with physical obstructions	1.20–1.50
4. Very Poor - "Blind" work conditions, cramped and restricted motions, including physical obstructions, especially visual	2.00–5.00

access situations, the weights of heavy LRUs, and the length of test classification. The fourth column lists the times, in hours, taken from Tables 15.3 through 15.5 and their sum.

The same procedure is followed for each element of the test set. The data thus accumulated are then transferred to a maintainability prediction worksheet, Fig. 15.11, which also lists the failure rates for each element. The equations at the bottom of the form are used to calculate the predicted MTTR and maximum corrective maintenance time at the 95th percentile [CMAX(95)].

Note that Fig. 15.11 reflects the fact that each type of multiprogrammer I/O card has a different failure rate and must therefore be listed separately, and that the equations used for MTTR and CMAX(95) are based on the normal distribution rather than on the lognormal distribution. The reason for the normal distribution is that fault isolation is accomplished using BIT and is extremely fast in relation to other maintenance times, and because these fault isolation times are not highly variable for any component.

The final step is to compare the predicted values to the allocated values (in this case shown in Fig. 15.9) or to stated MTTR and MaxCT(95) requirements to determine whether the design meets its maintainability requirements.

15.9.2 Methods Time Measurement

Another method for estimating maintenance time elements is by methods time measurement (MTM). This method has been under development over a long

TABLE 15.5 Verification Test Time Conversion Table

Verification Test Time Standards		
	Time range	
	Minutes	Hours
1. Short test	$T \leqslant 1$	$T \leqslant 0.0167$
2. Medium length test	$1 < T \leqslant 5$	$0.0167 < T \leqslant 0.0833$
3. Lengthy test	$35 \leqslant T \leqslant 120$	$0.5833 \leqslant T \leqslant 2.0$

ELEMENTAL ACTIVITY ANALYSIS

UNIT IDENT: *MULTIPROGRAMMER – REPLACE 1 I/O CARD*

STEP NO.	ELEMENTAL ACTIVITY (EA)	TIME BREAKDOWN DETAILS	EA TIME (HOURS)
1	WALK TO REAR OF CABINET	WALK 20' @ 0.0014/10'	0.0028
2	OPEN DOOR	HINGED DOOR WITH 1/4 TURN LATCH & "T" HANDLE	0.0028
3	DISCONNECT POWER CABLE	3 PRONG CONNECTOR	0.0028
4	REMOVE REAR COVER PLATE	1 SCREW @ 0.0111 3 SCREWS @ 0.0028	0.0195
5	DISCONNECT CONNECTOR FROM FAILED I/O CARD	CIRCUIT CARD CONNECTOR	0.0042
6	REMOVE & REPLACE FAILED CARD	REMOVE @ 0.0028 PLUG IN @ 0.0028 × 1.5 ASIDE @ 0.0028	0.0098
7	CONNECT CARD CONNECTOR	CIRCUIT CARD CONNECTOR	0.0069
8	INSTALL REAR COVER PLATE	1 SCREW @ 0.0125 3 SCREWS @ 0.0042	0.0254
9	CONNECT POWER CABLE	3 PRONG CONNECTOR	0.0056
10	CLOSE REAR DOOR	AS ABOVE	0.0028
11	WALK TO FRONT OF CABINET	" "	0.0028
12	VERIFICATION TEST	"LONG TEST"	1.0000
		SUBTOTAL	1.0854
		TOTAL (INCLUDING 1.2 DELAY FACTOR)	1.302

FIG. 15.10   Maintainability prediction worksheet.

period of time for estimating the time to perform manufacturing operations. Its application began in earnest in the early 1950s, and a well-established body of statistically derived time elements exists. The armed forces have adapted this system of predicting maintenance times and have a computerized data bank which

LRU & QUANTITY EA.	FAILURE RATE $\lambda + 10^6$	MAINTENANCE TASK TIMES (HOURS)				MTTR	λ(MTTR)	(MTTR)^2	λ(MTTR)^2
		DISASS'Y	REMOVE AND REPLACE	REASS'Y	TEST VERIFY				
Universal Counter (1)	132	0.0672	0.0160	0.0822	0.0833	0.288	38.02	0.0829	10.949
Function Generator (1)	167	0.0672	0.0160	0.0822	0.0833	0.288	48.10	0.0829	13.844
Multiprogrammer (1)									
Mainframe (1)	120	0.1390	0.2230	0.2068	1.0	1.878	225.36	3.527	423.230
Card A (1)	33	0.0321	0.0098	0.0366	1.0	1.302	42.96	1.695	55.942
B (2)	26	0.0321	0.0098	0.0366	1.0	1.302	33.85	1.695	44.075
C (10)	150	0.0321	0.0098	0.0366	1.0	1.302	195.30	1.695	254.281
D (1)	21	0.0321	0.0098	0.0366	1.0	1.302	27.34	1.695	35.599
E (1)	76	0.0321	0.0098	0.0366	1.0	1.302	98.95	1.695	128.836
AC Power Supply (1)	100	0.0626	0.0104	0.3058	0.0167	0.475	47.45	0.225	22.54
Data Acq./Control (1)									
Mainframe	76	0.2174	0.0795	0.2409	0.5	1.25	95		
Option 010 5	7.5								
TOTAL	5022						4793		5391

MEAN CORRECTIVE MAINTENANCE TIME

$$MTTR = \frac{\Sigma \lambda_i MTTR_i}{\Sigma \lambda_i} = \frac{4793}{5022} = 0.95$$

MAXIMUM CORRECTIVE MAINTENANCE TIME

$$CMAX(95) = MTTR + 1.645 \sqrt{\frac{\Sigma \lambda (MTTR)^2}{\Sigma \lambda} - \left(\frac{\Sigma \lambda MTTR}{\Sigma \lambda}\right)^2}$$

$$= 0.95 + 1.645 \sqrt{\frac{5391}{5022} - \left(\frac{4793}{5022}\right)^2} = 1.6 \; HRS$$

FIG. 15.11 Maintainability prediction summary worksheet.

has been named Computer-Aided Time Standards (CATS). There is controversy about the validity of this method for deriving maintenance times because MTM was developed for repetitive manufacturing operations, and maintenance operations are seldom repetitive to the same degree. The uses of MTM methods have, however, been greatly expanded. MTM has been successfully applied to such nonrepetitive applications as precision tool rooms, and should be at least as reliable as other less structured methods. The use of a computer-based data bank speeds the application significantly and tends toward greater consistency.

15.9.3 Predicting Corrective, Preventive, and Active Maintenance Parameters

The most useful method of prediction found in MIL-HDBK-472, *Maintainability Prediction,* is Method 2, Part B, which is "used to predict Corrective, Preventive and Active Maintenance parameters." Method 2, Part B does not use the tabulated task times, now largely obsolete, in the handbook, but "utilizes estimates of man-hours required to perform a maintenance task which are based on past experience or an analysis of the design with respect to maintenance." The tables described above or the CATS computer program are good sources of labor-hour estimates.

This method considers only active corrective maintenance time, excluding administrative and logistics times, and only preventive maintenance downtime—

time when the equipment is down to permit performance of preventive maintenance. Active maintenance time combines corrective and preventive maintenance times as defined above.

Method 2 requires acceptance of these assumptions:

1. Reasonably good estimates of the tasks required to maintain a system or equipment can be made by studying their design.
2. The elapsed time in hours or the labor-hours to perform these tasks can be estimated from the same study.
3. The magnitude of the repair time, for a discrete repair, is the sum of the individual maintenance task times which are required for its completion.

This prediction technique identifies seven different maintenance tasks which may be required for a discrete repair:

Localization	Isolation
Disassembly	Interchange
Reassembly	Alignment
Checkout	

These seven tasks no longer have the same relative importance in terms of time consumed as they did when the handbook was issued. At that time, one of the most time-consuming tasks was fault isolation, because it was frequently performed on a trial-and-error basis. Automatic testing has taken much of the guesswork out of fault detection, localization, and isolation, with the result that repair times are often normally distributed rather than lognormally as in the past.

15.10 MAINTAINABILITY AND AVAILABILITY ANALYSIS

Maintainability and availability analyses are performed throughout the design and development phases of a system to evaluate that system's ability to be maintained and to determine that the probability that the system will be ready when needed is sufficient to assure that it will fulfill its purpose.

15.10.1 Maintainability Analysis

Maintainability analyses are performed to (1) establish the most cost-effective ways to achieve required maintainability, (2) quantify maintainability requirements at the design level, (3) evaluate the design for its conformance with both qualitative and quantitative maintainability requirements, and (4) generate maintainability data for use in maintenance planning and logistics support analyses (LSAs). Major maintainability analyses, maintainability allocation, and maintainability prediction have been covered in Secs. 15.7 and 15.9, respectively. Equations for other maintainability analyses are found in Sec. 15.3, Maintainability Parameters.

15.10.2 Availability Analysis

Availability analyses are performed to ensure that the system being designed has a satisfactory probability of being operational, as opposed to being down, so that it will fulfill the objectives for which it is being developed. For example, an antiballistic missile (ABM) system must have enough ABMs in operating condition and ready to be fired at any point in time to intercept incoming enemy missiles.

Maintainability and reliability determine the availability of systems and equipment. Availability is the measure of the fraction of time that the system or equipment is in operating condition in relation to total or calendar time. There are several commonly used measures of availability: inherent availability A_i, *achieved availability* A_a and operational availability A_o. These are defined in *Maintainability: Principles and Practices*[1]:

Availability (inherent):   The probability that a system or equipment, when used under stated conditions, without consideration for any scheduled or preventive action, in an ideal support environment (i.e., available tools, spares, personnel, data, etc.), shall operate satisfactorily at a given point in time. It excludes ready time, preventive maintenance downtime, logistics time, and waiting or administrative downtime. It may be expressed as

$$A_i = \frac{MTBF}{MTBF + MTTR} \qquad (15.14)$$

Inherent availability is entirely a product of design.

Availability (achieved):   The probability that a system or equipment, when used under stated conditions in an ideal support environment (i.e., available tools, spares, personnel, data, etc.), shall operate satisfactorily at a given point in time. It excludes logistics time and waiting or administrative downtime. It includes active preventive- and corrective-maintenance downtime. It may be expressed as

$$A_a = \frac{MTBM}{MTBM + \overline{M}} \qquad (15.15)$$

Availability (operational):   The probability that a system or equipment, when used under stated conditions in an actual operational environment, shall operate satisfactorily at a given point in time. It includes ready time, logistics time, and waiting or administrative downtime. It may be expressed as

$$A_o = \frac{MTBM + \text{ready time}}{(MTBM + \text{ready time}) + MDT} \qquad (15.16)$$

Ready time, while not defined in Ref. 1, means the time during which the system is in a functionally acceptable condition but is inactive or off.

15.10.3 Availability Model

The availability and reliability of complex military systems are represented by more complex models. These models are developed from basic input parameters through intermediate relationships to define the final model. The model is computerized for use in sensitivity studies as well as to estimate system availability and reliability.

The following model was developed for a complex weapons system which was designed to be on standby status most of its design life with many of its components dormant and thus not in testable condition. It was expected to be on alert or active status periodically to test the condition of normally dormant components, or for combat. The steps in its development are shown as an example of how such models are developed.

Input Data

Peacetime Failure Rates

L_1 = Failure rate of dormant components
L_2 = Failure rate of continuously active components
A = False alarm rate

Alert Failure Rates

L_3 = Full alert failure rate
L_4 = Standby alert failure rate

Engagement Failure Rates

L_5 = Weapon launch failure rate
L_6 = Weapon free flight failure rate

Testability (% of L_1)

E_0 = Testability at organizational maintenance level
E_1 = Testability at intermediate maintenance level

Single-Event Failure Probabilities

P_1 = During peacetime or alert mode
P_2 = During engagement

Downtimes (Hours)

M_1 = MTTR
M_2 = Operational delay time to repair continuously active components
M_3 = Remove, replace, and maintenance transport time
M_4 = Annual preventive maintenance time

Operational Times (Hours)

T_1 = Total deployment time
T_2 = Time between tests
T_3 = Total alert time

T_4 = Standby alert time
T_5 = Missile launch time
T_6 = Missile flight time
N = Power-up/power-down factor

Intermediate Calculations (Hours)

MTBF of dormant components testable at organizational level:

$$\lambda_1 = \frac{1}{\lambda_1 E_0(1 + N/T_2) + \alpha} \tag{15.17}$$

MTBF of dormant components testable only at the intermediate level:

$$\theta_2 = \frac{1}{\lambda_1(E_1 - E_0)(1 + N/T_2)} \tag{15.18}$$

MTBF of continuously active components:

$$\theta_3 = \frac{1}{\lambda_2} \tag{15.19}$$

Mean time to detect system critical failures testable at

1. Organizational level

$$\text{MTD}_1 = \frac{T_2}{1 - e^{-T_2/\theta_1}} - \theta_1 \tag{15.20}$$

2. Intermediate level

$$\text{MTD}_2 = \frac{\theta_4}{1 - e^{-\theta_4/\theta_2}} - \theta_2 \tag{15.21}$$

Mean time between maintenance actions of all testable dormant and continuously active components:

$$\theta_4 = \frac{1}{1/(\theta_1 + \text{MTD}_1) + 1/(\theta_2 + M_2)} \tag{15.22}$$

Availability Calculations

Inherent Availability

1. *Dormant:* Critical failures detected at operational level:

$$A_1 = \frac{\theta_1}{\theta_1 + \text{MTD}_1} \tag{15.23}$$

Additional failures detected at the intermediate level:

$$A_2 = \frac{\theta_2}{\theta_2 + \text{MTD}_2} \tag{15.24}$$

Total dormancy availability:

$$A_d = A_1 + A_2 \tag{15.25}$$

2. Availability of continuously active components

$$A_c = \frac{\theta_2}{\theta_2 + M_2} \tag{15.26}$$

3. Availability during alert

$$A_a = e^{-x_3 (T^3 - T^4)} - \lambda_4 T_4 \tag{15.27}$$

4. Total inherent availability

$$A_i = A_d \cdot A_c \cdot A_a \tag{15.28}$$

System Operational Effects

1. Maintenance-related transportation

$$A_4 = \frac{\theta_4}{\theta_4 + M_3} \tag{15.29}$$

2. Repair actions

$$A_5 = \frac{\theta_4}{\theta_4 + M_1} \tag{15.30}$$

3. Preventive maintenance (per year)

$$A_6 = \frac{8760}{8760 + M_4} \tag{15.31}$$

4. Availability based on system operational effects

$$A_o = A_4 \cdot A_5 \cdot A_6 \tag{15.32}$$

System Operational Availability

$$A = A_1 \cdot A_o \tag{15.33}$$

The model was generated by first defining the input parameters, then combining them in successive steps until the system operational availability was defined in the last equation. The purpose of displaying this model is to show the approach to its generation. The model itself is applicable only to the system for which it was developed.

15.11   MAINTAINABILITY VERIFICATION, TEST, AND DEMONSTRATION

The usual method of verifying maintainability is by analysis. The analysis most usually used to verify that quantitative maintainability requirements are being met is maintainability prediction. The most common methods of predicting maintainability are described in Sec. 15.9.

Qualitative requirements are verified by analysis of the design. This is frequently accomplished by the use of a checklist based on maintainability design criteria (see Sec. 15.8) when the design criteria are referenced in the product specification as requirements. Verification takes place before hardware and software are available.

Maintainability tests are performed on development hardware and are usually planned as part of the overall development testing. The major problem encountered during this period is that when a failure occurs, the first step is to perform a failure analysis and not to repair the failed part. The sequence of events starting with the failure until the item is repaired is not that normally encountered during the operational phase and great care is required to get valid maintainability data. Also, the small number of like items subjected to development testing makes it very difficult, if not impossible, to acquire enough data to establish confidence in the results. At best, only major maintainability problems are likely to be identified through this type of testing unless careful plans are laid and agreements are reached with the test conductors to carry them out.

Maintainability demonstrations are performed with dedicated tests and are therefore well controlled to produce the data necessary to determine whether the design meets its maintainability requirements. Demonstrations are frequently performed on high-fidelity mock-ups or on preproduction or production hardware and software at the customer's facility by customer personnel under conditions closely approximating those expected in the field. When carefully planned, maintainability demonstrations are the most accurate method of verifying that the maintainability requirements have (or have not) been met.

The usual steps required to prepare for, perform, and evaluate the results of maintainability demonstrations are as follows:

1. Selection of the specific method(s) to be used. This is dependent on the system or equipment characteristics (e.g., is it made up of many duplicate subsystems or subassemblies or only a few?), and the parameters to be demonstrated. These may include one or more of the following:
   a. Mean or median time to repair
   b. Maximum corrective maintenance time at a given percentile
   c. Preventive maintenance time
   d. Maintenance labor-hours per specified unit of time
   e. Failure detection capability
   f. Failure isolation capability
   g. False alarm rate
   h. Failure isolation ambiguity level (a measure of the ability of BIT/BITE to isolate a fault to a specific failed item)

2. Establishment of accept-reject criteria and of retest procedures should the accept criteria not be met.

3. Preparation of the maintainability demonstration plan and of the detailed test procedure.

4. Selection of the population of maintenance tasks from which the test sample will be selected.

5. Pretest preparation, including assembly of all test hardware, test support equipment, facilities, and identification of the personnel who will perform

the test, and the test monitors. If induced failures are to be isolated and repaired, they should be induced before the test personnel arrive, and provisions should be made for restoring the equipment to an acceptable condition after the test. Training in special maintenance skills, if required, is given prior to starting the tests.

6. Performance of the demonstration test(s).

7. Performance of posttest tasks, such as restoring test hardware to its original condition and verifying that it is acceptable for use on production items, if applicable, and returning test equipment and facilities to the pretest condition.

8. Analysis of test data.

9. Recommendations for corrective action, if required, and retests according to item 2 above.

10 Preparation of the demonstration test report.

Detailed information on specific methods used for performing demonstrations is found in MIL-STD-471 as well as other, more specialized documents, such as RADC-TR-81-320 (see References).

15.12 GLOSSARY OF TERMS AND ACRONYMS

15.12.1 Glossary of Terms*

Access To gain entry to part of a system. (1)

Accessibility A measure of the relative ease of admission to the various areas of an item for the purpose of operation or maintenance. (2)

Achieved Obtained as a result of measurement. (2)

Alignment Performing the adjustments that are necessary to return an item to specified operation. (2)

Allocation The process by which a top-level quantitative requirement is distributed among lower hardware items in relation to design characteristics, reliability, and maintainability features. (1)

Availability A measure of the degree to which an item is in an operable and committable state at the start of a mission when the mission is called for at an unknown (random) time. (Item state at start of a mission includes the combined effects of the readiness-related system R&M parameters but excludes mission time.) (2)

Availability, achieved The probability that a system or equipment, when used under stated conditions in an ideal support environment (i.e., available tools, spares, personnel, data, etc.), shall operate satisfactorily at a given point in time. It excludes logistics time and waiting or administrative downtime. It includes active preventive- and corrective-maintenance downtime. (17)

Availability, inherent The probability that a system or equipment, when used under stated conditions, without consideration for any scheduled or preventive action, in an

---

* Numbers in parentheses are source codes. Sources are listed at end of Sec. 15.12.1.

ideal support environment (i.e., available tools, spares, personnel, data, etc.), shall operate satisfactorily at a given point in time. It excludes ready time, preventive-maintenance downtime, logistics time, and waiting or administrative downtime. (17)

Availability, operational The probability that a system or equipment, when used under stated conditions in an actual operational environment, shall operate satisfactorily at a given point in time. It includes ready time, logistics time, and waiting or administrative downtime. (17)

Built-in test (BIT) The self-test hardware and software which is internal to a unit to test the unit. (5)

Built-in test equipment (BITE) A unit which is part of a system and is used for the express purpose of testing the system. BITE is an identifiable unit of a system. (5)

Checkout Tests or observations of an item to determine its condition or status. (2)

Component An assembly or any combination of parts, subassemblies, and assemblies mounted together in manufacture, assembly, maintenance, or rebuild. (1)

Condition monitor Technique used to determine when a family of devices is beginning to incur failures which need attention. (3)

Cost effectiveness A measure of the value received (effectiveness) for the resources expended (cost). (4)

Criticality A relative measure of the consequences of a failure mode and its frequency of occurrences. (2)

Degradation A gradual impairment in ability to perform. (2)

Demonstration, maintainability The joint contractor and procuring activity effort to determine whether specified maintainability contractual requirements have been achieved. (6) (''Maintainability'' added to term.)

Depot maintenance Maintenance performed at customer's major maintenance facility or at the item manufacturer's facility. (1)

Detection, fault An indication that an item is not operating within its specified operating limits. (1)

Diagnostics The action required to identify the location of a fault to a lower level of hardware than that at which the fault was detected. (7)

Direct maintenance labor-hours per maintenance action (DMLH/MA) A measure of the maintainability parameter related to item demand for maintenance personnel: The sum of direct maintenance labor-hours, divided by the total number of maintenance actions (preventive and corrective) during a stated period of time. (2)

Duty cycle A specified operating time of an item, followed by a specified time of nonoperation. (This often is expressed as the fraction of operating time for the cycle, e.g., the duty cycle is 15 percent.) (4)

Evaluation (as pertains to maintainability) (a) A joint contractor and procuring activity effort to determine, at all specific levels of maintenance, the impact of the operational, maintenance, and support environment on the maintainability parameters of the item. This may include measurement of depot-level maintenance tasks. (b) The joint contractor and procuring activity effort to assess inherent maintainability or fault detection and isolation design capabilities. This would take place at the termination of validation or full-scale development phases when no specific maintainability requirements have been levied (i.e., goals or situations where just an assessment is required). (6)

Fail-safe design One in which a failure will not adversely affect the safe operation of the system, equipment, or facility. (8)

Fault isolation   The process of determining the location of a fault to the extent necessary to effect repair. (8)

Fault localization   The process of determining the approximate location of a fault. (2)

General support equipment (GSE)   Equipment that has maintenance application to more than a single model or type of equipment. (1)

Inherent   Achievable under ideal conditions, generally derived by analysis, and potentially present in the design. (2)

Item   A nonspecific term used to denote any product, including systems, materials, parts, subassemblies, sets, accessories, etc. (10)

Item, life limited   An item having a limited and predictable useful life, which for reliability, safety, or economic reasons is replaced on a preplanned basis. (11)

Life cycle   All phases through which an item passes from conception through disposition. (2)

Line replaceable unit (LRU)   A unit which is identified for removal and replacement by organizational-level maintenance personnel. (1)

Logistics support   The materials and services required to operate, maintain, and repair a system. Logistics support includes the identification, selection, procurement, scheduling, stocking, and distribution of spares, repair parts, facilities, support equipment, trainers, technical publications, contractor engineering and technical services, and personnel training necessary to provide the capabilities required to keep the system in a functioning status.

Logistics support analysis (LSA)   A formal analytical technique to identify, define, analyze, quantify, and process logistics support requirements which requires data inputs from design, reliability, system safety, and maintainability. (1)

Maintainability   The measure of the ability of an item to be retained in or restored to specific condition when maintenance is performed by personnel having specified skill levels, using prescribed procedures and resources at each prescribed level of maintenance and repair. (2)

Maintainability apportionment   The assignment of maintainability subgoals to subsystems and elements thereof within a system which will result in meeting the overall maintainability goal for the system if each of these subgoals is attained. (1)

Maintainability design criteria   A body of detailed design characteristics which, as requirements or goals, are to be incorporated in a system or equipment design to assure that it will meet overall maintainability objectives. (1)

Maintainability guidelines   The recommended course of action applied toward the accomplishment of the maintainability goal for a specific system or item of equipment. (7)

Maintainability parameters   A group of factors or environmental, human, and design features that affect the performance of maintenance of an equipment. (7)

Maintainability prediction   The forecasting of quantitative maintainability characteristics of an item based on analysis of available information such as specifications, design guidelines, drawings, breadboard models, mock-ups, engineering models, pilot models, development equipment, maintenance environment, and experience with similar items, depending on the availability of information at the time the prediction is made. (7)

Maintainability requirement   A comprehensive statement of required maintenance characteristics (expressed in qualitative and quantitative terms) to be achieved in design and demonstrated in development. (7)

Maintenance   All actions necessary for retaining an item in, or restoring it to, a specified condition. (2)

Maintenance concept   A description of the planned general scheme for maintenance and support of an item in the operational environment. The maintenance concept provides the practical basis for design, layout, and packaging of the system and its test equipment and establishes the scope of maintenance responsibility for each level of maintenance and the personnel resources required to maintain the system. (7)

Maintenance, condition monitoring   The establishment of a maintenance requirement by observing that a component has failed, or the detection of an impending failure by the operator through route monitoring during normal operation. (12)

Maintenance, corrective   All actions performed as a result of failure to restore an item to a specified condition. Corrective maintenance can include any or all of the following steps: localization, isolation, disassembly, interchange, reassembly, alignment, and checkout. (2)

Maintenance levels   The three basic levels of maintenance which describe the organization or location at which maintenance is performed. These are generally called the organizational, intermediate, and depot levels. (1)

Maintenance, on-condition   Elimination of scheduled replacement in favor of periodic or continuous assessment to determine whether the item still functions within acceptable limits. (13)

Maintenance, preventive   All actions performed in an attempt to retain an item in specified condition by providing systematic inspection, detection, and prevention of incipient failures. (2)

Maintenance, scheduled   Preventive maintenance performed at prescribed points in an item's life. (2)

Maintenance, unscheduled   Corrective maintenance required by item conditions. (2)

Maximum time to repair   That time below which a specified percentage of all corrective maintenance tasks must be completed. (14)

Mean time between maintenance actions (MTBMA)   A measure of the system reliability parameter related to the demand for maintenance personnel: The total number of system life units, divided by the total number of maintenance actions (preventive and corrective) during a stated period of time. (2)

Mean time between removals (MTBR)   A measure of the system reliability parameter related to demand for logistic support: The total number of life units divided by the total number of items removed from that system during a stated period of time. This term is defined to exclude removals performed to facilitate other maintenance and removals for product improvement. (2)

Mean time to repair (MTTR)   A basic measure of maintainability: The sum of corrective maintenance times at any specific level of repair, divided by the total number of failures within an item repaired at that level, during a particular interval under stated conditions. (2)

Mean time to restore system (MTTRS)   A measure of the system maintainability parameter related to availability and readiness: The total corrective maintenance time, associated with downing events, divided by the total number of downing events, during a stated period of time. (Excludes time for off-system maintenance and repair of detached components.) (2)

Repair parts   Individual parts or assemblies required for the maintenance or repair of equipment, systems, or spares. Such repair parts may be repairable or nonrepairable assemblies or one-piece items. Consumable supplies used in maintenance, such as wipe rags, solvent, and lubricants, are not considered repair parts. (3)

Replaceable unit   Any unit that is designed and packaged to be readily removed and replaced in an equipment system without unnecessary calibration or adjustment. (15)

Support equipment   Items required to maintain systems in effective operating condition under various environments. Support equipment includes general and special-purpose vehicles, power units, stands, test equipment, tools, or test benches needed to facilitate or sustain maintenance action, to detect or diagnose malfunctions, and to monitor the operational status of equipment and systems. (3)

Testability   A characteristic of an item's design which allows the status (operable, inoperable, or degraded) of that item to be confidently determined in a timely manner. (5)

Time, administrative   That element of delay time, not included in the supply delay time. (2)

Time, delay   That element of downtime during which no maintenance is being accomplished on the item because of either supply or administrative delay. (2)

Time, supply delay   That element of delay time during which a needed replacement item is being obtained. (2)

Undetectable failure   A postulated failure mode in the FMEA for which there is no failure detection method by which the operator is made aware of the failure. (16)

Unit   An assembly of any combination of parts, subassemblies, and assemblies mounted together, normally capable of independent operation in a variety of situations. (10)

Useful life   The number of life units from manufacture to when the item has an unrepairable failure or unacceptable failure rate. (2)

Source Codes.

1. General accepted definition.
2. MIL-STD-721C, *Definitions of Terms for Reliability and Maintainability*.
3. Joseph D. Patton, Jr., *Maintainability and Maintenance Management,* Instrument Society of America, 1980.
4. AMCP 706-200, *Engineering Design Handbook: Development Guide for Reliability,* Part 6, Mathematical Appendix and Glossary.
5. MIL-STD-2084(AS), *General Requirements for Maintainability of Avionic and Electronic Systems and Equipment.*
6. MIL-STD-471A, *Maintainability, Verification/Demonstration/Evaluation.*
7. RADC-TR-74-308.
8. MIL-STD-1472, *Human Engineering Design Criteria for Military Systems, Equipment and Facilities.*
9. MIL-STD-882, *System Safety Program.*
10. MIL-STD-280, *Definitions of Item Levels, Item Exchangeability, Models and Related Terms.*
11. ARMP-1, Allied Reliability and Maintainability Publication, *NATO Requirements for Reliability and Maintainability* (draft 2).
12. AMCP 750-16, *AMC (DARCOM) Guide to Logistics Support Analysis.*
13. AFSC DH 1-9, *Design Handbook—Maintainability (for Ground Electronic Systems).*

14. MIL STD-785, *Reliability Program for Systems and Equipment Development and Production.*
15. MIL-STD-415, *Test Provisions for Electronic Systems and Associated Equipment, Design Criteria for.*
16. MIL-STD-1629, *Procedures for Performing a FMECA.*
17. Benjamin S. Blanchard, Jr., and E. Edward Lowery, *Maintainability, Principles and Practices,* New York, McGraw-Hill, 1969.

15.12.2 Acronyms

A_a	Achieved availability
A_i	Inherent availability
A_o	Operational availability
ABM	Antiballistic missile
BIT	Built-in test
BITE	Built-in test equipment
CATS	Computer-aided time standards
CEI	Contract end item
CMAX(X%)	Maximum corrective maintenance time at the Xth percentile
DOD	Department of Defense
EA	Elemental activity
FFBD	Functional-flow block diagram
FMEA	Failure mode and effects analysis
I/O	Input-output
ILS	Integrated logistic support
LCC	Life-cycle support
LRU	Line replaceable unit
LSA	Logistics support analysis
MdCT	Median corrective maintenance time
MLH/OH	Mean labor hours per operating hour
MTD	Mean time to detect
MTM	Methods time measurement
MTTR	Mean time to repair
ORU	Orbital replacement unit
RAM	Reliability, availability, maintainability

15.13 REFERENCES

1. Blanchard, Benjamin S., Jr., and Lowery, E. Edward, *Maintainability Principles and Practices,* New York, McGraw-Hill, 1969.
2. Goldman, A. S., and Slattery, T. B., *Maintainability: A Major Element of System Effectiveness,* New York, John Wiley, 1964.
3. Henley, Ernest J., and Kumamoto, Hiromitsu, *Reliability Engineering and Risk Assessment,* Englewood Cliffs, N.J., Prentice-Hall, 1981.

4. Locks, Mitchell O., *Reliability, Maintainability, and Availability Assessment,* Rochelle Park, N.J., Hayden, 1973.

5. RADC-TR-81-320, *Analysis of Built-In-Test False Alarm Conditions,* Rome Air Development Center, Air Force Systems Command, Griffis Air Force Base, NY 13441, August 1981.

6. *Computer-Aided Time Standards (CATS),* Office of the Assistant Secretary of Defense, Defense Productivity Program Office, Falls Church, VA 22041 (unpublished).

7. MDC-H0025, *The Reliability and Maintainability Handbook,* McDonnell-Douglas Astronautics Company, West, Huntington Beach, CA 92647, April 1982 (internal document).

CHAPTER 16
SOFTWARE RELIABILITY

Sally Dudley
Hewlett Packard Corporation

Software reliability is a function of defects and severity. Reliable software performs its stated operations over time without failure. Although software does not degrade over time through wear or environmental conditions, its logical behavior may be altered as a result of a variety of circumstances, including changes to other operating software in its environment.

16.1 PRIMARY RELIABILITY FACTORS

Reliability of software is affected by three primary factors:

1. Defects included in the code
2. Defects in interfaces with other code
3. Operational defects which cause changes to previously defect-free code

16.1.1 Defects Included in the Code

Defects found in code can be created in many ways. Design errors can be directly translated into code errors. If the design indicates use of a table of telephone numbers for the local area and the program is intended to be used in another area, it will be a problem if the programmer translates the design literally.

Defects can also be created through omission. When coding, the programmer may unintentionally omit a block of processing unless the design document is used as a tight control, checking off each design component as it is coded.

Another way defects may be created is through misuse of a tool. Inadequate understanding of the language being used can result in incorrect usage. Other tools such as file systems, graphics subroutines, and external data sources may not be incorporated correctly by the programmer.

A large class of defects can be attributed to simple coding errors. A typographical error such as a spelling error or transposition of characters may not be detected by the programmer or the compiler and produce an operating defect.

16.1.2 Defects in Interfaces with Other Code

It is often the case that individual software components are coded correctly to their design specifications, but the passing of data between modules or the accessing of external tables is incorrect. An example of this is when a payroll program updates a file sequenced by social security number and the following program prints a report sequenced by zip code. The data must be sorted between the two steps. If the sort step is omitted, however, the defect in the system cannot be attributed to inadequate code.

16.1.3 Operational Defects

A program can be affected by other programs in a variety of ways. If an incorrect edit procedure in one program allows faulty data to be used in another program, an error may occur. If a table of values is searched using a faulty index, an incorrect location may be updated. If that location is in some other program's storage space, unpredictable results can occur.

16.1 SOFTWARE QUALITY

Reliability is one aspect of software quality. Software quality from the user's perspective can include many attributes. Software which does what is expected, is easy to use, performs quickly and correctly, and adapts smoothly to new equipment will be considered by the user to be of high quality. Therefore, it is important to consider a variety of attributes for software under development.

16.2.1 Quality Attributes

The attributes used to describe software vary greatly. Such terms as usability, consistency, capability, recoverability, and maintainability are examples of a few. In order to provide some structure to reliability and quality planning and evaluation, it is useful to group these attributes. In Table 16.1, some of the many terms used to evaluate software are listed. These terms are grouped under headings: functionality, usability, performance, reliability, supportability, and manageability. The first four describe quality from the perspective of the user, the others from the perspective of the software supplier.

TABLE 16.1 Software Quality Attributes

Functionality	Features, capabilities	Performance	Speed
	Generality		Response time
	Security		Resource consumption
	Comprehensiveness		Throughput
	Completeness		Efficiency
			Environmental degradation
Usability	Ease of use	Supportability	Serviceability
	Consistency		Upgradeability
	Adaptability to user		Installability
	Human factors		Cost of upgrade
	Aesthetics		Cost of support
	Documentation		Configurability
	Embedded help or direction		
Reliability	Frequency of failure	Manageability	Integratability
	Severity of failure		Testability
	Recoverability		Enhanceability
	Predictability		Adaptability
	Accuracy		Maintainability
	Mean time to failure		Compatibility
	Protection from failure		Mean time to repair
	Failure modes		Localizability

16.3   ATTRIBUTE USE

16.3.1   Phases of Development

Software development includes seven phases (Table 16.2). The problem investigation is followed by a detailed specification of the planned solution. A prototype or model of the software may be produced in conjunction with, or instead of, a detailed specification. Once the specification is complete, the structure of the product is designed. The architecture of the product is presented in hierarchical structure diagrams, data flow diagrams, condition tables, flowcharts, and other symbolic design documents.

After design, the software is coded. Code modules are produced, tested, and integrated with others. Testing of the integrated product takes place within its planned operational environment, and finally the product is released for testing by its targeted user.

Throughout these phases, documentation is produced: both user documentation and documentation for the use of software maintenance programmers. The quality of the documentation requires the same attention as that of the code itself.

TABLE 16.2 Use of Attributes in Software Development Phases

Phase	Quality attribute planning	Quality attribute evaluation

	Setting objectives
	Defining plan to
	  achieve objectives
	Prioritizing conflict-
	  ing objectives
		Verification that
		  objectives were
		  met

Investigation	F, U, R, P, S, M	
Specification	F, U	F, U
Design	R, P, S, M	R, P, S, M
Construction		F,R,S,M
System/user testing		F, U, R, P, S, M
Release		R, P, M
Support	F, U, R	F, U, R, P, S, M

Note: F = functionality; U = usability; R = reliability; P = performance; S = supportability; M = manageability.

16.3.2 Attribute Planning

Planning and prioritizing of quality attributes begins at the time of investigation and the results are verified throughout the development cycle. Different attributes can be evaluated at different stages. (See Table 16.2.)

16.3.3 Reliability as an Attribute

Reliability is defined as operation over time without failure. The reliability attribute of software quality combines an aspect of failure (frequency, severity, and mode) with an aspect of recovery (protection of data, ability to restart, degree of restoration required). The user's perspective of reliability is primarily based on the perceived failure, perceived downtime until repair, and perceived loss of data or processing. The supplier's perspective includes these aspects, but potential liability and customer satisfaction are added.

16.4 SOFTWARE RELIABILITY TERMINOLOGY

The terminology used to describe software reliability includes terms to classify defects—their existence, their definition, their mode; terms to normalize defect counts to provide a means of comparison; and terms to describe severity or impact of defects.

16.4.1 Defects

Bugs. The term *bug* has long been used for software malfunctions. This commonly used term usually refers to an error in coding, a logic error, or a typographical error. The term is not used when describing an error state which involves a number of logical conditions which result in inaccurate results or nonconformance to specifications. Such conditions can be investigated to discover a combination of design errors, incorrect understanding of the specification, and classic bugs.

Nonconformance to Specification.   When the software does not do what it was intended to do, or does not do it consistently, a defect exists. This definition of defect may be difficult to understand for a variety of reasons:

1. Some software is not specified in writing because the solution is considered simple or straightforward (e.g., a rewrite of existing code).
2. Some specifications are limited to processing logic only (e.g., software whose user interface is provided by a prototype).
3. Even when very detailed formalized specifications are produced, it is very difficult to represent complex logical states completely enough to be used as a sole reference for designers. Also, large, complete documentation may not be read and may not be updated to reflect change.

From the user's perspective, nonconformance to the definition provided in user manuals or embedded help text will be considered a defect. Ultimately, nonconformance to user expectations will also be considered a defect.

Failure.   The term *failure,* when applied to software, is usually used to describe a state which does not allow the user to continue operations. Thus, a failure usually requires a restart of some kind, with associated data recovery activities. Failure may also be termed *hard failure* or shutdown.

16.4.2   Normalizing Factors

If reliability is defined as encountering defects over time, it is useful to provide normalizing factors in order to understand whether the product is improving or degrading and to be able to provide comparison among products. Table 16.3 lists normalization factor uses.

1. *Defects/KLOC (defects per thousand lines of code):*   Lines of code is the most commonly used normalizing factor. By providing normalization based on a sizing factor that is easy to determine, some degree of comparison is possible. Problems with using lines of code include:
   a. Different computer languages produce widely different program sizes
   b. The amount of documentation provided in the code in the form of comment lines can distort the measure
   c. Blank lines are included in many programs for readability

2. *Defects/KNCSS (defects per thousand noncomment source statements):* This measure uses program size with more precision, since comments and blank lines are not included. Special utilities or special compiler options are necessary to obtain this number. The problem with this measure is that it may discourage the writing of comments in code, thus degrading maintainability.

TABLE 16.3 Normalization Factor Uses

	Useful in assessing current quality	Useful in assessing current quality in comparison to previous or another product	Useful in comparing to prestated objectives	Useful in making personnel decisions (fixing, integrating, testing, supporting)	Useful in providing data for future objective setting	Useful as measure of customer satisfaction
Absolute number of defects	Somewhat	Somewhat	Yes	Yes	Somewhat	No
Defects/ KLOC	Somewhat	Yes (if comparable)	Yes	No	Yes	Somewhat
Defects/ KNCSS	Yes	Yes	Yes	No	Yes	Somewhat
Defects/ User	No	Somewhat	Somewhat	No	No	Yes
Defects/K hours test	No	Somewhat	Yes	No	Yes	Yes
Defects/ K hours test with coverage measure	Yes	Yes	Yes	No	Yes	Yes

3. *Defects per user*: Normalization based on usage may be used to measure the impact of reliability on the user community and the necessity to provide replacement software. It may also be used to predict effort required to support the product over its life. The problem with this measure is that it may be difficult to determine how many users are using the product and to what degree it is used.

4. *Defects per thousand hours of testing*: Prior to release to users, the defect tracking during testing can provide useful reliability data, as well as providing data for prediction of future failure rates and support needs. Problems with using hours of testing include:

   a. The results can be widely skewed based on the rigor, variety, and intensity of the testing (for example, repeating a simple stand-alone test for 1000 hours will not produce the same results as running numerous different tests simultaneously).

   b. The amount of program code executed during testing will vary based on the test types. Thus, it is important to couple this measure with a test coverage measure.

   c. When testing is done by people attempting to run the product manually, the results may be skewed by the knowledge and inventiveness of the tester.

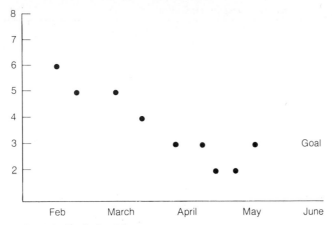

FIG. 16.1 Predicting failure rates

This measure can be very helpful, however, when estimating how much testing remains to be done. As defects are removed, the number remaining can be compared to a goal which was defined in the quality objectives stated in the product investigation phase. When combined with a mean time to failure (MTTF) measure, this measure can help predict failure rates in customer usage (see Fig. 16.1).

5. *Choice of a normalizing factor:* The choice of a normalizing factor depends on what reliability information is desired. Table 16.3 summarizes each factor's usefulness. Each can be helpful when used to compare to prestated objectives which use the same factors. However, some address only reliability, whereas others may be more useful in forecasting support staffing required or future quality plans. All are dependent on criteria which must be comparable, such as project schedule, skill levels of the programmer, etc.

16.4.3 Severity

Software defects can be classified by severity. Classification should be based on the severity of the problem from the user's perspective. A critical defect in a text-editing program used by a scientist may be quite different from a critical defect in software which directs a robot in polishing metal. Classification examples includes:

Critical: Causes loss of data, interrupts operation, requires restart, requires manual intervention, destroys network connection, endangers operators, less workaround

Severe: Degrades performance significantly, produces inconsistent results, requires manual monitoring, forces user to use different approach, has difficult workaround

Average: Adequate workaround, software recovers automatically, can be avoided, results require interpretation, performance is slowed somewhat, extra training is required, error message is difficult to interpret

TABLE 16.4   Influences on Severity Definition

Types of software	Types of users	Types of evaluators
Embedded control logic	Computer-literate professionals	Programmers
Data storage and manipulation	Office workers	Testers
Operating system	Other programmers	End users
Computer language compiler	Non-computer-literate persons	Support personnel
Data communications	Recreational users	Sales personnel
Business application	Scientists	Industry critics
Office application	Secretaries	
Recreational	Instrumentation	

Low: Cosmetic error, misspelling of error message, incorrectly linked user interface, inconvenient dialogue

When the reliability evaluation is being planned, it is important to define severity levels based on the type of product, the type of user, and the types of evaluators (see Table 16.4). For each of these, the definition of a critical or low severity defect may vary widely. It is important to agree on a standard set of criteria for the software to be evaluated in order to avoid conflict as defects are discovered.

16.5   CAUSES OF DEFECTS OVER TIME

Defects in software originate throughout the life cycle. Each phase in development and support has the potential to cause defects.

Imprecise specification: If the details of program logic are inadequately specified, the designer and coder may make decisions about logic based on intuition rather than the spec. This is often the case when a specification emphasizes the expected cases while omitting exception conditions or error handling. In software of moderate to extreme complexity, it is not always possible to specify every potential logic condition; in such cases, it is important to identify areas which should receive thorough evaluation when designed and coded.

Incomplete design: The translation of a narrative specification to a hierarchical design or flowchart may result in omission of parts or misinterpretation of meaning.

Incorrect coding: After the design is complete, the translation of design to code is subject to the same error insertion as design omission or misinterpretation. In addition, coding errors such as typing errors and transposition of numbers are possible.

Integration errors or version control: Reliability can be affected if software modules are omitted when the product's components are integrated or if the wrong versions of modules are used.

Unreliable tools: If a compiler, test package, or version-checking tool has defects, the code created may be defective as a result. Each tool has the potential to insert faults or mask the result of existing faults.

Other causes of defects are

Inadvertent alteration of stored data: Uncontrolled loops, inaccurate counting of buffer sizes, incorrect overlays can cause this problem.

Inadvertent alteration of logic: Same causes.

Unanticipated state: A set of conditions within the program or in the operational environment which was not planned.

Inadequate controls, validation: Data entering the program are not sufficiently edited to ensure the program will only be processing specified data types and formats.

16.6   RELIABILITY PLANNING

During initial investigation, it is important to explicitly state reliability objectives because subsequent specifications, designs, tests, and code must reflect these objectives (see Table 16.5). If the code must be defect-free prior to production use, numerous plans must be made throughout development. If a standard defect rate is acceptable, the approach need not be as thorough. Over time, defect rates have been shown to cluster around certain values for various types of software. Less than one defect/KLOC is expected in business application software; three to four defects/KLOC is more reasonable for compilers and subsystems such as sorts. Operating systems and data communications software may exhibit rates between 10 and 25 defects/KLOC.

Examples of objectives may be stated thus:

   0 Critical and severe defects prior to release

   10 Average and low defects prior to release

or

Fewer than twenty known defects with 80 percent of the product tested prior to customer testing

or

Zero defects after 100 percent code inspection of five most complex modules

Choosing a goal is very subjective at first, but as experience grows, the goals may be based on history of previous projects, demonstrated user expectations, or a planned approach to continuous improvement which would drive increasingly aggressive objectives.

16.7   DESIGNING FOR RELIABILITY

There are many software design approaches that can affect the reliability of the resulting code. These include:

TABLE 16.5   Reliability Objectives and Life-Cycle Phases

Phase	Objectives
Investigation	Existence of stated reliability objectives
	Priority of conflicting objectives stated
Specification	Specification of recovery modes, error checking, and handling
	Percent of specification to receive detailed inspection
	Number of defects found in specification
	Number of rewrites required
Design	Design inspections to be held
	Number of problems uncovered in design
	Percent of design to receive detailed inspection
	Number of module interfaces to be inspected
	Amount of logic or user interface to be prototyped
	Tools to be used to analyze design
Code	Code inspections to be held
	Number of problems uncovered in code
	Percent of code to receive detailed inspection
	Defect levels acceptable at module test
	Style and complexity analysis to be done
	Degree of test coverage expected at unit test
Test	Defect levels acceptable at integration test, user acceptance test, release to customers
Release	Defect level remaining
	Test coverage achieved
	Number of test sites
	Results of test site surveys
	MTTF at test sites
	MTTR at test sites
Support	Reported defect level from users
	Test coverage of regression testing
	Defect levels of update software
	Test usage of update software

Modular design: If the program is designed such that each function has its own module and no module has more than one function, then the risk of change is minimized.

Local data definition: If data are defined as local, they cannot be inadvertently modified by another module.

Data edit routines: The design can include logic which verifies that incoming data are in acceptable form and within acceptable value limits. If the data cannot be handled, error routines are designed to correct the data or terminate the processing to avoid a hard failure.

Soft failure: The program recognizes error states and stores all data, saves logical state information, and returns control to the higher process or the operator. Depending on the problem, the program may be able to be restarted after adjustments are made.

Redundancy: Programs which must be fail-safe can be designed with redundancy so that transactions are processed on two parallel systems. Utilities are provided which allow automatic switching to a second machine if the one in operation fails.

Design inspections: Inspections and reviews are part of the design process. The design documents—hierarchical diagrams, flowcharts, state diagrams, etc.—are reviewed in detail by a peer group. All defects are recorded. After correction, the design may be reinspected.

16.8 PLANNING FOR TESTING

Early in the design phase an overall test plan should be made. It addresses each kind of testing to be done and specifies the separate test plans that will be needed.

Structure: This overall plan is structured by phase, with plans to address each quality attribute. It should include the objective, the plan to achieve the objective, and the verification that the objective has been met.

Schedule: The testing schedule is based on the test plans and the schedules for the software components. Estimating testing time is very difficult. Some inputs that are useful are:

Test schedules planned for prior similar projects
Actual testing history of prior similar projects
Completion history of previous steps in current project
Separate test estimates plotted on a PERT (program evaluation and review technique) chart
Identification of key modules whose test expectations are uncertain (complex modules, modules with dependencies on external unreliable code, complicated interface modules, etc.)
The phases to be included in the schedule are unit (module) test, integration, system testing, user acceptance testing, stress testing, and regression testing.

As the testing progresses, the schedule for subsequent test phases should be reevaluated based on results achieved. If code is proving unreliable at unit test, it will be necessary to test it more prior to integration, and it may be necessary to increase subsequent test efforts.

Testing activities should be estimated at 40 to 60 percent of the total project schedule. As problems are found, their corrections must be unit-tested prior to reintegration, so the process is time consuming. Control over fix integration is essential to ensure that problems can be isolated to new changes (see Fig. 16.2).

16.9 PRIORITY OF OBJECTIVES

Test planning must consider the priority of the quality objectives stated early in the product investigation. If reliability is a higher priority than usability or

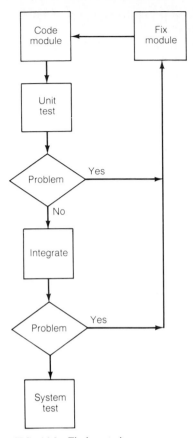

FIG. 16.2   Fix integration.

supportability, for example, the planning must emphasize reliability. Studies (Myers, 1976) have shown that a software product can be optimized for one of many attributes (size, documentation, reliability, performance, etc.) but that the others will be affected to varying degrees. It is therefore imperative that the designers and programmers consider these priorities.

16.9.1   Designing for Testing

When the product is designed, its testability must be considered. For the software to be testable, it may need special utilities, or changes to existing software tools; examples include:

Hooks in code: Ability to call test subroutines, ability to "trigger" exception conditions, ability to display interim results

Logging: Ability to log results to an external file for later analysis

Compiler options: Ability to specify modes of compilation to insert triggers, to enable logging, to display procedures executed, etc.

TABLE 16.6 Testing Tools

Type of tool	Degree of generality	Description	User of tool
Preparation tools	Very product specific and compiler dependent	Insert probes, log routines	Programmer creating product
Execution tools	↓	Enabled conditions, result logs, exception files	Test suite
Comparison and evaluation tools		Detailed results showing exceptions	Test suite
Summery reporting tools	Very general; all results in same format	Simple to evaluate, either pass or fail	Operator

Comparison tools: Utilities to compare actual results to previous results or a standard results file. These may compare sequential files, hierarchical data bases, log files, and so on.

Reporting tools: Programs which analyze the results of the execution and comparison tools and summarize the results (see Table 16.6).These should report simply a pass or fail condition and, in the failing case, provide specific aid in locating the failure and failure mode.

Coverage tools: Programs which determine what portions of the code have been executed by the test package. These tools provide the information to write the fewest possible test cases while providing the highest degree of coverage.

16.10 CODING FOR RELIABILITY

Writing code is a translation of a design to a sequential series of steps a computer can execute. As such, it has high potential for human error. To minimize such error, automation is an aid wherever possible. When compilers and linkers have strong error-checking abilities, many of the faults can be found by the computer. Beyond these, the most effective quality aid is desk checking followed by group reviews.

The programmer has the primary influence over the reliability of the code. There are two major components to the approach, understanding the problem and attention to detail.

16.10.1 Understanding the Problem

1. The specifications and design documentation must be complete and accurate.
2. Coding standards must be provided.
3. Testing tools and test strategies must be used.

4. The reliability objectives must be stated. An example might be failure-free operation 24 hours a day for medical applications, and one-hour downtime per week for an accounting system.

16.10.2 Attention to Detail

1. Checking for typos, reversals, unchecked limits, etc., is a part of individual programmer desk checking.
2. Checking for errors, inconsistencies, incorrect interfaces, etc., is a part of group inspections.
3. Checking for invalid constructs, incorrect module linkage, and exceeding boundaries is a part of the compiler and linker software.

There is a direct correlation between the complexity of the code and the difficulty of avoiding the creation of bugs, finding bugs, and fixing bugs. Therefore, to optimize for reliability it is useful to:

- Simplify the code itself:
  One statement per line
  Minimize control logic—use simple constructs
  Consistent meaningful naming conventions
  Avoid elegant constructs which require explanation
- Make the code visually simple:
  Use of blank lines and spaces for separations
  Avoid nonstandard symbols
  Embed comments separated from code by blank lines
  Indent to indicate subordinate control
  Group data definitions with comments in easy to locate area of code
- Plan for maintenance:
  Hints for maintainer in comments
  Indicators of related processes (other modules affected by changes)
  Minimize limits (table sizes, codes for data types, etc.)
  Where limits must exist, make them visible

16.11 TESTING

16.11.1 Testing Terminology

The following terms describe types of software testing.

Black box: Testing based only on the external attributes of the product. The internal structure and technique of the code are not considered. The external specifications are the basis for test development.

White box: Testing based on the internals of the product. The design and code itself are the basis for test development.

Static: Testing done without execution of the product. Static testing includes inspection and review, code analysis, style analysis, etc.

Dynamic: Testing by means of program execution. Running the program using test scenarios to find defects and weak points.

Functionality: Testing to ensure the appropriate features are provided and they perform as described.

Reliability: Testing to ensure the product continues to operate without failure over time.

Regression: Testing to ensure that changes made to the product have not caused it to "regress," i.e., to perform less reliably than previous versions.

System: Testing the completely integrated product in the intended environment. Types of system testing include performance testing, installation testing, and system resource utilization testing.

Stress: Once module testing is complete and system testing is producing stable results, it is useful to set up a stress test of the software. The intent is to stress the software until it fails, then analyze the failure mode. The result is data to be used in estimation of mean time before failure (MTBF). Stress tests should emulate a user's environment. In multiuser, multiprogram environments, this would mean the test bed would start and stop a variety of programs repeatedly to cause varied conditions over time.

Automation: Test automation is provided through test scripts which are in machine-readable form being executed by the computer with the test results summarized at the end. Automated testing provides repeatable tests which are required for regression tests. Automation also removes the human factor from testing so that the test results do not depend on the ingenuity of one tester over another (see Table 16.7).

16.11.2   Test Case Development

The design and coding of test cases require the same discipline as that of the product itself. Test cases should be modular, testing one function or aspect of the product. The test cases should be gathered into a library, called a test package or test suite. They should be easily modified as the product is modified. Test cases must be uniquely identifiable to provide traceability to the problem if failure occurs.

16.11.3   Test Case Inspection

Static inspection of test cases is similar to inspection of a design or code. The test case listing is reviewed line by line to determine clarity and correctness. The entire package is reviewed for completeness. The test package should be developed in parallel with the code. At release of the product, the test package becomes the test set used for regression testing.

TABLE 16.7 Testing during Development Phases

Phase	Static testing	Dynamic testing
Investigation	Document inspection	
Specification	Document inspection	
	Cross-reference check to investigation documents	
Design specification	Design inspection	
	Cross-reference check to documents	
	Design analysis tools	
Code	Code inspection	Functional tests
	Style analyzers	Reliability tests
	Cross-reference check to design documents	
System and user testing		Performance tests
		Configuration tests
		Installation tests
Release		Reliability tests
		Regression tests
Support	As above, by activity	As above, by activity

16.11.4 Prediction of Test Results

It is a useful exercise to predict defect rates expected in testing. There is little proven experience in doing this, but many people have experimented with different approaches, and the experimentation invariably results in a greater understanding of the process and improvements in the next product's testing.

Prediction can be based on history. If the same programmers who produced a previous product are producing the one to be tested, and if the product itself is similar in structure and use to the previous product, results achieved on the previous product will be useful in prediction. For example, a group that writes compilers repeatedly learns a great deal about the testing of compilers and can often extrapolate from previous testing history how long it will take to test the next version.

Prediction can also be based on sampling. If the product consists of many modules of code which are fairly similar in structure and function, the results found in testing the first few may be used to extrapolate the prediction for the rest. Care must be taken here, however. One of the primary variables among software components is the different coding styles and levels of experience of programmers. If the product was coded by many different people, the reliability of the different components may vary widely. So, prediction may be very misleading.

Prediction can also be based on industry experience. Data can be gathered on the product type (the examples listed above for defects/KLOC by product type may be a place to start) and a prediction made for the amount of variability from the standard.

Tracking the Prediction. Software developers find it very useful to graph a prediction and subsequently graph actual results during testing. At first, the results may be widely different from the prediction, but the programmer learns a great deal and predictions improve in the future. In addition, the time required to test becomes better understood as actual data are analyzed.

16.11.5 User Test Sites

Usability and *supportability* of end-user products is best performed in a target-user environment. Such testing will not be necessary for some embedded software such as peripheral drivers or data extraction routines; but end-user software, however, such as text processors or manufacturing applications are often designed and coded by people with minimal experience as users of such products. It is important therefore to expose the product to a controlled use by a real target user to evaluate its usefulness and its need for support.

User test sites will often encounter reliability problems in their test activity. It is useful to measure MTBF in a variety of user environments to identify weak aspects of the product and to calculate expected failure rates after general release.

16.12   RELIABILITY DURING SUPPORT

A great deal of ongoing work may be put into a software product. It may require extensions: new capabilities or new linkages to other products which increase its usefulness. It also may require modification to work with new hardware products, and it may be necessary to fix bugs found in customer usage. Maintaining high reliability while modifying a software product requires planning and testing comparable to that used in original development. Insertion of new bugs while fixing old bugs is very common.

16.12.1 Regression Testing

Regression tests are tests performed on modified software to ensure that the modifications have not caused the product to become less reliable than before. Complete regression test packages are a requirement during maintenance. Execution of these tests is necessary after each major modification to ensure reliability.

16.12.2 Version Control

Reliability problems often occur when large programs are incorrectly integrated, i.e., when versions of modules are integrated with incorrect versions of other modules. Precise version control, provided manually or through a software library management tool, can avoid this problem. Another term for version control is *configuration control.*

16.12.3   Update Management

A large software product may be worked on by many people with different schedules and plans. Changes made to one module by more than one person can cause a variety of logic and data manipulation problems. If different software versions exist simultaneously, it may be difficult to track a defect to the version of software it exists within.

Update management is required to control and manage updates so that test results and problem reports can be tracked back to causes.

16.12.4   Test Update

As software changes, it is often necessary to update the input to tests or the test logic itself; otherwise, the test suite (set of test cases) will become obsolete and useless. Test update requires version control in parallel with the code versions under test.

16.12.5   Patches

The usual way to modify software is to change the source code and recompile, producing new object code or machine language instructions. In this way, the machine instructions can always be re-created by recompiling the source.

Another modification technique is patching the machine code. Machine code is written and the actual stored instructions are modified. This technique is quicker than recompiling and can be done even when the source code is not available. Patching involves significant reliability risks, however. It is a risk because

- Changes to the source may conflict with existing patches.
- Patches can be created and applied anywhere, so even if rigid version control is maintained at the development site, patches applied at different user sites may conflict with one another.
- Patches are impossible to document internally.
- Removing patches can cause unpredictable results.

Reliability risks such as patches should be forbidden or, if unavoidable, controlled tightly. Source changes should be submitted wherever patches are created, so that subsequent compilations will include the changed logic.

16.13   OTHER RELIABILITY ISSUES

16.13.1   Reused Code

In order to minimize reliability problems resulting from coding errors, reuse of existing, proven code is a good practice. Code can be reused in a variety of forms:

- Data definitions in libraries

- Parameter lists in tables
- Libraries of logic routines
- Functional modules generalized for standard use
- Callable subroutines for error handling

16.13.2   Standards

Design and code standards can be vehicles to improve reliability. By providing consistency and a model, they limit the variety of ways a program can be written. They also provide a standard format for the maintenance programmer so that changes can be made with a minimum of confusion. Standards can be defined for

Design formats	Parameter passing
Data definition	Logic structures
Module interface formats	Code formats
Error handling	Instructions
Termination control	Documentation

16.14   GLOSSARY

Configuration management   Controlled archival and distribution of all software-related components and their interfaces. Includes source control, object code control, update management, version control, and documentation control.

External tables   List of information in a standard format kept outside the program to allow update without impacting the program. An example is a list of zip codes which might exist in one computer location for use by many different programs.

Failure   A state that does not allow the user to continue operation of the software.

Functionality   The feature set provided by the program; the program capabilities and behaviors; measured against user expectations, what the program does.

Hard failure   A failure which cannot be recovered without restarting the hardware.

Integration   Collecting the software components (module, programs) and establishing the appropriate interfaces among them to produce a working system.

Model   A representation of the program through simulation, prototyping, or diagrams.

Noncomment source statements   Code statements which do not consist entirely of text describing the program. Statements which include executable logic or data definition.

Performance   The speed at which the program executes and its impact on system resources (space utilization, processor time, etc.).

Prototyping   Providing a representation of the proposed software which behaves as the final program will, but is not in its final, supportable form. May involve experimental code or incomplete modules.

Regression testing   Testing to ensure the software has not regressed (gotten worse)

since its last version and to ensure that functions that worked prior to this version continue to work.

Reliability   The probability that the software will execute for a particular period of time without failure.

Simulation   Modeling a program, or a portion of a program, by artificially presenting an interface to the user which looks like the proposed interface.

Usability   The attributes of the software which affect the ability of the user to achieve expected results. Includes human factors, ergonomics, ease of learning, clarity of messages, intuitive paths through screens, and all user documentation.

BIBLIOGRAPHY

1. Arthur, Lowell, Jr., *Measuring Programmer Productivity and Software Quality,* Wiley, New York, 1985.

2. Beizer, Boris, *Software Testing Techniques,* Van Nostrand Reinhold, New York, 1983.

3. DeMarco, Tom, *Controlling Software Projects,* Yourdon Press, 1982.

4. Dunn, Robert, and Richard Ullman, *Quality Assurance for Computer Software,* McGraw-Hill, New York, 1982.

5. Mil-Hdbk-189, *Reliability Growth Management,* Department of Defense, 1981.

6. Myers, Glenford J., *The Art of Software Testing,* Wiley, New York, 1979.

7. Myers, Glenford J., *Software Reliability Principles and Practices,* Wiley, New York, 1976.

MATHEMATICS OF RELIABILITY

CHAPTER 17
LIFE DISTRIBUTIONS AND CONCEPTS*

Wayne Nelson
Private Reliability Consultant

Almost every major company yearly spends millions of dollars on product reliability. Much management and engineering effort goes into evaluating risks and liabilities, predicting warranty costs, evaluating replacement policies, assessing design changes, identifying causes of failure, and comparing alternate designs, vendors, materials, manufacturing methods, and the like. Major decisions are based on product life data, often from a few units. This chapter presents modern methods for extracting from life test and field data the information needed to make sound decisions. Such methods are successfully used on a great variety of products by many who have just a working knowledge of basic statistics from a first course.

This chapter presents basic concepts and theory for product life distributions, used as models for the life of products, materials, people, television programs, and many other things. The commonly used exponential, normal, lognormal, and Weibull distributions are presented and are used to analyze data graphically. This chapter also presents the Poisson and binomial distributions, which are probability models for the observed numbers of failures or defectives.

17.1 BASIC CONCEPTS AND THE EXPONENTIAL DISTRIBUTION

The cumulative distribution function F(y) for a continuous distribution represents the population fraction failing by age y. Any such $F(y)$ has the mathematical properties:

1. it is a continuous function for all y,

---

* Reprinted from Dr. Wayne Nelson, *How to Analyze Reliability Data,* vol. 6. Copyright American Society for Quality Control, Inc. 1983, Milwaukee, WI. Reprinted by permission.

2. $\lim_{y \to -\infty} F(y) = 0$ and $\lim_{y \to +\infty} F(y) = 1$, and

3. $F(y) \leqslant F(y')$ for all $y < y'$.

The exponential cumulative distribution function for the population fraction failing by age y is

$$F(y) = 1 - e^{-y/\theta}, \, y \geqslant 0$$

$\theta > 0$ is the mean time to failure. θ is in the same measurement units as y, for example, hours, months, cycles, etc. Figure 17.1a shows this cumulative distribution function. In terms of the "failure rate" $\lambda \equiv 1/\theta$,

$$F(y) = 1 - e^{-\lambda y}, \, y \geqslant 0$$

Engine fan example. The exponential distribution with a mean of $\lambda = 28,700$ hours was used to describe the hours to failure of a fan on diesel engines. The failure rate is $\lambda = 1/28,700 = 34.9$ failures per million hours. For the engine fans, the population fraction failing on an 8,000 hour warranty is $F(8,000) = 1 - \exp(-8,000/28,700) = 0.24$.

The reliability function $R(y)$ for a life distribution is the probability of survival beyond age y, namely,

$$R(y) \equiv 1 - F(y)$$

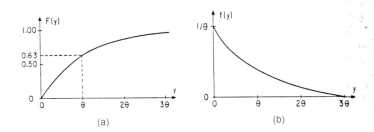

(a)   (b)

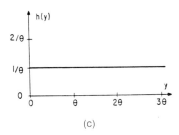

(c)

FIG. 17.1 (*a*) Exponential cumulative distribution; (*b*) exponential probability density; (*c*) exponential hazard function.

The exponential reliability function is

$$R(y) = e^{-y/\theta} , \ y \geqslant 0$$

For the engine fans, reliability for 8,000 hours is $R(8,000)$ = exp($-8,000/28,700$) = 0.76. That is, 76% of such fans survive warranty.

The 100Pth percentile of a distribution $F(\ )$ is the age y_P by which a proportion P of the population fails. It is the solution of

$$P = F(y_P)$$

In life data work, one often wants to know low percentiles such as the 1% and 10% points, which correspond to early failure. The 50% point is called the *median* and is commonly used as a "typical' life.

The 100Pth exponential percentile is

$$y_P = -\theta \ln(1 - P)$$

For example, the mean θ is roughly the 63rd percentile of the exponential distribution. For the diesel engine fans, median life is $y_{.50} = -28,700 \ln(1 - 0.50)$ = 19,900 hours. The 1st percentile is $y_{.01} = -28,700 \ln(1 - 0.01)$ = 288 hours.

The probability density of a cumulative distribution function is

$$f(y) \equiv \frac{dF(y)}{dy}$$

It corresponds to a histogram of the population life times.

The exponential probability density is

$$f(y) = (1/\theta) \, e^{-y/\theta}, \ y \geqslant 0$$

Figure 17.1*b* depicts this probability density. Also,

$$f(y) = \lambda \, e^{-\lambda y}, \ y \geqslant 0$$

The mean μ of a distribution with probability density $f(y)$ is

$$\mu \equiv \int_{-\infty}^{\infty} yf(y) \, dy$$

The integral runs over all possible outcomes y. The mean is also called the *average* or *expected life*. It corresponds to the arithmetic average of the lives of all units in a population. It is used as still another "typical" life.

The mean of the exponential distribution is

$$\mu = \int_{0}^{\infty} y(1/\theta)e^{-y/\theta} \, dy = \theta$$

This shows why θ is called the mean time to failure (MTTF). Also, $\mu = 1/\lambda$. For the diesel engine fans, the mean life is μ = 28,700 hours. Some repairable equipments have exponentially distributed time *between* failures, particularly after most components have been replaced a few times. Then θ is called the mean time between failures (MTBF).

The hazard function h (y) of a distribution is defined as

$$h(y) \equiv f(y)/[1 - F(y)] = f(y)/R(y)$$

It is the *(instantaneous) failure rate* at age y. That is, in the short time Δ from y to $y + \Delta$, a proportion $h(y)$ of the population that reached age y fails. $h(y)$ is a measure of proneness to failure as a function of age. It is also called the *hazard rate* and the *force of mortality*. In many applications, one wants to know whether the failure rate of a product increases or decreases with product age.

The exponential hazard function is

$$h(y) = [(1/\theta e^{-y/\theta}]/e^{-y/\theta} = 1/\theta, \ y \geqslant 0$$

Figure 17.1c shows this constant hazard function. Also, $h(y) = \lambda$, $y \geqslant 0$. Only the exponential distribution has a constant failure rate, a key characteristic. That is, for this distribution only, an old unit and a new unit have the same chance of failing over a future time interval Δ. For example, engine fans of any age will fail at a constant rate of $h(y) = 34.8$ failures per million hours.

A decreasing hazard function during the early life of a product is said to correspond to *infant mortality*. Figure 17.2 shows this near time zero. Such a failure rate often indicates that the product is poorly designed or suffers from manufacturing defects. Some products, such as some semiconductor devices, have a decreasing failure rate over their observed life.

An increasing hazard function during later life of a product is said to correspond to *wear-out* failure. This often indicates that failures are due to the product wearing out. Figure 17.2 shows this feature in the later part of the curve. Many products have an increasing failure rate over the entire range of life.

The bathtub curve. A few products show a decreasing failure rate in the early life and an increasing failure rate in later life. Figure 17.2 shows such a hazard function, called a "bathtub curve." Some products, such as high-reliability capacitors and semiconductor devices, are subjected to a burn-in. This weeds out early failures before units are put into service. Also, units are removed from service before wearout starts. Thus units are in service only in the low failure rate

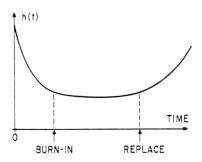

FIG. 17.2 Bathtub curve.

portion of their life. This increases their reliability in service. Jensen and Petersen (1982) comprehensively treat planning and analysis of burn-in procedures, including the economics.

17.1.1 Distributions for Special Situations

Distributions with failure at time zero. A fraction of a population may already be failed at time zero. Consumers may encounter a product that does not work when purchased. The model for this consists of the proportion p failed at time zero and a continuous life distribution for the rest. Such a cumulative distribution appears in Fig. 17.3a. The sample proportion failed at time zero is used to estimate p, and the failure times in the remainder of the sample are used to estimate the continuous distribution.

   Distributions with eternal survivors. Some units may never fail. This applies to (1) the time to death from a disease when some individuals are immune, (2) the time to redemption of trading stamps (some stamps are lost and never redeemed), (3) the time to product failure from a particular defect when some units lack that defect, and (4) time to warranty claim on a product whose warranty applies only to original owners, some of which sell the product before failure. Figure 17.3b depicts this situation.

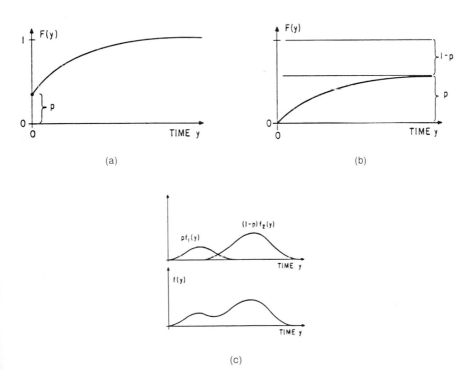

(a)

(b)

(c)

FIG. 17.3 (*a*) A cumulative distribution with a fraction failed at time zero; (*b*) a cumulative distribution with eternal survivors; (*c*) a mixture of distributions.

Mixtures of distributions. A population may consist of two or more subpopulations. Figure 17.3c depicts this situation. Units from different production periods may have different life distributions due to differences in design, raw materials, environment, etc. It is often important to identify such a situation and the production period, customer, environment, etc., that has poor units. Then suitable action may be taken on that portion of the population. A mixture should be distinguished from competing failure modes, described in Section 17.7.

17.2 NORMAL DISTRIBUTION

This section presents the normal distribution. Its hazard function increases. Thus it may describe products with wear-out failure.
   The normal probability density is

$$f(y) = (2\pi \sigma^2)^{-1/2}\exp[-(y - \mu)^2/(2\sigma^2)], \; -\infty < y < \infty$$

μ is the population mean and may have any value. σ is the population standard deviation and must be positive. μ and σ are in the same measurement units as y, for example, hours, months, cycles, etc. Figure 17.4a depicts this probability

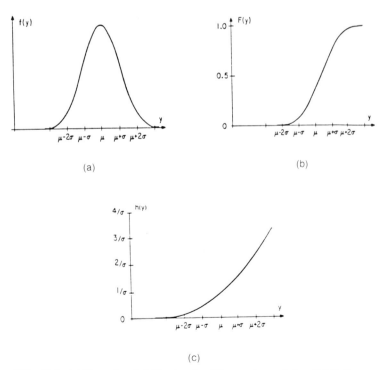

FIG. 17.4 (a) Normal probability density; (b) normal cumulative distribution; (c) normal hazard function.

density, which is symmetric about the mean μ. The figure shows that μ is the distribution median and σ determines the spread.

The range of y is from $-\infty$ to $+\infty$. Life must, of course, be positive. Thus the distribution fraction below zero must be small for this distribution to be a satisfactory approximation in practice.

The normal cumulative distribution function for the population fraction failing by age y is

$$F(y) = \int_{-\infty}^{y} (2\pi\sigma^2)^{-1/2} \exp[-(x - \mu)^2/(2\sigma^2)] \, dx, \qquad -\infty < y < \infty$$

Figure 17.4b depicts this function. This can be expressed in terms of the standard normal cumulative distribution function Φ ( ) as

$$F(y) = \Phi\left[(y - \mu)/\sigma\right], \quad -\infty < y < \infty$$

Many tables of $\Phi(z)$ give values only for $z \geq 0$. One then uses $\Phi(-z) = 1 - \Phi(z)$.

Transformer example. A normal life distribution with $\mu = 6250$ hours and $\sigma = 2600$ hours was used to represent life of a transformer. The fraction of the distribution with negative life times is $F(0) = \Phi[(0 - 6250)/2600] = \Phi(-2.40) = 0.0082$. This small fraction is ignored hereafter.

The 100Pth normal percentile is

$$y_P = \mu + z_P\sigma$$

Here z_P is the 100Pth standard normal percentile and is tabled below. The *median* (50th percentile) of the normal distribution is $y_{0.50} = \mu$, since $z_{0.50} = 0$.

Some standard percentiles are:

100P%:	0.1	1	2.5	5	10	50	90	97.5	99
z_P:	-3.090	-2.326	-1.960	-1.645	-1.282	0	1.282	1.960	2.326

Median transformer life is $y_{0.50} = 6250$ hours, and the 10th percentile is $y_{0.10} = 6250 + (-1.282)2600 = 2920$ hours.

The normal hazard function appears in Fig. 17.4c, which shows that the normal distribution has an *increasing failure rate* (wear out) with age.

A key question was: does transformer failure rate increase with age? If so, older units should be replaced first. The increasing failure rate of the normal distribution indicates that older units are more failure prone.

17.3 LOGNORMAL DISTRIBUTION

The lognormal distribution is used for certain types of life data, for example, metal fatigue and electrical insulation life. The lognormal and normal distributions are related, a fact used to analyze lognormal data with methods for normal data.

The lognormal probability density is

$$f(y) = \{0.4343/[(2\pi)^{1/2}y\sigma]\} \exp\{ - [\log(y)-\mu]^2/(2\sigma^2)\}, y > 0$$

μ is called the *log mean* and may have any value; it is the mean of the *log* of life—not of life. σ is called the *log standard deviation* and must be positive; it is the standard deviation of the *log* of life—not of life. μ and σ are not "times" like y; instead they are unitless pure numbers. $0.4343 \cong 1/\ln(10)$. Here log( ) denotes the common (base 10) logarithm. Some authors use the natural (base e) logarithm, denoted by ln( ). Figure 17.5a shows probability densities, which have a variety of shapes. The value of σ determines the shape of the distribution, and the value of μ determines the 50% point and the spread.

The lognormal cumulative distribution function for the population fraction failing by age y is

$$F(y) = \Phi \{[\log(y) - \mu]/\sigma\}, y > 0$$

Figure 17.5b shows lognormal cumulative distribution functions.

Locomotive control example. The life (in thousand miles) of an electronic control for locomotives was approximated by a lognormal distribution where $\mu = 2.236$ and $\mu = 0.320$. The population fraction failing on an 80 thousand mile warranty is $F(80) = \Phi \{[\log(80) - 2.236]/0.320\} = \Phi [-1.04] = 0.15$. This percentage was too high, and the control was redesigned.

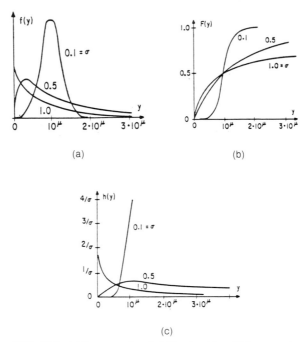

(a)   (b)

(c)

FIG. 17.5 (*a*) Lognormal probability densities; (*b*) lognormal cumulative distributions; (*c*) lognormal hazard functions.

The 100Pth lognormal percentile is

$$y_P = \text{antilog}\,[\mu + z_P\sigma]$$

Here z_P is the 100Pth standard normal percentile. The *median* (50th percentile) is $y_{0.50} = \text{antilog}\,[\mu]$.

For the locomotive control, $y_{0.50} = \text{antilog}\,[2.236] = 172$ thousand miles, regarded as a typical life. The 1% life is $y_{0.01} = \text{antilog}[2.236 + (-2.326)0.320] = 31$ thousand miles.

Lognormal hazard functions appear in Fig. 17.5c. For $\sigma \simeq 0.5$, $h(y)$ is essentially constant. For $\sigma \leq 0.2$, $h(y)$ increases and is much like that of a normal distribution. For $\sigma \geq 0.8$, $h(y)$ decreases. This flexibility makes the lognormal distribution popular and suitable for many products. The lognormal hazard function has a property seldom seen in products. It is zero at time zero, increases to a maximum, and then decreases to zero with increasing age. However, over most of its range, the lognormal distribution does fit life data on many products.

For the locomotive control, $\sigma = 0.320$. So the behavior of $h(y)$ is midway between the increasing and roughly constant hazard functions in Fig. 17.5c.

The relationship between the lognormal and normal distributions helps one understand the lognormal distribution in terms of the simpler normal distribution. The (base 10) log of a variable with a lognormal distribution with parameters μ and σ has a normal distribution with mean μ and standard deviation σ. Thus the analysis methods for normal data can be used for the logarithms of lognormal data.

17.4 WEIBULL DISTRIBUTION

The Weibull distribution is often used for product life, because it describes increasing and decreasing failure rates. It may be suitable for a "weakest link" product; i.e., the product consists of many parts with comparable life distributions and the product fails with the first part failure. For example, the life of a capacitor is determined by the shortest-lived portion of its dielectric.

The Weibull probability density function is

$$f(y) = (\beta/\alpha^\beta)\,y^{\beta-1}\,\exp[-(y/\alpha)^\beta], \ y > 0$$

The *shape parameter* β and the *scale parameter* α are positive. α is called the *characteristic life,* as it is always 63.2th percentile. α has the same units as y, for example, hours, months, cycles, etc. β is a unitless pure number. The Weibull probability densities in Fig. 17.6a show that β determines the shape of the distribution and α determines the spread. For $\beta = 1$, the Weibull distribution is the exponential distribution. For much life data, the Weibull distribution is more suitable than the exponential, normal, and lognormal distributions.

The Weibull cumulative distribution function for the population fraction failing by age y is

$$F(y) = 1 - \exp[-(y/\alpha)^\beta], \ y > 0$$

Figure 17.6b shows Weibull cumulative distribution functions.

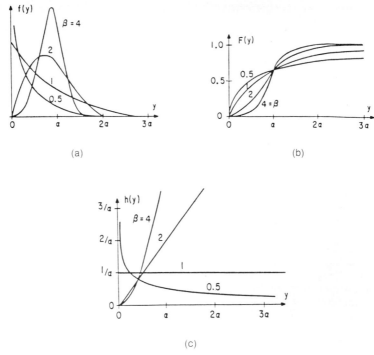

FIG. 17.6 (*a*) Weibull probability densities; (*b*) Weibull cumulative distributions; (*c*) Weibull hazard functions.

Winding example. The life of generator field windings was approximated by a Weibull distribution where $\alpha = 13$ years and $\beta = 2$. The population fraction of windings failing on a two-year warranty is $F(2.0) = 1 - \exp[-(2.0/13)^2] = 0.023$ or 2.3%.

The Weibull reliability function for the population fraction surviving beyond age y is

$$R(y) = \exp[-(y/\alpha)^\beta], \quad y > 0$$

For the windings, the population reliability for two years is $R(2.0) = \exp[-(2.0/13)^2] = 0.977$ or 97.7%

The 100Pth Weibull percentile is

$$y_P = \alpha[-\ln(1 - P)]^{1/\beta}$$

Here ln( ) is the natural logarithm. For example, $y_{0.632} \cong \alpha$ for any Weibull distribution. This may be seen in Fig. 17.6*b*.

For the windings, $y_{0.632} = 13[-\ln(1 - 0.632)]^{1/2} = 13$ years, the characteristic life. The 10th percentile is $y_{0.10} = 13[-\ln(1 - 0.10)]^{1/2} = 4.2$ years.

The Weibull hazard function is

$$h(y) = (\beta/\alpha)(y/\alpha)^{\beta-1}, \ y > 0$$

Figure 17.6c shows Weibull hazard functions. A power function of time, $h(y)$ increases for $\beta > 1$ and decreases for $\beta < 1$. For $\beta = 1$ (the exponential distribution), the failure rate is constant. With increasing or decreasing failure rates, the Weibull distribution flexibly describes product life.

For the windings, $\beta = 2$, and their failure rate increases with age, wear-out behavior. This tells utilities that preventive replacement of old windings will avoid costly failures in service.

17.5 POISSON DISTRIBUTION

The Poisson distribution is used for the number of occurrences of some event within some observed time, area, volume, etc. For example, it has been used to describe the yearly number of soldiers of a Prussian regiment kicked to death by horses, the number of flaws in a length of wire or computer tape, the number of failures of a repairable product over a certain period, and many other phenomena. It is appropriate if (1) the occurrences occur independently of each other over time (area, volume, etc.), (2) the chance of an occurrence is the same for each point in time (area, volume, etc.), and (3) the potential number of occurrences is unlimited.

The Poisson probability of y occurrences is

$$f(y) = (1/y!)(\lambda \ t)^y \ e^{-\lambda t}, \ y = 0,1,2,\ldots$$

Here t is the amount of exposure or observation; it may be a time, length, area, volume, etc. For example, for a power line, t is the product of length and time in thousand-ft-years. The *occurrence rate* λ must be positive; it is the expected number of occurrences per unit time, length, area, volume, etc. Figure 17.7a depicts Poisson probability functions.

Power line. For a power line, the yearly number of failures is assumed to have a Poisson distribution with $\lambda = 0.0256$ failures per year per thousand feet. For $t = 515.8$ thousand feet of line, the probability of no failures in a year is $f(0) = (1/0!)(0.0256 \times 515.8)^0 \exp(-0.0256 \times 515.8) = \exp(-13.2) = 1.8 \times 10^{-6}$.

The Poisson cumulative distribution function for the probability of y or fewer occurrences is

$$F(y) = \sum_{i=0}^{y} (1/i!)(\lambda t)^i e^{-\lambda t}$$

The Thorndike chart in Fig. 17.7b provides $F(y)$ as follows. Enter the chart on the horizontal axis at the value $\mu = \lambda t$. Go up to the curve labeled y. Then go horizontally to the vertical scale to read $F(y)$. For the power line, $\lambda t = 13.2$, and the probability of 15 or fewer failures is $F(15) = 0.75$ from the chart. $F(y)$ is tabulated in most textbooks.

The Poisson mean of the number Y of occurrences is

$$\mu = \lambda \ t$$

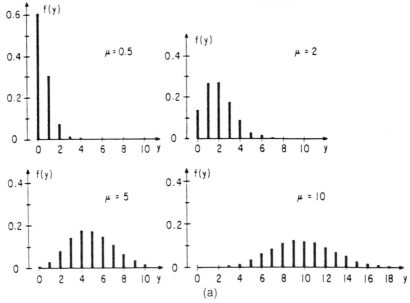

FIG. 17.7a　Poisson probability functions.

For the power line, the expected (mean) number of failures in a year is $\mu = 0.0256(515.8) = 13.2$ failures, useful in maintenance planning.

The Poisson standard deviation of the number Y of occurrences is

$$\sigma(Y) = (\lambda t)^{1/2}$$

For the power line, $\sigma(Y) = [13.2]^{1/2} = 3.63$ failures in a year.

A normal approximation to the Poisson $F(y)$ is

$$F(y) \cong \Phi[(y + 0.5 - \lambda t)/(\lambda t)^{1/2}]$$

Here $\Phi[\ ]$ is the standard normal cumulative distribution function. This approximation is satisfactory for most practical purposes if $\lambda t \geqslant 10$.

For the power line, the approximate probability of 15 or fewer failures in a year is $F(15) \cong \Phi[(15 + 0.5 - 13.2)/3.63] = \Phi\ [0.63] = 0.74$. The exact probability is 0.75.

Demonstration testing commonly involves the Poisson distribution. Repairable hardware "demonstrates" its reliability if units run a specified total time t with y or fewer failures. Units that fail are repaired and kept on test. A manufacturer designs the hardware to achieve a λ that assures passing the test with a desired high probability $100(1 - \alpha)\%$. The hardware can fail the test with $100\alpha\%$ probability, called the *producer's risk*.

Electronic system example. An electronic system was required to run $t = 10,000$ hours with $y = 2$ or fewer failures. For the electronic system, the producer's risk was to be 10%. To obtain the desired design λ, one must find $\lambda\ t$ such that the Poisson probability $F_{\lambda t}(y) = 1 - \alpha$. To do this, enter Fig. 17.7b on the vertical axis at $1 - \alpha$, go horizontally to the curve for y or fewer failures, and then go down to the horizontal axis to read the appropriate μ value. Then the

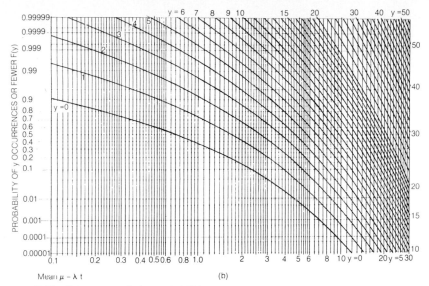

Mean $\mu - \lambda t$

(b)

FIG. 17.7b Poisson cumulative probabilities. (*From H. F. Dodge and H. G. Romig, Sampling Inspection Tables, Wiley, New York, Fig. 6. Copyright 1959, Bell Telephone Laboratories, Inc., reprinted by permission.*)

desired design failure rate is $\lambda = \mu/t$. For the electronic system, $1 - \alpha = 1 - 0.10 = 0.90$, $\mu = 1.15$, and $\lambda = 1.15/10{,}000 = 0.115$ failures per thousand hours.

Relationship of Poisson and exponential distributions. For a repairable product, suppose that times between failures are statistically independent and have an exponential distribution with failure rate λ. Then the *number* of failures in a total running time t over any number of units has a Poisson distribution with mean λt.

17.6 BINOMIAL DISTRIBUTION

The binomial distribution is used as a model for the number of sample units that fall in a specified category. For example, it is used for the number of defective units in samples from shipments and production, the number of units that fail on warranty, and the number of one-shot devices (used once) that work properly.

Its assumptions are (a) each sample unit has the same chance p of being in the category and (b) the outcomes of the n sample units are statistically independent.

The binomial probability of getting y category units in a sample of n units is

$$f(y) = \frac{n!}{y!(n-y)!}\, p^y(1-p)^{n-y}, \qquad y = 0, 1, 2, \ldots, n$$

p is the population proportion in the category ($0 \le p \le 1$). Figure 17.8 depicts binomial probability functions.

In reliability work, if the category is "failure" of a device, the proportion p is the *failure probability*, sometimes incorrectly called the failure rate. The proportion p is expressed as a percentage and differs from the Poisson failure rate λ,

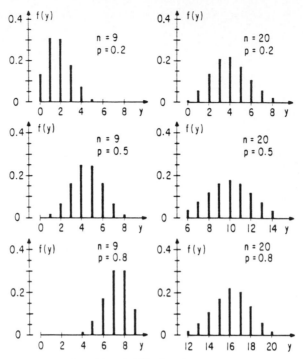

FIG. 17.8   Binomial probability functions.

which has the dimensions of failures per unit time. If the category is "successful operation" of a device, the proportion p is called the *reliability* of the device.

A *locomotive control* under development was assumed to fail on warranty with probability $p = 0.156$. $n = 96$ such controls were field tested, and $y = 15$ failures occurred on warranty. The probability is $f(15) = 96![15!(96 - 15)!]^{-1} (0.156)^{15} \times (1 - 0.156)^{96-15} = 0.111$.

The binomial cumulative distribution function for the probability of y or fewer sample items in the category is

$$F(y) = \sum_{i=0}^{y} \frac{n!}{i!(n - i)!} p^i(1 - p)^{n - i}, \qquad y = 0, 1, 2, \ldots, n$$

$F(y)$ is widely tabulated. For example, the probability of 15 or fewer warranty failures of the 96 controls is $F(15) = 0.571$ from a binomial table.

A *normal approximation* is

$$F(y) \cong \Phi\{(y + 0.5 - np)/[np(1 - p)]^{1/2}\}$$

here $\Phi[\ ]$ is the standard normal cumulative distribution function. This approximation is usually satisfactory if $np \geq 10$ and $n(1 - p) \geq 10$.

For example, for the controls, $F(15) \cong \Phi\{(15 + 0.5 - 96 \times 0.156)/[96 \times 0.156 (1 - 0.156)]^{1/2}\} = 0.556$. Similarly, the approximate probability of 15 failures is $f(15) = F(15) - F(14) \cong 0.556 - \Phi\{(14 + 0.5 - 96 \times 0.156)/[96 \times 0.156 \times (1 - 0.156)]^{1/2}\} = 0.112$ (0.111 exact). Here $96(0.156) = 15 > 10$ and $96(1 - 0.156) = 81 > 10$.

The mean of the number Y of sample items in the category is

$$\mu = np$$

This is the number n of sample units times the population proportion p in the category. For example, the mean number of failures in samples of 96 locomotive controls is $\mu = 96 \times 0.156 = 15.0$ failures.

Acceptance sampling plans based on the binomial distribution appear in MIL-STD-105D and in quality control books, for example, Grant and Leavenworth (1980) and Schilling (1982).

An acceptance sampling plan specifies the number n of sample units and the acceptable number of y of defective units in the sample. If the sample contains y or fewer defective units the product passes; otherwise, it fails. A plan had $n = 20$ and $y = 1$. If the product has a proportion defective of $p = 0.01$, the chance it passes inspection is $F(1) = f(0) + f(1) = [20!/0!(20-0)!] \, 0.01^0 \, 0.99^{20} + [20!/1!(20-1)!] \, 0.01^1 \, 0.99^{19} = 0.983$, which could be read from a binomial table. The chance of passing as a function of p is called the *operating characteristic* (OC) *curve* of the plan (n,y). The OC curve for $n = 20$ and $y = 1$ appears in Fig. 17.9.

17.7 SERIES SYSTEMS AND MULTIPLE CAUSES OF FAILURE

Many products fail from more than one cause. For example, any part in an appliance may fail and cause the appliance to fail. Also, humans may die from accidents, various diseases, etc. The series-system model represents the relationship between the product life distribution and those of its parts. Graphical analyses of data with a number of causes of failure follow later.

This section presents the series-system model, the product rule for reliability, the addition law for failure rates, and the resulting distribution when some failure modes are eliminated.

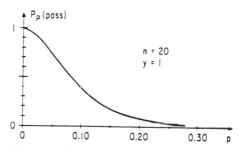

FIG. 17.9 OC curve of acceptance sampling plan.

Series systems and the product rule. Suppose that a product has a potential time to failure from each of M causes (also called competing risks or failure modes). Such a product is called a *series system* if its life is the smallest of those M potential times to failure. That is, the first part failure produces system failure.

Let $R(y)$ denote the system reliability function and let $R_1(y),\ldots,R_M(y)$ denote the reliability functions of the M causes (each in the absence of all other causes). It is assumed that the M potential times to failure of a system are statistically *independent*. Such systems are said to have *independent competing risks* or to be *series systems* with independent causes of failure. For such systems, it can be shown that

$$R(y) = R_1(y)R_2(y) \cdots R_M(y)$$

This key result is the *product rule* for reliability of series systems (with independent components).

Three-way bulb. By engineering definition, a three-way lightbulb fails if either filament fails. Filament 1 (2) has a normal life distribution with a mean of 1500 (1200) hours and a standard deviation of 300 (240) hours. Filament reliability functions are depicted as straight lines on normal probability paper in Fig. 17.10. The life distribution of such bulbs was needed, in particular, the median life. Filament lives are assumed independent; so the bulb reliability is $R(y) = \{1 - \Phi[(y - 1500)/300]\} \times \{1 - \Phi[(y - 1200/240]\}$; here $\Phi[\ ]$ is the standard normal cumulative distribution. For example, $R(1200) = \{1 - \Phi[(1200 - 1500)/300]\} \times \{1 - \Phi[(1200 - 1200)/240]\} = 0.421$. $R(y)$ is plotted in Fig. 17.10 and is not quite a straight line (not a normal distribution). The median life is obtained by solving $R(y_{0.50}) = 0.50$ to get $y_{0.50} = 1160$ hours; this also can be obtained from the plot.

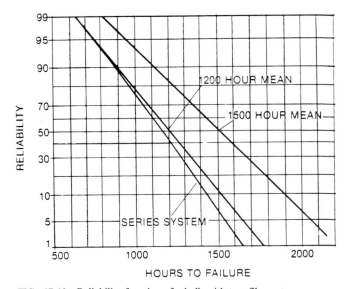

FIG. 17.10 Reliability function of a bulb with two filaments.

Addition law for failure rates. Denote the system hazard function by $h(y)$ and those for the failure causes by $h_1(y), \ldots, h_M(y)$. Then it can be shown that

$$h(y) = h_1(y) + h_2(y) + \cdots + h_M(y)$$

This is called the *addition law for failure rates* for *independent* failure modes (or competing risks). This law is depicted in Fig. 17.11, which shows the hazard functions of the two components of a series system and the system hazard function.

A pronounced increase in the failure rate of a product may occur at some age. This may indicate that a new failure cause with an increasing failure rate is becoming dominant at that age, as in Fig. 17.11.

Exponential causes. Suppose that M independent causes have *exponential* life distributions with failure rates $\lambda_1, \ldots, \lambda_M$. Then series systems consisting of such components have an exponential life distribution with a constant failure rate

$$\lambda = \lambda_1 + \cdots + \lambda_M$$

This simple relationship is often *incorrectly* used for reliability analysis of systems with components that do not have constant failure rates. Then the previous equation is correct.

Freight train. A high-priority freight train was required to have its three locomotives all complete a one-day run. If a locomotive failed, the railroad had to pay a large penalty. The railroad needed to know the reliability of such trains. Time to failure for such locomotives has an exponential distribution with $\lambda_o = 0.023$ failures per day. A series system, the train has an exponential distribution of time to delay with $\lambda = 0.023 + 0.023 + 0.023 = 0.069$ delays per day. Train reliability on a one-day run is $R(1) = \exp(-0.069 \times 1) = 0.933$.

Elimination of failure modes. Often it is important to know how elimination of some causes of failure will improve product life. Suppose that cause 1 is eliminated (this may be a collection of causes.) Then $R_1(y) = 1$, $h_1(y) = 0$, and the

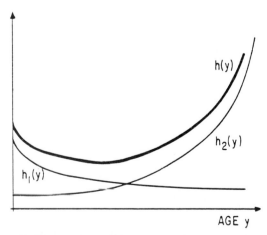

FIG. 17.11 Hazard functions of a series system and its components.

life distribution for the remaining causes has

$$R^*(y) = R_2(y) \times \ \cdots \ \times R_M(y), \ h^*(y) = h_2(y) + \cdots + h_M(y)$$

If the 1500-hour filament were replaced by one with essentially unlimited life, the bulb would have the life distribution of the 1200-hour filament.

Series systems with dependence. Some series-system products contain parts with statistically *dependent* lifetimes. For example, adjoining segments of a cable may have positively correlated lives, that is, have similar lives. Models for dependent part lives are complicated; David and Moeschberger (1979) and Block and Savits (1981) comprehensively survey multivariate distributions with dependence.

17.8   HAZARD PLOTTING OF LIFE DATA

Data plots are used for display and interpretation of data because they are simple and effective, as described by Nelson (1979). Hazard plots are widely used to analyze field and life test data on products consisting of electronic and mechanical parts and ranging from small electrical appliances through heavy industrial equipment. This section presents hazard plots to estimate a life distribution from multiply censored life data. Such plots do not apply to failures found on inspection, since the failure occurred earlier at an unknown time.

Appliance component. Data that illustrate hazard plotting appear in Table 17.1, which shows the cycles (number of times used) to failure of a component of a small appliance in a development program. Each survival time has a + to indicate that the failure time of the unfailed component is beyond. Failure times are unmarked. Engineering wanted an estimate of the percentage failing on warranty (500 cycles) and an estimate of median life.

17.8.1   Steps to Make a Hazard Plot

1. Order the *n* times from smallest to largest as shown in Table 17.1 without regard to which are survival or failure times. Label the times with reverse ranks; that is, label the first time with *n*, the second with *n*−1,..., and the *n*th with 1. There are *n*=54 appliance components in the example.

2. Calculate a hazard value for each *failure* as 100/*k*, where *k* is its reverse rank, as shown in Table 17.1. For example, the failure at 145 cycles has reverse rank 46, and its hazard value is 100/46 = 2.2%. Hazard values are tabulated in Table 17.2.

3. Calculate the cumulative hazard value for each *failure* as the sum of its hazard value and the cumulative hazard value of the preceding failure. For example, for the failure at 145 cycles, the cumulative hazard value of 6.0 is the hazard value 2.2 plus the previous cumulative hazard value 3.8. Cumulative hazard values appear in Table 17.1. Cumulative hazard values have no physical meaning and may exceed 100%.

4. Choose a hazard paper. There are hazard papers* for the exponential, Weibull, extreme value, normal, and lognormal distributions. The distribution is often chosen from engineering knowledge of the product.

---

* Offered in the catalog of TEAM, Box 25, Tamworth, N.H. 03886.

TABLE 17.1   Hazard Calculations for the Appliance Component

Cycles	Reverse rank k	Hazard 100/k	Cum. hazard	Cycles	Reverse rank k	Hazard 100/k	Cum. hazard
45+	54			608+	27		
47	53	1.9	1.8	608+	26		
73	52	1.9	3.8	608+	25		
136+	51			608+	24		
136+	50			608+	23		
136+	49			608	22	4.6	24.8
136+	48			608+	21		
136+	47			608+	20		
145	46	2.2	6.0	630	19	5.3	30.1
190+	45			670	18	5.6	35.7
190+	44			670	17	5.9	41.6
281+	43			731+	16		
311	42	2.4	8.4	838	15	6.7	48.3
417+	41			964	14	7.1	55.4
485 ⊦	40			964	13	7.7	63.1
485+	39			1164+	12		
490	38	2.6	11.0	1164+	11		
569+	37			1164+	10		
571+	36			1164+	9		
571	35	2.9	13.9	1164+	8		
575	34	2.9	16.8	1164+	7		
608+	33			1164+	6		
608+	32			1198 ⊦	5		
608+	31			1198	4	25.0	88.1
608+	30			1300+	3		
608	29	3.4	20.2	1300+	2		
608+	28			1300+	1		

5. On the vertical axis of the hazard paper, mark a time scale that brackets the data. For the component data, normal hazard paper was chosen, and marked from 0 to 1200 cycles as shown in Fig. 17.12.

6. On the paper, plot each failure time vertically against its cumulative hazard value on the horizontal axis as shown in Fig. 17.12. Survival times are *not* plotted; hazard and cumulative hazard values are not calculated for them. However, the survival times do determine the proper plotting positions of the failure times through the reverse ranks. STATPAC of Nelson and others (1983) and other computer programs do such calculations and make such plots.

7. If the plot of failure times is roughly straight, the theoretical distribution fits the data. By eye fit a straight line through the data points. Also, one can just use the plotted points without a line.

The line estimates the cumulative percentage failing (read from the horizontal probability scale at the top of the grid) as a function of age. The straight line, as explained below,

TABLE 17.2   Hazard Values 100/k for k = 1 to 200

1	100.00	51	1.96	101	.99	151	.66	26	3.85	76	1.32	126	.79	176	.57
2	50.00	52	1.92	102	.98	152	.66	27	3.70	77	1.30	127	.79	177	.56
3	33.33	53	1.89	103	.97	153	.65	28	3.57	78	1.28	128	.78	178	.56
4	25.00	54	1.85	104	.96	154	.65	29	3.45	79	1.27	129	.78	179	.56
5	20.00	55	1.82	105	.95	155	.65	30	3.33	80	1.25	130	.77	180	.56
6	16.67	56	1.79	106	.94	156	.64	31	3.23	81	1.23	131	.76	181	.55
7	14.29	57	1.75	107	.93	157	.64	32	3.12	82	1.22	132	.76	182	.55
8	12.50	58	1.72	108	.93	158	.63	33	3.03	83	1.20	133	.75	183	.55
9	11.11	59	1.69	109	.92	159	.63	34	2.94	84	1.19	134	.75	184	.54
10	10.00	60	1.67	110	.91	160	.62	35	2.86	85	1.18	135	.74	185	.54
11	9.09	61	1.64	111	.90	161	.62	36	2.78	86	1.16	136	.74	186	.54
12	8.33	62	1.61	112	.89	162	.62	37	2.70	87	1.15	137	.73	187	.53
13	7.69	63	1.59	113	.88	163	.61	38	2.63	88	1.14	138	.72	188	.53
14	7.14	64	1.56	114	.88	164	.61	39	2.56	89	1.12	139	.72	189	.53
15	6.67	65	1.54	115	.87	165	.61	40	2.50	90	1.11	140	.71	190	.53
16	6.25	66	1.52	116	.86	166	.60	41	2.44	91	1.10	141	.71	191	.52
17	5.88	67	1.49	117	.85	167	.60	42	2.38	92	1.09	142	.70	192	.52
18	5.56	68	1.47	118	.85	168	.60	43	2.33	93	1.08	143	.70	193	.52
19	5.26	69	1.45	119	.84	169	.59	44	2.27	94	1.06	144	.69	194	.52
20	5.00	70	1.43	120	.83	170	.59	45	2.22	95	1.05	145	.69	195	.51
21	4.76	71	1.41	121	.83	171	.58	46	2.17	96	1.04	146	.68	196	.51
22	4.55	72	1.39	122	.82	172	.58	47	2.13	97	1.03	147	.68	197	.51
23	4.35	73	1.37	123	.81	173	.58	48	2.08	98	1.02	148	.68	198	.51
24	4.17	74	1.35	124	.81	174	.57	49	2.04	99	1.01	149	.67	199	.50
25	4.00	75	1.33	125	.80	175	.57	50	2.00	100	1.00	150	.67	200	.50

yields information on the life distribution. If the plot is curved, plot the data on another hazard paper. If no hazard paper yields a straight enough plot, draw a smooth curve through the plotted data. Then, as described below, use the curve in the same way as a straight line to estimate percentiles and failure probabilities.

The basic assumption. Hazard plotting is valid if the life distribution of units censored at a given age is the same as the life distribution of units that run beyond that age. For example, this assumption is not satisfied if units are removed from service when they look like they are about to fail.

17.8.2   How to Interpret a Hazard Plot

The probability and data scales on a hazard paper are exactly the same as those on the corresponding probability paper. Thus, a hazard plot is interpreted the same way as a probability plot, and the scales on hazard paper are used like those on probability paper. The cumulative hazard scale is only an aid for plotting multiply censored data.

Estimate of the percentage failing. The population percentage failing by a given age is estimated from the fitted line or curve as follows. Enter the plot on the time scale at the given age, go to the fitted line, and then go to the corresponding

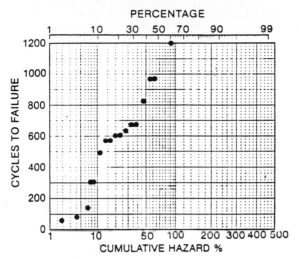

FIG. 17.12 Normal hazard plot of appliance component data.

point on the probability scale to read the percentage. For example, the estimate of the percentage of components failing by 500 cycles (warranty) is 12%; this answers a basic question.

Percentile estimate. To estimate a percentile, enter the plot on the probability scale at the given percentage, go to the fitted line, and then go to the corresponding point on the time scale to read the percentile. For example, the estimate of the 50th percentile, nominal component life, is 1000 cycles.

17.9 LIFE DISTRIBUTION WITH FAILURE MODES ELIMINATED

Hazard plotting also provides an estimate of the life distribution that would result if certain failure modes were eliminated by proposed design changes. It is costly and time consuming to change a design and collect and analyze data to determine the value of design changes. Instead this can be done using past data. It is assumed that the cause of each failure is identified. The following example illustrates the method.

The method. Suppose that a proposed design change of the appliance would eliminate mode 11 failures and leave other failure modes unchanged. Past data including mode 11 and other failure modes are given in Table 17.3 and are used to predict the resulting life distribution. Hazard calculations for the life distribution without mode 11 are shown in Table 17.3. Each failure time by mode 11 is treated as a censoring time, since the new design would have run that long without failure. The failure times for the remaining modes are plotted against their cumulative hazard values as shown in Fig. 17.13 on Weibull paper. About 10% would fail on warranty (500 cycles) with mode 11 eliminated.

Failure rate behavior. Often it is useful to know how the failure rate depends on product age. A failure rate that increases with age usually indicates that

TABLE 17.3 Hazard Calculations with Mode 11 Eliminated

Cycles	Failure mode	Cum. hazard	Cycles	Failure mode	Cum. hazard
45	1	1.85	608+		
47	11		608+		
73	11		608+		
136+			608+		
136	6	3.85	608+		
136+			608	11	
136+			608+		
136+			608+		
145	11		630	11	
190+			670	11	
190+			670	11	
281	12	6.18	731+		
311	11		838	11	
417	12	8.62	964	11	
485+			964	11	
485+			1164+		
490	11		1164+		
569	1	11.32	1164+		
571+			1164+		
571	11		1164+		
575	11		1164+		
608+			1164+		
608+			1198	9	31.32
608+			1198	11	
608+			1300+		
608+	11		1300+		
608+			1300+		

failures are due to wearout. A failure rate that decreases with age usually indicates that failures are due to manufacturing or design defects that cause early failures.

For data plotted on Weibull hazard paper, the following assesses the nature of the failure rate. A Weibull failure rate increases (decreases) if the shape parameter is greater (less) than 1. To estimate the Weibull shape parameter, draw a straight line parallel to the plotted data, so it passes through the "origin" of the Weibull hazard paper and through the shape parameter scale, as in Fig. 17.13. The value on that scale is the estimate of the shape parameter; it is 0.7 in Fig. 17.13, indicating a decreasing failure rate (design or manufacturing defects).

Assumptions. The hazard plotting method above is based on four assumptions: (1) Each unit has a potential failure time for each failure mode. (2) The observed time to failure for a unit is the smallest of its potential times to failure. (3) Potential times to failure for different failure modes are statistically independent. (4) The mode of each failure is identified. Thus the product is regarded to be a series system.

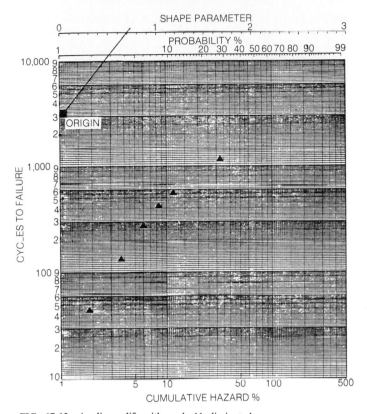

FIG. 17.13   Appliance life with mode 11 eliminated.

17.10   LIFE DISTRIBUTION OF A FAILURE MODE

Information is sometimes desired on the distribution of time to failure for a particular failure mode. An estimate of its distribution provides information on the nature of the failure mode and on the effect of design changes on that mode.

The method. An example of the method involves the data with a mix of failure modes in Table 17.3. Hazard calculations for mode 11 are shown in Table 17.1. In these calculations, each failure time for another mode is treated as a censoring time for mode 11, that is, as if those units were removed from test before they failed by mode 11. The failure times for mode 11 are plotted against their cumulative hazard values in Fig. 17.12.

REFERENCES

Bain, L. J. (1978), *Statistical Analysis of Reliability and Life-Testing Models: Theory and Methods,* Marcel Dekker, New York.

Block, H. W., and Savits, T. H. (1981), "Multivariate Distributions in Reliability Theory and Life Testing," Technical Report No. 81-13, Inst. for Statistics and Applications, Dept. of Math. and Statistics, Univ. of Pittsburgh, Pittsburgh, PA 15260.

David, H. A., and Moeschberger, M. L. (1979), *The Theory of Competing Risks*, Griffin's Statistical Monograph No. 39, Methuen, London.

Grant, E. L., and Leavenworth, R. S. (1980), *Statistical Quality Control*, 5th ed., McGraw-Hill, New York.

Hahn, G. J., and Shapiro, S. S. (1967), *Statistical Models in Engineering*, Wiley, New York.

Jensen, F., and Petersen, N. E. (1982), *Burn-in: An Engineering Approach to the Design and Analysis of Burn-in Procedures*, Wiley, New York.

Locks, M. O. (1973), *Reliability, Maintainability, and Availability Assessment*, Hayden, Rochelle Park, NJ.

Mann, N. R., Schafer, R. E., and Singpurwalla, N. D. (1974), *Methods for Statistical Analysis of Reliability and Life Data*, Wiley, New York.

Martz, H. F., and Waller, R. A. (1982), *Bayesian Reliability Analysis*, Wiley, New York.

Meeker, W. Q., and Duke, S. (1979), "CENSOR—A User-Oriented Computer Program for Life Data Analysis," Department of Statistics, Iowa State University, Ames, IA 50011.

Nelson, W. (1979), "How to Analyze Data with Simple Plots," Volume 1 of the ASQC Basic References in Quality Control: Statistics Techniques, J. Dudewicz, Editor. For sale from the Amer. Soc. for Quality Control, 310 W. Wisconsin Ave., Milwaukee, WI 53203.

Nelson, Wayne (1982), *Applied Life Data Analysis*, Wiley, New York.

Nelson, W. B., Morgan, C. B., and Caporal, P. "(1983)" 1983 STATPAC Simplified—A Short Introduction to How to Run STATPAC, a General Statistical Package for Data Analysis," General Electric Company Corporate Research & Development TIS Report 83CRD146.* Outside General Electric, STATPAC may be obtained on license through Mr. M. Keith Burk, Technology Marketing Operation, GE Corporate Research & Development, 120 Erie Blvd., Schenectady, NY 12305. Phone: (518) 385-3801.

Schilling, E. G. (1982), *Acceptance Sampling in Quality Control*, Marcel Dekker, New York.

CHAPTER 18
TECHNIQUES OF ESTIMATING RELIABILITY AT DESIGN STAGE

Kailash C. Kapur, Ph.D., P.E.
Professor, Department of Industrial Engineering and
Operations Research, Wayne State University

18.1   INTRODUCTION

Reliability is basically a design parameter and must be incorporated into the system at the design stage. Reliability is an inherent attribute of a system resulting from design, just as is the system's capacity, performance, or power rating. The reliability level must be established at the design phase, and subsequent testing and production will not raise the reliability without a basic design change. With increasing system complexity, reliability becomes an elusive and difficult design parameter to define and achieve. It also becomes more difficult to control, demonstrate, and ensure as an operational characteristic under the projected conditions of use by the customer. However, past history has demonstrated that where reliability was recognized as a necessary program development component, with the practice of various reliability engineering methods throughout the evolutionary life cycle of the system, reliability can be quantified during the specification of design requirements, can be predicted by testing, can be controlled during production, and can be sustained in the field.

The term *system effectiveness* is often used to describe the overall capability of a system to accomplish its mission or perform its intended function. System effectiveness is defined as the probability that the system can successfully meet an operational demand within a given time when operating under specified conditions.[1] For consumer products, system effectiveness is related to customer satisfaction. Our objective is to design and manufacture products that will meet the needs and expectations of the customer. Effectiveness is influenced by the way the system is designed, manufactured, used, and maintained, and thus is a function of all the life-cycle activities as well as system attributes such as design adequacy, performance measures, safety, reliability, quality, producibility, maintainability, and availability (see Fig. 18.1). Reliability is one of the major attributes determining system effectiveness.

The purpose of this chapter is to present some of the reliability methodologies and philosophies that are applicable during the design stage in the life cycle of a system. Figure 18.2 represents the complexity of balancing different design requirements[2] and trade-offs between reliability, maintainability, cost, etc.

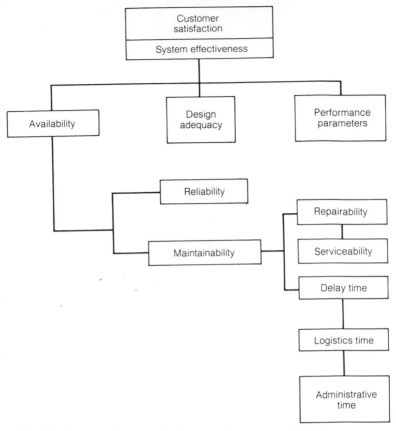

FIG. 18.1 Concept of system effectiveness and customer satisfaction.

18.2 SOURCES OF FAILURE DATA

This section briefly explains some of the sources of failure data which designers can use for reliability analysis. Relatively few sources for failure rate data are available. There are some data on electronic components, particularly in military applications. However, on mechanical components practically no good data are commercially available. The use of any existing data bank to obtain failure rate data on any particular design should be done with great care. The applicability of any past history on failure rate to a current design depends on the degree of similarities in the design, the environment, and the definition of failure. With these words of caution, three sources of failure rate information are covered in this section.

MIL-HDBK-217C. This handbook is concerned with reliability prediction for electronic systems.[3] It contains two methods of reliability prediction, the *part stress analysis* and the *parts count methods*. These methods vary in complexity and in the degree of information needed to apply them. The part stress analysis

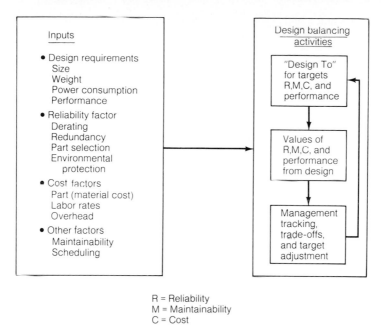

R = Reliability
M = Maintainability
C = Cost

FIG. 18.2   Design balancing activities.

method requires the greatest amount of information and is applicable during the later design stages when actual hardware and circuits are being designed. To apply this method, a detailed parts list including part stresses must be available.

The parts count method requires less information and is relatively easy to apply. The information needed to apply this method is (1) generic part type (resistor, capacitor, etc.) and quantity, (2) part quality levels, and (3) equipment environment. Tables are provided in the handbook to determine the factors in a failure rate model that can be used to predict the overall failure rate. The parts count method is obviously easier to apply and, one would assume, less accurate than the stress analysis method.

Nonelectronic Parts Reliability Data.   The *Nonelectronic Parts Reliability Data Handbook*[4] (1981) was prepared under the supervision of the Reliability Analysis Center at Griffiss AFB. This handbook is intended to complement MIL-HDBK-217C in that it contains information on some mechanical components. Specifically, the handbook provides failure rate and failure mode information for mechanical, electrical, pneumatic, hydraulic, and rotating parts. Again, little is known about the environment, design specifics, etc., that produced the failure rate data in this handbook.

GIDEP.   The Government-Industry Data Exchange Program (GIDEP)[5] (1976) is a cooperative venture between government and industry that provides a means to exchange certain types of technical data. Participants in GIDEP are provided with access to various data banks. The failure experience data bank contains failure information generated when significant problems are identified on systems in the field. These data are reported to a central data bank by the participants. The reliability-maintainability data bank contains failure rate and failure mode data on parts, components, and systems based on field operations. The data are reported by the participants. Here again, the problem is one of identifying the exact operating conditions that caused failure.

18.3 DESIGN REVIEW

The design review, a formal and documented review of a system design, is conducted by a committee of senior company personnel who are experienced in various pertinent aspects of product design, reliability, manufacturing, materials, stress analysis, human factors, safety, logistics, maintenance, etc. The design review extends over all phases of product development from conception to production. In each phase previous work is updated, and the review is based on current information.

A mature design rquires trade-offs between many conflicting factors such as performance, manufacturability, reliability, and maintainability. These trade-offs depend heavily on experienced judgment and require continuous communication between experienced reviewers. The design review committee approach has been found to be extremely beneficial to this process. The committee adopts the system's point of view and considers all conceivable phases of design and system use to ensure that the best trade-offs have been made for the particular situation (see Fig. 18.2). A complete design review procedure must be multiphased in order to follow the design cycle until the system is released for production. The product development cycle is presented in Table 18.1. A typical review committee is shown in Table 18.2 where the review phases are keyed to the product development cycle given in Table 18.1. Here the review process has been subdivided into six phases, and each phase is an update or more detailed analysis based on the latest knowledge.

Ultimately the design engineer has the responsibility of investigating and incorporating the ideas and suggestions posed by the design review committee. The design review committee chairperson is responsible for adequately reporting all suggestions by way of a formal and documented summary. The design engineer then can accept or reject various points in the summary; however, he or she must formally report back to the committee stating reasons for any actions.

The basic philosophy of a design review is being presented here, and it should be recognized that considerably more thought and detail must go into developing the management structure and procedures for conduct in order to have a successful review procedure. It should be noted that this review procedure not only considers reliability but also all important factors in order to ensure that a mature design will result from the design effort.

A design review technique has proved effective in identifying failure situations early in the design cycle and before product testing. With proper organization and appropriate management support, the design review procedure is an effective means for promoting early product maturity. Further discussion on design review can be found in Refs. 6, 7, and 8. A brief discussion of some of the components for design review are given below.

18.3.1 Policy

An endeavor should be made to conduct design reviews on all new products and on major revisions of existing products which seriously affect cost, interchange ability, function, performance, or appearance.

TABLE 18.1   Product Design Cycle

I. Marketing Research
   Needs analysis
   Forecast sales
   Set broad performance objectives
   Establish program cost objectives
   Establish technical feasibility
   Establish manufacturing capacity

II. Concepts
   Formulate project teams
   Refine project—broadly outline product
   Develop rough ideas of product
   Develop and consider alternatives

III. Design

   A. Preliminary design
   Design calculations
   Rough drawings and sketches
   Pursue different alternatives
   Obtain manufacturing engineering cost estimates
      of design approaches

   B. Detailed design
   Design calculations
   Stress analysis
   Vibration considerations
   Complete and detailed design
      package (drawings)

IV. Manufacturing Engineering

   A. Process planning

   B. Quality system planning

V. Prototype Program

   A. Build components and prototypes

   B. Write test plans
   1. Component/subsystem tests
   2. System tests

VI. Finalized Design

   A. Design changes due to manufacturing engineering input and
      test input

   B. Freeze design

   C. Release to manufacturing

18.3.2   Appointment of Chairperson

The task of the chairperson requires a high-level of tact, broad understanding of the design requirements, and technical knowledge of the various disciplines involved. Each review committee will have its designated chair, and the person selected should neither be the designer of the product nor be associated with the design.

TABLE 18.2   Design Review Committee

Member	I	II	III	IV	V	VI	Responsibility
Chairperson (manager of product design function)	X	X	X	X	X	X	Ensure that review is conducted in an efficient fashion. Issue major reports and monitor follow-up.
Design engineer (of this product)		X	X	X	X	X	Prepare and present design approaches with calculations and supporting data.
Design engineer (not of this product)		X	X	X	X	X	Review and verify adequacy of design (may require more than one specialist).
Customer and/or marketing representative	X	X	X	X	X	X	Ensure that customer's viewpoint is adequately presented (especially at the design trade-offs' stage)
Reliability engineer	X	X	X	X	X	X	Evaluate design for reliability consistent with system goals.
Manufacturing engineer			X	X	X	X	Ensure manufacturability at reasonable cost. Check for tooling adequacy and assembly problems.
Materials engineer			X				Ensure optimum material usage considering application and environment.
Stress analysis			X				Review and verify stress calculations.
Quality control engineer			X	X	X	X	Review tolerancing problems, manufacturing capability, inspection, and testing problems.
Human factors engineer			X	X			Ensure adequate consideration to human operator, identification of potential human-induced problems, and person-machine interface.

18.3.3   Participants

The various functional activities represented in a design review vary with the type of review. Generally, the participants will include personnel from engineering,

TABLE 18.2 Design Review Committee (*Continued*)

Member	| Review phase |						Responsibility
	I	II	III	IV	V	VI	
Safety engineer			X				Ensure safety to operating and auxiliary personnel.
Maintainability engineer		X	X				Analyze for ease of maintenance, repair, and handling of field service problems.
Test engineer			X		X	X	Present test procedures and results.
Logistics			X				Evaluate and specify logistic support. Identify logistics problems.

manufacturing, tooling, marketing, purchasing, reliability, quality control, and cost reduction. In addition, it may be desirable to invite specialists from other activities to participate. A design review committee should consist of about a dozen people.

18.3.4 Agenda

The chairperson will have the agenda prepared for the meeting and will distribute it with the advanced information.

18.3.5 Checklists

Checklists will be developed to remind participants of all items which should be considered in each review. Such a list also aids the designer by reminding him or her of the things that should not be overlooked. Some of the points to consider during design review are given below.

- Quality of products is defined by customers.
- Review adequately the needs and expectation of the customer.
- Review customer performance and environmental requirements.
- Confirmation on use of approved parts in an approved manner.
- For high reliability make conservative choices of materials and components.
- Simplify design wherever possible.
- Wherever possible avoid new state-of-the-art use of materials and manufacturing processes.
- Wherever possible use previously proven designs and components.
- Look for misapplication of components and materials.
- Check for the influence of environmental extremes on the system.

- Make provision for vibration and shock.
- Make provision for heat transfer.
- Search for potential sources of fluid leaks.
- Check all subsystem interfaces for failure problems.
- Minimize the requirement for assembly adjustments and selective fit requirements in production.
- Provisions for testability and inspectability where required.
- Determine serviceability index, particularly for components with highest failure rates. Good maintainability may compensate for poor reliability.
- Analyze potential failure modes and their effect.
- Perform a worst-case analysis.
- Make sure that review checklists are based on product experience in design, production, and field use.

18.3.6   Conduct of Design Review Meetings

Introductory comments by the chairperson should set a constructive tone and climate for the meeting. The specific objectives of the design review should be stated and should relate to the overall objectives, namely, achieving optimum product design from the standpoint of reliability, function, cost, appearance, and other requirements of the customer.

An attendee will be appointed secretary and will take notes on useful ideas submitted and other pertinent comments. The person will also record when additional action is required and by whom. It may be a good idea to tape the entire design review committee meetings to make sure that the secretary has correct information for inclusion in the record. It may be useful to save the tapes and use them to resolve arguments in the future.

The design engineer or product manager should describe adequately the product being reviewed and include a comparison of customer requirements versus the expected performance of the product.

The chairperson should make sure that the discussion follows a systematic plan so that no major subject areas are omitted. The discussion should follow the prepared agenda. Checklists should be used to prevent omission of important design considerations.

18.3.7   Timing and Duration of Reviews

It is usually desirable to have design reviews at more than one point in the design and development cycle of a new product; e.g., a design review may be held when

1. Marketing or customer requirements are completed
2. Specifications and drawings are completed
3. After testing the prototypes and prior to manufacturing release
4. After production assessment tests

In addition, tool drawings should be similarly reviewed prior to their manufacture.

18.3.8 Scheduling of Reviews

The reliability manager and engineering manager will schedule all design reviews with the appropriate attendees and advise them of time, place, and subject. Specific time for the design reviews will be incorporated into the product design and development schedule and approved so that all participating functions can plan their efforts efficiently.

18.3.9 Advance Information

At least ten days before the actual date of the review, information should be distributed or made available to the designated participants. This may be in the form of specification, a competitive cost analysis, preliminary layouts, etc. Distributing this information beforehand will help to assure that the participants are well prepared to contribute constructively to the objectives of the review.

18.3.10 Follow-up

Documentation: The secretary is responsible for preparing and distributing minutes of the meeting, indicating the ideas generated and the action to be taken. The minutes should also note by whom the action is to be taken and when.

Utilization of ideas: The designer is responsible for investigating and incorporating into the design those ideas which will aid in achieving optimum product design. The design review chairperson will have the responsibility of following up on the utilization of the ideas proposed and of the assigned action items.

Final report: A final report should be issued covering the investigation of suggestions made and the reasons for their adoption or rejection.

18.3.11 Reliability Design Guidelines

There are some basic principles of reliability in design that are useful for the designer.[9] Each concept is briefly discussed in terms of its role in the design of reliable systems.

Simplicity. Simplification of system configuration contributes to reliability improvement by reducing the number of failure modes. A common approach is called *component integration,* which is the use of a single part to perform multiple functions.

Use of Proven Components and Preferred Designs
1. If working within time and cost constraints, use proven components because this minimizes analysis and testing to verify reliability.
2. Mechanical and fluid system design concepts can be categorized and proven configurations given first preference.

Stress and Strength Design.   Use various sources of data on strength of materials and strength degradation with time as it relates to fatigue. The traditional and common use of safety factors does not address reliability, and new techniques like the probabilistic design approach should be used. The probabilistic design approach is explained in a following section.

Redundancy.   Redundancy sometimes may be the only cost-effective way to design a reliable system from less reliable components.

Local Environmental Control.   Severe local environment sometimes prevents achievement of required component reliability. The environment should be modified to achieve high reliability. Some typical environmental problems are as follows:

1. Shock and vibration
2. Heat
3. Corrosion

Identification and Elimination of Critical Failure Modes.   This is accomplished through failure mode, effect, and criticality analysis (FMECA) and also by fault-tree analysis. These are discussed later in this chapter.

Self-Healing.   A design approach which has possibilities for future development is the use of self-healing devices. Automatic sensing and switching devices represent a form of self-healing.

Detection of Impending Failures.   Achieved reliability in the field can be improved by the introduction of methods and/or devices for detecting impending failures. Some of the examples are

1. Screening of parts and components
2. Periodic maintenance schedules
3. Monitoring of operations

Preventive Maintenance.   Preventive maintenance procedures can enhance the achieved reliability, but the procedures are sometimes difficult to implement. Hence, effective preventive maintenance procedures must be considered at the design stage.

Tolerance Evaluation.   In a complex system, it is necessary to consider the expected range of manufacturing process tolerances, operational environment, and all stresses, as well as the effect of time. Two tolerance evaluation methods, worst-case tolerance analysis and statistical tolerance analysis, are discussed later in this chapter.

Human Engineering.   Human activities and limitations can be very important to system reliability. The design engineer must consider factors which directly refer to human aspects, such as

1. Human factors
2. Person-machine interface
3. Evaluation of the person in the system
4. Human reliability

18.4 FAILURE MODE AND EFFECT ANALYSIS

Failure mode and effect analysis (FMEA) is a design evaluation procedure used to identify all conceivable and potential failure modes and to determine the effect of each on system performance (see Fig. 18.3). This procedure is accomplished by formal documentation which serves (1) to standardize the procedure, (2) as a means of historical documentation, and (3) as a basis for future improvement. The FMEA procedure consists of a sequence of logical steps starting with the analysis of lower-level subsystems or components. The analysis assumes a failure point of view and identifies all potential modes of failure along with the causative agent, which is termed the *failure mechanism*. The effect of each failure mode is then traced up to the system level. A criticality rating is developed for each failure mode and resulting effect. The rating is based on the probability of occurrence, severity, and detectability. For failures scoring high on this rating, design changes to reduce criticality are recommended. This procedure is aimed at providing a better design from a reliability standpoint. Correct usage of the FMEA process will result in two improvements, (1) an improvement in the reliability of the product through the anticipation of problems and institution of corrections prior to

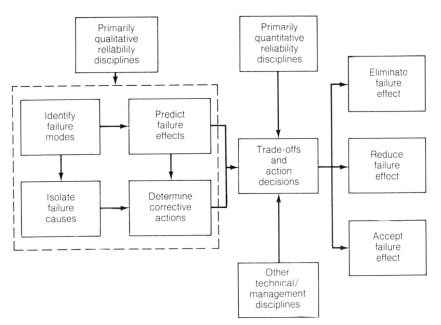

FIG. 18.3 FMEA methodology.

going into production, and (2) an improvement in the validity of the analytical method itself through the strict documentation that allows the rationale for every step in the decision process to be seen.

18.4.1 Steps for FMEA Study[10,11]

A. System Description and Block Diagrams. The first step of the FMEA is the description of the item to be analyzed and the definition of its function. These are used to develop the functional block diagram and to generate a "function and output" list. The description, functional block diagram, listing of functions, and output of each element in the block diagram are the basic data required to initiate the FMEA. Functional and reliability block diagrams which illustrate the operation, interrelationships, and interdependencies of functional entities should be developed and all system interfaces should be indicated. Guidelines should be developed to pick the indenture level at which failures are postulated. FMEA shall be initiated at the initial indenture level selected and shall proceed through decreasing indenture levels. The lower the indenture level, the greater the detail level of analysis required.

Guidelines. Use the following guidelines to select the indenture level.[10]

1. The initial analysis shall be performed on the initial indenture level as soon as the design concept is defined and details of output requirements are specified. As the design progresses, refinements to the initial analysis shall be made at progressively greater detail levels in an interactive process.

2. The lowest indenture level at which the analysis is effective is the level for which information is available to establish definition and description of functions.

3. The results of the initial indenture level analysis shall be used to determine the requirements for the next and subsequent analysis levels since the analysis will identify both the critical functional elements for which an assembly or part-oriented analysis is required and the noncritical functional elements that do not require further analysis.

4. The level of detail is influenced by previous experience. A lesser level of detail can be justified for a subsystem having a good reliability record. Conversely, a greater level of detail is indicated for a subsystem having a questionable reliability history or for an unproven design. Regardless of the reliability history, an item having a known high-severity failure mode should be investigated.

5. A deeper level of detail is necessary for an item performing a critical function, whereas a lesser level of detail may be indicated for an ancillary item which does not influence any significant part of the function.

B. Failure Modes. The next step is to identify failure modes. A failure mode is the manner in which a component or system fails and thus does not meet design intent and customer requirements. By examining the outputs of all the items identified in the applicable block diagrams, all realistically probable failure modes of the items shall be identified, described, and evaluated. Failure modes of the individual output functions of each block diagram element are postulated from the stated requirements contained in the equipment specifications as tabulated in the function and output list. Therefore, a "realistically probable" failure mode is one which causes a deviation from specified required output function requirements.

C. Failure Cause(s).   The next step is to develop a concise description of causes for each potential failure mode. The causes of each failure mode are identified in order to estimate the probability of occurrence, to formulate recommended corrective action, and to uncover secondary effects. A failure mode can have more than one cause, and all possible independent causes for each failure mode shall be identified and described. The failure causes within the next lower indenture level shall be considered. For example, failure causes at the third indenture level shall be considered when conducting a second indenture level analysis.

D. Effect(s) of Failure.   The consequences of each potential failure mode on item operation, function, or status shall be identified, evaluated, and recorded. The failure effect is the consequence of failure and focuses on the specific element in the block diagram being analyzed which is affected by the failure under consideration. The failure effect also impacts on the next higher indenture level and ultimately can affect the initial indenture level under analysis. Therefore, both a "local" effect and an "end" effect shall be evaluated.

Local effects concentrate specifically on the impact of the postulated failure mode on the operation and function of the item under consideration. The consequences of each postulated failure on the output of the item are described along with the second-order effects. The purpose of defining the local effects is to provide a base for judgment when evaluating existing compensating provisions or formulating recommended corrective actions. It should be noted that in certain instances there may not be a local effect beyond the description of the failure mode itself.

End-effect analysis evaluates and defines the effect of the postulated failure on the operation, function, and status of the next higher indenture level and eventually on the initial indenture level. The end effect described may be the result of a double failure. For instance, the failure of a safety device to function results in a catastrophic end effect only in the event that both the safety device fails and the prime function goes beyond the limit for which the safety device is set. These end effects resulting from a double failure should be so indicated in the FMEA.

E. Symptoms and Detection.   Symptoms and detection features shall be identified and described. Symptoms indicate the recognized item behavior pattern(s) which signify that a malfunction has occurred. Symptoms may be confined to the operation of the specific item under consideration (local) or may result in both local and overall system or equipment evidence of malfunction. Detection is made possible by the features incorporated in the decision through which occurrence of a failure mode is recognized. Where reliability and maintainability characteristics and criticality dictate, special provisions may have to be included for detection of specific malfunctions. Performance-monitoring devices can be included at various levels related to the criticality of the function and associated reliability and maintainability.

Descriptions of indications which are evident to an operator that a system has malfunctioned or failed, other than the identified warning devices, shall be recorded. Proper correlation of a system malfunction or failure may require identification of normal as well as abnormal indications. If no indication exists, determine if the undetected failure will jeopardize the mission objectives or personnel safety. If the undetected failure allows the system to remain in a safe state, a second failure situation should be explored to determine whether or not an indication will be evident to an operator. Indications to the operator should be described as follows:

1. *Normal:*   An indication that is evident to an operator when the system or equipment is operating normally
2. *Abnormal:*   An indication that is evident to an operator when the system has malfunctioned or failed
3. *Incorrect:*   An erroneous indication to an operator due to the malfunction or failure of an indicator (i.e., instruments, sensing devices, visual or audible warning devices, etc.)

Describe the most direct procedure that allows an operator to isolate the malfunction or failure. An operator will know only the initial symptoms until further specific action is taken such as performing a more detailed built-in test (BIT). The failure being considered in the analysis may be of lesser importance or likelihood than another failure that could produce the same symptoms, and this must be considered. Fault isolation procedures require a specific action or series of actions by an operator followed by a check or cross reference either to instruments, control devices, circuit breakers, or combinations thereof. This procedure is followed until a satisfactory course of action is determined.

F. Severity Classification.   Severity classifications are assigned to provide a qualitative measure of the worst potential consequences resulting from design error or the failure. A severity classification shall be assigned to each identified failure mode. Classification of level of severity may be established as follows:[10]

Category I. Catastrophic: A failure that may cause death or mission loss

Category II. Critical: A failure that may cause severe injury or major system damage

Category III. Marginal: A failure that may cause minor injury or degradation in mission performance

Category IV. Minor: A failure that does not cause injury or system damage but may result in system failure and unscheduled maintenance

Table 18.3 gives another example of developing a severity index ranking based on a scale of 1 to 10, where 10 indicates the highest ranking or catastrophic failure.

G.. Probability of Failure Occurrence.   The probability of occurrence of each postulated, identified, and potential failure mode shall be assessed in quantitative terms by means of an analytically derived estimate. This probability shall provide a measure of the expected number of occurrences of each identified failure mode during a specific time interval. This interval can be based on expected time between overhaul, item mission time, or any other interval deemed appropriate. The probability of occurrence of each postulated failure mode is the combined probability of occurrence of all the identified causes for that failure mode. The probability of occurrence of each failure mode is then used to establish the appropriate failure probability level.

TABLE 18.3   Severity Index Ranking

Ranking	Criteria
1	*Unreasonable* to expect that the minor nature of this failure will degrade the performance of the system.
2–3	*Minor* nature of failure will cause slight annoyance to the customer.
4–6	*Moderate* failure will cause customer dissatisfaction.
7–8	*High* degree of customer dissatisfaction and inoperation of the system.
9–10	*Very high* severity ranking in terms of safety-related failures and non-conformance to regulations and standards.

Failure Probability Levels.   The individual failure mode probabilities of occurrence shall be grouped into distinct, logically defined levels. The individual levels shall be described and defined in qualitative or quantitative terms.

Probability of occurrence levels may be defined as follows:

Level A—Frequent

Level B—Reasonably probable

Level C—Occasional

Level D—Remote

Level E—Extremely unlikely

Failure rate data sources can be used for the quantitative assessment of failure rates. A ranking system can be developed based on the above guidelines using both qualitative as well as quantitative information.

H. Criticality Analysis.   Criticality is the combination of probability of occurrence and the level of severity. The true impact of a failure mode on the reliability of an item is best defined by the combination of these two characteristics. For instance, a specific failure mode with a high probability of occurrence may have a negligible effect on required item function. This means that the level of severity is negligible and its criticality, therefore, is relatively low. Another failure mode with low probability of occurrence may result in complete loss of item function. This failure mode is identified with a high level of severity but a relatively low criticality. An approach for criticality analysis is given below based on Ref. 10.

Failure Effect Probability β.   The β values are the conditional probability that the failure effect will result in the identified criticality classification, given that the failure mode occurs. The β values represent the analyst's judgment of conditional probability that the loss will occur and may be quantified as follows.

Failure effect	β value
Actual loss	1.00
Probable loss	$0.10 < \beta < 1.0$
Possible loss	$0 < \beta \le 0.10$
No effect	0

Failure Mode Ratio α.   The fraction of the part failure rate λ_p related to the particular failure mode under consideration shall be evaluated by the analyst and recorded. The failure mode ratio is the probability, expressed as a decimal fraction, that the part or item will fail in the identified mode. If all potential failure modes of a particular part or item are listed, the sum of the values for that part or item will equal one. Individual failure mode multipliers can be derived from failure rate source data or from test and operational data. If failure mode data are not available, the values shall represent the analyst's judgment based upon an analysis of the item's functions.

Part Failure Rate λ_p.   The part failure rate λ_p from the appropriate reliability prediction or as calculated using the procedure described in MIL-HDBK-217C[3] are developed. Where appropriate, application factors π_A, environmental factors π_E, and other π factors as may be required shall be applied to the base failure rates λ_b obtained from handbooks or other reference material to adjust for differences in operating stresses. Values of π factors utilized in computing λ_p shall be listed.

Operating Time t.   The operating time in hours or the number of operating cycles of the item shall be derived from the system definition.

Failure Mode Criticality Number C_m.   The value of the failure mode criticality number C_m shall be calculated. C_m is the portion of the criticality number for the item due to one of its failure modes under a particular severity classification. For a particular severity classification and operational phase, the C_m for a failure mode can be calculated with the following equation:

$$C_m = \beta \, \alpha \, \lambda_p t \tag{18.1}$$

where C_m = criticality number for failure mode
β = conditional probability
α = failure mode ratio
λ_p = part failure rate
t = duration of time, usually expressed in hours or number of operating cycles

Item Criticality Number C_r.   The second criticality number calculation is for the item under analysis. A criticality number for an item is the number of system failures of a specific type expected due to the item's failure modes. The specific type of system failure is expressed by the severity classification for the item's failure modes. For a particular severity classification, the C_r for an item is the sum of the failure mode criticality numbers C_m under the severity classification.

Based on the criticality analysis, a criticality matrix can be developed which provides a means of identifying and comparing each failure mode to all other failure modes with respect to severity.

I. **Corrective Action and Follow-up.**   A well-developed FMEA will be of limited value unless positive corrective actions, actions recommended by other activities, and all recommended actions arising from the FMEA are taken and followed up on. Based on the FMEA analysis, design actions can be used for the following:

- Eliminate the "cause of failure."
- Reduce the probability of "occurrence" by reducing the probability that the cause of failure will result in the failure mode.

- Reduce the severity of failure by redesign to produce a fail-safe, system redundancy, etc.
- Increase the probabilty of detection.

18.5 FAULT-TREE ANALYSIS

Fault-tree analysis is a method of system reliability and safety analysis. It provides an objective basis for analyzing system design, justifying system changes, performing trade-off studies, analyzing common failure modes, and demonstrating compliance with safety requirements. The concept of fault-tree analysis was originated in 1961 by H. A. Watson of Bell Telephone Laboratories to evaluate the safety of the Minuteman missile launch control system.[12,13] Many reliability techniques are inductive and are concerned primarily with ensuring that hardware will accomplish its intended functions. Fault-tree analysis is a detailed deductive analysis that usually requires considerable information about the system. It is concerned with ensuring that all critical aspects of a system are identified and controlled. It is a graphical representation of Boolean logic associated with the development of a particular system failure (consequence), called the *top event,* to basic failures (causes), called *primary events.* These top events can be broad, all-encompassing events, such as "release of radioactivity from a nuclear power plant" or "inadvertent launch of an ICBM missile," or they can be specific events, such as "failure to insert control rods" or "energizing power available to ordnance ignition line."

Fault-tree analysis is of value in

1. Providing options for qualitative and quantitative reliability analysis
2. Helping the analyst to understand system failures deductively
3. Pointing out the aspects of a system that are important with respect to the failure of interest
4. Providing the analyst with insight into system behavior

A fault tree is a model that graphically and logically represents the various combinations of possible events, both fault and normal, occurring in a system that leads to the top event. A fault event is an abnormal system state. A normal event is an event that is expected to occur. The term *event* denotes a dynamic change of state that occurs in a system element. System elements include hardware, software, human, and environmental factors.

Steps for fault-tree analysis are given in Fig. 18.4.[12] The structure of a tree is shown in Fig. 18.5. The undesired event appears as the top event and is likened to more basic fault events by event statements and logic gates. Fault-tree analysis is different from FMEA because it is restricted only to the identification of the system elements and events that lead to one particular undesired event.

18.5.1 Fault-Tree Building Blocks[12]

Fault trees are developed using two building blocks, gate symbols and event symbols.

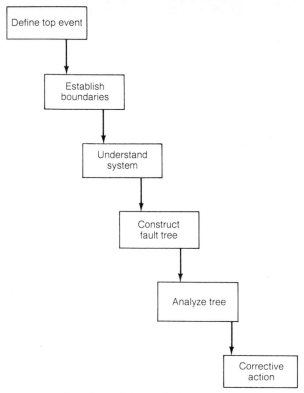

FIG. 18.4   Steps for fault-tree analysis.

Gate Symbols.   Gate symbols connect events according to their causal relations. A gate may have one or more input events but only one output event. Table 18.4 shows some of the gate symbols. The output events of AND gates occur if all input events occur simultaneously. The output events of OR gates happen if any one of the input events occurs. The causal relation expressed by AND gate or OR gate is deterministic. The inhibit gate represented by a hexagon can be used to represent a probabilistic causal relation. The event on the side of the gate is called a conditional event and the output event on the top happens if the input event at the bottom and the conditional event occur simultaneously. The priority AND gate requires the input events to occur in a specific order. The exclusive OR gate can be represented by a combination of AND gate and an OR gate. An m-out-of-n voting gate requires that at least m out of the n input events ($m \leq n$) occur.

Event Symbol.   Some of the event symbols are given in Table 18.5. The rectangle defines an event that is the output of a logic gate and is dependent on the type of the logic gate and the inputs to the logic gate. The circle defines a basic inherent failure of a system when operated under specified conditions. It is therefore a primary failure and is also referred to as a generic failure. The diamond represents a failure other than a primary failure that is purposely not developed further. The

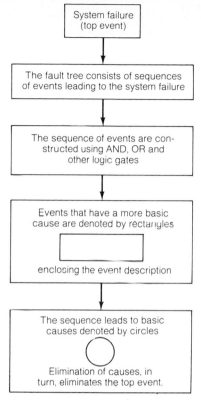

FIG. 18.5 Structuring a fault tree.

switch or house event represents an event that is expected to occur or to never occur because of design and normal conditions, such as phase change in a system. Thus, the house event is used to examine special cases of a fault tree by forcing some events to occur and other events not to occur. When we turn on the house event, the fault tree presumes the occurrence of the event, and vice versa when we turn it off. Table 18.5 represents a transfer-out triangle (has a line to its side from a gate) and transfer-in triangle (has a line from its apex to another gate). The transfer triangles are used to simplify the representation of fault trees.

Some Guidelines for the Development of Fault Trees. The fault tree is so structured that the sequence of events that leads to the undesired event is shown below the top event and is logically related to the undesired event by OR and AND gates. The input events to each logic gate that are also output of other logic gates at a lower level are shown as rectangles. These events are developed further until the sequence of events leads to basic causes of interest, called *basic events*. The basic events appear as circles. Diamonds at the bottom of the fault tree represent the limit of resolution of the fault tree.

The structuring process for the construction of a fault tree identifies three failure mechanisms or causes that can contribute to a fault state.[14]

TABLE 18.4 Gate Symbols

Gate symbol	Gate name	Casual relations
	AND gate	Output event occurs if all the input events occur simultaneously
	OR gate	Output event occurs if any one of the input events occurs
	Inhibit gate	Input produces output when conditional event occurs
	Priority AND gate	Output event occurs if all input events occur in the order from left to right
	Exclusive OR gate	Output event occurs if one, but not both, of the input events occur
m-out-of-n gate (voting or sample gate)		Output event occurs if m out of n input events occur

TABLE 18.5   Event Symbols

Event symbol	Meaning
Rectangle	Event represented by a gate
Circle	Basic event with sufficient data
Diamond	Undeveloped event
Switch or House	Either occurring or not occurring
Oval	Conditional event used with inhibit gate
Triangles	Transfer symbol

1. A primary failure is due to the internal characteristics of the system under consideration.

2. A secondary failure is due to the excessive environmental or operational stresses.

3. A command fault is inadvertent operation or nonoperation of a system element due to failure(s) of elements that can control or limit the flow of energy [called initiating element(s)] to respond as initiated to system conditions.

Example for Fault-Tree Construction[15]. This example explains some of the fundamental aspects of fault-tree construction. Figure 18.6 shows a sample system for the operation of an electric motor. The system boundary conditions are:

$$\begin{aligned}
\text{Top event} &= \text{motor overheats} \\
\text{Initial condition} &= \text{switch closed} \\
\text{Not-allowed events} &= \text{failures due to effects external to system} \\
\text{Existing events} &= \text{switch closed} \\
\text{Treetop} &= \text{shown in Fig. 18.7}
\end{aligned}$$

Figure 18.7 reflects the inductive reasoning that the motor overheats if an electrical overload is supplied to the motor or a primary failure within the motor causes the overheating; for example, bearings lose their lubrication or a wiring failure occurs within the motor.

From a knowledge of the components, the fault tree shown in Fig. 18.8 is constructed. The event "excessive current to motor" occurs if excessive current is present in the circuit and the fuse fails to open. The event "excessive current in circuit" occurs if the wire fails, shorts, or the power supply surges. The fault tree is now complete to the level of primary failures.

For the same sample system but with different system boundary conditions, a second example illustrates the treatment of secondary failures, i.e., failures possibly caused by failure feedback between components. For this example, the system boundary conditions are

$$\begin{aligned}
\text{Top event} &= \text{motor does not operate} \\
\text{Initial condition} &= \text{switch closed} \\
\text{Not-allowed events} &= \text{failures due to effects external to} \\
&\quad \text{system (operator failures not included)} \\
\text{Existing events} &= \text{none} \\
\text{Treetop} &= \text{shown in Fig. 18.9}
\end{aligned}$$

The completed fault tree is shown in Fig. 18.10. Here the diamond symbol is used to indicate that the event "switch open" is not developed to its causes. An open switch is a failure external to the system bounds and, in this analysis, insufficient information is available for developing the event.

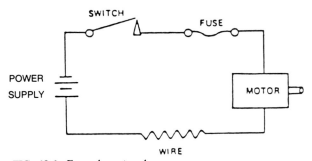

FIG. 18.6 Example system 1.

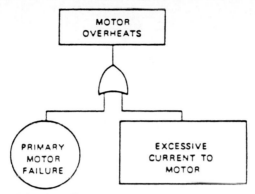

FIG. 18.7 First treetop system boundary for example system 1.

The event "fuse fails open" occurs if a primary or second fuse failure occurs. Secondary fuse failure can occur if an overload in the circuit occurs, because an overload can cause the fuse to open. The fuse does not open, however, every time an overload is present in the circuit because all conditions of an overload do not

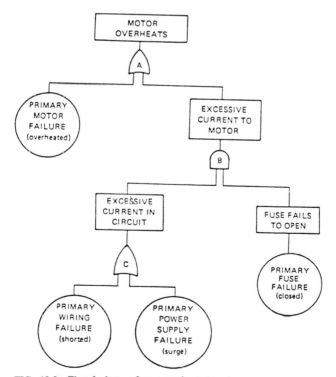

FIG. 18.8 First fault tree for example system 1.

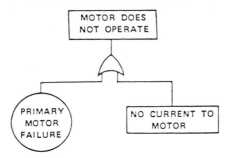

FIG. 18.9 Second treetop system boundary for example system 1.

result in sufficient overcurrent to open the fuse. The inhibit condition then is used as a weighting factor applied to all the fault events in the domain of the inhibit condition. Because the inhibit condition is treated as an AND logic gate in a probabilistic analysis, it is a probabilistic weighting factor. The inhibit condition has many variations in fault-tree analysis, but in all cases it represents a probabilistic weighting factor.

Even though the generation and analysis of fault trees nominally are separate tasks, there is a great deal of interaction between the two. During the course of analysis, engineers become aware of things they had forgotten or not realized while the tree was being generated.

Let us look at another example shown in Fig. 18.11. The purpose of the system is to provide light from a bulb. When the switch is closed, the relay contacts close and the contacts of the circuit breaker, a normally closed relay, open. If the relay contacts open, the light goes out and the operator immediately opens the switch, which in turn causes the circuit breaker contacts to close and restore the light. The system boundary conditions include:

$$
\begin{aligned}
\text{Top event} &= \text{no light}\\
\text{Initial condition} &= \text{switch closed, relay contacts closed, or circuit breaker}\\
&\ \text{contacts open}\\
\text{Not-allowed events} &= \text{Operator failures, wiring failures, secondary failures}
\end{aligned}
$$

Operator failures, wiring failures, and secondary failures are neglected to simplify the fault tree (see Fig. 18.11). The fault tree for this example is shown in Fig. 18.12.

18.6   PROBABILISTIC APPROACH TO DESIGN

Reliability is basically a design parameter and must be incorporated in the system at the design stage. One way to quantify reliability during design and to design for reliability is the probabilistic approach to design.[15-18] The design variables and parameters are random variables and, hence, the design methodology must consider them as random variables. The reliability of any system is a function of the reliabilities of its components. In order to analyze the reliability of the system, we have to first understand how to compute the reliabilities of the components. The basic idea in reliability analysis from the probabilistic design methodology

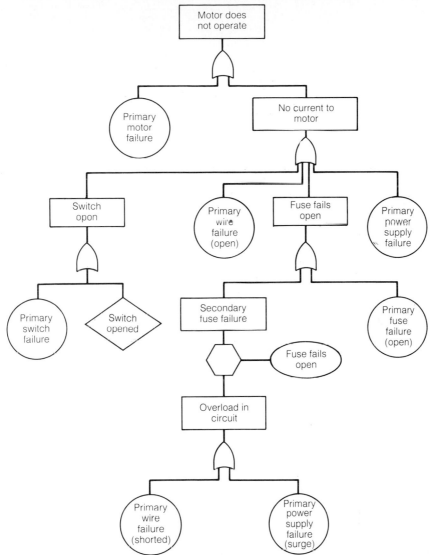

FIG. 18.10 Second fault tree for example system 1.

viewpoint is that a given component has certain strength which, if exceeded, will result in the failure of the component. The factors which determine the strength of the component are random variables as are the factors which determine the stresses or loads acting on the component. *Stress* is used to indicate any agency that tends to induce failure, while *strength* indicates any agency resisting failure. *Failure* itself is taken to mean failure to function as intended. It is defined to have occurred when the actual stress exceeds the actual strength for the first time.

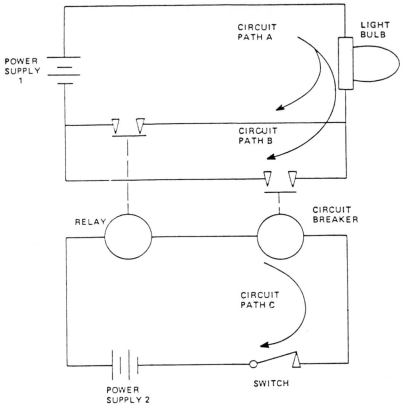

FIG. 18.11 Example system 2.

Let $f(x)$ and $g(y)$ be the probability density functions for the stress random variable X and the strength random variable Y, respectively, for a certain mode of failure. Also, let $F(x)$ and $G(y)$ be the cumulative distribution functions for the random variables X and Y, respectively. Then the reliability R of the component for the failure mode under consideration with the assumption that the stress and the strength are independent random variables is given by (see Fig. 18.13)[17–19]

$$R = P\{Y > X\}$$

$$= \int_{-\infty}^{\infty} g(y)\left\{\int_{-\infty}^{y} f(x)\,dx\right\} dy$$

$$= \int_{-\infty}^{\infty} g(y)F(y)\,dy$$

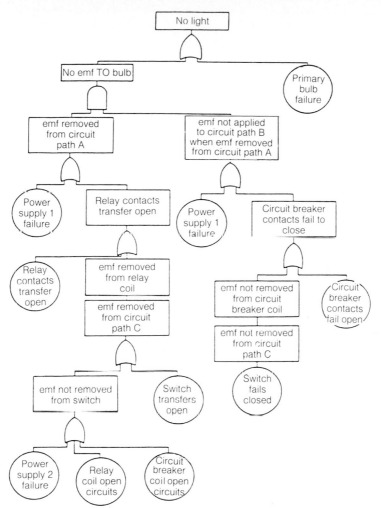

FIG. 18.12   Fault tree to example system 2.

$$= \int_{-\infty}^{\infty} f(x) \left\{ \int_{x}^{\infty} g(y) \, dy \right\} dx$$

$$= \int_{-\infty}^{\infty} f(x) \{1 - G(x)\} \, dx \qquad (18.2)$$

For example, suppose the stress random variable X is normally distributed with a mean value of μ_X and standard deviation of σ_X and the strength random variable is also normally distributed with parameters μ_Y and σ_Y, then the reliability R is given by

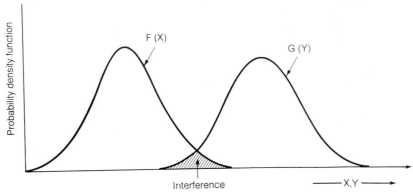

FIG. 18.13   Stress-strength interference.

$$R = \Phi\left[\frac{\mu_Y - \mu_X}{\sqrt{\sigma_Y^2 + \sigma_X^2}}\right] \tag{18.3}$$

where $\Phi[.]$ is the cumulative distribution function for the standard normal variable.

Example

$$\begin{array}{ll} \mu_Y = 40{,}000 & \sigma_Y = 4{,}000 \\ \mu_X = 30{,}000 & \sigma_X = 3{,}000 \end{array}$$

Then, factor of safety = 40,000/30,000 = 1.33, and

$$R = \Phi\left[\frac{40{,}000 - 30{,}000}{\sqrt{(4{,}000)^2 + (3{,}000)^2}}\right] = \Phi[2]$$

$$= 0.97725$$

If we change μ_X to 20,000, increasing the factor of safety to 2, we have,

$$R = \Phi\left[\frac{40{,}000 - 20{,}000}{\sqrt{(4{,}000)^2 + (3{,}000)^2}}\right] = \Phi[4]$$

$$= 0.99997$$

We now develop the relationship between reliability, factor of safety and variability of the stress, and the strength random variables. Let

V_Y = coefficient of variation for the strength random variable Y

$$= \frac{\sigma_Y}{\mu_Y}$$

V_X = coefficient of variation for the stress random variable X

$$= \frac{\sigma_X}{\mu_X}$$

n = factor of safety

$$= \frac{\mu_Y}{\mu_X}$$

Substituting these values in Eq. (18.3) for reliability R, when X and Y are normally distributed, we have

$$R = \Phi\left[\frac{n-1}{\sqrt{V_Y^2 n^2 + V_X^2}}\right] \tag{18.4}$$

Thus, the above relation can be used to relate reliability, factor of safety, coefficient of variation for stress, and strength random variable. For example, let $n = 2.0$, $V_Y = 0.25$, $V_X = 0.15$. Then

$$R = \Phi\left[\frac{2-1}{\sqrt{(0.25)^2(2)^2 + (0.15)^2}}\right]$$

$$= \Phi[1.91] = 0.972$$

There are four basic ways in which the designer can increase reliability:

1. *Increase mean strength:*   Can be achieved by increasing size, weight, using stronger material, etc.
2. *Decrease average stress:*   Controlling loads, using higher dimensions.
3. *Decrease stress variations:*   This variation is harder to control, but can be effectively truncated by putting limitations on use conditions.
4. *Decrease strength variation:*   The inherent part-to-part variation can be reduced by improving the basic process, controlling the process, and utilizing tests to eliminate the less desirable parts.

The probabilistic design methodology is illustrated in Fig. 18.14.

If we consider the total *design reliability program,* the steps can be summarized below:

1. Define the design problem.
2. Identify the design variables and parameters involved.

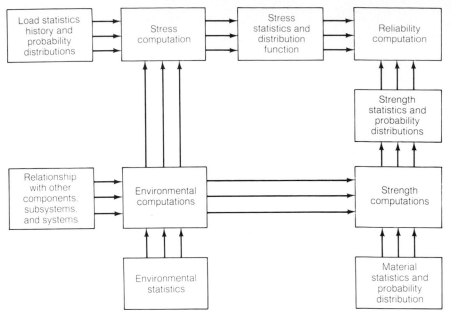

FIG. 18.14   Probabilistic design methodology.

3. Conduct a failure modes, effects, and criticality analysis.
4. Verify the significant design parameter selection.
5. Formulate the relationship between the critical parameters and the failure-governing criteria involved.
6. Determine the failure-governing stress function.
7. Determine the failure-governing stress distribution.
8. Determine the failure-governing strength function.
9. Determine the failure-governing strength distribution.
10. Calculate the reliability associated with these failure-governing distributions for each critical failure mode.
11. Iterate the design to obtain the design reliability goal.
12. Optimize the design in terms of performance, cost, weight, etc.
13. Repeat optimization for each critical component.
14. Calculate system reliability.
15. Iterate to optimize system reliability.

18.7 STATISTICAL VARIATION AND TOLERANCE ANALYSIS

The reliability of an engineering design is a function of several design parameters and random variables. The design performance can be expressed as a function of these design variables and parameters. In this section we discuss how to combine several random variables. For example, a voltage distribution can be derived from the resistance and the current probability distributions. In a mechanical system, the stress distribution can be derived from the force and cross-sectional-area distributions. We have to develop mathematical variability models for the behavior of the system. The model must be accurate enough to simulate the behavior of the system over the range of operation. We discuss some of the techniques for variation analysis.

Let us assume that the performance of the system Y is a function of n random variables and/or parameters $X_1, X_2, \ldots, X_n$, i.e.,

$$Y = f(X_1, X_2, \ldots, X_n) \tag{18.5}$$

We wish to determine the properties of the random variable Y. If we know the probability density functions of the random variables $X_1, \ldots, X_n$, we may be able to find the probability density function of random variable Y, but this may be difficult. In many design situations, only the first few moments of the random variables $X_1, X_2, \ldots, X_n$ are known, and it is necessary to find the corresponding moments of the random variable Y. Computation of the moments of Y using the Taylor series approximation is discussed below.

18.7.1   Taylor Series Approximation to Find Moments

Let $Y = f(X_1, X_2, \ldots, X_n)$ represent a general equation where a design variable Y is a function of other design variables $X_1, X_2, \ldots, X_n$. Given

$$E[X_i] = \mu_i \qquad i = 1, 2, \ldots, n$$
$$V[X_i] = \sigma_i \qquad i = 1, 2, \ldots, n$$

We have the approximate values for μ_Y and σ_Y using Taylor series approximations as follows:

$$\mu_y \cong f(\mu_1, \mu_2, \ldots, \mu_n) + \frac{1}{2} \sum_{i=1}^{n} \left. \frac{\partial^2 f(X)}{\partial X_i^2} \right|_{x=\mu} V(X_i) \tag{18.6}$$

$$V[Y] \cong \sum_{i=1}^{n} \left\{ \left. \frac{\partial f(X)}{\partial X_i} \right|_{x=\mu} \right\}^2 V(X_i) \tag{18.7}$$

where $X = (X_1, \ldots, X_n)$
$\mu = (\mu_1, \ldots, \mu_n)$

Example.   Let us consider two resistances in parallel as shown in Fig. 18.15.[18] The mean and standard deviation for each resistance R_1 and R_2 are given, and we wish to approximately find the mean and standard deviation of R_T, the terminal resistance.

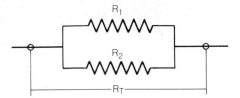

FIG. 18.15 Two resistances in parallel.

We have

$$R_T = f(R_1, R_2) = \frac{R_1 R_2}{R_1 + R_2}$$

$$\mu_{R_1} = 100\ \Omega \qquad \sigma_{R_1} = 10\ \Omega$$

$$\mu_{R_2} = 200\ \Omega \qquad \sigma_{R_2} = 15\ \Omega$$

Then

$$E\ [R_T] \cong f(100, 200) = \frac{100 \times 200}{100 + 200} = 66.7\ \Omega$$

$$\frac{\partial f}{\partial R_1} = \frac{R_2^2}{(R_1 + R_2)^2} \qquad \frac{\partial f}{\partial R_2} = \frac{R_1^2}{(R_1 + R_2)^2}$$

$$\left.\frac{\partial f}{\partial R_1}\right|_{R = \mu} = 0.444 \qquad \left.\frac{\partial f}{\partial R_2}\right|_{R = \mu} = 0.111$$

Hence, using the Taylor series approximation, Eq. (18.7), we have

$$\sigma^2{}_{RT} \cong (0.444)^2(10)^2 + (0.111)^2(15)^2$$
$$= 22.4858$$

or

$$\sigma_{RT} = 4.74\ \Omega$$

If we use 3σ tolerance limits, we have $R_1 = 100 \pm 30$, $R_2 = 200 \pm 45$ and then $R_T = 66.7 \pm 14.2$.

If the random variables $X_1, ..., X_n$ are *not independent*, Eq. (18.7) for $V[Y]$ using the Taylor series approximation is given by

$$V(y) = \sum_{i=1}^{n} \left\{ \left.\frac{\partial f(X)}{\partial X_i}\right|_{x = \mu} \right\}^2 V(X_i)$$

$$+ 2 \sum_{i=1}^{n-1} \sum_{j=i+1}^{n} \left(\left.\frac{\partial f}{\partial X_i}\right|_{x = \mu} \right) \left(\left.\frac{\partial f}{\partial X_j}\right|_{x = \mu} \right) \text{cov}\ (X_i, X_j) \qquad (18.8)$$

18.7.2 Worst-Case Methods[15]

The worst-case method of variability analysis is a nonstatistical approach that can be used to determine whether it is possible, with given parameter tolerance limits, for the system performance characteristics to fall outside specifications. The answer is obtained by using system models in which parameters are set at either their upper or lower tolerance limits. Parameter values are chosen to cause each performance characteristic to assume first its maximum and then its minimum expected value. If the performance characteristic values fall within specifications, the designer can be confident that the system has high drift reliability. If specifications are exceeded, drift-type failures are possible, but the probability of their occurrence remains unknown.

Worst-case analysis is based on expressing the model for the performance variable Y as a function of design parameters $X_1, \ldots, X_n$ and expanding these functions in Taylor series about the nominal values. The design parameters include all pertinent part characteristics, inputs, loads, and environmental factors. Let the model for the performance variable Y_i be

$$Y_i = f_i X_1, \ldots, X_n \tag{18.9}$$

The linear relation which relates changes in Y_i to changes in design parameters X_1, $X_2, \ldots, X_n$ is

$$\Delta Y_i = \sum_{j=1}^{n} \left(\frac{\partial f_i}{\partial X_j} \bigg|_{x=\mu} \Delta X_j \right) \tag{18.10}$$

where vector μ represents the nominal value of the design parameter vector X and ΔX_j is the variation of design parameter X_j, $\Delta X_j = \mu_j - X_{j\min}$ or $\Delta X_{j\max} - \mu_j$, where μ_j is the nominal value for parameter j, $j = 1, \ldots, n$.

A set of these equations must be derived to relate all performance factors to all design variables. The partial derivatives of f with respect to each dependent variable X_j must be computed. One of the most important steps in a worst-case analysis is to decide whether to use a high or low parameter tolerance limit for each component part when analyzing a specific performance characteristic. If the slope of the function that relates a parameter to a performance characteristic is known, the selection of parameter limit is easy: when the slope of the parameter function is positive, the upper tolerance limit is chosen if the maximum value of the performance characteristic is desired. For parameter functions with negative slopes, the lower tolerance limit corresponds to the maximum performance characteristic value.

An important part of worst-case analysis is to determine the sensitivity of system performance to variations in input parameters. The sensitivity of a system essentially is measured as the effect of parameter variations on the system performance. In equation form, sensitivity can be expressed by

$$S_{ij} = \frac{\partial Y_i}{\partial X_j} \bigg|_{x=\mu} \quad \text{or} \quad \frac{\partial Y_i}{\partial X_j} \bigg|_{o} \tag{18.11}$$

where S_{ij} = the sensitivity of the performance measure Y_i to the variation in
         the system design parameter X_j
       o = evaluated at nominal conditions, usually the mean values

Another equation that is used is

$$S_{ij} = \left. \frac{\partial \ln Y_i}{\partial \ln X_j} \right|_{x=\mu} \approx \frac{\Delta Y_i / Y_{io}}{\Delta X_j / X_{jo}} \tag{18.12}$$

The forms of the variation equation which correspond to the two sensitivities are

$$\Delta Y_i = \sum_{j=1}^{n} S_{ij}\, \Delta X_j \tag{18.13}$$

$$\frac{\Delta Y_i}{Y_{io}} = \sum_{j=1}^{n} S_{ij} \frac{\Delta X_j}{X_{jo}} \tag{18.14}$$

If a design fails the worst-case analysis, look at the absolute values of the individual terms in Eq. (18.11) or (18.12). The ones which contribute the most ought to be reduced—they are the bottlenecks. It does little good to reduce the small terms because they have so little effect on the total variation. It is not unusual to have well over half the variation due to one or two parameters. If several performance parameters have too much variation, the major contributors ought to be listed for each. If a few parameters are causing most of the difficulty, attention can be devoted to them. If not, an extensive redesign might be necessary. Figure 18.16 shows a block diagram of the worst-case method.

18.7.3   Monte Carlo Method[15,20,21]

In the Monte Carlo method, a large number of replicas of the system are simulated by mathematical models. The values of the variables and parameters are randomly selected based on their probability distributions. The performance of the overall system can be compared to a given set of specifications. The large number of values of the system performance are used to develop a frequency distribution for the performance of the system. Figure 18.17 is a block diagram of the Monte Carlo method. Monte Carlo methods are very good to simulate the performance of the system but give very little help in identifying and correcting problems.

18.8   ALLOCATION OF RELIABILITY REQUIREMENTS

Reliability and design engineers must translate overall system performance, including reliability, into component performance, including reliability. The

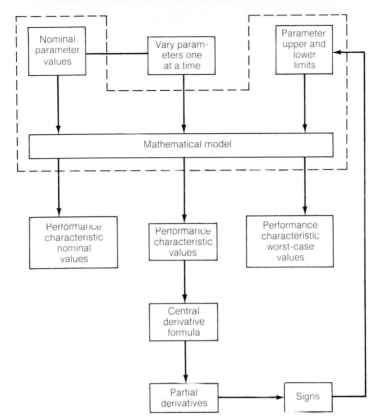

FIG. 18.16   Worst-case method.

process of assigning reliability requirements to individual components to attain the specified system reliability is called *reliability allocation*. There exist many different ways that reliability can be allocated to components in order to achieve a specific system reliability.[1] The allocation problem is complex for several reasons, among which are: (1) the role a component plays for the functioning of the system, (2) the methods available for accomplishing this function, (3) the complexity of the component, and (4) the changeable reliability of the component with the type of function to be performed. The problem is further complicated by the lack of detailed information on many of these factors early in the system design phase. However, a tentative reliability allocation must be accomplished in order to guide the design engineer. The typical decision process from a reliability allocation standpoint is illustrated in Fig. 18.18. A process such as this attempts to force all concerned to make decisions in an orderly and knowledgeable fashion rather than on an ad hoc basis.

Some of the advantages of the reliability allocation program are

1. The reliability allocation program forces system design and development personnel to understand and develop the relationships between component, subsystem, and system reliabilities. This leads to an understanding of the basic reliability problems inherent in the design.

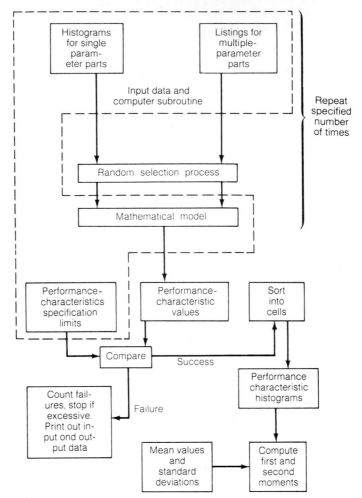

FIG. 18.17 The Monte Carlo method.

2. The design engineer is obliged to consider reliability equally with other system parameters such as weight, cost, and performance characteristics.

3. Reliability allocation program ensures adequate design, manufacturing methods, and testing procedures.

The allocation process is approximate and the system effectiveness parameters, such as reliability and maintainability apportioned to the subsystems, are used as guidelines to determine design feasibility. If the allocated parameters for a system cannot be achieved using the current technology, then the system must be modified and the allocations reassigned. This procedure is repeated until an allocation is achieved that satisfies the system requirements.

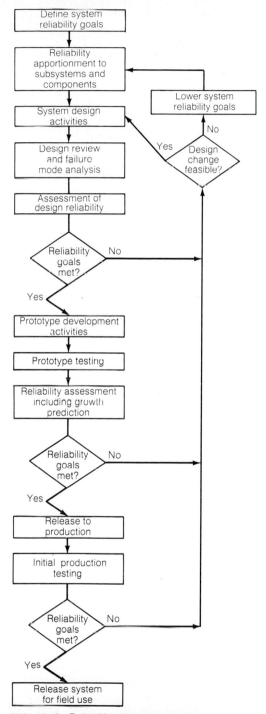

FIG. 18.18 Reliability allocation process.

18.37

The allocation of specified system reliability R^* to the component reliability requires solving the following inequality

$$f(R_1, R_2, \ldots, R_n) \geq R^* \tag{18.15}$$

where R_i = the reliability allocated to the ith unit
$\quad\quad f$ = the functional relationship between the components and the system

For series and parallel systems, the functional relationship f is well known. This relationship is complex for other system configurations. If we are interested in reliability as a function of time, Eq. (18.15) may be generalized by considering R^* and R_i, $i = 1, \ldots, n$ as functions of time t.

Most of the basic reliability allocation models are based on the assumption that component failures are independent, the failure of any component results in system failure (i.e., the system is composed of components in series), and the failure rates of the components are constant. These assumptions lead to the following equation as a special case of Eq. (18.15)

$$R_1(t)R_2(t)\ldots R_n(t) \geq R^*(t) \tag{18.16}$$

Let λ_i = failure rate of the ith component and λ^* = failure rate of the system; then Eq. (18.16) becomes, using the exponential distribution,

$$e^{-\lambda_1 t} e^{-\lambda_2 t} \ldots e^{-\lambda_n t} \geq e^{-\lambda^* t}$$

or

$$\lambda_1 + \lambda_2 + \cdots + \lambda_n \leq \lambda^*$$

The purpose of the allocation model is to allocate $R^*(t)$ or λ^* to the components of the system. We will discuss some of the reliability allocation methods.

In the ARINC method,[1] we assume that the subsystems are in series with constant failure rates, that any subsystem failure causes a system failure, and that the subsystem mission times equal the system mission time. This apportionment technique requires the expression of the required reliability in terms of the failure rates. The objective is to choose λ_i^*'s such that

$$\sum_{i=1}^{n} \lambda_i^* \leq \lambda^* \tag{18.17}$$

where λ_i^* = the failure rate allocated to subsystem i
$\quad\quad i$ = $1, \ldots, n$
$\quad\quad \lambda^*$ = the required system failure rate
The following steps summarize this technique:

1. Determine the subsystem failure rates λ_i from the past data, observed or estimated.

2. Assign a weighting factor ω_i to each subsystem according to the failure rates determined in step 1, where ω_i is given by

$$\omega_i = \frac{\lambda_i}{\sum_{i=1}^{n} \lambda_i} \qquad i = 1, \ldots, n \qquad (18.18)$$

Thus ω_i represents the relative failure vulnerability of the ith component and

$$\sum_{i=1}^{n} \omega_i = 1$$

3. Compute the subsystem failure rate requirements using

$$\lambda_i = \omega_i \lambda^* \qquad i = 1, \ldots, n$$

assuming equality holds in Eq. (18.17).

It is clear that this method allocates the new failure rates based on relative weighting factors that are functions of the past failure rates of the subsystems.

Example[18]. Consider a system composed of three subsystems with the estimated failure rates of $\lambda_1 = 0.005$, $\lambda_2 = 0.003$, and $\lambda_3 = 0.001$ failure per hour, respectively. The system has a mission time of 20 hours. A system reliability of 0.95 is required. Find the reliability requirements for the subsystems.

Using Eq. (18.18), we compute the weighting factors:

$$\omega_1 = \frac{0.005}{0.005 + 0.003 + 0.001} = 0.555$$

$$\omega_2 = \frac{0.003}{0.005 + 0.003 + 0.001} = 0.333$$

$$\omega_3 = \frac{0.001}{0.005 + 0.003 + 0.001} = 0.111$$

We know that

$$R^*(20) = \exp[-\lambda^*(20)] = 0.95$$

or

$$\lambda^* = 0.00256 \text{ failure per hour}$$

Hence the failure rates for the subsystems are

$$\lambda_1 = \omega_1 \lambda^* = 0.555 \times 0.00256 = 0.00142$$

and similarly,

$$\lambda_2^* = 0.333 \times 0.00256 = 0.000852$$

$$\lambda_3^* - 0.111 \times 0.00256 = 0.000284$$

The corresponding apportioned reliabilities for the subsystem are

$$R_1^*(20) = \exp[-0.00142(20)] = 0.97$$

$$R_2^* (20) = \exp [-0.000852 (20)] = 0.98$$
$$R_3^* (20) = \exp [-0.000284 (20)] = 0.99$$

The AGREE allocation method[22] is more sophisticated than the previous method. This method is based on component or subsystem complexity and explicitly considers the relationship between component and system failure. The AGREE formula is used to determine the minimum mean time between failures (MTBF) for each component required to meet the system reliability. The components are supposed to have constant failure rates that are independent of each other and they operate in series with respect to their effect on system success.

Component complexity is defined in terms of modules and their associated circuitry. Examples of a module are an electron tube, a transistor, or a magnetic tape; a diode is considered a half-module. It is recommended that for digital computers (where the module count is high), the count should be reduced because failure rates for digital parts are generally far lower than for radio-radar types. The importance factor of a unit or subsystem is defined in terms of the probability of system failure if the particular subsystem fails. The importance factor of 1 means that the subsystem must operate for the system to operate successfully, and the importance factor of 0 means that the failure of the subsystem has no effect on system operation.

The allocation assumes that each module makes an equal contribution to system success. An equivalent requirement is that each module have the same failure rate. Making the observation that $e^{-x} \approx 1 - x$ when x is very small, the allocated failure rate to the ith unit is given by

$$\lambda_i = \frac{N_i[-\ln R^*(t)]}{N \omega_i t_i} \qquad i = 1, 2, \ldots, n \tag{18.19}$$

where t = mission time, or the required system operation time
  t_i = time units for which the ith subsystem will be required to operate during t units of system operation $(0 < t_i \leqslant t)$
  N_i = number of modules in ith subsystem
  N = total number of modules in the system = ΣN_i
  ω_i = importance factor for the ith subsystem
   = P [system failure|subsystem i fails]
  $R^*(t)$ = required system reliability for operation time t

The allocated reliability for the ith subsystem for t_i operating time units is given by

$$R_i(t_i) = 1 - \frac{1 - [R^*(t)]^{N_i/N}}{\omega_i} \tag{18.20}$$

The AGREE formula will lead to distorted allocation if the importance factor for a certain unit is very low. It is a good approximation if ω_i is close to one for each subsystem.

Example.[18] A system consisting of four subsystems is required to demonstrate a reliability level of 0.95 for 10 hours of continuous operation. Subsystems 1 and 3 are essential for the successful operation of the system. Subsystem 2 has to function for only 9 hours for the operation of the system, and its importance factor is 0.95. Subsystem 4 has an importance factor of 0.90 and must function for

8 hours for the system to function. Solve the reliability allocation problem by AGREE method using the data in Table 18.6. We have

$$N = \sum_{i=1}^{4} N_i = 20$$

The minimum acceptable failure rates for the subsystems are given by Eq. (18.19) and these are

$$\lambda^1 = \frac{15(-\ln 0.95)}{(210)(1.0)(10)} = 0.000366$$

$$\lambda^2 = \frac{25(-\ln 0.95)}{(210)(0.95)(9)} = 0.000714$$

$$\lambda^3 = \frac{100(-\ln 0.95)}{(210)(1.0)(10)} = 0.002442$$

$$\lambda^4 - \frac{70(-\ln 0.95)}{(210)(0.90)(8)} = 0.002377$$

Thus, the allocated subsystem reliabilities are, using Eq. (18.20)

$$R_1(10) = 1 - \frac{1 - (0.95)^{15/210}}{1} = 0.99635$$

$$R_2(9) = 1 - \frac{1 - (0.95)^{25/210}}{0.95} = 0.99274$$

$$R_3(10) = 1 - \frac{1 - (0.95)^{100/210}}{1} = 0.97587$$

$$R_4(8) = 1 - \frac{1 - (0.95)^{70/210}}{0.90} = 0.98116$$

As a check, we have the system reliability as

$$R^* = (0.99635)(0.99274)(0.97587)(0.98116)$$
$$= 0.94723$$

which is slightly less than the specified reliability. This is due to the approximate nature of the formula and because the importance factors for subsystems 2 and 4 are less than 1.

TABLE 18.6 Data for Reliability Allocation

Subsystem number	Number of modules, N_i	Importance factor, ω_i	Operating time, t_i
1	15	1.00	10
2	25	0.95	9
3	100	1.00	10
4	70	0.90	8

REFERENCES

1. Aeronautical Research Incorporated, Engineering and Statistical Staff, in W. H. Von Alven (ed.), *Reliability Engineering,* Prentice-Hall, Englewood Cliffs, N.J., 1964.

2. *Reliability Design Handbook,* Reliability Analysis Center, RDG-376, Griffiss Air Force Base, N.Y., March 1976.

3. Mil-Hdbk-217C, *Military Standardization Handbook, Reliability Prediction of Electronic Equipment,* U.S. Department of Defense, April 1979.

4. NPRD-Z, *Nonelectronic Parts Reliability Data Handbook,* Reliability Analysis Center, Rome Air Development Center, Griffiss AFB, N.Y., 1981.

5. Mil-Std-1556A, *Government-Industry Data Exchange Program (GIDEP),* U.S. Department of Defense, USAF, February 1976.

6. Juran, J. M., and F. M. Gryna, Jr., *Quality Planning and Analysis,* 2d ed., McGraw-Hill, New York, 1980.

7. Mil-Std-785B, *Reliability Program for Systems and Equipment Development and Production,* U.S. Department of Defense, 15 September 1980.

8. Mil-Std-1543A, *Reliability Program Requirements for Space and Missile Systems,* U.S. Department of Defense, 25 June 1982.

9. *Quality Assurance-Reliability Handbook,* AMC Pamphlet No. 702-3, Headquarters, U.S. Army Materiel Command, Alexandria, Va., October 1968.

10. Mil-Std-1629A, *Procedures for Performing a Failure Mode, Effects and Criticality Analysis,* U.S. Department of Defense, November 1980.

11. ARP-926, *Design Analysis Procedure for Failure Mode, Effects and Criticality Analysis (FMECA),* Society of Automotive Engineers, New York, September 1967.

12. Henley, E. J., and H. Kumamoto, *Reliability Engineering and Risk Assessment,* Prentice-Hall, Englewood Cliffs, N.J., 1981.

13. Fussell, J. E., G. J. Powers, and R. G. Bennetts, "Fault Trees—A State of the Art Discussion," *IEEE Transactions on Reliability,* vol. R-30, no. 1, April 1974.

14. R. E. Barlow, J. B. Fussel, and N. D. Singurwalla (eds.), *Reliability and Fault Tree Analysis,* Society of Industrial and Applied Mathematics, Pennsylvania, Pa., 1975.

15. *Engineering Design Handbook, Part 2, Design for Reliability,* AMC Pamphlet No. 706-196, U.S. Army Materiel Command, Alexandria, Va., January 1976.

16. Kececioglu, D., and D. Cormier, "Designing a Specified Reliability Directly into a Component," *Proceedings of the Third Annual Aerospace Reliability and Maintainability Conference,* pp. 520–530, 1968.

17. Haugen, E. G., *Probabilistic Approach to Design,* Wiley, New York, 1968.

18. Kapur, K. C., and L. R. Lamberson, *Reliability in Engineering Design,* Wiley, New York, 1977.

19. Kapur, K. C., "Reliability and Maintainability," *Industrial Engineering Handbook,* chap. 5, Wiley, New York, 1982.

20. Mark, D. G., and L. H. Stember, Jr., "Variability Analysis," *Electro-Technology,* July 1965.

21. NASA CR-1126, *Practical Reliability, Vol. I, Parameter Variation Analysis,* Research Triangle Institute, Research Triangle Park, N.C., July 1968.

22. Reliability of Military Electronic Equipment, Advisory Group on Reliability of Electronic Equipment (AGREE), Office of the Assistant Secretary of Defense, U.S. Government Printing Office, Washington, D.C., June 1957.

CHAPTER 19
MATHEMATICAL AND STATISTICAL METHODS AND MODELS IN RELIABILITY AND LIFE STUDIES

Kailash C. Kapur, Ph.D., P.E.
*Professor, Department of Industrial
Engineering and Operations Research,
Wayne State University*

TABLE 19.1 Glossary of Symbols and Abbreviations

$P(\cdot)$	probability of event $(\cdot)$
$\binom{n}{k}$	combination of size k from n objects
R	reliability
p.d.f.	probability density function
c.d.f.	cumulative distribution function
r.v.	random variable
$f(t)$	probability density function for random variable T
$F(t)$	cumulative distribution function for random variable
$R(t)$	reliability function for random variable T
$h(t)$	failure rate
$E[x]$	expected value for random variable x
μ	mean of a random variable
σ	standard deviation
ρ	correlation coefficient
z	standard normal variable
$\Phi(z)$	cumulative distribution function for standard normal variable
$\Gamma(\cdot)$	gamma function
MTBF	mean time between failures
MTTR	mean time to repair
$M(t)$	maintainability function
χ^2	chi-square distribution
F	F distribution
$\hat{\theta}$	point estimator of parameter θ
β	slope parameter for Weibull distribution

19.1 ELEMENTS OF PROBABILITY THEORY

Reliability is the probability of a product performing its intended function satisfactorily for its intended life under specified operating conditions. Hence study of probability theory is essential for understanding the reliability of a product. The purpose of this section is to provide a basic knowledge of probability theory.

Probability is a method to model and describe the random variation in systems. Probability theory provides the mathematical foundation and language of statistics and helps us deduce from a mathematical model the properties of a physical process, whereas statistical inference determines the properties of the model from observed data. Statistics is the art and science of gathering, analyzing, and making decisions from the data and will be discussed in Sec. 19.3.

Table 19.1 gives a glossary of the symbols and abbreviations to be used in this chapter.

19.1.1 Fundamentals of Set Theory

A *set* is a collection of objects viewed as a single entity. The individual objects of the set are called the *elements* of the set. Sets usually are denoted by capital letters: A, B, C, ..., Y, Z; elements are designated by lowercase letters: a, b, c, ..., y, z. If a is an element of the set A, we write $a \in A$, and we write $a \notin A$ for a is not an element of A. A set is called a *finite* set when it contains a finite number of elements and an *infinite* set otherwise. The *null set* $\emptyset$ is the set that contains no elements. The *total* or *universal* set Ω is the set which contains all the elements under consideration.

We say a set A is a subset of set B if each element of A is also an element of B and write as $A \subseteq B$. The relation $\subseteq$ is referred to as *set inclusion*.

The Algebra Sets

Definition.   The *union* of the two sets A and B, denoted by $A \cup B$, is the set of all elements of either set, that is, $c \in (A \cup B)$ means $c \in A$, or $c \in B$, or both.

Definition.   The *intersection* of the two sets A and B, denoted by $A \cap B$, is the set of all elements common to both A and B, that is, $c \in (A \cap B)$ means $c \in A$ and $c \in B$.

Definition.   The *complement* of a set A, denoted by $\bar{A}$, is the set of elements of the universal set that do not belong to A.

Definition.   Two sets are said to be *disjoint* or *mutually exclusive* if they have no elements in common, i.e., $A \cap B = \emptyset$.

When considering sets and operations on sets, *Venn diagrams* can be used to represent sets diagrammatically. Figure 19.1a shows a Venn diagram for $A \cap B$ and Fig. 19.1b shows a Venn diagram for $A \cup B$. Figure 19.1c shows a Venn diagram with three sets A, B, and C.

19.1.2 Probability Definitions

There is a natural relation between probability theory and set theory based on the concept of a random experiment for which it is impossible to state a particular outcome, but we can define the set of all possible outcomes.

Definition.   The *sample space* of an experiment, denoted by S, is the set of all possible outcomes of the experiment.

Definition.   An *event* is any collection of outcomes of the experiment or subset of

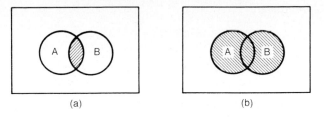

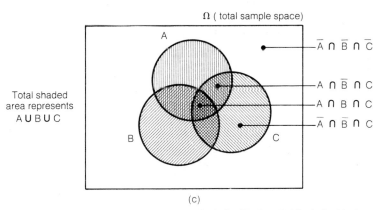

FIG. 19.1   Venn diagrams. (*a*) $A \cap B$ (shaded); (*b*) $A \cup B$ (shaded); (*c*) three sets A, B, and C.

the sample space S. An event is said to be *simple* if it consists of exactly one outcome, and *compound* if it consists of more than one outcome.

The objective of probability is to assign to each event A of the sample space S associated with an experiment a number $P(A)$, called the probability of event A, which will give a precise measure of the chance that A will occur. The function $P(\cdot)$ has the following properties:

1. $0 \le P(A) \le 1$ for each event A of S

2. $P(S) = 1$

3. For any finite number k of mutually exclusive events defined on S,

$$P\left(\bigcup_{i=1}^{k} A_i \right) = \sum_{i=1}^{k} P(A_i) \tag{19.1}$$

4. If $A_1, A_2, A_3, \dots$ is a denumerable or countably infinite sequence of mutually exclusive events defined on S, then

$$P(A_1 \cup A_2 \cup A_3 \cup \cdots) = P(A_1) + P(A_2) + P(A_3) + \cdots \tag{19.2}$$

We can also use the concept *relative frequency* to develop the function $P(\cdot)$. If we repeat an experiment n times and event A occurs n_A times, $0 \le n_A \le n$, then the value of the relative frequency $f_A = n_A/n$ approaches $P(A)$ as n increases to infinity.

Properties of Probability

1. If $\emptyset$ is the empty or null set, then $P(\emptyset) = 0$

2. $P(\bar{A}) = 1 - P(A)$ (19.3)

3. $P(A \cup B) = P(A) + P(B) - P(A \cap B)$ (19.4)

$$
\begin{aligned}
\textbf{4. } P(A_1 \cup A_2 \cup \cdots \cup A_n) = & \sum_{i=1}^{k} P(A_i) - \sum_{i=1}^{n-1} \sum_{j=i+1}^{n} P(A_i \cap A_j) \\
& + \sum_{i=1}^{n-2} \sum_{j=i+1}^{n-1} \sum_{k=j+1}^{n} P(A_i \cap A_j \cap A_k) \\
& + \cdots + (-1)^{n+1} P(A_1 \cap A_2 \cap \cdots \cap A_n)
\end{aligned}
$$

       (19.5)

Conditional Probability.   We will frequently be interested in evaluating the probability of events where the event is *conditioned* on some subset of the sample space.

     Definition.   The *conditional probability* of event A given event B is defined as

$$
P(A/B) = \frac{P(A \cap B)}{P(B)} \qquad \text{if } P(B) > 0
$$

       (19.6)

This statement can be restated to what is often called the multiplication rule, that is,

$$
P(A \cap B) = P(A/B)P(B) \qquad P(B) > 0
$$

       (19.7)

$$
P(A \cap B) = P(B/A)P(A) \qquad P(A) > 0
$$

       (19.8)

     Definition. A and B are called *independent* events if and only if

$$
P(A \cap B) = P(A)P(B)
$$

       (19.9)

This definition leads to the following statement. If A and B are independent events, then

$$
P(A/B) = P(A) \qquad \text{and} \qquad P(B/A) = P(B)
$$

Partitions, Total Probability, and Bayes' Theorem.   If $A_1, \ldots, A_n$ are disjoint subsets of S (mutually exclusive events) and if $A_1 \cup A_2 \cup \cdots \cup A_n = S$, then these subsets are said to form a *partition* of S. The *total probability* of any other event B is given by

$$
P(B) = \sum_{i=1}^{n} P(B/A_i)P(A_i)
$$

       (19.10)

Another important result of total probability is *Bayes' theorem*. If $A_1, A_2, \ldots, A_k$ constitute a partition of the sample space S and B is an arbitrary event, then Bayes' theorem states that

$$
P(A_i/B) = \frac{P(A_i \cap B)}{P(B)} = \frac{P(B/A_i)P(A_i)}{\sum_{i=1}^{n} P(B/A_i)P(A_i)} \qquad i = 1, 2, \ldots, n
$$

       (19.11)

where $P(A_i) > 0$ for $i = 1, \ldots, n$ and $P(B) > 0$.

Permutations and Combinations.   In some situations we will have to resort to the relative frequency concept and successive trials to *estimate* probabilities and hence have to enumerate the number of outcomes favorable to event A.

Definition. Suppose that a set consists of ordered collections of k elements (k-tuples) and that there are n possible choices for the first element; for each choice of first element there are n_2 possible choices of the second element, and so for each possible choice of the first $(k - 1)$ elements, there are n_k choices of the kth element. Then the product rule states that there are $n_1, n_2, \ldots, n_k$ possible k element (k-tuple) sets.

Definition. Any ordered sequence of k objects from n distinct objects is called a *permutation* of size k of the objects. The number of permutations of n objects taken k at a time is denoted by P_k^n and is given by

$$P_k^n = n(n - 1) \cdots (n - k + 1) = \frac{n!}{(n - k)!} \tag{19.12}$$

where

$$n! = n(n - 1)(n - 2) \cdots (2)(1) = \prod_{i=1}^{n} i$$

$n!$ is read n factorial; also $0! = 1$.

Definition. Given a set of n distinct objects, any unordered subset of size k of the objects is called a *combination.* The number of combinations of size k which can be formed from n distinct objects is denoted by $\binom{n}{k}$ and is given by

$$\binom{n}{k} = \frac{n!}{(n - k)! k!} \tag{19.13}$$

Further details on probability concepts can be found in references 1 and 2.

Example 19.1 System Reliability Computation. In this section applications of probability concepts discussed before are presented to compute reliability of a system given reliabilities of the components. A reliability block diagram has to be developed based on a careful analysis of the manner in which the system operates. Let R_i be the reliability of the ith component, $i = 1, \ldots, n$. We present reliability equations for different system configurations.

Series Structure. The structure is called a *series* structure when the system functions if and only if all the n components of the system function (see Fig. 19.2a). The components are assumed to fail or function independently of one another. Then the reliability R_S of the system is given by

$$R_S = \prod_{i=1}^{n} R_i \tag{19.14}$$

Parallel Structure. The structure is said to be *parallel* when the system functions if at least one of the n components of the system functions (see Fig. 19.2b). The reliability of the system is given by

$$R_S = 1 - \prod_{i=1}^{n} (1 - R_i) \tag{19.15}$$

Combination Structure. We can compute reliability of a system which consists of both series and parallel subassemblies. Figure 19.2c shows such a structure. The numbers in each block are the reliabilities of the components. The system reliability

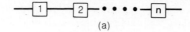

(a)

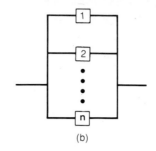

(b)

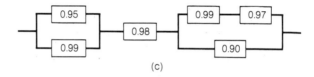

(c)

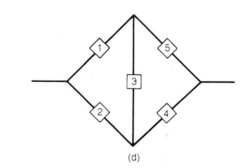

(d)

FIG. 19.2 Structures. (a) Series; (b) parallel; (c) combination; (d) bridge.

is given by

$$R_S = [1 - (1 - 0.95)(1 - 0.99)](0.98)[1 - (1 - 0.99 \times 0.97)(1 - 0.90)]$$

$$= 0.9995 \times 0.98 \times 0.99603$$

$$= 0.97562$$

The k-out-of-n Structure. For this structure, the system works if and only if at least k components out of the n components work; $1 \leq k \leq n$. For the case when $R_i = R$ for all i, we have

$$R_S = \sum_{i=k}^{n} \binom{n}{i} R^i (1 - R)^{n-i} \tag{19.16}$$

Let us now consider an example of the above system.

A military planner has eight helicopters available to perform a mission. At least six helicopters are required to perform the mission successfully. Each helicopter has a reliability of 0.80 for the duration of the mission. The helicopters work or fail independently of each other. What is the reliability of this mission, i.e., what is the probability that the mission will be a success? We have using Eq. (19.16)

$$R_{mission} = \sum_{k=6}^{8} \binom{8}{k}(0.8)^k(1 - 0.8)^{8-k}$$

$$= 0.79692$$

Coherent Systems. The reliability block diagrams for many systems cannot be represented by the above three configurations. In general, the concept of coherent systems can be used to determine the reliability of any system.[3]* The performance of each of the n components in the system is represented by a binary indicator variable x_i which takes the value 1 if the ith component functions and 0 if the ith component fails. Similarly, the binary variable ϕ indicates the state of the system, and ϕ is a function of $x = (x_1, \ldots, x_n)$ and $\phi(x)$ is called the *structure function* of the system. The structure function is represented by using the concept of minimal path and minimal cut. A *minimal path* is a minimal set of components whose functioning ensures the functioning of the system. A *minimal cut* is a minimal set of components whose failures cause the system to fail. Let $\alpha_j(x)$ be the jth minimal path series structure for path $A_j, j = 1, \ldots, p$ and $\beta_k(x)$ be the kth minimal parallel cut structure for cut $B_k, k = 1, \ldots, s$. Then, we have

$$\alpha_j(x) = \prod_{i \in A_j} x_i \tag{19.17}$$

$$\beta_k(x) = 1 - \prod_{i \in B_k} (1 - x_i) \tag{19.18}$$

and

$$\phi(x) = 1 - \prod_{j=1}^{p} [1 - \alpha_j(x)] \tag{19.19}$$

$$= \prod_{k=1}^{s} \beta_k(x) \tag{19.20}$$

For the bridge structure (Fig. 19.2d), we have

$$\alpha_1 = x_1 x_5 \qquad \beta_1 = 1 - (1 - x_1)(1 - x_2)$$
$$\alpha_2 = x_2 x_4 \qquad \beta_2 = 1 - (1 - x_4)(1 - x_5)$$
$$\alpha_3 = x_1 x_3 x_4 \qquad \beta_3 = 1 - (1 - x_1)(1 - x_3)(1 - x_4)$$
$$\alpha_4 = x_2 x_3 x_5 \qquad \beta_4 = 1 - (1 - x_2)(1 - x_3)(1 - x_5)$$

Then, the reliability of the system is given by $R_S = P[\phi(x) = 1]$.
The reliability R_S for the bridge structure is given by the following expression

$$R_S = R_1 R_5 + R_1 R_3 R_4 + R_2 R_3 R_5 + R_2 R_4$$
$$- R_1 R_3 R_4 R_5 - R_1 R_2 R_3 R_5 - R_1 R_2 R_4 R_5$$
$$- R_1 R_2 R_3 R_4 - R_2 R_3 R_4 R_5 + 2R_1 R_2 R_3 R_4 R_5$$

---

*Superscripts refer to numbered references at end of chapter.

If all $R_i = R = 0.9$, we have

$$R_S = 2R^2 + 2R^3 - 5R^4 + 2R^5$$

$$= 0.9785$$

The exact calculations for R_S are generally very tedious because the paths and the cuts are dependent, since they may contain a same component. Bounds on system reliability are given by

$$\prod_{k=1}^{s} P[\beta_k(x) = 1] \le P[\phi(x) = 1] \le 1 - \prod_{j=1}^{p} \{1 - P[\alpha_j(x) = 1]\}$$

Using these bounds for the bridge structure (Fig. 19.2d), we have when $R_i = R = 0.9$

$$\text{Upper bound on } R_S = 1 - (1 - R^2)^2(1 - R^3)^2$$

$$= 0.9973$$

$$\text{Lower bound on } R_S = [1 - (1 - R)^2]^2[1 - (1 - R)^3]^3$$

$$= 0.9781$$

The bounds on system reliability can be improved, and details are given in Barlow and Proschan.[3]

19.1.3   Discrete and Continuous Random Variables

Definition.   Let S be the sample space associated with experiment ε. Let X be a function that assigns a real number $X(e)$ to every outcome $e \in S$, then $X(e)$ is called a *random variable* (r.v.).

Discrete Random Variables.   If the range space R_X of the random variable X is either finite or countably infinite, then X is called a discrete random variable.

Continuous Random Variables.   If the range space R_X of the random variable X is an interval or a collection of intervals, then X is called a continuous random variable.

Probability Density Function.   For a continuous random variable X, we define

$$P[a \le X \le b] = \int_a^b f_X(x)\, dx \qquad (19.21)$$

where the function f_X, called the *probability density function* (p.d.f.), satisfies the following properties:

1. $f_X(x) \ge 0$ for all $x \in R_X$
2. $\int_{R_X} f_X(x)\, dx = 1$

Probability Mass Function.   For a discrete random variable, we associate a number $p_X(x_i) = P(X = x_i)$ for each outcome x_i in R_X, where the numbers $p_X(x_i)$ satisfy the following properties:

1. $p_X(x_i) \geq 0$ for all i
2. $\sum_{\text{all } i} p_X(x_i) = 1$

Cumulative Distribution Function. The cumulative distribution function (c.d.f.) of a random variable X is denoted by $F_X(\cdot)$ and is defined by

$$F_X(x) = P[X \leq x] \qquad \text{for all } x$$

$$F_X(x) = \sum_{\text{all } i \text{ such that } x_i \leq x} p(x_i) \qquad X \text{ discrete random variable} \qquad (19.22)$$

$$= \int_{-\infty}^{x} f(u)\, du \qquad X \text{ continuous random variable} \qquad (19.23)$$

Properties of the cumulative distribution function are

$$F(-\infty) = \lim_{x \to -\infty} F(x) = 0$$

$$F(+\infty) = \lim_{x \to +\infty} F(x) = 1$$

$$F(x_2) \leq F(x_1) \qquad \text{for all } x_2 \leq x_1$$

Example 19.2 Reliability Function. Let T be the time-to-failure random variable. Then reliability at time t, $R(t)$, is the probability that the system will not fail by time t, or

$$R(t) = P[T > t]$$

$$= 1 - P[T \leq t]$$

$$= 1 - F(t)$$

$$= 1 - \int_0^t f(\tau)\, d\tau \qquad (19.24)$$

where $f(t)$ and $F(t)$ are the probability density function and cumulative distribution function for the T, respectively. For example, if the time to failure T is exponentially distributed, then

$$f(t) = \lambda e^{-\lambda t} \qquad t \geq 0,\ \lambda > 0$$

$$F(t) = \int_0^t \lambda e^{-\lambda \tau}\, d\tau = 1 - e^{-\lambda t} \qquad t \geq 0$$

and

$$R(t) = e^{-\lambda t} \qquad t \geq 0$$

Figure 19.3 shows the shapes of the above functions.

Example 19.3 Failure Rate. If we have a large population of the items whose

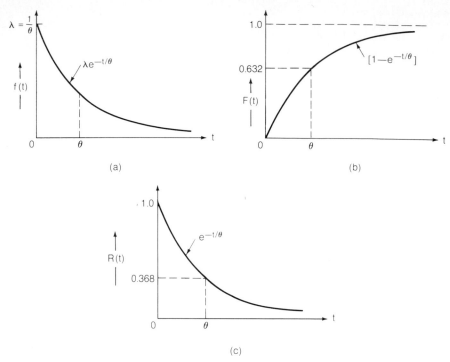

FIG. 19.3 (*a*) Exponential density function; (*b*) exponential distribution function; (*c*) reliability function for exponential distribution.

reliability we are interested in studying, then for replacement and maintenance purposes we are interested in the rate at which the items in the population, which have survived at any point in time, will fail. This is called the *failure rate* or *hazard rate* and is given by the following relationship:

$$h(t) = \frac{f(t)}{R(t)}$$ (19.25)

The failure rate for most components follows the curve shown in Fig. 19.4*a*, which is called the *life characteristic curve.*[4] Fig. 19.4*b* gives three types of failures, namely quality failures, stress-related failures, and wearout failures. The sum total of these failures gives the overall failure rate of Fig. 19.4*a*. Figure 19.4*b* is also referred to as the *bathtub curve*. The failure rate curve in Fig. 19.4*a* has three distinct periods. The initial decreasing failure rate is termed *infant mortality* and is due to the early failure of substandard products. Latent material defects, poor assembly methods, and poor quality control can contribute to a high initial failure rate. A short period of in-plant product testing, termed *burn-in*, is used by manufacturers to eliminate these early failures from the consumer market. The flat, middle portion of the failure-rate curve

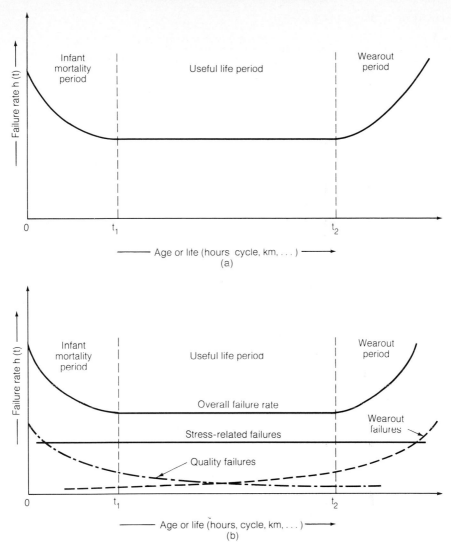

FIG. 19.4 (*a*) Failure rate-life characteristic curve; (*b*) failure rate based on components of failure.

represents the design failure rate for the specific product as used by the consumer market. During the useful-life portion, the failure rate is relatively constant. It might be decreased by redesign or restricting usage. Finally, as products age they reach a wearout phase characterized by an increasing failure rate.

The hazard rate is defined as the limit of the instantaneous failure rate given no failure up to time t, and is given by[5]

$$h(t) = \lim_{\Delta t \to 0} \frac{P[t < T \le t + \Delta t \mid T > t]}{\Delta t}$$

$$= \lim_{\Delta t \to 0} \frac{R(t) - R(t + \Delta t)}{\Delta t \cdot R(t)}$$

$$= \frac{1}{R(t)} \left[-\frac{d}{dt} R(t) \right]$$

$$= \frac{f(t)}{R(t)}$$

If the time-to-failure distribution is exponential, then using Example 19.2, we have

$$h(t) = \frac{\lambda e^{-\lambda t}}{e^{-\lambda t}} = \lambda$$

Thus, exponential distribution has a constant failure rate. Also,

$$f(t) = h(t) \exp\left[-\int_0^t h(\tau)\, d\tau \right] \qquad (19.26)$$

and thus

$$R(t) = \exp\left[-\int_0^t h(\tau)\, d\tau \right] \qquad (19.27)$$

The term *mission reliability* is the reliability of a product for a specified time period t^*, i.e.,

$$\text{Mission reliability} = R(t^*) = P[T > t^*] \qquad (19.28)$$

19.1.4 Expected Value

Definition. The expected value of a function $g(X)$ of a discrete or continuous random variable is

$$E[g(X)] = \sum_{\text{all } x} g(x)p(x) \qquad X \text{ is discrete r.v.}$$

$$= \int_{-\infty}^{\infty} g(x)f(x)\, dx \qquad X \text{ is continuous r.v.} \qquad (19.29)$$

Some of the properties of expected value are

$$E[c] = c \qquad c \text{ is a constant}$$

$$E[cg(X)] = cE[g(X)]$$

$$E[g_1(X) + g_2(X)] = E[g_1(X)] + E[g_2(X)]$$

Definition. The kth moment of the random variable X about origin is defined as

$$v_k = E[X^k] = \sum_{\text{all } x} x^k p(x) \qquad X \text{ is discrete random variable}$$

$$= \int_{-\infty}^{\infty} x^k f(x)\, dx \qquad X \text{ is continuous random variable} \tag{19.30}$$

The mean or expected value of the random variable X is

$$v_1 = E[X] \tag{19.31}$$

$E[X]$ is also denoted by μ or μ_1.

A measure that is frequently used as an indirect indicator of system reliability is called the *mean time to failure* (MTTF), which is the expected or mean value of the time-to-failure random variable. Thus, the MTTF is theoretically defined as

$$\text{MTTF} = E[T] = \int_0^{\infty} t f(t)\, dt = \int_0^{\infty} R(t)\, dt \tag{19.32}$$

Sometimes the term *mean time between failures* (MTBF) is also used to denote $E[T]$. For the exponential distribution,

$$E[T] = \int_0^{\infty} t \lambda e^{-\lambda t}\, dt = \frac{1}{\lambda} = \theta \tag{19.33}$$

Definition. The kth moment of the random variable X about mean v_1 is defined as

$$\mu_k = E[(x - v_1)^k] \qquad k \geq 2$$

$$= E[(x - \mu)^k] \qquad k \geq 2 \tag{19.34}$$

We have

$$\mu_k = \sum_{i=0}^{k} \binom{k}{i} (-1)^i v_{k-i}\, v_1^i \tag{19.35}$$

$\mu_2 = E[(x - \mu)^2] = \sigma^2$ is called the *variance* of the random variable X and *standard deviation* σ is the positive square root of the variance. Other measures[6] for a random variable are

$$\text{Coefficient of variation} = \eta = \frac{\sigma}{\mu}$$

$$\text{Coefficient of skewness} = \alpha_3 = \frac{\mu_3}{\mu_2^{3/2}}$$

$$\text{Coefficient of kurtosis} = \alpha_4 = \frac{\mu_4}{\mu_2^2}$$

Definition. The moment-generating function (MGF) of a random variable X is the expected value of e^{tX} and is given by

$$M_X(t) = E[e^{tX}] \tag{19.36}$$

Whenever MGF exists, it is unique and completely determines the distribution. MGF can be used to find all the moments about origin v_k of the random variable X.

The moments are given by

$$v_k = \frac{\partial^k M_X(t)}{\partial t^k}\bigg|_{t=0} \qquad (19.37)$$

Thus MGF is used to find moments and study the form of a probability distribution.

19.1.5   Bivariate Probability Distributions

If X_1 and X_2 are discrete random variables, joint probability distribution is given by

$$p(x_1, x_2) = P[X_1 = x_1, X_2 = x_2] \qquad (19.38)$$

where the function P has the properties:

$$p(x_1, x_2) \geq 0$$

and

$$\sum_{\text{all } x_1} \sum_{\text{all } x_2} p(x_1, x_2) = 1$$

For continuous random variables X_1 and X_2, the joint density function $f(x_1, x_2)$ has the properties:

$$f(x_1, x_2) \geq 0 \qquad \text{for all } x_1, x_2$$

$$\int_{x_2} \int_{x_1} f(x_1, x_2)\, dx_1\, dx_2 = 1$$

The marginal distributions of X_1 and X_2 are given by

$$f_1(x_1) = \int_{-\infty}^{\infty} f(x_1, x_2)\, dx_2 \qquad (19.39)$$

$$f_2(x_2) = \int_{-\infty}^{\infty} f(x_1, x_2)\, dx_1 \qquad (19.40)$$

Definition.   If $[X_1, X_2]$ is a two-dimensional random variable, the *covariance*, denoted by σ_{12} is

$$\text{cov}(X_1, X_2) = \sigma_{12} = E[(X_1 - E(X_1))(X_2 - E(X_2))]$$

and the correlation coefficient, denoted by ρ, is

$$\rho = \frac{\text{cov}(X_1, X_2)}{\sqrt{V(X_1)}\sqrt{V(X_2)}} = \frac{\sigma_{12}}{\sigma_1 \sigma_2} \qquad (19.41)$$

The value of ρ will always be in the interval $[-1, +1]$. X_1 and X_2 are independent random variables if and only if

$$f(x_1, x_2) = f_1(x_1) f_2(x_2) \qquad (19.42)$$

If X_1 and X_2 are independent, then $\rho = 0$.

19.2   DISTRIBUTIONS USED IN RELIABILITY

In this section we will summarize some of the important distributions used in reliability. Table 19.2 summarizes some of the discrete distributions. The table gives equations for mean, variance, and moment-generating functions for different distributions.

We will briefly discuss the applications of the discrete distributions given in Table 19.2. The *Bernoulli* distribution is extensively used for situations where an individual experiment, trial, or test has only two possible outcomes, such as success or failure, go or no-go, defective or nondefective, or pass or fail. The n Bernoulli trials are called a *Bernoulli process* if the trials are independent and each trial has only two possible outcomes, and probability of success remains constant from trial to trial. The number of successes in n Bernoulli trials has a *binomial* distribution. Applications of the binomial distribution will be discussed in this chapter. Some of the properties of the binomial distribution are given below:

1. $p(x)$ has maximum value for values of x for which $p(n + 1) - 1 \leq x \leq p(n + 1)$.

2. As $n \to \infty$, and $p \to 0$ in such a way that np remains constant, the binomial distribution approaches the Poisson distribution with parameter $\alpha = np$.

3. As $n \to \infty$, the binomial distribution approaches the normal distribution with $\mu = np$ and $\sigma^2 = np(1 - p)$. This approximation is good when $p = 1/2$ and poor for $p < 1/(n + 1)$, $p > n/(n + 1)$, and outside 3σ limits for the random variable.

The *geometric* distribution is related to a sequence of Bernoulli trials except that the number of trials is not fixed and the random variable of interest is defined to be the number of trials required to achieve the first success. An interesting and useful property of the geometric distribution is that it has no memory, that is,

$$P(X > x + k \mid X > k) = P(X > x)$$

This property is analogous to the memoryless property of the exponential distribution. The geometric distribution is the only discrete distribution having this property. The *Pascal* distribution is an extension of the geometric distribution where the random variable X is the Bernoulli trial on which the rth success occurs, where r is an integer. If r is not an integer, then the distribution given in Table 19.2 is called a *negative binomial distribution*. The *hypergeometric distribution* is used to model sampling without replacement from a population. Let the universe consist of N elements, D of which possess a given property. Then the probability that a random sample of size n, without replacement, will contain exactly x elements which possess this property is given by a hypergeometric distribution. As $n \to \infty$, the hypergeometric distribution approaches the binomial distribution with parameters n and $p = D/N$. The approximation is satisfactory for $10n < N$. The *Poisson distribution* is a useful approximation to the binomial and hypergeometric distributions. It also arises when the number of possible events is large but the probability of occurrence of the event over a given area or interval is small, e.g., number of defects, number of failures in a given time interval, and problems in maintainability and availability.

Now we summarize the probability density function, cumulative distribution function, reliability function, hazard function, and MTBF for some of the well-known continuous distributions that are used in reliability.

TABLE 19.2 Summary of Discrete Distributions

Distribution	Parameters	Probability function: $p(x)$	Mean	Variance	Moment-generating function
Bernoulli	$0 \le p \le 1$	$p(x) = p^x q^{1-x}$ $x = 0, 1$ $= 0$ otherwise	p	pq	$pe^t + q$
Binomial	$n = 1, 2, \dots$ $0 \le p \le 1$	$p(x) = \binom{n}{x} p^x q^{n-x}$ $x = 0, 1, 2, \dots, n$ $= 0$ otherwise	np	npq	$(pe^t + q)^n$
Geometric	$0 < p < 1$	$p(x) = pq^{x-1}$ $x = 1, 2, \dots$ $= 0$ otherwise	$1/p$	q/p^2	$pe^t/(1 - qe^t)$
Pascal (neg. binomial)	$0 < p < 1$ $r = 1, 2, \dots$ $(r > 0)$	$p(x) = \binom{x-1}{r-1} p^r q^{x-r}$ $x = r, r+1, r+2, \dots$ $= 0$ otherwise	r/p	rq/p^2	$\left[\dfrac{pe^t}{1 - qe^t}\right]^r$
Hypergeometric	$N = 1, 2, \dots$ $n = 1, 2, \dots, N$ $D = 1, 2, \dots, N$	$p(x) = \dfrac{\binom{D}{x}\binom{N-D}{n-x}}{\binom{N}{n}}$ $x = 0, 1, 2, \dots,$ $\min(n, D)$ $= 0$ otherwise	$n\left[\dfrac{D}{N}\right]$	$n\left[\dfrac{D}{N}\right]\left[1 - \dfrac{D}{N}\right]\left[\dfrac{N-n}{N-1}\right]$	—
Poisson	$\alpha > 0$	$p(x) = e^{-\alpha}(\alpha)^x/x!$ $x = 0, 1, 2, \dots$ $= 0$ otherwise	α	α	$\exp[\alpha(e^t - 1)]$

Source: From Ref. 1.

Exponential Distribution

$$f(t) = \lambda e^{-\lambda t} \qquad t \geq 0 \qquad\qquad (19.43)$$

$$F(t) = 1 - e^{-\lambda t} \qquad t \geq 0 \qquad\qquad (19.44)$$

$$R(t) = e^{-\lambda t} \qquad t \geq 0 \qquad\qquad (19.45)$$

$$h(t) = \lambda \qquad\qquad (19.46)$$

$$\text{MTBF} = \theta = \frac{1}{\lambda} \qquad\qquad (19.47)$$

Thus, the failure rate for the exponential distribution is always constant.

Normal Distribution

$$f(t) = \frac{1}{\sigma\sqrt{2\pi}} \exp\left[-\frac{1}{2}\left(\frac{t-\mu}{\sigma}\right)^2\right] \qquad -\infty < t < \infty \qquad (19.48)$$

$$F(t) = \Phi\left(\frac{t-\mu}{\sigma}\right) \qquad\qquad (19.49)$$

$$R(t) = 1 - \Phi\left(\frac{t-\mu}{\sigma}\right) \qquad\qquad (19.50)$$

$$h(t) = \frac{\phi[(t-\mu)/\sigma]/\sigma}{R(t)} \qquad\qquad (19.51)$$

$$\text{MTBF} = \mu$$

$\Phi(Z)$ is the cumulative distribution function and $\phi(Z)$ is the probability density function for the standard normal variate Z. The failure rate for the normal distribution is a monotonically increasing function.

Lognormal Distribution

$$f(t) = \frac{1}{\sigma t\sqrt{2\pi}} \exp\left[-\frac{1}{2}\left(\frac{\ln t - \mu}{\sigma}\right)^2\right] \qquad t \geq 0 \qquad (19.52)$$

$$F(t) = \Phi\left(\frac{\ln t - \mu}{\sigma}\right) \qquad\qquad (19.53)$$

$$R(t) = 1 - \Phi\left(\frac{\ln t - \mu}{\sigma}\right) \qquad\qquad (19.54)$$

$$h(t) = \frac{\phi[(\ln t - \mu)/\sigma]}{t\sigma R(t)} \qquad\qquad (19.55)$$

$$\text{MTBF} = \exp\left(\mu + \frac{\sigma^2}{2}\right)$$

The failure rate for the lognormal distribution is neither always increasing nor always decreasing. It takes different shapes depending on the parameters μ and σ.

Weibull Distribution

$$f(t) = \frac{\beta(t-\delta)^{\beta-1}}{(\theta-\delta)^{\beta}} \exp\left[-\left(\frac{t-\delta}{\theta-\delta}\right)^{\beta}\right] \qquad t \geq \delta \geq 0 \qquad (19.56)$$

$$F(t) = 1 - \exp\left[-\left(\frac{t-\delta}{\theta-\delta}\right)^{\beta}\right] \qquad\qquad (19.57)$$

$$R(t) = \exp\left[-\left(\frac{t-\delta}{\theta-\delta}\right)^{\beta}\right] \qquad\qquad (19.58)$$

$$h(t) = \frac{\beta(t-\delta)^{\beta-1}}{(\theta-\delta)^{\beta}} \qquad\qquad (19.59)$$

$$\text{MTBF} = \theta\Gamma\left(1+\frac{1}{\beta}\right) \qquad\qquad (19.60)$$

The failure rate for the Weibull distribution is decreasing when $\beta < 1$, is constant when $\beta = 1$ (same as the exponential distribution), and is increasing when $\beta > 1$.

Gamma Distribution

$$f(t) = \frac{\lambda^{\eta}}{\Gamma(\eta)} t^{\eta-1} e^{-\lambda t} \qquad t \geq 0 \qquad\qquad (19.61)$$

$$F(t) = \sum_{k=\eta}^{\infty} \frac{(\lambda t)^{k} e^{-\lambda t}}{k!} \qquad \text{when } \eta \text{ is an integer} \qquad (19.62)$$

$$R(t) = \sum_{k=0}^{\eta-1} \frac{(\lambda t)^{k} e^{-\lambda t}}{k!} \qquad \text{when } \eta \text{ is an integer} \qquad (19.63)$$

$$h(t) = \frac{f(t)}{R(t)} \qquad [\text{using Eqs. (19.61) and (19.63)}]$$

$$\text{MTBF} = \frac{\eta}{\lambda} \qquad\qquad (19.64)$$

We will now briefly discuss the application of some of the above continuous distributions in reliability analysis of products. *Exponential distribution* is a good model for the life of a complex system which has a large number of components. Because the exponential distribution has a constant failure rate, it is a good model for the useful life of many products after the end of the infant mortality period. Some applications for the exponential distribution are electrical and electronic systems, computer systems, and automobile transmissions. The *normal distribution* is used to model various physical, mechanical, electrical, or chemical properties of systems. Some examples are gas molecule velocity, wear, noise, chamber pressure from firing ammunition, tensile strength of aluminum alloy steel, capacity variation of electrical condensers, electrical power consumption in a given area, generator output voltage, and electrical resistance. The *lognormal distribution* is a skewed distribution and can be used to model situations where large occurrences are concentrated at the tail (left) end of the range. Some examples are amount of electricity used by different customers, downtime of systems, time to repair, light intensities of bulbs, concentration of chemical process residues, and automotive mileage accumulation by different customers. The *two-parameter Weibull distribution* can also be used to model skewed data. When $\beta < 1$, the failure rate for the Weibull distribution is decreasing and hence can be used to model infant mortality or debugging period or for situations when the reliability in terms of failure rate is improving or for reliability growth. When $\beta = 1$, the Weibull distribution is the same as the exponential distribution, and all of the previous comments for the exponential distribution are applicable. When $\beta > 1$, the failure rate is increasing, and hence it is a good model for determining wearout and end-of-useful life period. Some of the examples are corrosion life, fatigue life, antifriction bearings, transmission gears, and life of electronic tubes. The *three-parameter* Weibull distribution is a good model when we have a minimum life and the odds of the component failing before the minimum life are close to zero. Many strength characteristics of systems do have a minimum value significantly greater than zero. Some examples are electrical resistance, capacitance, and fatigue strength.

The failure rate for the gamma distribution is decreasing when $\eta < 1$, is constant when $\eta = 1$, and is increasing when $n > 1$.

Example 19.4 Normal Distribution. The time to failure for a component is normally distributed with an expected value of 20 hours and a standard deviation of 3 hours. We wish to answer the following questions:

1. What is the reliability of the component at 25 hours of operating time?
2. What is the probability that the component will fail between 25 hours and 28 hours?
3. What is the failure rate of the component at 25 hours of operating time?

For question 1, we have, using Eq. (19.50),

$$R(25) = 1 - F(25) = 1 - \Phi\left(\frac{25 - 20}{3}\right) = 1 - \Phi(1.667)$$

$$= 1 - 0.95221 = 0.04779$$

For question 2, we have, using Eq. (19.49),

$$F(28) - F(25) = \Phi(2.667) - \Phi(1.667)$$

$$= 0.99617 - 0.95221 = 0.04396$$

For question 3, we have, using Eq. (19.51),

$$h(25) = \frac{\phi[(25 - 20)/3]}{\sigma R(25)} = \frac{0.09949}{3 \times 0.04779} = 0.6939 \text{ failures/hour}$$

Example 19.5 Weibull Distribution.   A component has a Weibull failure distribution for the time to failure with the following parameters:

$$\delta = 1000 \qquad \beta = 4.5 \qquad \theta = 3000$$

One hundred components are put on test at time zero and we wish to answer the following questions.

1. What is the expected number of components functioning at 2000 time units?
2. What is the expected number of failures in the time interval (2000, 2100)?
3. What is the failure rate at 2000 time units?

We have using Eq. (19.58)

$$R(2000) = \exp\left[-\left(\frac{2000 - 1000}{3000 - 1000}\right)^{4.5} \right] = 0.9567$$

$$R(2100) = \exp\left[-\left(\frac{2100 - 1000}{3000 - 1000}\right)^{4.5} \right] = 0.9344$$

Hence the expected number of components functioning at 2000 is

$$E[N_S(2000)] = 95.67$$

Similarly

$$E[N_S(2100)] = 93.44$$

and hence the expected number of failures in the interval [2000, 2100] = 2.23. The failure rate is [using Eq. (19.59)]

$$h(2000) = \frac{4.5(1000)^{3.5}}{(2000)^{4.5}} = 0.00019887 \text{ failures/hour}$$

Example 19.6 Maintainability Analysis.   Maintainability is one of the system design parameters which has a great impact on the effectiveness of the system. Failures will occur no matter how reliable a system is made. The ability of a system to be maintained, i.e., retained in, or restored to, effective usable condition is often as important to system effectiveness as is its reliability. Maintainability is a characteristic of systems and it is designed just like reliability. It is concerned with such system attributes as accessibility to failed parts, diagnosis of failures, repairs, test points, test equipment and tools, maintenance manuals, displays, and safety. Maintainability can be defined as a characteristic of design and installation which imparts to a system a great inherent ability to be maintained, so as to lower the required maintenance labor-hours, skill levels, tools, test equipment, facilities, and logistics costs and thus achieve greater availability.

   Maintainability Measures.   Maintainability is the probability that a system in need of maintenance will be retained in, or restored to, a specified operational condition within a given period of time. Thus, the underlying random variable is the maintenance time. Let T be the repair time random variable. Then the maintainability function $M(t)$ is given by

$$M(t) = P[T \leq t]$$

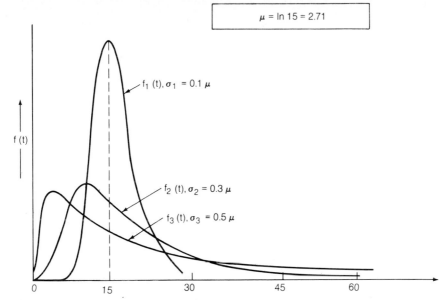

FIG. 19.5 Lognormal probability density function.

If the repair time T follows the exponential distribution with mean time to repair (MTTR) of $1/\mu$, where μ is the repair rate, then

$$M(t) = 1 - \exp\left(-\frac{t}{\text{MTTR}}\right)$$

Various other distributions such as lognormal, Weibull, and normal are used to model the repair time. In addition, other time-related indices such as median (50th percentile) and M_{max} (90th or 95th percentile) are used as maintainability measures. The lognormal probability density functions with a median time to repair of 15 minutes but with different values for standard deviation are given in Fig. 19.5, and Fig. 19.6 shows the associated maintainability functions. Distribution 1 has the least variability, and distribution 3 has the highest variability. From the maintainability function plot, different percentiles, such as 90th, can be easily read. Distribution 3 has the highest value for the 90th percentile. In other instances, the maintenance labor-hours per system operating hour or maintenance ratio (MR) can be specified and maintainability design goals then derived from such specifications.

MTTR, which is the mean of the distribution of system repair time, can be evaluated by

$$\text{MTTR} = \frac{\sum_{i=1}^{n} \lambda_i t_i}{\sum_{i=1}^{n} \lambda_i}$$

where n = number of components in the system
λ_i = failure rate of the ith repairable component
t_i = time required to repair the system when the ith component fails

In addition, other quantities, such as mean active corrective maintenance time and mean active preventive maintenance time, are used to measure maintainability.

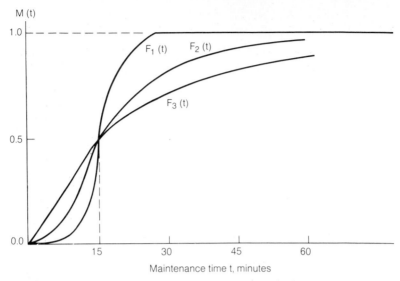

FIG. 19.6 Maintainability functions $M(t)$ based on lognormal distribution with median = 15 minutes.

Some components of the corrective maintenance tasks are

Localization: Determining the location of a failure to the extent possible, without using accessory test equipment.

Isolation: Determining the location of a failure by the use of accessory test equipment.

Disassembly: Disassembling the equipment to gain access to the item being replaced.

Interchange: Removing the failed item and installing the replacement.

Alignment: Performing any alignment, testing, and adjustment made necessary by the repair action.

Checkout: Performing checks or tests to verify that the equipment has been restored to a satisfactory operating condition.

As an example of MTTR computation, assume a communication system consists of five assemblies with the data given in Table 19.3. Column 2 gives the number of units n_i for assembly i. Column 3 indicates the failure rate per thousand hours for each unit. Thus, column 4 gives us the total failure rate for an assembly i. Column 5 gives the average time to perform all the maintenance actions discussed before. Then, MTTR is given by

$$MTTR = \frac{\sum n_i \lambda_i t_i}{\sum n_i \lambda_i}$$

$$= \frac{63.5}{161} = 0.394 \text{ hours}$$

TABLE 19.3 Worksheet for MTTR Prediction

1	2	3	4	5	6
					Repair time
		$\lambda_i,$	$n_i \lambda_i,$	$t_i,$	per 10^3 hours,
Assemblies	n_i	$\times 10^3$	$\times 10^3$	hours	$n_i \lambda_i t_i$
1	4	10	40	0.10	4.0
2	6	5	30	0.20	6.0
3	2	8	16	1.00	16.0
4	1	15	15	0.50	7.5
5	5	12	60	0.50	30.0
			$\Sigma = 161$		$\Sigma = 63.5$

Example 19.7 Gamma Distribution. The failure distribution of a component has the form of the gamma distribution with parameters $\eta = 2$ and $\lambda = 0.001$. We wish to determine the reliability of the component and the hazard function after an operation of 100 time units and also the mean life of the component. Here $\eta = 2$, $\lambda = 0.001$, $t = 100$, and hence $\lambda t = 0.1$. Using Eq. (19.61), we have

$$f(t) = \frac{(0.001)^2}{\Gamma(2)} \, t e^{-0.001t} \qquad t \geq 0$$

and

$$f(100) = \frac{(0.001)^2}{1} \times 100 e^{-0.1} = 9 \times 10^{-5}$$

Also from Eq. (19.63)

$$R(t) = \sum_{k=0}^{\eta-1} \frac{(\lambda t)^k e^{-\lambda t}}{k!}$$

Hence

$$R(100) = \frac{(0.1)^0 e^{-0.1}}{1} + \frac{(0.1)^1 e^{-0.1}}{1} = 0.995$$

Also, we have

$$h(100) = \frac{f(100)}{R(100)} = \frac{9 \times 10^{-5}}{0.995} = 9.09 \times 10^{-5} \text{ failures/time unit}$$

Mean life $= \dfrac{2}{0.001} = 2000$ time units [using Eq. (19.64)].

19.3 ELEMENTS OF STATISTICAL THEORY

In order to evaluate the reliability of a system, we have to observe and collect data. We only collect data for a part of the population. Statistics is the science of drawing conclusions about a population based on an analysis of the sample data from the population. A subset of observations selected from a population is called a *sample*. There are many different ways to take samples from a population. Furthermore, the conclusions that we can draw about the population will depend on how the sample is selected. We want the samples to be *representative* of the population and we want the sample to be random. A *random sample* is one such that every member of the population has an equal chance of being in the sample.

The process of drawing conclusions about populations based on sample data makes considerable use of statistics. A *statistic* is any function of the observations in a random sample that does not depend on any unknown parameters. For example, if $X_1, X_2, ..., X_n$ is a random sample of size n, then the sample mean $\bar{X}$ and the sample variance S^2 are statistics, where

$$\bar{X} = \frac{\sum_{i=1}^{n} X_i}{n} \qquad (19.65)$$

and

$$S^2 = \frac{\sum_{i=1}^{n} (X_i - \bar{X})^2}{n - 1} \qquad (19.66)$$

The statistical procedures require that we understand the probabilistic behavior of the statistic. In general, we call the probability distribution of a statistic a *sampling distribution*. There are several important sampling distributions used in reliability estimation and we briefly discuss some of them.

19.3.1 Sampling Distributions[6, 7]

For the sampling distribution of $\bar{X}$ (Eq. 19.65), we have

$$\mu_{\bar{X}} = \mu_X \qquad \text{and} \qquad \sigma_{\bar{X}}^2 = \frac{\sigma_X^2}{n} \qquad (19.67)$$

Central Limit Theorem.   If a population has a mean μ and a finite variance σ^2, then the sampling distribution of means approaches the normal distribution with mean μ and variance σ^2/n as n increases, i.e., the sampling distribution of mean is asymptotically normal.

Chi-Square Distribution.   Let $Z_1, Z_2, ..., Z_v$ be normally and independently distributed random variables, with mean $\mu = 0$ and variance $\sigma^2 = 1$. Then the random variable

$$\chi^2 = Z_1^2 + Z_2^2 + \cdots + Z_v^2 \qquad (19.68)$$

has chi-square distribution with v degrees of freedom with the probability density function

$$f_{\chi^2}(u) = \frac{1}{2^{v/2}\Gamma(v/2)} u^{(v/2)-1} e^{-u/2} \qquad u > 0 \qquad (19.69)$$

The percentage points of χ_v^2 distribution are given in the Appendix.

$$P[\chi_v^2 \geq \chi_{\alpha, v}^2] = \alpha \qquad (19.70)$$

For example, $P[\chi_{18}^2 \geq \chi_{0.05, 18}^2] = P[\chi_{18}^2 \geq 18.31] = 0.05$ When v is large, χ^2 can be approximated by normal distribution and the approximation is given by

$$\chi_{\alpha, \, v} \approx \frac{[Z_\alpha + \sqrt{2v - 1}]^2}{2} \tag{19.71}$$

For example, we wish to find $\chi^2_{0.05, \, 170}$. We have

$$\chi^2_{0.05, \, 170} \approx \frac{[1.645 + \sqrt{2(170) - 1}]^2}{2} = 201.10$$

Theorem. If $X_1, \ldots, X_n$ are random samples from a normal population with mean μ and variance σ^2, then $(n - 1)S^2/\sigma^2$ is distributed as chi-square with $(n - 1)$ degrees of freedom.

Goodness of Fit Test. The test statistic is given by

$$\chi^2 = \frac{\sum_{i=1}^k (f_{oi} - f_{ei})^2}{f_{ei}} \tag{19.72}$$

where k = number of classes
  f_{oi} = observed frequency of ith class
  f_{ei} = expected frequency of the ith class (should be greater than 5)

The larger the value of χ^2 (Eq. 19.72), the greater the discrepancy. The degrees of freedom are given by $v = k - 1 - m$, where m is the number of population parameters used in the computation of expected frequencies.

Student's t Distribution. Let Z be normally distributed with mean 0 and variance 1, and V be a chi-square with v degrees of freedom. If Z and V are independent, then the random variable

$$T = \frac{Z}{\sqrt{V/k}} \tag{19.73}$$

has the probability density function

$$f(t) = \frac{\Gamma[(v + 1)/2]}{\sqrt{\pi v} \, \Gamma(v/2)} \left(1 + \frac{t^2}{v}\right)^{-(v + 1)/2} \qquad -\infty < t < \infty \tag{19.74}$$

and is said to follow the t distribution with v degrees of freedom and abbreviated t_v.
  The cumulative distribution functions for the t distribution are given in the Appendix.
  If X_i, $i = 1, \ldots, n$ are independently normally distributed random variables with mean μ and variance σ^2, then $Z = (\bar{X} - \mu)/(\sigma/\sqrt{n})$ is $N(0, 1)$. Furthermore, $t = (\bar{X} - \mu)/(S/\sqrt{n})$ has a t distribution with $(n - 1)$ degrees of freedom.

F Distribution. Given two independently distributed chi-square variables χ_1^2 with v_1 degrees of freedom and χ_2^2 with v_2 degrees of freedom, the random variable

$$F = \frac{\chi_1^2/v_1}{\chi_2^2/v_2} \tag{19.75}$$

has an F distribution with v_1 and v_2 degrees of freedom. The probability density function of F is given by

$$f(F) = \frac{\Gamma[(v_1 + v_2)/2]}{\Gamma(v_1/2)\Gamma(v_2/2)} \left(\frac{v_1}{v_2}\right)^{v_1/2} \frac{F^{(v_1/2)-1}}{[1 + (v_1/v_2)F]^{(v_1+v_2)/2}} \qquad 0 < F < \infty \qquad (19.76)$$

The percent points of the F distribution are given in Appendix A.7. For example, $P[F \geq F_{0.05, 5, 10}] = P[F \geq 3.33] = 0.05$. We also have

$$F_{(1-\alpha), v_1, v_2} = \frac{1}{F_{\alpha, v_1, v_2}}$$

Consider two random samples, one from each of two normal populations; then the statistic $(s_1^2/\sigma_1^2)/(s_2^2/\sigma_2^2)$ is distributed as an F distribution with $(n_1 - 1)$ and $(n_2 - 1)$ degrees of freedom.

19.3.2 Statistical Inference

Statistical inference is the process by which information from sample data is used to draw conclusions about the population from which the sample was selected. The techniques of statistical inference can be divided into two major areas: parameter estimation and hypothesis testing.

Parameter Estimation.   One of the important problems in reliability is to estimate the parameters of the life distributions. For example, we wish to know the average life of a light bulb. An *estimator* is some function of the sample values which provides an estimate of the population parameter. A single-valued estimate of the population parameter is called a *point estimate*. Let $\hat{\theta} = h(X_1, \ldots, X_n)$ be a point estimator of the parameter θ. Some of the properties of the estimators are given below.

1. $\hat{\theta}$ is an *unbiased* estimator of θ if $E(\hat{\theta}) = \theta$.
2. $\hat{\theta}$ is a *consistent* estimator of θ if $\hat{\theta}$ converges to θ in probability as sample size increases.
3. If $\hat{\theta}_1$ and $\hat{\theta}_2$ are two different unbiased estimators of θ and if $E[(\hat{\theta}_1 - \theta)^2] < E[(\hat{\theta}_2 - \theta)^2]$ then $\hat{\theta}_1$ is a more *efficient* estimator of θ than $\hat{\theta}_2$.
4. $\hat{\theta}$ is a *sufficient* estimator if no other independent estimate based on the sample is able to yield any further information about the parameter which is being estimated.

To construct an interval estimator of the unknown parameter θ, we find two numbers θ_L and θ_U such that

$$P[\theta_L \leq \theta \leq \theta_U] = 1 - \alpha$$

and the resulting interval, $\theta_L \leq \theta \leq \theta_U$, is called a $100(1 - \alpha)$ percent confidence interval for θ. The interpretation of the confidence interval is that if many random samples are collected, and a $100(1 - \alpha)$ percent confidence interval on θ is computed from each sample, then $100(1 - \alpha)$ percent of these intervals will contain the true value of θ.

Tests of Hypotheses.   Many reliability problems require that we decide whether or not a statement about some parameter of the life distribution is true or false. This statement is usually called a *hypothesis*. For example, we want to say that the mean

TABLE 19.4   Decisions in Hypothesis Testing

	H_0 is true	H_0 is false
Accept H_0	No error $(1 - \alpha)$	Type II error or β error
Reject H_0	Type I error or α error	No error $(1 - \beta)$

life of a component which follows exponential distribution is at least 1000 hours. The *null hypothesis* (H_0) is the hypothesis that we are interested in. Any hypothesis which differs from this is called the *alternate hypothesis*, designated by H_1. Type I error, designated α (also called *producer's risk*), is the error made in rejecting a hypothesis that is true. 100α is the *significance* level of the test. Type II error (see Table 19.4), designated β (also called *consumer's risk*), is the error made in accepting a hypothesis which is false. $(1 - \beta)$ is called the *power* of the test.

A graph of $(1 - \beta)$ versus values of H_1 is called the *power curve*. A graph of β versus values of H_1 is called the *operating characteristic* (OC) curve. Thus the OC curve gives us a Type II error as a function of different values of the underlying parameter for which we have formulated a hypothesis. Since reliability is a function of time as well as of the parameters of the underlying distribution, it is not practical to provide OC curves in terms of actual reliability. OC curves are mostly provided in terms of the unknown parameter of the assumed distribution. In practice, the problem of selecting a test procedure on the basis of operating characteristics can be formulated as that of selecting two points on the OC curve and then finding a test plan which meets this specification. (See Sec. 19.9 for illustration.)

19.4   THE EXPONENTIAL DISTRIBUTION

The exponential distribution is a very popular and easy to use model to represent time to failure. Selection of the exponential distribution as an appropriate model implies that the failure rate is constant over the range of predictions. For certain failure situations and over certain portions of product life, the assumption of a constant failure rate may be appropriate. The probability density function for an exponentially distributed time-to-failure random variable T is given by (see Eq. 19.43)

$$f(t, \lambda) = \lambda e^{-\lambda t} \qquad t \geq 0 \qquad (19.77)$$

where the parameter λ is the failure rate. The reciprocal of the failure rate ($\theta = 1/\lambda$) is the mean or expected life. For products that are repairable, the parameter θ is referred to as the mean time between failures (MTBF) and for nonrepairable products θ is called the mean time to failure (MTTF). The parameter λ must be known (or estimated) for any specific application situation.

The reliability function is given by

$$R(t) = e^{-\lambda t} \qquad t \geq 0 \qquad (19.78)$$

The relationship between $f(t)$ and $R(t)$ is illustrated in Fig. 19.7. For any value of t the quantity $R(t)$ provides the chance of survival beyond time t. If the reliability function is evaluated at the MTBF, it should be noted that $R(\theta) = 0.368$, or there is only a 36.8 percent chance of surviving the mean life.

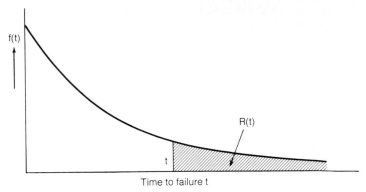

FIG. 19.7   The exponential distribution.

19.4.1   Point Estimation[5,7]

The estimator for the mean life parameter θ is given by

$$\hat{\theta} = \frac{T}{r} \tag{19.79}$$

where T = the total accumulated test time considering both failed and unfailed items, and r = the total number of failures. It is reasonable to take the estimator for λ as $\hat{\lambda} = 1/\hat{\theta}$. The estimator $\hat{\theta}$ is termed a *maximum likelihood* estimator. It has the properties of *unbiasedness, minimum variance, efficiency,* and *sufficiency.*
   The reliability is estimated by

$$\hat{R}(t) = e^{-\hat{\lambda}t} \qquad t \geq 0 \tag{19.80}$$

and if one wants to estimate a time for a given reliability level R, this is obtained by

$$\hat{t} = \hat{\theta} \ln (1/R) \tag{19.81}$$

Confidence-Interval Estimates.[8-11]   The confidence intervals for the mean life or for reliability depend on the testing situation. A life-testing situation where n devices are placed on test and it is agreed to terminate the test at the time of the rth failure $(r \leq n)$ is called *failure-censored* and is termed *Type II life testing.* A time-censored life test is one where the total accumulated time T is specified; this is termed *Type I life testing.* So, in Type I testing T is specified and r occurs as a result of testing, whereas in a Type II situation, r is specified and T results from testing. The confidence limits are slightly different for each situation.

Failure-Censored Life Tests.   In this situation the number of failures r at which the test will be terminated is specified with n items placed on test. The $100 (1 - \alpha)$ percent two-sided confidence interval for θ, the mean life, is

$$\frac{2T}{\chi^2_{\alpha/2,\, 2r}} \leq \theta \leq \frac{2T}{\chi^2_{1-\alpha/2,\, 2r}} \tag{19.82}$$

The quantities $\chi^2_{\delta,\, \nu}$ are the $(1 - \delta)$ percentiles of a chi-square distribution with ν degrees of freedom.

The one-sided lower $100(1 - \alpha)$ percent confidence limit is given by

$$\frac{2T}{\chi^2_{\alpha, \, 2r}} \leq \theta \tag{19.83}$$

The confidence limits on reliability for any specified time can be found by substituting the above limits on θ into the reliability function.

Time-Censored Life Tests.   In this situation the accumulated test time T is specified for the test and the test produces r failures. The $100(1 - \alpha)$ percent two-sided confidence interval on the mean life is given by

$$\frac{2T}{\chi^2_{\alpha/2, \, 2(r+1)}} \leq \theta \leq \frac{2T}{\chi^2_{1-\alpha/2, \, 2r}} \tag{19.84}$$

If only the one-sided $100(1 - \alpha)$ percent lower confidence interval is desired, this is given by

$$\frac{2T}{\chi^2_{\alpha, \, 2(r+1)}} \leq \theta \tag{19.85}$$

For a number of failure-censored life tests, the termination time T_r is a random variable with

$$E[T_r] = \theta \sum_{i=1}^{r} \frac{1}{n - i + 1} \tag{19.86}$$

and

$$V[T_r] = \theta^2 \sum_{i=1}^{r} \frac{1}{(n - i + 1)^2} \tag{19.87}$$

Example 19.8.   A designer wants to use a certain electronic counter in the design of a piece of equipment. Results from a previous test on 50 counters showed that none failed during a 3000-hour test. Based on this test and a 90 percent one-sided confidence limit, is it realistic to expect that the probability of survival of this counter will be at least 85 percent for a one-year period of operation? Assume exponential distribution as the underlying failure distribution. For this example, we have

$$\text{Total test time} = 50 \times 3000 = 150,000 \text{ hours}$$

Hence, using Eq. (19.85), we have

$$\frac{2T}{\chi^2_{0.1, \, 2}} \leq \theta \quad \text{or} \quad \frac{2 \times 150,000}{4.605} \leq \theta \quad \text{or} \quad 65,146 \leq \theta$$

Hence,

$$R(365 \times 24) = R(8760) = e^{-8760/65,146} = 0.8742$$

Hence it is realistic to expect that the probability of survival of this counter will be at least 85 percent for a one-year period of operation.

19.4.2   Hypothesis Testing

Hypothesis testing is another approach to statistical decision making. Whereas confidence limits offer some degree of protection against meager statistical knowledge, the hypothesis testing approach is easily misused. To use this approach one should be very familiar with the inherent errors involved in applying the hypothesis testing procedure.

Consider a Type II testing situation where n items are placed on test and the test is terminated at the time of the rth failure. An MTBF goal θ_g is to be verified. The hypotheses for this situation are

$$H_0: \quad \theta \leq \theta_g$$
$$H_1: \quad \theta > \theta_g \tag{19.88}$$

If H_0 is rejected, it will be concluded that the goal has been met.

The statistic calculated from test data is

$$\chi_c^2 = \frac{2T}{\theta_g} \tag{19.89}$$

and the decision criteria is to reject H_0 and assume that the goal has been met if $\chi_c^2 > \chi_{\alpha, 2r}^2$. Here the level of significance is taken as α.

The probability of accepting a design, that is, concluding that the goal has been met where the design has a true MTBF of θ_1, is

$$P_1 = P\left[\chi_{2r}^2 \geq \frac{\theta_g}{\theta_1} \chi_{\alpha, 2r}^2\right] \tag{19.90}$$

and can be looked up in χ^2 tables.

Example 19.9. The reliability of an electronic device is under investigation. The MTBF goal for this device for a particular application is 2000 hours.

Ten units are placed on test and the test is truncated at the time of the second failure. The failure times are

$$t_i: \quad 187 \text{ hours}; \quad 462 \text{ hours}$$

The estimated mean life would be

$$\hat{\theta} = \frac{187 + 462 + 8(462)}{2} = 2172.5 \text{ hours}$$

The calculated χ^2 value is

$$\chi_c^2 = \frac{2(4345)}{2000} = 4.345$$

Using $\alpha = 0.10$, the critical value as obtained from χ^2 tables is

$$\chi_{0.10, 4}^2 = 7.779$$

So, in this case we would not conclude that the goal has been met.

Now let us investigate the Type II error associated with this procedure. Suppose the true mean was $\theta_1 = 3000$ hours; then the probability of concluding that the goal was met by this testing procedure is

$$P_1 = P\left[\chi_4^2 \geq \frac{2000}{3000} (7.779)\right] = 0.24$$

Or, even though the goal is 2000 hours, the decision procedure has only a 24 percent chance of accepting a design with a true MTBF of 3000 hours. Obviously the decision procedure may not be acceptable in a practical situation. The procedure can be improved by increasing r, the number of failures. For example, if we used $r = 8$, then

$$P_1 = P\left[\chi^2_{16} \geq \frac{2000}{3000}(7.779)\right] = 0.995$$

Or, we now only have a 0.5 percent chance of concluding that the design has not met the MTBF goal.

19.5 THE WEIBULL DISTRIBUTION

The Weibull distribution[11-13] is considerably more versatile than the exponential distribution and can be expected to fit many different failure patterns. However, when applying a distribution, the failure pattern should be carefully studied and the mixture of failure modes noted. The selection and application of a distribution should then be based on this study and on any knowledge of the underlying physical failure phenomena.

Graphical procedures for the Weibull distribution are attractive in that they provide practitioners with a visual representation of the situation. Although the graphical approach will be covered in the following material, it should be recognized that there are better statistical estimation procedures; however, these procedures require the use of a computer. Different computer programs are available in the open literature and should be used wherever possible.

The reliability function for the three-parameter Weibull distribution is given by

$$R(t) = \exp\left[-\left(\frac{t - \delta}{\theta - \delta}\right)^{\beta}\right] \qquad t \geq \delta \qquad (19.91)$$

where δ = the minimum life ($\delta \geq 0$)
$\qquad \theta$ = the characteristic life ($\theta > \delta$)
$\qquad \beta$ = the Weibull slope or shape parameter ($\beta > 0$)

The two-parameter Weibull has a minimum life of zero and the reliability function is

$$R(t) = \exp\left[-\left(\frac{t}{\theta}\right)^{\beta}\right] \qquad t \geq 0 \qquad (19.92)$$

where θ and β are as previously defined ($\theta > 0$). The term characteristic life resulted from the fact that $R(\theta) = 0.368$; or, there is a 36.8 percent chance of surviving the characteristic life for any Weibull distribution.

The hazard function for the Weibull distribution is given by

$$h(t) = \frac{\beta}{\theta^{\beta}} t^{\beta - 1} \qquad t \geq 0 \qquad (19.93)$$

It can be seen that the hazard function will decrease for $\beta < 1$, increase for $\beta > 1$, and remain constant for $\beta = 1$.

The expected or mean life for the two-parameter Weibull is given by

$$\mu = \theta \Gamma \left(1 + \frac{1}{\beta} \right)$$ (19.94)

where $\Gamma(\cdot)$ is a gamma function and its value can be found in gamma tables. The standard deviation for the Weibull distribution is

$$\sigma = \theta \sqrt{\Gamma \left(1 + \frac{2}{\beta} \right) - \Gamma^2 \left(1 + \frac{1}{\beta} \right)}$$ (19.95)

19.5.1 Graphical Estimation[9, 11]

The Weibull distribution is very amenable to graphical estimation. This procedure will now be illustrated.

The cumulative distribution for the two-parameter Weibull is given by

$$F(t) = 1 - \exp\left[-(t/\theta)^\beta\right]$$ (19.96)

Hence by rearranging and taking logarithms one can obtain

$$\ln \left(\ln \frac{1}{1 - F(t)} \right) = \beta \ln t - \beta \ln \theta$$ (19.97)

Weibull paper is scaled such that t_j and $p_j = F(t_j)$ can be plotted directly and a straight line fitted to the data. A convenient way to assign values to p_j for plotting is to calculate

$$p_j = \frac{j - 0.3}{n + 0.4}$$ (19.98)

where j is the order of magnitude of the observation and n is the sample size. This is essentially median rank plotting and tables for median ranks are available[11] and can also be used.

Example 19.10.   The design for an aluminum flexible-drive hub on computer disk packs is under study. The failure mode of interest is fatigue. Data from 14 hubs placed on an accelerated life test follow (see Table 19.5). The plotted data with a visually fitted line are shown on the Weibull paper in Fig. 19.8. The slope of this line provides an estimate of β, which in this case is about 2.3. Most commercially available Weibull papers have a special scale for estimating β.

The characteristic life can be estimated by recalling that $R(\theta) = 0.368$; or that $F(\theta) = 0.632$. So, one can locate 63.2 percent on the cumulative probability scale, project across to the plotted line, and then project down to the time-to-failure axis. In Fig. 19.8, the characteristic life is about 330,000 cycles.

Weibull paper offers a quick and convenient method for analyzing a failure situation. The population line plotted on the paper can be used to estimate either percent failure at a given time or the time at which a given percentage will fail. Also, a concave plot is indicative of a nonzero minimum life.[9-11]

Confidence Limits for Graphical Analysis.   Lower and upper confidence limits for graphical analysis can be computed using the following equations.[11]

TABLE 19.5 Data for Aluminum Flexible-Drive Hub

j	Cycles to failure	p_j^*
1	93,000	5.6
2	147,000	13.5
3	192,000	21.7
4	214,000	29.8
5	260,000	37.9
6	278,000	46.0
7	297,000	54.0
8	319,000	62.1
9	349,000	70.1
10	388,000	78.2
11	460,000	86.3
12	510,000	94.4

*The p_j values were calculated using Eq. (19.98).

Lower Limit

$$w_\alpha = \frac{j/(n-j+1)}{F_{1-\alpha,\ 2(n-j+1),\ 2j} + j/(n-j+1)} \qquad \alpha \geq 0.50 \qquad (19.99)$$

Upper Limit

$$w_\alpha = \frac{[j/(n-j+1)]F_{\alpha,\ 2j,\ 2(n-j+1)}}{1 + [j/(n-j+1)]F_{\alpha,\ 2j,\ 2(n-j+1)}}, \qquad \alpha < 0.50 \qquad (19.100)$$

Use $\alpha/2$ for two-sided limits.

19.6 SUCCESS-FAILURE TESTING

Success-failure testing[9,10] describes a situation where a product (component, subsystem, etc.) is subjected to a test for a specified length of time T (or cycles, stress reversals, miles, etc.). The product either survives to time T (i.e., survives the test) or fails prior to time T.

Testing of this type can frequently be found in engineering laboratories where a test "bogey" or target has been established and new designs are tested against this bogey. The bogey will specify a set number of cycles in a certain test environment and at a predetermined stress level.

The probability model for this testing situation is the binomial distribution given by

$$p(y) = \binom{n}{y} R^y (1 - R)^{n-y} \qquad y = 0, 1, 2, \ldots, n \qquad (19.101)$$

where R = the probability of surviving the test
 n = the number of items placed on test
 y = the number of survivors

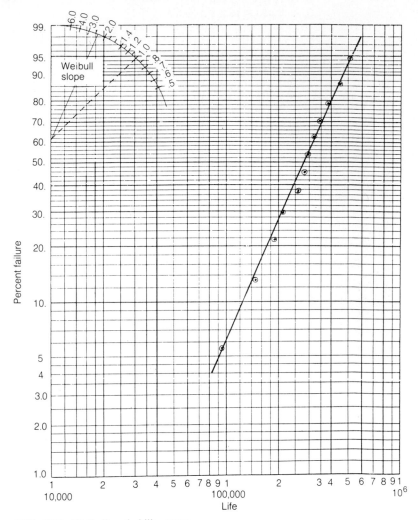

FIG. 19.8 Weibull probability paper.

The value R is the reliability, which is the probability of surviving the test. Procedures for estimating product reliability R based on this testing situation will now be covered.

19.6.1 Point Estimate

The point estimate of reliability is simply calculated as

$$\hat{R} = \frac{y}{n}$$ (19.102)

19.6.2 Confidence Limit Estimate

The $100(1 - \alpha)$ percent lower confidence limit on the reliability R is calculated by

$$R_L = \frac{y}{y + (n - y + 1)F_{\alpha,\,2(n-y+1),\,2y}} \qquad (19.103)$$

where $F_{\alpha,\,2(n-y+1),\,2y}$ is obtained from F tables. Here again n is the number of items placed on test and y is the number of survivors. The $100(1 - \alpha)$ percent upper confidence limit on R is given by

$$R_U = \frac{(y + 1)F_{\alpha,\,2(y+1),\,2(n-y)}}{(n - y) + (y + 1)F_{\alpha,\,2(y+1),\,2(n-y)}} \qquad (19.104)$$

The F tables that are usually available are somewhat limited. Therefore, it is convenient to have an approximation for the lower confidence limit that uses the standard normal distribution. The lower confidence on reliability can be approximated by

$$R_L = \frac{y - 1}{n + z_\alpha \sqrt{n(n - y + 1)/(y - 2)}} \qquad (19.105)$$

where z_α = the standard normal variate as given in Table 19.6
$\quad\ y$ = the number of successes
$\quad\ n$ = the sample size

It should be noted that $P[Z \geq z_\alpha] = \alpha$ where Z is the standard normal variable. Values given in Table 19.6 can be read from cumulative distribution tables for standard normal variable given in Appendix A.4.

Example 19.11   A weapon system has completed a test schedule. The test is equivalent to 60 missions. Dividing the test schedule up into 60 missions results in seven failed missions. Let us estimate the mission reliability.
    In this case the number of successes (y) is

$$y = 60 - 7 = 53 \text{ successful missions}$$

out of $n = 60$ missions. Then the point estimate for mission reliability is

$$\hat{R}_m = 53/60 = 0.883$$

Let us now find a 75 percent lower confidence limit. The exact lower 75 percent limit is found by using an F value of

$$F_{0.25,\,8,\,106} = 1.31$$

TABLE 19.6   Standard Normal Variates

Confidence level, $1 - \alpha$, %	Confidence factor, z_α
95	1.645
90	1.281
80	0.841
70	0.525
50	0.0

and substituting into the confidence limit Eq. (19.103) gives

$$R_L = \frac{53}{53 + (8 \times 1.31)} = 0.835$$

Or, the 75 percent lower confidence limit on mission reliability is $0.835 \le R_m$. If the normal approximation was used, the lower limit's value would be

$$R_L = \frac{52}{60 + 0.675\sqrt{[60(60 - 53 + 1)/51]}} = 0.838$$

As can be seen, this approximation provides limits that are reasonably close to the exact values.

19.6.3　Success Testing

In receiving inspection and sometimes in engineering test labs one encounters a situation where a no-failure ($r = 0$) test is to be specified. The concern is usually to ensure that a reliability level has been achieved at a specified confidence level. A special adaptation of the confidence limit formula can be derived for this situation.

For the special case where $r = 0$ (i.e., no failures), the lower $100(1 - \alpha)$ percent confidence limit on the reliability is

$$R_L = \alpha^{1/n} \tag{19.106}$$

where α is the level of significance and n is the sample size (i.e., number placed on test). Then with $100(1 - \alpha)$ percent confidence, we can say that

$$R_L \le R$$

where R is the true reliability.

If we let $C = 1 - \alpha$ be the desired confidence level (i.e., 0.80, 0.90, etc.), then the necessary sample size to demonstrate a desired reliability level R is

$$n = \frac{\ln (1 - C)}{\ln R} \tag{19.107}$$

For example, if $R = 0.80$ is to be demonstrated with 90 percent confidence, we have

$$n = \frac{\ln 0.10}{\ln 0.80} = 11$$

Thus, we must place 11 items on test and allow no failures. This is frequently referred to as success testing.

19.6.4　Conversion to the Exponential Distribution

In some cases, where it is reasonable to assume an exponential time-to-failure distribution, it is desirable to obtain the reliability to predict failure over time. If

$$n = \text{number placed on test}$$

$$r = \text{number failing}$$

$$y = \text{number of successes}$$

then a point estimate of the fraction failing the fixed length T test is

$$\hat{p} = \frac{r}{n}$$

Or, the reliability is estimated as

$$\hat{R} = \frac{y}{n}$$

From the exponential distribution, the probability of failing prior to time T is given by

$$F(T) = 1 - e^{-T/\theta}$$

So, an estimate of the MTBF can be obtained from

$$\frac{r}{n} = 1 - e^{-T/\theta}$$

or

$$\theta = \frac{T}{\ln [1/(1 - r/n)]} \tag{19.108}$$

Confidence limits can also be placed on the MTBF. If we denote the lower confidence limits on the reliability by

$$R_L \leq R$$

then, one can also say

$$R_L \leq e^{-T/\theta}$$

or

$$\frac{T}{\ln (1/R_L)} \leq \theta \tag{19.109}$$

In the case of success testing where no failures are allowed, the lower confidence limit on the MTBF is (using Eq. (19.106))

$$-\frac{nT}{\ln \alpha} \leq \theta$$

Or, for any specified time t, the $100(1 - \alpha)$ percent lower confidence limit on reliability is

$$\exp \left[-\frac{t(-\ln \alpha)}{nT} \right] \leq R(t) \tag{19.110}$$

19.7 SYSTEM RELIABILITY MODELS

Dynamic system reliability models are an extension of the static models (Example 19.1) where time-dependent reliability functions are used for each subsystem. Consider the following notation:

$R_i(t)$ = the reliability function for the ith subsystem

$R_s(t)$ = the system reliability function

These will be used to reformulate series and parallel systems considered in Example 19.1.

19.7.1 Series Configuration

The system reliability is given by

$$R_s(t) = \prod_{i=1}^{n} R_i(t) \qquad (19.111)$$

where n = the number of subsystems.
The failure rate $h_s(t)$ for the series system is given by

$$h_s(t) = \sum_{i=1}^{n} h_i(t) \qquad (19.112)$$

where $h_i(t)$ is the failure rate for the ith subsystem.
If all subsystems have an exponentially distributed time to failure, then

$$h_s(t) = \sum_{i=1}^{n} \lambda_i \qquad (19.113)$$

where λ_i is the failure rate for the ith subsystem.

19.7.2 Parallel Configuration

The system reliability for a parallel configuration where all parallel subsystems are activated when the system is turned on is given by

$$R_s(t) = 1 - \prod_{i=1}^{n} [1 - R_i(t)] \qquad (19.114)$$

If the time to failure for each subsystem is exponential, then

$$R_s(t) = 1 - \prod_{i=1}^{n} [1 - \exp{(-\lambda_i t)}] \qquad (19.115)$$

If $\lambda_i = \lambda = 1/\theta$ for all i, then the expected system life given by

$$E_s(T) = \sum_{i=1}^{n} \frac{\theta}{i} \qquad (19.116)$$

A standby parallel redundant system is depicted in Fig. 19.9. In this system the standby unit is activated by the switching mechanism S. For a system with two units (i.e., one standby unit), the system reliability is

$$R_s^2(t) = R_1(t) + \int_0^t f_1(t_1)R_2(t - t_1)\, dt_1 \qquad (19.117)$$

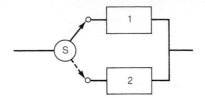

FIG. 19.9 Standby redundant configuration.

where it is assumed that the switch cannot fail. Here $f_1(t)$ is the probability density function for device 1 and $R_2(t)$ is the reliability function for device 2. For the special case where each subsystem is identical with an exponential time to failure, the system reliability is

$$R_a^2(t) = e^{-\lambda t}(1 + \lambda t) \qquad t \geq 0 \tag{19.118}$$

Similarly, if we have three subsystems, the reliability of the system is given by

$$R_s^3(t) = e^{-\lambda t}\left[1 + \lambda t + \frac{(\lambda t)^2}{2}\right], \qquad t \geq 0 \tag{19.119}$$

For the general case, when we have n subsystems,

$$R_s^n(t) = e^{-\lambda t}\sum_{i=0}^{n-1} \frac{(\lambda t)^i}{i!} \tag{19.120}$$

If the switch has a reliability of P_s, then the reliability for the two-unit system is

$$R_s^2(t) = R_1(t) + P_s \int_0^t f_1(t_1)R_2(t - t_1)\, dt_1 \tag{19.121}$$

For identical subsystems we have an exponential time to failure of

$$R_s^2(t) = e^{-\lambda t}(1 + P_s \lambda t) \qquad t \geq 0 \tag{19.122}$$

19.8 BAYESIAN RELIABILITY IN DESIGN AND TESTING

Probability is a mathematical concept used in conjunction with random events. The concept of relative frequency is widely used to define probability. Suppose we are interested in the probability of an event A associated with a random experiment ε. We perform the experiment N times and event A occurs N_A times, $0 \leq N_A \leq N$. Then, the probability of event A is defined by

$$P(A) = \lim_{N \to \infty} \frac{N_A}{N}$$

Thus probability is associated with "long-run" percentages. Another popular use of probability is as "degree of belief." The term *Bayes' probability* has become associated with degree-of-belief probability. The degree-of-belief probability is also called *subjective* probability. If a person is prudent, the degree-of-belief probability is the same as the long-run percentage, when that percentage becomes known.

Bayes' formula provides a means of converting the degree-of-belief probability we had before any test data or objective data were obtained to the degree-of-belief probability after the test data or objective data are obtained. Thus, Bayes' formula provides the mathematics by which a rational person has the degree-of-belief probability changed by evidence or some outcomes of a random experiment. Thus, the degree-of-belief or subjective probabilities are dynamic in nature. The Bayesian approach is based on the work by Reverend Thomas Bayes in the eighteenth century. His work was republished in *Biometrika*.[14] The Bayesian approach is illustrated in Fig. 19.10.

The mathematics underlying the Bayesian approach are not controversial. The approach is controversial in terms of interpretation of equations concerning knowledge as "prior" and what constitutes reasonable prior knowledge. Also, it is difficult to develop experimental or analytical methods for the quantification of belief in the performance of new systems based on past experience.

Example 19.12.   We are concerned with the reliability level of a new but as yet untested system. We feel that the system might have one of two possible reliability levels denoted by R_1 and R_2. (It can be easily generalized for a continuous range of reliability.) Based on past experience we believe that the system may have a reliability level $R_1 = 0.95$, but if the design engineer has miscalculated on a certain factor in question, then the reliability level will be at a lower level $R_2 = 0.75$.

We will express our confidence in the system designer by assigning an 80 percent chance that level R_1 has been attained, thus leaving a 20 percent chance that level R_2 has been attained.

In this example, the confidence probabilities are subjective in nature and are based on past experience with similar systems. Objective data are obtained by testing. Now suppose that we test one system and find that it operates successfully. We now wish to find the probability that level R_1 has been attained. New information should modify our prior belief in the reliability of the system.

Bayes' theorem is based on the concept of conditional probabilities. We have

$$P(A \mid B) = P(A) \, \frac{P(B \mid A)}{P(B)} \qquad (19.123)$$

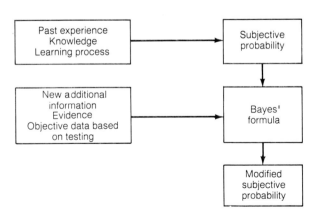

FIG. 19.10 Bayesian approach.

$P(A)$ is the *prior* probability of the event A before information about B becomes available. In the Bayesian approach, $P(A)$ is generally a degree-of-belief or subjective probability. $P(A \mid B)$ is called the *posterior* probability, which is the probability of event A modified by the objective data about event B.

The conditional probability can be extended to the following version of Bayes' theorem. Let $A_1, A_2, \ldots, A_k$ be a partition of the sample space associated with the random experiment which is used to collect objective data. Then, Bayes' theorem is [see also Eq. (19.11)]

$$P(A_i \mid B) = \frac{P(A_i)P(B \mid A_i)}{\sum_{i=1}^{k} P(A_i)P(B \mid A_i)} \tag{19.124}$$

Applications to Example 19.12. Let us apply the previous result to Example 19.12. Let us define

R_i = the event that reliability level R_i has been attained

S_i = the event that the *l*th system results in a success

We want to find

$$P(R_1 \mid S_1) = \frac{P(R_1)P(S_1 \mid R_1)}{P(R_1)P(S_1 \mid R_1) + P(R_2)P(S_1 \mid R_2)}$$

$$= \frac{0.80 \times 0.95}{0.80 \times 0.95 + 0.20 \times 0.75}$$

$$= 0.835$$

Thus, our confidence that level R_1 has been achieved goes up from 80 to 83.5 percent.

Let us assume that the second system is tested, and it is also successful. Then we wish to compute

$$P(R_1 \mid S_1 \cap S_2) = \frac{P(R_1)P(S_1 \cap S_2 \mid R_1)}{P(R_1)P(S_1 \cap S_2 \mid R_1) + P(R_2)P(S_1 \cap S_2 \mid R_2)}$$

$$= \frac{0.80 \times (0.95)^2}{(0.80)(0.95)^2 + (0.20)(0.75)^2}$$

$$= 0.865$$

We can see how the probability of events R_1 is updated by application of Bayes' theorem as new information becomes available. If we had tested one system and it resulted in a failure, then, as expected, our confidence would go down.

$$P(R_1 \mid F_1) = \frac{P(R_1)P(F_1 \mid R_1)}{P(R_1)P(F_1 \mid R_1) + P(R_2)P(F_1 \mid R_2)}$$

$$= \frac{0.80 \times 0.05}{0.80 \times 0.05 + 0.20 \times 0.25}$$

$$= 0.4444$$

19.8.1   Bayes' Theorem for Continuous Random Variables

This theorem enables prior knowledge about an unknown parameter θ (for example, true reliability or failure rate or MTBF) to be combined with subsequent test data to produce an updated or posterior knowledge about the unknown parameter θ about which we have some prior knowledge in the form of a probability density function.

The Bayes' theorem is

$$k(\theta \mid y) = \frac{h(\theta)g(y \mid \theta)}{\int_\theta h(\theta)g(y \mid \theta)\, d\theta} \tag{19.125}$$

where $h(\theta)$ = prior probability density function that expresses our degree of belief in θ

$g(y \mid \theta)$ = conditional probability density function for y (hard or actual test data or statistic) given θ

$k(\theta \mid y)$ = posterior probability density function for θ given y

Example 19.13.   Let us consider the continuous version of Example 19.12, and say that our belief in the reliability R of the system is expressed by a prior probability density function.

$$h(R) = 4R^3 \qquad 0 \le R \le 1$$

Thus, the expected value of reliability for the system is

$$E[R] = \int_0^1 R \cdot 4R^3 \, dR = 0.80$$

Let us assume that we test one system and it is a success. Then

$$k(R \mid S_1) = \frac{h(R)g(S_1 \mid R)}{\int_0^1 h(R)g(S_1 \mid R)\, dR}$$

$$= \frac{4R^3 \cdot R}{\int_0^1 4R^3 \cdot R \, dR}$$

$$= 5R^4 \qquad 0 \le R \le 1$$

Thus, the posterior expected value for reliability is

$$E[R] = \int_0^1 R \cdot 5R^4 \, dR = \frac{5}{6}$$

Similarly, we have the posterior reliability as given below when we test one system and it fails.

$$k(R \mid F_1) = \frac{h(R)g(F_1 \mid R)}{\int_0^1 h(R)g(F_1 \mid R)\, dR}$$

$$= \frac{4R^3(1 - R)}{\int_0^1 4R^3(1 - R)\, dR}$$

$$= 5(4R^3 - 4R^4) \qquad 0 \le R \le 1$$

The posterior expected value for reliability now is less than before, as expected, and is given by

$$E[R] = \int_0^1 R \cdot 5(4R^3 - 4R^4)\, dR = \frac{2}{3}$$

19.9   SEQUENTIAL LIFE TESTING

Sequential testing[15] is a situation where one continually reassesses test results to arrive at a decision with the minimum amount of testing. So, at decision points in the test, the alternatives that are available are

1. Accept the product
2. Reject the product
3. Continue testing—not enough information to make a decision

The OC curve (see Sec. 19.3.2) for a sequential test must be determined beforehand by specifying four values. These values are illustrated in the OC curve in Fig. 19.11). The values R_0, R_1, α, β must be specified in order to design the sequential test. Once the OC curve has been determined, then a sequential graph can be constructed that will have three regions as shown in Fig. 19.12.

19.9.1   Success-Failure Testing—Sequential Life Testing for Binomial Distribution[11]

In this situation each product tested either survives or fails the test. Let p be the probability of failure for a product. Let the null hypothesis be

$$H_0: \quad p \leq p_0$$

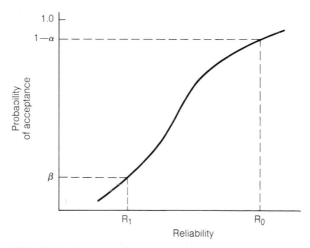

FIG. 19.11   Operating characteristic curve.

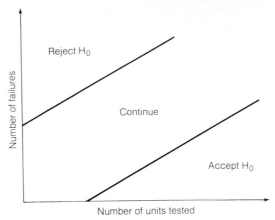

FIG. 19.12   Sequential life testing.

and the alternate hypothesis be

$$H_1: \quad p > p_0$$

Here p_0 is the value of p such that if $p = p_0$, then the probability of accepting H_0 is $(1 - \alpha)$. Also let p_1 be a value of p such that $p_1 > p_0$, and for $p = p_1$, the probability of accepting H_0 is β. These quantities (α, p_0, p_1, and β) define the sequential test.

Let $y =$ the total number of failures in n trials. Then the sequential test plan is given by (the continue region)

$$\frac{n}{D} \ln \left(\frac{1 - p_0}{1 - p_1} \right) - \frac{1}{D} \ln \left(\frac{1 - \alpha}{\beta} \right) < y < \frac{n}{D} \ln \left(\frac{1 - p_0}{1 - p_1} \right) + \frac{1}{D} \ln \left(\frac{1 - \beta}{\alpha} \right) \quad (19.126)$$

where

$$D = \ln \left[\left(\frac{p_1}{p_0} \right) \left(\frac{1 - p_0}{1 - p_1} \right) \right] \quad (19.127)$$

Let the inequality in Eq. (19.126) be represented as

$$A_n < y < B_n \quad (19.128)$$

Now by our procedural rules we accept H_0 if $y \leq A_n$, reject H_0 if $y \geq B_n$, and take an additional observation if $A_n < y < B_n$.

The boundaries of Eq. (19.128) graph as parallel straight lines, and a sequential testing graph can be constructed similar to Fig. 19.12. The test results can then be plotted on this graph to provide a visual representation of the test progress.

The entire OC curve for this test can also be determined. The probability of accepting H_0 when p is the true fraction failing is

$$Pa(p) = \frac{B^h - 1}{B^h - A^h} \quad (19.129)$$

where

$$p = \frac{1 - [(1 - p_1)/(1 - p_0)]^h}{(p_1/p_0)^h - [(1 - p_1)/(1 - p_0)]^h} \qquad A = \frac{\beta}{1 - \alpha}, \quad B = \frac{1 - \beta}{\alpha} \qquad (19.130)$$

To obtain the OC curve for this test, Eq. (19.130) is first used with an arbitrarily selected value of h to compute p. Then Eq. (19.129) is used to calculate the probability of acceptance for the value of p. Of course, for $p = p_0$, $Pa(p_0) = \alpha$ and for $p = p_1$, $Pa(p_1) = \beta$.

The expected number of observations needed to reach a decision is given by

$$E(p, n) = \frac{Pa(p) \ln A + [1 - Pa(p)] \ln B}{p \ln (p_1/p_0) + (1 - p) \ln [(1 - p_1)/(1 - p_0)]} \qquad (19.131)$$

where $Pa(p)$ is given by Eq. (19.129).

19.9.2 Exponential Distribution

Let us consider the sequential test procedure for the following test of a hypothesis:

$$H_0: \quad \theta = \theta_0$$
$$H_1: \quad \theta = \theta_1 \qquad \theta_1 < \theta_0$$

In order to design the test, the following requirements must be specified:

1. Determine an acceptable mean life θ_0.
2. Determine the magnitude of an α risk which may be tolerated, where α is the probability of rejecting H_0 if $\theta = \theta_0$ or $P(H_1/H_0) = \alpha$.
3. Determine an unacceptable or limiting mean life θ_1 where $\theta_1 < \theta_0$.
4. Determine the magnitude of a β risk which may be tolerated, where $P(H_0/H_1) = \beta$ or β is the probability of accepting H_0 if $\theta = \theta_1 < \theta_0$.

Let us consider the situation where items are tested to failure in a sequential (one-at-a-time) fashion. Let t_i be the time to failure of the ith item. The sequential decision criteria are

If $h_1 + ns < V(t) < h_0 + ns$, continue the test
If $V(t) \geq h_0 + ns$, stop the test and accept H_0
If $V(t) \leq h_1 + ns$, stop the test and reject H_0

where $V(t)$ is the total time on test, $\sum_{i=1}^{n} t_i$, and n is the number of failures by total time $V(t)$ and where

$$h_0 = \frac{\ln [(1 - \alpha)/\beta]}{1/\theta_1 - 1/\theta_0} \qquad (19.132)$$

$$h_1 = \frac{\ln [\alpha/(1 - \beta)]}{1/\theta_1 - 1/\theta_0} \qquad (19.133)$$

$$s = \frac{\ln (\theta_0/\theta_1)}{1/\theta_1 - 1/\theta_0} \qquad (19.134)$$

The OC curve for the test is given by

$P(\theta) = $ probability of accepting H_0 when θ is the true parameter

$$= \frac{B^h - 1}{B^h - A^h} \tag{19.135}$$

where

$$\theta = \frac{(\theta_0/\theta_1)^h - 1}{h(1/\theta_1 - 1/\theta_0)}$$

$$A = \frac{\beta}{1-\alpha} \qquad B = \frac{1-\beta}{\alpha}$$

Parameter h can be any real number, and meaningful selections are made by trial and error.

The five points which can be easily found and enable us to sketch the OC curve are given in Table 19.7. The expected number of failures required to reach a decision, $E_\theta(R)$, is dependent upon the lot mean life θ and may be found as follows:

$$E_\theta(R) = \frac{-h_1 - (h_0 - h_1)P(\theta)}{s - \theta} \qquad s \neq \theta \tag{19.136}$$

$$E_\theta(R) = \frac{-h_0 h_1}{s^2} \qquad s = \theta \tag{19.137}$$

where the random variable R is the number of failures required to reach a decision. If $\theta = \theta_1$ we have

$$E_\theta(R) \approx \frac{\beta \ln [\beta/(1-\alpha)] + (1-\beta) \ln [(1-\beta)/\alpha]}{\ln (\theta_0/\theta_1) - [1 - (\theta_1/\theta_0)]} \tag{19.138}$$

and when $\theta = \theta_0$, we have

$$E_\theta(R) \approx \frac{(1-\alpha) \ln [\beta/(1-\alpha)] + \alpha \ln [(1-\beta)/\alpha]}{\ln (\theta_0/\theta_1) - [(\theta_0/\theta_1) - 1]} \tag{19.139}$$

The expected waiting time to reach a decision, $E_\theta(T)$, is a function of θ and is found by (n is the number of items put on test at a time)

$$E_\theta(T) = \frac{\theta}{n} E_\theta(R) \qquad \text{for replacement test} \tag{19.140}$$

TABLE 19.7   OC Curve for Sequential Plan

θ	$P(\theta)$
0	0
θ_1	β
s	$\dfrac{\ln[(1-\beta)/\alpha]}{\ln[(1-\beta)/\alpha] - \ln[\beta/(1-\alpha)]}$
θ_0	$1 - \alpha$
α	1

or

$$E_\theta(T) \approx \theta \ln \left[\frac{n}{n - E_\theta(R)} \right] \qquad \text{for a nonreplacement test} \qquad (19.141)$$

Example 19.14 We want to test a new product to see if it meets a standard of $\theta_0 = 1000$ hours for its MTBF with $\alpha = 0.05$. We decide $\theta_1 = 500$ hours with $\beta = 0.10$. Then, we have

$$h_0 = \frac{\ln\,[(1 - 0.05)/0.10]}{1/500 - 1/1000} = 2251.29$$

$$h_1 = \frac{\ln\,[0.05/(1 - 0.10)]}{1/500 - 1/1000} = -2890.37$$

$$s = \frac{\ln\,(1000/500)}{1/500 - 1/1000} = 693.15$$

Hence, continue region is given by

$$693n - 2890 < V(t) < 693n + 2251$$

Accept H_0 when $V(t) \geq 693n + 2251$
Reject H_0 when $V(t) \leq 693n - 2890$

The OC curve can be plotted using Table 19.7 and we have

$$P(500) = 0.10$$
$$P(1000) = 0.95$$

and

$$P[\theta = 693] = \frac{\ln\,[(1 - 0.1)/0.05]}{\ln\,[(1 - 0.1)/0.05] - \ln\,[0.10/(1 - 0.05)]}$$

$$= \frac{2.89037}{2.89037 - (-2.25129)}$$

$$= 0.56214$$

The expected number of failures to reach a decision is given by [using Eqs. (19.137) to (19.139)]

$$E_{500}(R) = 12.31 \qquad E_{693}(R) = 13.54 \qquad E_{1000}(R) = 6.50$$

Similarly, the expected time to reach a decision is given by [using Eqs. (19.140) and (19.141)]

$$E_{500}(T) = 500 \times 12.31 = 6155 \text{ hours}$$
$$E_{693}(T) = 693 \times 12.54 = 9385 \text{ hours}$$
$$E_{1000}(T) = 1000 \times 6.50 = 6500 \text{ hours}$$

19.10 CONCLUSION

In this chapter we have presented probability theory and statistical methods that can help us model and estimate reliability of the product. Reliability is defined in terms of probability, and hence probability theory plays an important role in reliability. Probability theory is concerned with the methods of analysis that can be used to study random phenomena associated with the life of a product. A failure distribution describes and models mathematically the life of a product. There are several physical or other causes that may be responsible for the failure of a device and generally it is not possible to isolate these causes and mathematically account for all of them. Thus the selection of a failure distribution is based on statistical studies about the life of a product and the data accumulated on the product. The failure data and an understanding of the physical causes for failure should complement each other in the selection of a failure distribution. Some of the statistical procedures to estimate the parameters of the failure distribution are given in this chapter. There is tremendous interplay between probability, statistics, and reliability.

REFERENCES

1. Hines, W. W., and D. C. Montgomery, *Probability and Statistics in Engineering and Management Science*, 2d ed., Wiley, New York, 1980.
2. Feller, W., *An Introduction to Probability Theory and its Applications*, 3d ed., Wiley, New York, 1968.
3. Barlow, R. E., and F. Proschan, *Statistical Theory of Reliability and Life Testing*, Holt, Rinehart and Winston, New York, 1975.
4. *Reliability Design Handbook*, Reliability Analysis Center, RDG-376, Griffiss Air Force Base, New York, March 1976.
5. Mann, N. R., R. E. Schafer, and N. D. Singpurwalla, *Methods for Statistical Analysis of Reliability and Life Data*, Wiley, New York, 1974.
6. Bowker, A. H., and G. J. Lieberman, *Handbook of Industrial Statistics*, Prentice-Hall, Englewood Cliffs, N.J., 1955.
7. Duncan, A. J., *Quality Control and Industrial Statistics*, 4th ed., Irwin, Homewood, Ill., 1974.
8. Kapur, K. C., "Reliability and Maintainability," Chap. 8.5 in G. Salvendy (ed.), *Industrial Engineering Handbook*, Wiley, New York, 1982.
9. Kapur, K. C., and L. R. Lamberson, "Reliability," in H. A. Rothbart, (ed.), *Mechanical Design and Systems Handbook*, McGraw-Hill, New York.
10. Kapur, K. C., and L. R. Lamberson, "Reliability in Product Design and Testing," Lecture Notes, Detroit, Michigan, copyrighted 1982.
11. Kapur, K. C., and L. R. Lamberson, *Reliability in Engineering Design*, Wiley, New York, 1977.
12. Weibull, W., "A Statistical Distribution Function of Wide Applicability," *Journal of Applied Mechanics*, pp. 293–296, 1951.
13. Weibull, W., *Fatigue Testing and Analysis of Results*, Macmillan, New York, 1961.
14. Bayes, T., "An Essay Towards a Problem in the Doctrine of Chances," *Biometrika*, vol. 45, 1958.
15. Wald, A., *Sequential Analysis*, Wiley, New York, 1947.

APPENDIXES

APPENDIX A: TABLES

Table A.1   Summation of Terms of Poisson's Exponential Binomial Limit*
1,000 × probability of c or fewer occurrences of event that has average number of occurrences equal to c' or np'

c' or np' \ c	0	1	2	3	4	5	6	7	8	9
0.02	980	1,000								
0.04	961	999	1,000							
0.06	942	998	1,000							
0.08	923	997	1,000							
0.10	905	995	1,000							
0.15	861	990	999	1,000						
0.20	819	982	999	1,000						
0.25	779	974	998	1,000						
0.30	741	963	996	1,000						
0.35	705	951	994	1,000						
0.40	670	938	992	999	1,000					
0.45	638	925	989	999	1,000					
0.50	607	910	986	998	1,000					
0.55	577	894	982	998	1,000					
0.60	549	878	977	997	1,000					
0.65	522	861	972	996	999	1,000				
0.70	497	844	966	994	999	1,000				
0.75	472	827	959	993	999	1,000				
0.80	449	809	953	991	999	1,000				
0.85	427	791	945	989	998	1,000				
0.90	407	772	937	987	998	1,000				
0.95	387	754	929	984	997	1,000				
1.00	368	736	920	981	996	999	1,000			
1.1	333	699	900	974	995	999	1,000			
1.2	301	663	879	966	992	998	1,000			
1.3	273	627	857	957	989	998	1,000			
1.4	247	592	833	946	986	997	999	1,000		
1.5	223	558	809	934	981	996	999	1,000		
1.6	202	525	783	921	976	994	999	1,000		
1.7	183	493	757	907	970	992	998	1,000		
1.8	165	463	731	891	964	990	997	999	1,000	
1.9	150	434	704	875	956	987	997	999	1,000	
2.0	135	406	677	857	947	983	995	999	1,000	

Table A.1 Summation of Terms of Poisson's Exponential Binomial Limit
(Continued)

c'or np' \ c	0	1	2	3	4	5	6	7	8	9
2.2	111	355	623	819	928	975	993	998	1,000	
2.4	091	308	570	779	904	964	988	997	999	1,000
2.6	074	267	518	736	877	951	983	995	999	1,000
2.8	061	231	469	692	848	935	976	992	998	999
3.0	050	199	423	647	815	916	966	988	996	999
3.2	041	171	380	603	781	895	955	983	994	998
3.4	033	147	340	558	744	871	942	977	992	997
3.6	027	126	303	515	706	844	927	969	988	996
3.8	022	107	269	473	668	816	909	960	984	994
4.0	018	092	238	433	629	785	889	949	979	992
4.2	015	078	210	395	590	753	807	936	972	080
4.4	012	066	185	359	551	720	844	921	964	985
4.6	010	056	163	326	513	686	818	905	955	980
4.8	008	048	143	294	476	651	791	887	944	975
5.0	007	040	125	265	440	616	762	867	932	968
5.2	006	034	109	238	406	581	732	845	918	960
5.4	005	029	095	213	373	546	702	822	903	951
5.6	004	024	082	191	342	512	670	797	886	941
5.8	003	021	072	170	313	478	638	771	867	929
6.0	002	017	062	151	285	446	606	744	847	916

c'or np' \ c	10	11	12	13	14	15	16
2.8	1,000						
3.0	1,000						
3.2	1,000						
3.4	999	1,000					
3.6	999	1,000					
3.8	998	999	1,000				
4.0	997	999	1,000				
4.2	996	999	1,000				
4.4	994	998	999	1,000			
4.6	992	997	999	1,000			
4.8	990	996	999	1,000			
5.0	986	995	998	999	1,000		
5.2	982	993	997	999	1,000		
5.4	977	990	996	999	1,000		
5.6	972	988	995	998	999	1,000	
5.8	965	984	993	997	999	1,000	
6.0	957	980	991	996	999	999	1,000

Table A.1   Summation of Terms of Poisson's Exponential Binomial Limit
(Continued)

c / c' or np'	0	1	2	3	4	5	6	7	8	9
6.2	002	015	054	134	259	414	574	716	826	902
6.4	002	012	046	119	235	384	542	687	803	886
6.6	001	010	040	105	213	355	511	658	780	869
6.8	001	009	034	093	192	327	480	628	755	850
7.0	001	007	030	082	173	301	450	599	729	830
7.2	001	006	025	072	156	276	420	569	703	810
7.4	001	005	022	063	140	253	392	539	676	788
7.6	001	004	019	055	125	231	365	510	648	765
7.8	000	004	016	048	112	210	338	481	620	741
8.0	000	003	014	042	100	191	313	453	593	717
8.5	000	002	009	030	074	150	256	386	523	653
9.0	000	001	006	021	055	116	207	324	456	587
9.5	000	001	004	015	040	089	165	269	392	522
10.0	000	000	003	010	029	067	130	220	333	458

c' or np'	10	11	12	13	14	15	16	17	18	19
6.2	949	975	989	995	998	999	1,000			
6.4	939	969	986	994	997	999	1,000			
6.6	927	963	982	992	997	999	999	1,000		
6.8	915	955	978	990	996	998	999	1,000		
7.0	901	947	973	987	994	998	999	1,000		
7.2	887	937	967	984	993	997	999	999	1,000	
7.4	871	926	961	980	991	996	998	999	1,000	
7.6	854	915	954	976	989	995	998	999	1,000	
7.8	835	902	945	971	986	993	997	999	1,000	
8.0	816	888	936	966	983	992	996	998	999	1,000
8.5	763	849	909	949	973	986	993	997	999	999
9.0	706	803	876	926	959	978	989	995	998	999
9.5	645	752	836	898	940	967	982	991	996	998
10.0	583	697	792	864	917	951	973	986	993	997

c' or np'	20	21	22
8.5	1,000		
9.0	1,000		
9.5	999	1,000	
10.0	998	999	1,000

Table A.1 Summation of Terms of Poisson's Exponential Binomial Limit
(*Continued*)

c' or np' \ c	0	1	2	3	4	5	6	7	8	9
10.5	000	000	002	007	021	050	102	179	279	397
11.0	000	000	001	005	015	038	079	143	232	341
11.5	000	000	001	003	011	028	060	114	191	289
12.0	000	000	001	002	008	020	046	090	155	242
12.5	000	000	000	002	005	015	035	070	125	201
13.0	000	000	000	001	004	011	026	054	100	166
13.5	000	000	000	001	003	008	019	041	079	135
14.0	000	000	000	000	002	006	014	032	062	109
14.5	000	000	000	000	001	004	010	024	048	088
15.0	000	000	000	000	001	003	008	018	037	070

	10	11	12	13	14	15	16	17	18	19
10.5	521	639	742	825	888	932	960	978	988	994
11.0	460	579	689	781	854	907	944	968	982	991
11.5	402	520	633	733	815	878	924	954	974	986
12.0	347	462	576	682	772	844	899	937	963	979
12.5	297	406	519	628	725	806	869	916	948	969
13.0	252	353	463	573	675	764	835	890	930	957
13.5	211	304	409	518	623	718	798	861	908	942
14.0	176	260	358	464	570	669	756	827	883	923
14.5	145	220	311	413	518	619	711	790	853	901
15.0	118	185	268	363	466	568	664	749	819	875

	20	21	22	23	24	25	26	27	28	29
10.5	997	999	999	1,000						
11.0	995	998	999	1,000						
11.5	992	996	998	999	1,000					
12.0	988	994	997	999	999	1,000				
12.5	983	991	995	998	999	999	1,000			
13.0	975	986	992	996	998	999	1,000			
13.5	965	980	989	994	997	998	999	1,000		
14.0	952	971	983	991	995	997	999	999	1,000	
14.5	936	960	976	986	992	996	998	999	999	1,000
15.0	917	947	967	981	989	994	997	998	999	1,000

Appendix A: Tables

Table A.1 Summation of Terms of Poisson's Exponential Binomial Limit
(*Continued*)

c' or np' \ c	4	5	6	7	8	9	10	11	12	13
16	000	001	004	010	022	043	077	127	193	275
17	000	001	002	005	013	026	049	085	135	201
18	000	000	001	003	007	015	030	055	092	143
19	000	000	001	002	004	009	018	035	061	098
20	000	000	000	001	002	005	011	021	039	066
21	000	000	000	000	001	003	006	013	025	043
22	000	000	000	000	001	002	004	008	015	028
23	000	000	000	000	000	001	002	004	009	017
24	000	000	000	000	000	000	001	003	005	011
25	000	000	000	000	000	000	001	001	003	006

	14	15	16	17	18	19	20	21	22	23
16	368	467	566	659	742	812	868	911	942	963
17	281	371	468	564	655	736	805	861	905	937
18	208	287	375	469	562	651	731	799	855	899
19	150	215	292	378	469	561	647	725	793	849
20	105	157	221	297	381	470	559	644	721	787
21	072	111	163	227	302	384	471	558	640	716
22	048	077	117	169	232	306	387	472	556	637
23	031	052	082	123	175	238	310	389	472	555
24	020	034	056	087	128	180	243	314	392	473
25	012	022	038	060	092	134	185	247	318	394

	24	25	26	27	28	29	30	31	32	33
16	978	987	993	996	998	999	999	1,000		
17	959	975	985	991	995	997	999	999	1,000	
18	932	955	972	983	990	994	997	998	999	1,000
19	893	927	951	969	980	988	993	996	998	999
20	843	888	922	948	966	978	987	992	995	997
21	782	838	883	917	944	963	976	985	991	994
22	712	777	832	877	913	940	959	973	983	989
23	635	708	772	827	873	908	936	956	971	981
24	554	632	704	768	823	868	904	932	953	969
25	473	553	629	700	763	818	863	900	929	950

	34	35	36	37	38	39	40	41	42	43
19	999	1,000								
20	999	999	1,000							
21	997	998	999	999	1,000					
22	994	996	998	999	999	1,000				
23	988	993	996	997	999	999	1,000			
24	979	987	992	995	997	998	999	999	1,000	
25	966	978	985	991	994	997	998	999	999	1,000

Table A.2 Confidence Limits for a Proportion (One-sided), $n \leq 30$

(For confidence limits for $n > 30$, see Charts B.1 to B.4)

If the observed proportion is r/n, enter the table with n and r for an upper one-sided limit. For a lower one-sided limit, enter the table with n and $n - r$ and subtract the table entry from 1.

r	90%	95%	99%	r	90%	95%	99%	r	90%	95%	99%
	$n = 2$				$n = 3$				$n = 4$		
0	0.684	0.776	0.900	0	0.536	0.632	0.785−	0	0.438	0.527	0.684
1	0.949	0.975−	0.995−	1	0.804	0.865−	0.941	1	0.680	0.751	0.859
				2	0.965+	0.983	0.997	2	0.857	0.902	0.958
								3	0.974	0.987	0.997
	$n = 5$				$n = 6$				$n = 7$		
0	0.369	0.451	0.602	0	0.319	0.393	0.536	0	0.280	0.348	0.482
1	0.584	0.657	0.778	1	0.510	0.582	0.706	1	0.453	0.521	0.643
2	0.753	0.811	0.894	2	0.667	0.729	0.827	2	0.596	0.659	0.764
3	0.888	0.924	0.967	3	0.799	0.847	0.915+	3	0.721	0.775−	0.858
4	0.979	0.990	0.998	4	0.907	0.937	0.973	4	0.830	0.871	0.929
				5	0.983	0.991	0.998	5	0.921	0.947	0.977
								6	0.985+	0.993	0.999
	$n = 8$				$n = 9$				$n = 10$		
0	0.250	0.312	0.438	0	0.226	0.283	0.401	0	0.206	0.259	0.369
1	0.406	0.471	0.590	1	0.368	0.429	0.544	1	0.337	0.394	0.504
2	0.538	0.600	0.707	2	0.490	0.550	0.656	2	0.450	0.507	0.612
3	0.655+	0.711	0.802	3	0.599	0.655+	0.750	3	0.552	0.607	0.703
4	0.760	0.807	0.879	4	0.699	0.749	0.829	4	0.646	0.696	0.782
5	0.853	0.889	0.939	5	0.790	0.831	0.895−	5	0.733	0.778	0.850
6	0.931	0.954	0.980	6	0.871	0.902	0.947	6	0.812	0.850	0.907
7	0.987	0.994	0.999	7	0.939	0.959	0.983	7	0.884	0.913	0.952
				8	0.988	0.994	0.999	8	0.945+	0.963	0.984
								9	0.990	0.995−	0.999
	$n = 11$				$n = 12$				$n = 13$		
0	0.189	0.238	0.342	0	0.175−	0.221	0.319	0	0.162	0.206	0.298
1	0.310	0.364	0.470	1	0.287	0.339	0.440	1	0.268	0.316	0.413
2	0.415+	0.470	0.572	2	0.386	0.438	0.537	2	0.360	0.410	0.506
3	0.511	0.564	0.660	3	0.475+	0.527	0.622	3	0.444	0.495−	0.588
4	0.599	0.650	0.738	4	0.559	0.609	0.698	4	0.523	0.573	0.661
5	0.682	0.729	0.806	5	0.638	0.685−	0.765+	5	0.598	0.645+	0.727
6	0.759	0.800	0.866	6	0.712	0.755−	0.825+	6	0.669	0.713	0.787
7	0.831	0.865−	0.916	7	0.781	0.819	0.879	7	0.736	0.776	0.841
8	0.895+	0.921	0.957	8	0.846	0.877	0.924	8	0.799	0.834	0.889
9	0.951	0.967	0.986	9	0.904	0.928	0.961	9	0.858	0.887	0.931
10	0.990	0.995+	0.999	10	0.955−	0.970	0.987	10	0.912	0.934	0.964
				11	0.991	0.996	0.999	11	0.958	0.972	0.988
								12	0.992	0.996	0.999
	$n = 14$				$n = 15$				$n = 16$		
0	0.152	0.193	0.280	0	0.142	0.181	0.264	0	0.134	0.171	0.250
1	0.251	0.297	0.389	1	0.236	0.279	0.368	1	0.222	0.264	0.349
2	0.337	0.385+	0.478	2	0.317	0.363	0.453	2	0.300	0.344	0.430
3	0.417	0.466	0.557	3	0.393	0.440	0.529	3	0.371	0.417	0.503
4	0.492	0.540	0.627	4	0.464	0.511	0.597	4	0.439	0.484	0.569
5	0.563	0.610	0.692	5	0.532	0.577	0.660	5	0.504	0.548	0.630

This table was computed by Robert S. Gardner and is reproduced, by permission of the author and the publisher, from E. L. Crow, F. A. Davis, and M. W. Maxfield, *Statistics Manual*, China Lake, Calif., U.S. Naval Ordnance Test Station, NAVORD Report 3369 (NOTS 948), pp. 262–265.

Table A.2 Confidence Limits for a Proportion (One-sided), $n \leq 30$
(Continued)

r	90%	95%	99%	r	90%	95%	99%	r	90%	95%	99%
	n = 14 (Continued)				n = 15 (Continued)				n = 16 (Continued)		
6	0.631	0.675−	0.751	6	0.596	0.640	0.718	6	0.565+	0.609	0.687
7	0.695+	0.736	0.805+	7	0.658	0.700	0.771	7	0.625−	0.667	0.739
8	0.757	0.794	0.854	8	0.718	0.756	0.821	8	0.682	0.721	0.788
9	0.815−	0.847	0.898	9	0.774	0.809	0.865+	9	0.737	0.773	0.834
10	0.869	0.896	0.936	10	0.828	0.858	0.906	10	0.790	0.822	0.875−
11	0.919	0.939	0.967	11	0.878	0.903	0.941	11	0.839	0.868	0.912
12	0.961	0.974	0.989	12	0.924	0.943	0.969	12	0.886	0.910	0.945−
13	0.993	0.996	0.999	13	0.964	0.976	0.990	13	0.929	0.947	0.971
				14	0.993	0.997	0.999	14	0.966	0.977	0.990
								15	0.993	0.997	0.999

r	90%	95%	99%	r	90%	95%	99%	r	90%	95%	99%
	n = 17				n = 18				n = 19		
0	0.127	0.162	0.237	0	0.120	0.153	0.226	0	0.114	0.146	0.215+
1	0.210	0.250	0.332	1	0.199	0.238	0.316	1	0.190	0.226	0.302
2	0.284	0.326	0.410	2	0.269	0.310	0.391	2	0.257	0.296	0.374
3	0.352	0.396	0.480	3	0.334	0.377	0.458	3	0.319	0.359	0.439
4	0.416	0.461	0.543	4	0.396	0.439	0.520	4	0.378	0.419	0.498
5	0.478	0.522	0.603	5	0.455+	0.498	0.577	5	0.434	0.476	0.554
6	0.537	0.580	0.658	6	0.512	0.554	0.631	6	0.489	0.530	0.606
7	0.594	0.636	0.709	7	0.567	0.608	0.681	7	0.541	0.582	0.655+
8	0.650	0.689	0.758	8	0.620	0.659	0.729	8	0.592	0.632	0.702
9	0.703	0.740	0.803	9	0.671	0.709	0.774	9	0.642	0.680	0.746
10	0.754	0.788	0.845−	10	0.721	0.756	0.816	10	0.690	0.726	0.788
11	0.803	0.834	0.883	11	0.769	0.801	0.855−	11	0.737	0.770	0.827
12	0.849	0.876	0.918	12	0.815−	0.844	0.890	12	0.782	0.812	0.863
13	0.893	0.915+	0.948	13	0.858	0.884	0.923	13	0.825−	0.853	0.897
14	0.933	0.950	0.973	14	0.899	0.920	0.951	14	0.866	0.890	0.927
15	0.968	0.979	0.991	15	0.937	0.953	0.975−	15	0.905−	0.925−	0.954
16	0.994	0.997	0.999	16	0.970	0.980	0.992	16	0.941	0.956	0.976
				17	0.994	0.997	0.999	17	0.972	0.981	0.992
								18	0.994	0.997	0.999

r	90%	95%	99%	r	90%	95%	99%	r	90%	95%	99%
	n = 20				n = 21				n = 22		
0	0.109	0.139	0.206	0	0.104	0.133	0.197	0	0.099	0.127	0.189
1	0.181	0.216	0.289	1	0.173	0.207	0.277	1	0.166	0.198	0.266
2	0.245−	0.283	0.358	2	0.234	0.271	0.344	2	0.224	0.259	0.330
3	0.304	0.344	0.421	3	0.291	0.329	0.404	3	0.279	0.316	0.389
4	0.361	0.401	0.478	4	0.345+	0.384	0.460	4	0.331	0.369	0.443
5	0.415−	0.456	0.532	5	0.397	0.437	0.512	5	0.381	0.420	0.493
6	0.467	0.508	0.583	6	0.448	0.487	0.561	6	0.430	0.468	0.541
7	0.518	0.558	0.631	7	0.497	0.536	0.608	7	0.477	0.515+	0.587
8	0.567	0.606	0.677	8	0.544	0.583	0.653	8	0.523	0.561	0.630
9	0.615+	0.653	0.720	9	0.590	0.628	0.695+	9	0.568	0.605−	0.672
10	0.662	0.698	0.761	10	0.636	0.672	0.736	10	0.611	0.647	0.712
11	0.707	0.741	0.800	11	0.679	0.714	0.774	11	0.654	0.689	0.750
12	0.751	0.783	0.837	12	0.722	0.755+	0.811	12	0.695+	0.729	0.786
13	0.793	0.823	0.871	13	0.764	0.794	0.845+	13	0.736	0.767	0.821
14	0.834	0.860	0.902	14	0.804	0.832	0.878	14	0.775+	0.804	0.853
15	0.873	0.896	0.931	15	0.842	0.868	0.908	15	0.813	0.840	0.884
16	0.910	0.929	0.956	16	0.879	0.901	0.935−	16	0.850	0.874	0.912
17	0.944	0.958	0.977	17	0.914	0.932	0.959	17	0.885+	0.906	0.938
18	0.973	0.982	0.992	18	0.946	0.960	0.978	18	0.918	0.935+	0.961
19	0.995−	0.997	0.999	19	0.974	0.983	0.993	19	0.949	0.962	0.979
				20	0.995−	0.988	1.000	20	0.976	0.984	0.993
								21	0.995+	0.998	1.000

Table A.2   Confidence Limits for a Proportion (One-sided), $n \leq 30$
(Continued)

r	90%	95%	99%	r	90%	95%	99%	r	90%	95%	99%
	n = 23				n = 24				n = 25		
0	0.095+	0.122	0.181	0	0.091	0.117	0.175−	0	0.088	0.113	0.168
1	0.159	0.190	0.256	1	0.153	0.183	0.246	1	0.147	0.176	0.237
2	0.215+	0.249	0.318	2	0.207	0.240	0.307	2	0.199	0.231	0.296
3	0.268	0.304	0.374	3	0.258	0.292	0.361	3	0.248	0.282	0.349
4	0.318	0.355−	0.427	4	0.306	0.342	0.412	4	0.295−	0.330	0.398
5	0.366	0.404	0.476	5	0.352	0.389	0.460	5	0.340	0.375+	0.444
6	0.413	0.451	0.522	6	0.398	0.435−	0.505−	6	0.383	0.420	0.488
7	0.459	0.496	0.567	7	0.442	0.479	0.548	7	0.426	0.462	0.531
8	0.503	0.540	0.609	8	0.484	0.521	0.590	8	0.467	0.504	0.571
9	0.546	0.583	0.650	9	0.526	0.563	0.630	9	0.508	0.544	0.610
10	0.589	0.625−	0.689	10	0.567	0.603	0.668	10	0.548	0.583	0.648
11	0.630	0.665−	0.727	11	0.608	0.642	0.705−	11	0.587	0.621	0.684
12	0.670	0.704	0.763	12	0.647	0.681	0.740	12	0.625−	0.659	0.719
13	0.710	0.742	0.797	13	0.685+	0.718	0.774	13	0.662	0.695−	0.752
14	0.748	0.778	0.829	14	0.723	0.754	0.806	14	0.699	0.730	0.784
15	0.786	0.814	0.860	15	0.759	0.788	0.837	15	0.735−	0.764	0.815+
16	0.822	0.848	0.889	16	0.795+	0.822	0.867	16	0.770	0.798	0.845+
17	0.857	0.880	0.916	17	0.830	0.854	0.894	17	0.804	0.830	0.873
18	0.890	0.910	0.941	18	0.863	0.885+	0.920	18	0.837	0.861	0.899
19	0.922	0.938	0.962	19	0.895+	0.914	0.943	19	0.869	0.890	0.923
20	0.951	0.963	0.980	20	0.925+	0.941	0.964	20	0.899	0.918	0.946
21	0.977	0.984	0.993	21	0.953	0.965+	0.981	21	0.928	0.943	0.966
22	0.995+	0.998	1.000	22	0.978	0.985−	0.994	22	0.955+	0.966	0.982
				23	0.996	0.998	1.000	23	0.979	0.986	0.994
								24	0.996	0.998	1.000
	n = 26				n = 27				n = 28		
0	0.085−	0.109	0.162	0	0.082	0.105+	0.157	0	0.079	0.101	0.152
1	0.142	0.170	0.229	1	0.137	0.164	0.222	1	0.132	0.159	0.215−
2	0.192	0.223	0.286	2	0.185+	0.215+	0.277	2	0.179	0.208	0.268
3	0.239	0.272	0.337	3	0.231	0.263	0.326	3	0.223	0.254	0.316
4	0.284	0.318	0.385−	4	0.275−	0.308	0.373	4	0.265+	0.298	0.361
5	0.328	0.363	0.430	5	0.317	0.351	0.417	5	0.306	0.339	0.404
6	0.370	0.405+	0.473	6	0.358	0.392	0.458	6	0.346	0.380	0.445−
7	0.411	0.447	0.514	7	0.397	0.432	0.498	7	0.385−	0.419	0.484
8	0.451	0.487	0.554	8	0.436	0.471	0.537	8	0.422	0.457	0.521
9	0.491	0.526	0.592	9	0.475−	0.509	0.574	9	0.459	0.494	0.558
10	0.529	0.564	0.628	10	0.512	0.547	0.610	10	0.496	0.530	0.593
11	0.567	0.602	0.664	11	0.549	0.583	0.645+	11	0.532	0.565−	0.627
12	0.604	0.638	0.698	12	0.585−	0.618	0.679	12	0.567	0.600	0.660
13	0.641	0.673	0.731	13	0.620	0.653	0.711	13	0.601	0.634	0.692
14	0.676	0.708	0.763	14	0.655+	0.687	0.743	14	0.635+	0.667	0.723
15	0.711	0.742	0.794	15	0.689	0.720	0.773	15	0.669	0.699	0.753
16	0.746	0.774	0.823	16	0.723	0.752	0.802	16	0.701	0.731	0.782
17	0.779	0.806	0.851	17	0.756	0.783	0.831	17	0.733	0.762	0.810
18	0.812	0.837	0.878	18	0.788	0.814	0.857	18	0.765−	0.792	0.837
19	0.843	0.866	0.903	19	0.819	0.843	0.883	19	0.796	0.821	0.863
20	0.874	0.894	0.927	20	0.849	0.871	0.907	20	0.826	0.849	0.888
21	0.903	0.921	0.948	21	0.879	0.899	0.930	21	0.855+	0.876	0.911
22	0.931	0.946	0.967	22	0.907	0.924	0.950	22	0.883	0.902	0.932
23	0.957	0.968	0.983	23	0.934	0.948	0.968	23	0.911	0.927	0.952
24	0.979	0.986	0.994	24	0.958	0.969	0.983	24	0.936	0.950	0.969
25	0.996	0.998	1.000	25	0.980	0.987	0.994	25	0.960	0.970	0.984
				26	0.996	0.998	1.000	26	0.981	0.987	0.995−
								27	0.996	0.998	1.000

Table A.2   Confidence Limits for a Proportion (One-sided), $n \leq 30$
(Continued)

r	90 %	95 %	99 %	r	90 %	95 %	99 %
		$n = 29$				$n = 30$	
0	0.076	0.098	0.147	0	0.074	0.095 +	0.142
1	0.128	0.153	0.208	1	0.124	0.149	0.202
2	0.173	0.202	0.260	2	0.168	0.195 +	0.252
3	0.216	0.246	0.307	3	0.209	0.239	0.298
4	0.257	0.288	0.350	4	0.249	0.280	0.340
5	0.297	0.329	0.392	5	0.287	0.319	0.381
6	0.335 −	0.368	0.432	6	0.325 −	0.357	0.420
7	0.372	0.406	0.470	7	0.361	0.394	0.457
8	0.409	0.443	0.507	8	0.397	0.430	0.493
9	0.445 +	0.479	0.542	9	0.432	0.465 +	0.527
10	0.481	0.514	0.577	10	0.466	0.499	0.561
11	0.515 +	0.549	0.610	11	0.500	0.533	0.594
12	0.550	0.583	0.643	12	0.533	0.566	0.626
13	0.583	0.616	0.674	13	0.566	0.598	0.657
14	0.616	0.648	0.705 −	14	0.599	0.630	0.687
15	0.649	0.680	0.734	15	0.630	0.661	0.716
16	0.681	0.711	0.763	16	0.662	0.692	0.744
17	0.712	0.741	0.791	17	0.692	0.721	0.772
18	0.743	0.771	0.818	18	0.723	0.750	0.799
19	0.774	0.800	0.843	19	0.752	0.779	0.824
20	0.803	0.828	0.868	20	0.782	0.807	0.849
21	0.832	0.855 −	0.892	21	0.810	0.834	0.873
22	0.860	0.881	0.914	22	0.838	0.860	0.896
23	0.888	0.906	0.935 −	23	0.865 +	0.885 +	0.917
24	0.914	0.930	0.954	24	0.891	0.909	0.937
25	0.938	0.951	0.970	25	0.917	0.932	0.955 +
26	0.961	0.971	0.985	26	0.941	0.953	0.972
27	0.982	0.988	0.995 −	27	0.963	0.972	0.985 +
28	0.996	0.998	1.000	28	0.982	0.988	0.995 +
				29	0.996	0.998	1.000

Table A.3 Confidence Limits for a Proportion (Two-sided), $n \leq 30$*

(For confidence limits for $n > 30$, see Charts B.1 to B.4)

Upper limits are in boldface.   The observed proportion in a random sample is r/n.

$n = 1$ (left) / **$n = 2$** (right)

r	90% lower	90% upper	95% lower	95% upper	99% lower	99% upper	r	90% lower	90% upper	95% lower	95% upper	99% lower	99% upper
0	0	**0.900**	0	**0.950**	0	**0.990**	0	0	**0.684**	0	**0.776**	0	**0.900**
1	0.100	**1**	0.050	**1**	0.010	**1**	1	0.051	**0.949**	0.025+	**0.975−**	0.005+	**0.995−**
							2	0.316	**1**	0.224	**1**	0.100	**1**

$n = 3$ (left) / **$n = 4$** (right)

r	90% lower	90% upper	95% lower	95% upper	99% lower	99% upper	r	90% lower	90% upper	95% lower	95% upper	99% lower	99% upper
0	0	**0.536**	0	**0.632**	0	**0.785−**	0	0	**0.500**	0	**0.527**	0	**0.684**
1	0.035−	**0.804**	0.017	**0.865−**	0.003	**0.941**	1	0.026	**0.680**	0.013	**0.751**	0.003	**0.859**
2	0.196	**0.965+**	0.135−	**0.983**	0.059	**0.997**	2	0.143	**0.857**	0.098	**0.902**	0.042	**0.958**
3	0.464	**1**	0.368	**1**	0.215+	**1**	3	0.320	**0.974**	0.249	**0.987**	0.141	**0.997**
							4	0.500	**1**	0.473	**1**	0.316	**1**

$n = 5$ (left) / **$n = 6$** (right)

r	90% lower	90% upper	95% lower	95% upper	99% lower	99% upper	r	90% lower	90% upper	95% lower	95% upper	99% lower	99% upper
0	0	**0.379**	0	**0.500**	0	**0.602**	0	0	**0.345−**	0	**0.402**	0	**0.536**
1	0.021	**0.621**	0.010	**0.657**	0.002	**0.778**	1	0.017	**0.542**	0.009	**0.598**	0.002	**0.706**
2	0.112	**0.753**	0.076	**0.811**	0.033	**0.894**	2	0.093	**0.667**	0.063	**0.729**	0.027	**0.827**
3	0.247	**0.888**	0.189	**0.924**	0.106	**0.967**	3	0.201	**0.799**	0.153	**0.847**	0.085−	**0.915+**
4	0.379	**0.979**	0.343	**0.990**	0.222	**0.998**	4	0.333	**0.907**	0.271	**0.937**	0.173	**0.973**
5	0.621	**1**	0.500	**1**	0.398	**1**	5	0.458	**0.983**	0.402	**0.991**	0.294	**0.998**
							6	0.655+	**1**	0.598	**1**	0.464	**1**

$n = 7$ (left) / **$n = 8$** (right)

r	90% lower	90% upper	95% lower	95% upper	99% lower	99% upper	r	90% lower	90% upper	95% lower	95% upper	99% lower	99% upper
0	0	**0.316**	0	**0.377**	0	**0.500**	0	0	**0.255−**	0	**0.315+**	0	**0.451**
1	0.015	**0.500**	0.007	**0.554**	0.001	**0.643**	1	0.013	**0.418**	0.006	**0.500**	0.001	**0.590**
2	0.079	**0.684**	0.053	**0.659**	0.023	**0.764**	2	0.069	**0.582**	0.046	**0.685−**	0.020	**0.707**
3	0.170	**0.721**	0.129	**0.775**	0.071	**0.858**	3	0.147	**0.745+**	0.111	**0.711**	0.061	**0.802**
4	0.279	**0.830**	0.225+	**0.871**	0.142	**0.929**	4	0.240	**0.760**	0.193	**0.807**	0.121	**0.879**
5	0.316	**0.921**	0.341	**0.947**	0.236	**0.977**	5	0.255−	**0.853**	0.289	**0.889**	0.198	**0.939**
6	0.500	**0.985+**	0.446	**0.993**	0.357	**0.999**	6	0.418	**0.931**	0.315+	**0.954**	0.293	**0.980**
7	0.684	**1**	0.623	**1**	0.500	**1**	7	0.582	**0.987**	0.500	**0.994**	0.410	**0.999**
							8	0.745+	**1**	0.685−	**1**	0.549	**1**

$n = 9$ (left) / **$n = 10$** (right)

r	90% lower	90% upper	95% lower	95% upper	99% lower	99% upper	r	90% lower	90% upper	95% lower	95% upper	99% lower	99% upper
0	0	**0.232**	0	**0.289**	0	**0.402**	0	0	**0.222**	0	**0.267**	0	**0.376**
1	0.012	**0.391**	0.006	**0.443**	0.001	**0.598**	1	0.010	**0.352**	0.005+	**0.397**	0.001	**0.512**
2	0.061	**0.515+**	0.041	**0.558**	0.017	**0.656**	2	0.055−	**0.500**	0.037	**0.603**	0.016	**0.624**
3	0.129	**0.610**	0.098	**0.711**	0.053	**0.750**	3	0.116	**0.648**	0.087	**0.619**	0.048	**0.703**
4	0.210	**0.768**	0.169	**0.749**	0.105+	**0.829**	4	0.188	**0.659**	0.150	**0.733**	0.093	**0.782**
5	0.232	**0.790**	0.251	**0.831**	0.171	**0.895−**	5	0.222	**0.778**	0.222	**0.778**	0.150	**0.850**
6	0.390	**0.871**	0.289	**0.902**	0.250	**0.947**	6	0.341	**0.812**	0.267	**0.850**	0.218	**0.907**
7	0.485−	**0.939**	0.442	**0.959**	0.344	**0.983**	7	0.352	**0.884**	0.381	**0.913**	0.297	**0.952**
8	0.609	**0.988**	0.557	**0.994**	0.402	**0.999**	8	0.500	**0.945+**	0.397	**0.963**	0.376	**0.984**
9	0.768	**1**	0.711	**1**	0.598	**1**	9	0.648	**0.990**	0.603	**0.995−**	0.488	**0.999**
							10	0.778	**1**	0.733	**1**	0.624	**1**

$n = 11$ (left) / **$n = 12$** (right)

r	90% lower	90% upper	95% lower	95% upper	99% lower	99% upper	r	90% lower	90% upper	95% lower	95% upper	99% lower	99% upper
0	0	**0.197**	0	**0.250**	0	**0.359**	0	0	**0.184**	0	**0.236**	0	**0.321**
1	0.010	**0.315+**	0.005−	**0.369**	0.001	**0.509**	1	0.009	**0.294**	0.004	**0.346**	0.001	**0.445+**
2	0.061	**0.423**	0.033	**0.500**	0.014	**0.593**	2	0.045+	**0.398**	0.030	**0.450**	0.013	**0.555−**
3	0.105−	**0.577**	0.079	**0.631**	0.043	**0.660**	3	0.096	**0.500**	0.072	**0.550**	0.039	**0.679**
4	0.169	**0.685+**	0.135+	**0.667**	0.084	**0.738**	4	0.154	**0.602**	0.123	**0.654**	0.076	**0.698**
5	0.197	**0.698**	0.200	**0.750**	0.134	**0.806**	5	0.184	**0.706**	0.181	**0.706**	0.121	**0.765+**
6	0.302	**0.803**	0.250	**0.800**	0.194	**0.866**	6	0.271	**0.729**	0.236	**0.764**	0.175−	**0.825+**
7	0.315+	**0.831**	0.333	**0.865−**	0.262	**0.916**	7	0.294	**0.816**	0.294	**0.819**	0.235−	**0.879**
8	0.423	**0.895+**	0.369	**0.921**	0.340	**0.957**	8	0.398	**0.846**	0.346	**0.877**	0.302	**0.924**
9	0.577	**0.951**	0.500	**0.967**	0.407	**0.986**	9	0.500	**0.904**	0.450	**0.928**	0.321	**0.961**
10	0.685−	**0.990**	0.631	**0.995+**	0.500	**0.999**	10	0.602	**0.955−**	0.550	**0.970**	0.445+	**0.987**
11	0.803	**1**	0.750	**1**	0.641	**1**	11	0.706	**0.991**	0.654	**0.996**	0.555−	**0.999**
							12	0.816	**1**	0.764	**1**	0.679	**1**

* This table was calculated by Edwin L. Crow, Eleanor G. Crow, and Robert S. Gardner and is reproduced, by permission of the authors and the publisher, from E. L. Crow, F. A. Davis, and M. W. Maxfield, *Statistics Manual*, China Lake, Calif., U.S. Naval Ordnance Test Station, NAVORD Report 3369 (NOTS 948), 1955, pp. 257–261.

Table A.3　Confidence Limits for a Proportion (Two-sided), $n \le 30$
(Continued)

$n = 13$

r	90%		95%		99%	
0	0	0.173	0	0.225+	0	0.302
1	0.008	0.276	0.004	0.327	0.001	0.429
2	0.042	0.379	0.028	0.434	0.012	0.523
3	0.088	0.470	0.066	0.520	0.036	0.594
4	0.142	0.545-	0.113	0.587	0.069	0.698
5	0.173	0.621	0.166	0.673	0.111	0.727
6	0.246	0.724	0.224	0.740	0.159	0.787
7	0.276	0.754	0.260	0.776	0.213	0.841
8	0.379	0.827	0.327	0.834	0.273	0.889
9	0.455+	0.858	0.413	0.887	0.302	0.931
10	0.530	0.912	0.480	0.934	0.406	0.964
11	0.621	0.958	0.566	0.972	0.477	0.988
12	0.724	0.992	0.673	0.996	0.571	0.999
13	0.827	1	0.775-	1	0.698	1

$n = 14$

r	90%		95%		99%	
0	0	0.163	0	0.207	0	0.286
1	0.007	0.261	0.004	0.312	0.001	0.392
2	0.039	0.365+	0.026	0.389	0.011	0.500
3	0.081	0.422	0.061	0.500	0.033	0.608
4	0.131	0.578	0.104	0.611	0.064	0.636
5	0.163	0.594	0.153	0.629	0.102	0.714
6	0.224	0.645+	0.206	0.688	0.146	0.751
7	0.261	0.739	0.207	0.793	0.195-	0.805+
8	0.355-	0.776	0.312	0.794	0.249	0.854
9	0.406	0.837	0.371	0.847	0.286	0.898
10	0.422	0.869	0.389	0.896	0.364	0.936
11	0.578	0.919	0.500	0.939	0.392	0.967
12	0.635-	0.961	0.611	0.974	0.500	0.989
13	0.739	0.993	0.688	0.996	0.608	0.999
14	0.837	1	0.793	1	0.714	1

$n = 15$

r	90%		95%		99%	
0	0	0.154	0	0.191	0	0.273
1	0.007	0.247	0.003	0.302	0.001	0.373
2	0.036	0.326	0.024	0.369	0.010	0.461
3	0.076	0.400	0.057	0.448	0.031	0.539
4	0.122	0.500	0.097	0.552	0.059	0.627
5	0.154	0.600	0.142	0.631	0.094	0.672
6	0.205+	0.674	0.191	0.668	0.135-	0.727
7	0.247	0.675-	0.192	0.706	0.179	0.771
8	0.325+	0.753	0.294	0.808	0.229	0.821
9	0.326	0.795-	0.332	0.809	0.273	0.865+
10	0.400	0.846	0.369	0.838	0.328	0.906
11	0.500	0.878	0.448	0.903	0.373	0.941
12	0.600	0.924	0.552	0.943	0.461	0.969
13	0.674	0.964	0.631	0.976	0.539	0.990
14	0.753	0.993	0.698	0.997	0.627	0.999
15	0.846	1	0.809	1	0.727	1

$n = 16$

r	90%		95%		99%	
0	0	0.147	0	0.179	0	0.264
1	0.007	0.235+	0.003	0.273	0.001	0.357
2	0.034	0.305+	0.023	0.352	0.010	0.451
3	0.071	0.381	0.053	0.429	0.029	0.525-
4	0.114	0.450	0.090	0.500	0.055+	0.579
5	0.147	0.550	0.132	0.571	0.088	0.643
6	0.189	0.619	0.178	0.648	0.125+	0.705-
7	0.235+	0.695-	0.179	0.727	0.166	0.739
8	0.299	0.701	0.272	0.728	0.212	0.788
9	0.305+	0.765-	0.273	0.821	0.261	0.834
10	0.381	0.811	0.352	0.822	0.295+	0.875-
11	0.450	0.853	0.429	0.868	0.357	0.912
12	0.550	0.886	0.500	0.910	0.421	0.945-
13	0.619	0.929	0.571	0.947	0.475+	0.971
14	0.695-	0.966	0.648	0.977	0.549	0.990
15	0.765-	0.993	0.727	0.997	0.643	0.999
16	0.853	1	0.821	1	0.736	1

$n = 17$

r	90%		95%		99%	
0	0	0.140	0	0.167	0	0.243
1	0.006	0.225+	0.003	0.254	0.001	0.346
2	0.032	0.290	0.021	0.337	0.009	0.413
3	0.067	0.364	0.050	0.417	0.027	0.500
4	0.107	0.432	0.085-	0.489	0.052	0.587
5	0.140	0.500	0.124	0.544	0.082	0.620
6	0.175+	0.568	0.166	0.594	0.117	0.662
7	0.225+	0.636	0.167	0.663	0.155+	0.757
8	0.277	0.710	0.253	0.746	0.197	0.758
9	0.290	0.723	0.254	0.747	0.242	0.803
10	0.364	0.775-	0.338	0.833	0.243	0.845
11	0.432	0.825-	0.406	0.834	0.338	0.883
12	0.500	0.860	0.456	0.876	0.380	0.918
13	0.568	0.893	0.511	0.915+	0.413	0.948
14	0.636	0.933	0.583	0.950	0.500	0.973
15	0.710	0.968	0.663	0.979	0.587	0.991
16	0.775-	0.994	0.746	0.997	0.654	0.999
17	0.860	1	0.833	1	0.757	1

$n = 18$

r	90%		95%		99%	
0	0	0.135-	0	0.157	0	0.228
1	0.006	0.216	0.003	0.242	0.001	0.318
2	0.030	0.277	0.020	0.325-	0.008	0.397
3	0.063	0.349	0.047	0.381	0.025+	0.466
4	0.101	0.419	0.080	0.444	0.049	0.534
5	0.135-	0.482	0.116	0.556	0.077	0.603
6	0.163	0.536	0.156	0.619	0.110	0.682
7	0.216	0.584	0.157	0.625+	0.145+	0.686
8	0.257	0.651	0.236	0.675+	0.184	0.772
9	0.277	0.723	0.242	0.758	0.226	0.774
10	0.349	0.743	0.325-	0.764	0.228	0.816
11	0.416	0.784	0.375-	0.843	0.314	0.855-
12	0.464	0.837	0.381	0.844	0.318	0.890
13	0.518	0.865+	0.444	0.884	0.397	0.923
14	0.581	0.899	0.556	0.920	0.466	0.951
15	0.651	0.937	0.619	0.953	0.534	0.975-
16	0.723	0.970	0.675+	0.980	0.603	0.992
17	0.784	0.994	0.758	0.997	0.682	0.999
18	0.865+	1	0.843	1	0.772	1

Table A.3 Confidence Limits for a Proportion (Two-sided), $n \leq 30$
(Continued)

n = 19

r	90%		95%		99%	
0	0	0.130	0	0.150	0	0.218
1	0.006	0.209	0.303	0.232	0.001	0.305+
2	0.028	0.265+	0.019	0.316	0.008	0.383
3	0.059	0.337	0.044	0.365-	0.024	0.455+
4	0.095+	0.387	0.075+	0.426	0.046	0.515+
5	0.130	0.440	0.110	0.500	0.073	0.564
6	0.151	0.560	0.147	0.574	0.103	0.617
7	0.209	0.613	0.150	0.635+	0.137	0.695-
8	0.238	0.614	0.222	0.655+	0.173	0.707
9	0.265+	0.663	0.232	0.688	0.212	0.782
10	0.337	0.735-	0.312	0.768	0.218	0.788
11	0.386	0.762	0.345-	0.778	0.293	0.827
12	0.387	0.791	0.365-	0.850	0.305+	0.863
13	0.440	0.849	0.426	0.853	0.383	0.897
14	0.560	0.870	0.500	0.890	0.436	0.927
15	0.613	0.905-	0.574	0.925+	0.485-	0.954
16	0.668	0.941	0.635+	0.956	0.545-	0.976
17	0.735-	0.972	0.684	0.981	0.617	0.992
18	0.791	0.994	0.768	0.997	0.695-	0.999
19	0.870	1	0.850	1	0.782	1

n = 20

r	90%		95%		99%	
0	0	0.126	0	0.143	0	0.209
1	0.005+	0.203	0.003	0.222	0.001	0.293
2	0.027	0.255-	0.018	0.294	0.008	0.375-
3	0.056	0.328	0.042	0.351	0.023	0.424
4	0.090	0.367	0.071	0.411	0.044	0.500
5	0.126	0.422	0.104	0.467	0.069	0.576
6	0.141	0.500	0.140	0.533	0.098	0.601
7	0.201	0.578	0.143	0.589	0.129	0.637
8	0.221	0.633	0.209	0.649	0.163	0.707
9	0.255-	0.642	0.222	0.706	0.200	0.726
10	0.325	0.675+	0.293	0.707	0.209	0.791
11	0.358	0.745+	0.294	0.778	0.274	0.800
12	0.367	0.779	0.351	0.791	0.293	0.837
13	0.422	0.799	0.411	0.857	0.363	0.871
14	0.500	0.859	0.467	0.860	0.399	0.902
15	0.578	0.874	0.533	0.896	0.424	0.931
16	0.633	0.910	0.589	0.929	0.500	0.956
17	0.672	0.944	0.649	0.958	0.576	0.977
18	0.745+	0.973	0.706	0.982	0.625+	0.992
19	0.797	0.995-	0.778	0.997	0.707	0.999
20	0.874	1	0.857	1	0.791	1

n = 21

r	90%		95%		99%	
0	0	0.123	0	0.137	0	0.201
1	0.005+	0.197	0.002	0.213	0.000	0.283
2	0.026	0.245-	0.017	0.277	0.007	0.347
3	0.054	0.307	0.040	0.338	0.022	0.409
4	0.086	0.353	0.068	0.398	0.041	0.466
5	0.121	0.407	0.099	0.455+	0.065+	0.534
6	0.130	0.458	0.132	0.506	0.092	0.591
7	0.191	0.542	0.137	0.551	0.122	0.653
8	0.192	0.593	0.197	0.602	0.155-	0.661
9	0.245+	0.647	0.213	0.662	0.189	0.717
10	0.306	0.693	0.276	0.723	0.201	0.743
11	0.307	0.694	0.277	0.724	0.257	0.799
12	0.353	0.755+	0.338	0.787	0.283	0.811
13	0.407	0.808	0.398	0.803	0.339	0.845+
14	0.458	0.809	0.449	0.863	0.347	0.878
15	0.542	0.870	0.494	0.868	0.409	0.908
16	0.593	0.879	0.545-	0.901	0.466	0.935-
17	0.647	0.914	0.602	0.932	0.534	0.959
18	0.693	0.946	0.662	0.960	0.591	0.978
19	0.755+	0.974	0.723	0.983	0.653	0.993
20	0.808	0.995-	0.787	0.998	0.717	1.000
21	0.877	1	0.863	1	0.799	1

n = 22

r	90%		95%		99%	
0	0	0.116	0	0.132	0	0.194
1	0.005-	0.182	0.002	0.205+	0.000	0.273
2	0.024	0.236	0.016	0.264	0.007	0.334
3	0.051	0.289	0.038	0.326	0.021	0.396
4	0.082	0.340	0.065-	0.389	0.039	0.454
5	0.115-	0.393	0.094	0.424	0.062	0.505-
6	0.116	0.444	0.126	0.500	0.088	0.550
7	0.181	0.500	0.132	0.576	0.116	0.604
8	0.182	0.556	0.187	0.582	0.147	0.666
9	0.236	0.607	0.205+	0.617	0.179	0.682
10	0.289	0.660	0.260	0.674	0.194	0.727
11	0.290	0.710	0.264	0.736	0.242	0.758
12	0.340	0.711	0.326	0.740	0.273	0.806
13	0.393	0.764	0.383	0.795-	0.318	0.821
14	0.444	0.818	0.418	0.813	0.334	0.853
15	0.500	0.819	0.424	0.868	0.396	0.884
16	0.556	0.884	0.500	0.874	0.450	0.912
17	0.607	0.885+	0.576	0.906	0.495+	0.938
18	0.660	0.918	0.611	0.935+	0.546	0.961
19	0.711	0.949	0.674	0.962	0.604	0.979
20	0.764	0.976	0.736	0.984	0.666	0.993
21	0.818	0.995+	0.795-	0.998	0.727	1.000
22	0.884	1	0.868	1	0.806	1

Table A.3   Confidence Limits for a Proportion (Two-sided), $n \leq 30$
(Continued)

$n = 23$

r	90%		95%		99%	
0	0	0.111	0	0.127	0	0.187
1	0.005−	0.174	0.002	0.198	0.000	0.365+
2	0.023	0.228	0.016	0.255	0.007	0.323
3	0.049	0.274	0.037	0.317	0.020	0.386
4	0.078	0.328	0.062	0.361	0.038	0.429
5	0.110	0.381	0.090	0.409	0.059	0.500
6	0.111	0.431	0.120	0.457	0.084	0.571
7	0.173	0.479	0.127	0.543	0.111	0.580
8	0.174	0.522	0.178	0.591	0.140	0.616
9	0.228	0.569	0.198	0.639	0.171	0.677
10	0.273	0.619	0.247	0.640	0.187	0.702
11	0.274	0.672	0.255−	0.683	0.229	0.735−
12	0.328	0.726	0.317	0.745+	0.265+	0.771
13	0.381	0.727	0.360	0.753	0.298	0.813
14	0.431	0.772	0.361	0.802	0.323	0.829
15	0.478	0.826	0.409	0.822	0.384	0.860
16	0.521	0.827	0.457	0.873	0.420	0.889
17	0.569	0.889	0.543	0.880	0.429	0.916
18	0.619	0.890	0.591	0.910	0.500	0.941
19	0.672	0.922	0.639	0.938	0.571	0.962
20	0.726	0.951	0.683	0.963	0.614	0.980
21	0.772	0.977	0.745+	0.984	0.677	0.993
22	0.826	0.995+	0.802	0.998	0.735−	1.000
23	0.889	1	0.873	1	0.813	1

$n = 24$

r	90%		95%		99%	
0	0	0.105+	0	0.122	0	0.181
1	0.004	0.165+	0.002	0.191	0.000	0.259
2	0.022	0.221	0.015+	0.246	0.006	0.313
3	0.047	0.264	0.035−	0.308	0.019	0.364
4	0.075−	0.317	0.059	0.347	0.036	0.416
5	0.105−	0.370	0.086	0.396	0.057	0.464
6	0.105+	0.423	0.115−	0.443	0.080	0.536
7	0.165−	0.448	0.122	0.500	0.106	0.584
8	0.165+	0.552	0.169	0.557	0.133	0.636
9	0.221	0.553	0.191	0.604	0.163	0.638
10	0.259	0.587	0.234	0.653	0.181	0.687
11	0.264	0.630	0.246	0.661	0.216	0.720
12	0.317	0.683	0.308	0.692	0.257	0.743
13	0.370	0.736	0.339	0.754	0.280	0.784
14	0.413	0.741	0.347	0.766	0.313	0.819
15	0.447	0.779	0.396	0.809	0.362	0.837
16	0.448	0.835−	0.443	0.831	0.364	0.867
17	0.552	0.835+	0.500	0.878	0.416	0.894
18	0.577	0.895−	0.557	0.885+	0.464	0.920
19	0.630	0.895+	0.604	0.914	0.536	0.943
20	0.683	0.925+	0.653	0.941	0.584	0.964
21	0.736	0.953	0.692	0.965+	0.636	0.981
22	0.779	0.978	0.754	0.985−	0.687	0.994
23	0.835−	0.996	0.809	0.998	0.741	1.000
24	0.895−	1	0.878	1	0.819	1

$n = 25$

r	90%		95%		99%	
0	0	0.102	0	0.118	0	0.175+
1	0.004	0.159	0.002	0.185+	0.000	0.246
2	0.021	0.214	0.014	0.238	0.006	0.305−
3	0.045−	0.255−	0.034	0.303	0.018	0.352
4	0.072	0.307	0.057	0.336	0.034	0.403
5	0.101	0.362	0.082	0.384	0.054	0.451
6	0.102	0.390	0.110	0.431	0.077	0.500
7	0.158	0.432	0.118	0.475−	0.101	0.549
8	0.159	0.500	0.161	0.525+	0.127	0.597
9	0.214	0.568	0.185+	0.569	0.155+	0.648
10	0.246	0.610	0.222	0.616	0.175+	0.658
11	0.255	0.611	0.238	0.664	0.205+	0.695+
12	0.307	0.640	0.296	0.683	0.245+	0.754
13	0.360	0.693	0.317	0.704	0.246	0.755−
14	0.389	0.745+	0.336	0.762	0.305−	0.795−
15	0.390	0.754	0.384	0.778	0.342	0.825−
16	0.432	0.786	0.431	0.815−	0.352	0.845−
17	0.500	0.841	0.475−	0.839	0.403	0.873
18	0.568	0.842	0.525+	0.882	0.451	0.899
19	0.610	0.898	0.569	0.890	0.500	0.923
20	0.638	0.899	0.616	0.918	0.549	0.946
21	0.693	0.928	0.664	0.943	0.597	0.966
22	0.745+	0.955+	0.697	0.966	0.648	0.982
23	0.786	0.979	0.762	0.986	0.695+	0.994
24	0.841	0.996	0.815−	0.998	0.751	1.000
25	0.898	1	0.882	1	0.825−	1

$n = 26$

r	90%		95%		99%	
0	0	0.098	0	0.114	0	0.170
1	0.004	0.152	0.002	0.180	0.000	0.235−
2	0.021	0.209	0.014	0.230	0.006	0.298
3	0.043	0.247	0.032	0.283	0.017	0.342
4	0.069	0.299	0.054	0.325+	0.033	0.393
5	0.097	0.343	0.079	0.374	0.052	0.442
6	0.098	0.377	0.106	0.421	0.073	0.487
7	0.151	0.419	0.114	0.465−	0.097	0.526
8	0.152	0.460	0.154	0.506	0.122	0.562
9	0.209	0.540	0.180	0.542	0.149	0.607
10	0.233	0.581	0.212	0.579	0.170	0.658
11	0.247	0.623	0.230	0.626	0.195−	0.678
12	0.299	0.657	0.282	0.675−	0.234	0.702
13	0.342	0.658	0.283	0.717	0.235−	0.765+
14	0.343	0.701	0.325+	0.718	0.298	0.766
15	0.377	0.753	0.374	0.770	0.322	0.805+
16	0.419	0.767	0.421	0.788	0.342	0.830
17	0.460	0.791	0.458	0.820	0.393	0.851
18	0.540	0.848	0.494	0.846	0.438	0.878
19	0.581	0.849	0.535−	0.886	0.474	0.903
20	0.623	0.902	0.579	0.894	0.513	0.927
21	0.657	0.903	0.626	0.921	0.558	0.948
22	0.701	0.931	0.675−	0.946	0.607	0.967
23	0.753	0.957	0.717	0.968	0.658	0.983
24	0.791	0.979	0.770	0.986	0.702	0.994
25	0.848	0.996	0.820	0.998	0.765+	1.000
26	0.902	1	0.886	1	0.830	1

Table A.3 Confidence Limits for a Proportion (Two-sided), $n \leq 30$
(Continued)

$n = 27$

r	90%		95%		99%	
0	0	0.093	0	0.110	0	0.166
1	0.004	0.146	0.002	0.175-	0.000	0.225-
2	0.020	0.204	0.013	0.223	0.006	0.297
3	0.042	0.239	0.031	0.270	0.017	0.332
4	0.066	0.291	0.052	0.316	0.032	0.384
5	0.093	0.327	0.076	0.364	0.050	0.419
6	0.094	0.365+	0.101	0.415-	0.070	0.461
7	0.145+	0.407	0.110	0.437	0.093	0.539
8	0.146	0.447	0.148	0.500	0.117	0.581
9	0.204	0.500	0.175-	0.563	0.143	0.587
10	0.221	0.553	0.202	0.570	0.166	0.617
11	0.239	0.593	0.223	0.598	0.185-	0.668
12	0.291	0.635-	0.269	0.636 ·	0.224	0.702
13	0.326	0.673	0.270	0.684	0.225-	0.716
14	0.327	0.674	0.316	0.730	0.284	0.775+
15	0.365+	0.709	0.364	0.731	0.298	0.776
16	0.407	0.761	0.402	0.777	0.332	0.815+
17	0.447	0.779	0.430	0.798	0.383	0.834
18	0.500	0.796	0.437	0.825+	0.413	0.857
19	0.553	0.854	0.500	0.852	0.419	0.883
20	0.593	0.855-	0.563	0.890	0.461	0.907
21	0.635-	0.906	0.585+	0.899	0.539	0.930
22	0.673	0.907	0.636	0.924	0.581	0.950
23	0.709	0.934	0.684	0.948	0.616	0.968
24	0.761	0.958	0.730	0.969	0.668	0.983
25	0.796	0.980	0.777	0.987	0.703	0.994
26	0.854	0.996	0.825+	0.998	0.775+	1.000
27	0.907	1	0.890	1	0.834	1

$n = 28$

r	90%		95%		99%	
0	0	0.090	0	0.106	0	0.162
1	0.004	0.140	0.002	0.170	0.000	0.218
2	0.019	0.201	0.013	0.217	0.005+	0.273
3	0.040	0.232	0.030	0.259	0.016	0.323
4	0.064	0.284	0.050	0.307	0.031	0.365-
5	0.089	0.312	0.073	0.357	0.048	0.408
6	0.090	0.355-	0.098	0.384	0.068	0.449
7	0.139	0.396	0.106	0.424	0.089	0.500
8	0.140	0.435+	0.142	0.463	0.112	0.551
9	0.197	0.473	0.170	0.537	0.137	0.592
10	0.208	0.527	0.192	0.576	0.162	0.635+
11	0.232	0.565-	0.217	0.616	0.175+	0.636
12	0.284	0.604	0.258	0.619	0.214	0.677
13	0.310	0.645+	0.259	0.645+	0.218	0.727
14	0.312	0.688	0.307	0.693	0.272	0.728
15	0.335-	0.690	0.355-	0.741	0.273	0.782
16	0.396	0.716	0.381	0.742	0.323	0.786
17	0.435+	0.768	0.384	0.783	0.364	0.825-
18	0.473	0.792	0.424	0.808	0.365-	0.838
19	0.527	0.803	0.463	0.830	0.408	0.863
20	0.565-	0.860	0.537	0.858	0.449	0.888
21	0.604	0.861	0.576	0.894	0.500	0.911
22	0.645+	0.910	0.616	0.902	0.551	0.932
23	0.688	0.911	0.643	0.927	0.592	0.952
24	0.716	0.936	0.693	0.950	0.635+	0.969
25	0.768	0.960	0.741	0.970	0.677	0.984
26	0.799	0.981	0.783	0.987	0.727	0.995-
27	0.860	0.996	0.830	0.998	0.782	1.000
28	0.910	1	0.894	1	0.838	1

$n = 29$

r	90%		95%		99%	
0	0	0.087	0	0.103	0	0.160
1	0.004	0.135-	0.002	0.166	0.000	0.211
2	0.018	0.190	0.012	0.211	0.005	0.263
3	0.039	0.225-	0.029	0.251	0.015+	0.316
4	0.062	0.279	0.049	0.299	0.030	0.354
5	0.086	0.303	0.070	0.340	0.046	0.397
6	0.087	0.345-	0.094	0.374	0.065+	0.438
7	0.134	0.385+	0.103	0.413	0.086	0.477
8	0.135-	0.425-	0.136	0.451	0.108	0.523
9	0.189	0.463	0.166	0.500	0.132	0.562
10	0.190	0.500	0.184	0.549	0.157	0.603
11	0.225-	0.537	0.211	0.587	0.165+	0.646
12	0.276	0.575+	0.247	0.626	0.206	0.654
13	0.294	0.615-	0.251	0.660	0.211	0.684
14	0.308	0.655+	0.299	0.661	0.260	0.737
15	0.345-	0.697	0.339	0.701	0.263	0.740
16	0.385+	0.706	0.340	0.749	0.316	0.789
17	0.425-	0.724	0.374	0.753	0.346	0.794
18	0.463	0.775+	0.413	0.789	0.354	0.835-
19	0.500	0.810	0.451	0.816	0.397	0.843
20	0.537	0.811	0.500	0.834	0.438	0.868
21	0.575+	0.865+	0.549	0.864	0.477	0.892
22	0.615-	0.866	0.587	0.897	0.523	0.914
23	0.655	0.913	0.626	0.906	0.562	0.935-
24	0.697	0.914	0.660	0.930	0.603	0.954
25	0.721	0.938	0.701	0.951	0.646	0.970
26	0.775+	0.961	0.749	0.971	0.684	0.985-
27	0.810	0.982	0.789	0.988	0.737	0.995-
28	0.865+	0.996	0.834	0.998	0.789	1.000
29	0.913	1	0.897	1	0.840	1

$n = 30$

r	90%		95%		99%	
0	0	0.084	0	0.100	0	0.152
1	0.004	0.130	0.002	0.163	0.000	0.206
2	0.018	0.183	0.012	0.205+	0.005+	0.256
3	0.037	0.219	0.028	0.244	0.015-	0.310
4	0.059	0.266	0.047	0.292	0.028	0.345-
5	0.083	0.295-	0.068	0.325-	0.045-	0.388
6	0.084	0.336	0.091	0.364	0.063	0.430
7	0.129	0.376	0.100	0.403	0.083	0.469
8	0.130	0.416	0.131	0.440	0.104	0.505+
9	0.182	0.455+	0.163	0.476	0.127	0.538
10	0.183	0.492	0.175+	0.524	0.151	0.570
11	0.219	0.524	0.205+	0.560	0.152	0.612
12	0.265-	0.554	0.236	0.597	0.198	0.655+
13	0.266	0.584	0.244	0.636	0.206	0.671
14	0.295-	0.624	0.292	0.675+	0.294	0.692
15	0.336	0.664	0.324	0.676	0.256	0.744
16	0.376	0.705+	0.325-	0.708	0.308	0.751
17	0.416	0.734	0.364	0.756	0.329	0.794
18	0.446	0.735+	0.403	0.764	0.345-	0.802
19	0.476	0.781	0.440	0.795-	0.388	0.848
20	0.508	0.817	0.476	0.825-	0.430	0.849
21	0.545-	0.818	0.524	0.837	0.462	0.873
22	0.584	0.870	0.560	0.869	0.495-	0.896
23	0.624	0.871	0.597	0.900	0.531	0.917
24	0.664	0.916	0.636	0.909	0.570	0.937
25	0.705+	0.917	0.675+	0.932	0.612	0.955+
26	0.734	0.941	0.708	0.953	0.655+	0.972
27	0.781	0.963	0.756	0.972	0.690	0.985+
28	0.817	0.982	0.795-	0.988	0.744	0.995-
29	0.870	0.996	0.837	0.998	0.794	1.000
30	0.916	1	0.900	1	0.848	1

Appendix A: Tables

Table A.4 Areas under the Normal Curve*

Proportion of total area under the curve that is under the portion of the curve from $-\infty$ to $(X_i - \bar{X}')/\sigma'$. (X_i represents any desired value of the variable X.)

$\dfrac{X_i - \bar{X}'}{\sigma'}$	0.09	0.08	0.07	0.06	0.05	0.04	0.03	0.02	0.01	0.00
−3.5	0.00017	0.00017	0.00018	0.00019	0.00019	0.00020	0.00021	0.00022	0.00022	0.00023
−3.4	0.00024	0.00025	0.00026	0.00027	0.00028	0.00029	0.00030	0.00031	0.00033	0.00034
−3.3	0.00035	0.00036	0.00038	0.00039	0.00040	0.00042	0.00043	0.00045	0.00047	0.00048
−3.2	0.00050	0.00052	0.00054	0.00056	0.00058	0.00060	0.00062	0.00064	0.00066	0.00069
−3.1	0.00071	0.00074	0.00076	0.00079	0.00082	0.00085	0.00087	0.00090	0.00094	0.00097
−3.0	0.00100	0.00104	0.00107	0.00111	0.00114	0.00118	0.00122	0.00126	0.00131	0.00135
−2.9	0.0014	0.0014	0.0015	0.0015	0.0016	0.0016	0.0017	0.0017	0.0018	0.0019
−2.8	0.0019	0.0020	0.0021	0.0021	0.0022	0.0023	0.0023	0.0024	0.0025	0.0026
−2.7	0.0026	0.0027	0.0028	0.0029	0.0030	0.0031	0.0032	0.0033	0.0034	0.0035
−2.6	0.0036	0.0037	0.0038	0.0039	0.0040	0.0041	0.0043	0.0044	0.0045	0.0047
−2.5	0.0048	0.0049	0.0051	0.0052	0.0054	0.0055	0.0057	0.0059	0.0060	0.0062
−2.4	0.0064	0.0066	0.0068	0.0069	0.0071	0.0073	0.0075	0.0078	0.0080	0.0082
−2.3	0.0084	0.0087	0.0089	0.0091	0.0094	0.0096	0.0099	0.0102	0.0104	0.0107
−2.2	0.0110	0.0113	0.0116	0.0119	0.0122	0.0125	0.0129	0.0132	0.0136	0.0139
−2.1	0.0143	0.0146	0.0150	0.0154	0.0158	0.0162	0.0166	0.0170	0.0174	0.0179
−2.0	0.0183	0.0188	0.0192	0.0197	0.0202	0.0207	0.0212	0.0217	0.0222	0.0228
−1.9	0.0233	0.0239	0.0244	0.0250	0.0256	0.0262	0.0268	0.0274	0.0281	0.0287
−1.8	0.0294	0.0301	0.0307	0.0314	0.0322	0.0329	0.0336	0.0344	0.0351	0.0359
−1.7	0.0367	0.0375	0.0384	0.0392	0.0401	0.0409	0.0418	0.0427	0.0436	0.0446
−1.6	0.0455	0.0465	0.0475	0.0485	0.0495	0.0505	0.0516	0.0526	0.0537	0.0548
−1.5	0.0559	0.0571	0.0582	0.0594	0.0606	0.0618	0.0630	0.0643	0.0655	0.0668
−1.4	0.0681	0.0694	0.0708	0.0721	0.0735	0.0749	0.0764	0.0778	0.0793	0.0808
−1.3	0.0823	0.0838	0.0853	0.0869	0.0885	0.0901	0.0918	0.0934	0.0951	0.0968
−1.2	0.0985	0.1003	0.1020	0.1038	0.1057	0.1075	0.1093	0.1112	0.1131	0.1151
−1.1	0.1170	0.1190	0.1210	0.1230	0.1251	0.1271	0.1292	0.1314	0.1335	0.1357
−1.0	0.1379	0.1401	0.1423	0.1446	0.1469	0.1492	0.1515	0.1539	0.1562	0.1587
−0.9	0.1611	0.1635	0.1660	0.1685	0.1711	0.1736	0.1762	0.1788	0.1814	0.1841
−0.8	0.1867	0.1894	0.1922	0.1949	0.1977	0.2005	0.2033	0.2061	0.2090	0.2119
−0.7	0.2148	0.2177	0.2207	0.2236	0.2266	0.2297	0.2327	0.2358	0.2389	0.2420
−0.6	0.2451	0.2483	0.2514	0.2546	0.2578	0.2611	0.2643	0.2676	0.2709	0.2743
−0.5	0.2776	0.2810	0.2843	0.2877	0.2912	0.2946	0.2981	0.3015	0.3050	0.3085
−0.4	0.3121	0.3156	0.3192	0.3228	0.3264	0.3300	0.3336	0.3372	0.3409	0.3446
−0.3	0.3483	0.3520	0.3557	0.3594	0.3632	0.3669	0.3707	0.3745	0.3783	0.3821
−0.2	0.3859	0.3897	0.3936	0.3974	0.4013	0.4052	0.4090	0.4129	0.4168	0.4207
−0.1	0.4247	0.4286	0.4325	0.4364	0.4404	0.4443	0.4483	0.4522	0.4562	0.4602
−0.0	0.4641	0.4681	0.4721	0.4761	0.4801	0.4840	0.4880	0.4920	0.4960	0.5000

* From *Statistical Quality Control*, 3d ed., by Eugene L. Grant. Copyright 1964. McGraw-Hill Book Company. Used by permission.

Table A.4   Areas under the Normal Curve (*Continued*)

$\frac{X_i-\bar{X}'}{\sigma'}$	0.00	0.01	0.02	0.03	0.04	0.05	0.06	0.07	0.08	0.09
+0.0	0.5000	0.5040	0.5080	0.5120	0.5160	0.5199	0.5239	0.5279	0.5319	0.5359
+0.1	0.5398	0.5438	0.5478	0.5517	0.5557	0.5596	0.5636	0.5675	0.5714	0.5753
+0.2	0.5793	0.5832	0.5871	0.5910	0.5948	0.5987	0.6026	0.6064	0.6103	0.6141
+0.3	0.6179	0.6217	0.6255	0.6293	0.6331	0.6368	0.6406	0.6443	0.6480	0.6517
+0.4	0.6554	0.6591	0.6628	0.6664	0.6700	0.6736	0.6772	0.6808	0.6844	0.6879
+0.5	0.6915	0.6950	0.6985	0.7019	0.7054	0.7088	0.7123	0.7157	0.7190	0.7224
+0.6	0.7257	0.7291	0.7324	0.7357	0.7389	0.7422	0.7454	0.7486	0.7517	0.7549
+0.7	0.7580	0.7611	0.7642	0.7673	0.7704	0.7734	0.7764	0.7794	0.7823	0.7852
+0.8	0.7881	0.7910	0.7939	0.7967	0.7995	0.8023	0.8051	0.8079	0.8106	0.8133
+0.9	0.8159	0.8186	0.8212	0.8238	0.8264	0.8289	0.8315	0.8340	0.8365	0.8389
+1.0	0.8413	0.8438	0.8461	0.8485	0.8508	0.8531	0.8554	0.8577	0.8599	0.8621
+1.1	0.8643	0.8665	0.8686	0.8708	0.8729	0.8749	0.8770	0.8790	0.8810	0.8830
+1.2	0.8849	0.8869	0.8888	0.8907	0.8925	0.8944	0.8962	0.8980	0.8997	0.9015
+1.3	0.9032	0.9049	0.9066	0.9082	0.9099	0.9115	0.9131	0.9147	0.9162	0.9177
+1.4	0.9192	0.9207	0.9222	0.9236	0.9251	0.9265	0.9279	0.9292	0.9306	0.9319
+1.5	0.9332	0.9345	0.9357	0.9370	0.9382	0.9394	0.9406	0.9418	0.9429	0.9441
+1.6	0.9452	0.9463	0.9474	0.9484	0.9495	0.9505	0.9515	0.9525	0.9535	0.9545
+1.7	0.9554	0.9564	0.9573	0.9582	0.9591	0.9599	0.9608	0.9616	0.9625	0.9633
+1.8	0.9641	0.9649	0.9656	0.9664	0.9671	0.9678	0.9686	0.9693	0.9699	0.9706
+1.9	0.9713	0.9719	0.9726	0.9732	0.9738	0.9744	0.9750	0.9756	0.9761	0.9767
+2.0	0.9773	0.9778	0.9783	0.9788	0.9793	0.9798	0.9803	0.9808	0.9812	0.9817
+2.1	0.9821	0.9826	0.9830	0.9834	0.9838	0.9842	0.9846	0.9850	0.9854	0.9857
+2.2	0.9861	0.9864	0.9868	0.9871	0.9875	0.9878	0.9881	0.9884	0.9887	0.9890
+2.3	0.9893	0.9896	0.9898	0.9901	0.9904	0.9906	0.9909	0.9911	0.9913	0.9916
+2.4	0.9918	0.9920	0.9922	0.9925	0.9927	0.9929	0.9931	0.9932	0.9934	0.9936
+2.5	0.9938	0.9940	0.9941	0.9943	0.9945	0.9946	0.9948	0.9949	0.9951	0.9952
+2.6	0.9953	0.9955	0.9956	0.9957	0.9959	0.9960	0.9961	0.9962	0.9963	0.9964
+2.7	0.9965	0.9966	0.9967	0.9968	0.9969	0.9970	0.9971	0.9972	0.9973	0.9974
+2.8	0.9974	0.9975	0.9976	0.9977	0.9977	0.9978	0.9979	0.9979	0.9980	0.9981
+2.9	0.9981	0.9982	0.9983	0.9983	0.9984	0.9984	0.9985	0.9985	0.9986	0.9986
+3.0	0.99865	0.99869	0.99874	0.99878	0.99882	0.99886	0.99889	0.99893	0.99896	0.99900
+3.1	0.99903	0.99906	0.99910	0.99913	0.99915	0.99918	0.99921	0.99924	0.99926	0.99929
+3.2	0.99931	0.99934	0.99936	0.99938	0.99940	0.99942	0.99944	0.99946	0.99948	0.99950
+3.3	0.99952	0.99953	0.99955	0.99957	0.99958	0.99960	0.99961	0.99962	0.99964	0.99965
+3.4	0.99966	0.99967	0.99969	0.99970	0.99971	0.99972	0.99973	0.99974	0.99975	0.99976
+3.5	0.99977	0.99978	0.99978	0.99979	0.99980	0.99981	0.99981	0.99982	0.99983	0.99983

Table A.5   Percentiles of the *t* Distribution*

df	$t_{.60}$	$t_{.70}$	$t_{.80}$	$t_{.90}$	$t_{.95}$	$t_{.975}$	$t_{.99}$	$t_{.995}$
1	.325	.727	1.376	3.078	6.314	12.706	31.821	63.657
2	.289	.617	1.061	1.886	2.920	4.303	6.965	9.925
3	.277	.584	.978	1.638	2.353	3.182	4.541	5.841
4	.271	.569	.941	1.533	2.132	2.776	3.747	4.604
5	.267	.559	.920	1.476	2.015	2.571	3.365	4.032
6	.265	.553	.906	1.440	1.943	2.447	3.143	3.707
7	.263	.549	.896	1.415	1.895	2.365	2.998	3.499
8	.262	.546	.889	1.397	1.860	2.306	2.896	3.355
9	.261	.543	.883	1.383	1.833	2.262	2.821	3.250
10	.260	.542	.879	1.372	1.812	2.228	2.764	3.169
11	.260	.540	.876	1.363	1.796	2.201	2.718	3.106
12	.259	.539	.873	1.356	1.782	2.179	2.681	3.055
13	.259	.538	.870	1.350	1.771	2.160	2.650	3.012
14	.258	.537	.868	1.345	1.761	2.145	2.624	2.977
15	.258	.536	.866	1.341	1.753	2.131	2.602	2.947
16	.258	.535	.865	1.337	1.746	2.120	2.583	2.921
17	.257	.534	.863	1.333	1.740	2.110	2.567	2.898
18	.257	.534	.862	1.330	1.734	2.101	2.552	2.878
19	.257	.533	.861	1.328	1.729	2.093	2.539	2.861
20	.257	.533	.860	1.325	1.725	2.086	2.528	2.845
21	.257	.532	.859	1.323	1.721	2.080	2.518	2.831
22	.256	.532	.858	1.321	1.717	2.074	2.508	2.819
23	.256	.532	.858	1.319	1.714	2.069	2.500	2.807
24	.256	.531	.857	1.318	1.711	2.064	2.492	2.797
25	.256	.531	.856	1.316	1.708	2.060	2.485	2.787
26	.256	.531	.856	1.315	1.706	2.056	2.479	2.779
27	.256	.531	.855	1.314	1.703	2.052	2.473	2.771
28	.256	.530	.855	1.313	1.701	2.048	2.467	2.763
29	.256	.530	.854	1.311	1.699	2.045	2.462	2.756
30	.256	.530	.854	1.310	1.697	2.042	2.457	2.750
40	.255	.529	.851	1.303	1.684	2.021	2.423	2.704
60	.254	.527	.848	1.296	1.671	2.000	2.390	2.660
120	.254	.526	.845	1.289	1.658	1.980	2.358	2.617
∞	.253	.524	.842	1.282	1.645	1.960	2.326	2.576
df	$-t_{.40}$	$-t_{.30}$	$-t_{.20}$	$-t_{.10}$	$-t_{.05}$	$-t_{.025}$	$-t_{.01}$	$-t_{.005}$

When the table is read from the foot, the tabled values are to be prefixed with a negative sign. Interpolation should be performed using the reciprocals of the degrees of freedom.

*The data of this table are taken from Table III of Fischer and Yates: *Statistical Tables for Biological, Agricultural and Medical Research*, published by Longman Group U.K., Ltd., London (previously published by Oliver & Boyd, Ltd., Edinburgh and by permission of the author and publishers. From *Introduction to Statistical Analysis*, 2d ed., by W. J. Dixon and F. J. Massey, Jr. Copyright, 1957. McGraw-Hill Book Company. Used by permisson.

Table A.6   Percentiles of the χ^2 Distribution*

df	Per Cent									
	.5	1	2.5	5	10	90	95	97.5	99	99.5
1	.000039	.00016	.00098	.0039	.0158	2.71	3.84	5.02	6.63	7.88
2	.0100	.0201	.0506	.1026	.2107	4.61	5.99	7.38	9.21	10.60
3	.0717	.115	.216	.352	.584	6.25	7.81	9.35	11.34	12.84
4	.207	.297	.484	.711	1.064	7.78	9.49	11.14	13.28	14.86
5	.412	.554	.831	1.15	1.61	9.24	11.07	12.83	15.09	16.75
6	.676	.872	1.24	1.64	2.20	10.64	12.59	14.45	16.81	18.55
7	.989	1.24	1.69	2.17	2.83	12.02	14.07	16.01	18.48	20.28
8	1.34	1.65	2.18	2.73	3.49	13.36	15.51	17.53	20.09	21.96
9	1.73	2.09	2.70	3.33	4.17	14.68	16.92	19.02	21.67	23.59
10	2.16	2.56	3.25	3.94	4.87	15.99	18.31	20.48	23.21	25.19
11	2.60	3.05	3.82	4.57	5.58	17.28	19.68	21.92	24.73	26.76
12	3.07	3.57	4.40	5.23	6.30	18.55	21.03	23.34	26.22	28.30
13	3.57	4.11	5.01	5.89	7.04	19.81	22.36	24.74	27.69	29.82
14	4.07	4.66	5.63	6.57	7.79	21.06	23.68	26.12	29.14	31.32
15	4.60	5.23	6.26	7.26	8.55	22.31	25.00	27.49	30.58	32.80
16	5.14	5.81	6.91	7.96	9.31	23.54	26.30	28.85	32.00	34.27
18	6.26	7.01	8.23	9.39	10.86	25.99	28.87	31.53	34.81	37.16
20	7.43	8.26	9.59	10.85	12.44	28.41	31.41	34.17	37.57	40.00
24	9.89	10.86	12.40	13.85	15.66	33.20	36.42	39.36	42.98	45.56
30	13.79	14.95	16.79	18.49	20.60	40.26	43.77	46.98	50.89	53.67
40	20.71	22.16	24.43	26.51	29.05	51.81	55.76	59.34	63.69	66.77
60	35.53	37.48	40.48	43.19	46.46	74.40	79.08	83.30	88.38	91.95
120	83.85	86.92	91.58	95.70	100.62	140.23	146.57	152.21	158.95	163.64

For large values of degrees of freedom the approximate formula

$$\chi_\alpha^2 = n \left(1 - \frac{2}{9n} + z_\alpha \sqrt{\frac{2}{9n}} \right)^3$$

where z_α is the normal deviate and n is the number of degrees of freedom, may be used.   For example $\chi_{.99}^2 = 60[1 - .00370 + 2.326(.06086)]^3 = 60(1.1379)^3 = 88.4$ for the 99th percentile for 60 degrees of freedom.

* From *Introduction to Statistical Analysis*, 2d ed., by W. J. Dixon and F. J. Massey, Jr. Copyright, 1957.   McGraw-Hill Book Company.   Used by permission.

Table A.7   **Percentiles of the $F(\nu_1,\nu_2)$ Distribution with Degrees of Freedom ν_1 for the Numerator and ν_2 for the Denominator***

ν_2	Cum. Prop.	1	2	3	4	5	6	7	8	9	10	11	12	Cum. Prop.
1	.0005	$.0^662$	$.0^550$	$.0^338$	$.0^294$	.016	.022	.027	.032	.036	.039	.042	.045	.0005
	.001	$.0^525$	$.0^210$	$.0^260$	.013	.021	.028	.034	.039	.044	.048	.051	.054	.001
	.005	$.0^462$	$.0^251$	.018	.032	.044	.054	.062	.068	.073	.078	.082	.085	.005
	.010	$.0^325$	.010	.029	.047	.062	.073	.082	.089	.095	.100	.104	.107	.010
	.025	$.0^215$	.026	.057	.082	.100	.113	.124	.132	.139	.144	.149	.153	.025
	.05	$.0^262$	.054	.099	.130	.151	.167	.179	.188	.195	.201	.207	.211	.05
	.10	.025	.117	.181	.220	.246	.265	.279	.289	.298	.304	.310	.315	.10
	.25	.172	.389	.494	.553	.591	.617	.637	.650	.661	.670	.680	.684	.25
	.50	1.00	1.50	1.71	1.82	1.89	1.94	1.98	2.00	2.03	2.04	2.05	2.07	.50
	.75	5.83	7.50	8.20	8.58	8.82	8.98	9.10	9.19	9.26	9.32	9.36	9.41	.75
	.90	39.9	49.5	53.6	55.8	57.2	58.2	58.9	59.4	59.9	60.2	60.5	60.7	.90
	.95	161	200	216	225	230	234	237	239	241	242	243	244	.95
	.975	648	800	864	900	922	937	948	957	963	969	973	977	.975
	.99	405^1	500^1	540^1	562^1	576^1	586^1	593^1	598^1	602^1	606^1	608^1	611^1	.99
	.995	162^2	200^2	216^2	225^2	231^2	234^2	237^2	239^2	241^2	242^2	243^2	244^2	.995
	.999	406^3	500^3	540^3	562^3	576^3	586^3	593^3	598^3	602^3	606^3	609^3	611^3	.999
	.9995	162^4	200^4	216^4	225^4	231^4	234^4	237^4	239^4	241^4	242^4	243^4	244^4	.9995
2	.0005	$.0^650$	$.0^550$	$.0^342$	.011	.020	.029	.037	.044	.050	.056	.061	.065	.0005
	.001	$.0^520$	$.0^210$	$.0^268$	.016	.027	.037	.046	.054	.061	.067	.072	.077	.001
	.005	$.0^450$	$.0^250$	.020	.038	.055	.069	.081	.091	.099	.106	.112	.118	.005
	.01	$.0^320$	.010	.032	.056	.075	.092	.105	.116	.125	.132	.139	.144	.01
	.025	$.0^213$	.026	.062	.094	.119	.138	.153	.165	.175	.183	.190	.196	.025
	.05	$.0^250$	.053	.105	.144	.173	.194	.211	.224	.235	.244	.251	.257	.05
	.10	.020	.111	.183	.231	.265	.289	.307	.321	.333	.342	.350	.356	.10
	.25	.133	.333	.439	.500	.540	.568	.588	.604	.616	.626	.633	.641	.25
	.50	.667	1.00	1.13	1.21	1.25	1.28	1.30	1.32	1.33	1.34	1.35	1.36	.50
	.75	2.57	3.00	3.15	3.23	3.28	3.31	3.34	3.35	3.37	3.38	3.39	3.39	.75
	.90	8.53	9.00	9.16	9.24	9.29	9.33	9.35	9.37	9.38	9.39	9.40	9.41	.90
	.95	18.5	19.0	19.2	19.2	19.3	19.3	19.4	19.4	19.4	19.4	19.4	19.4	.95
	.975	38.5	39.0	39.2	39.2	39.3	39.3	39.4	39.4	39.4	39.4	39.4	39.4	.975
	.99	98.5	99.0	99.2	99.2	99.3	99.3	99.4	99.4	99.4	99.4	99.4	99.4	.99
	.995	198	199	199	199	199	199	199	199	199	199	199	199	.995
	.999	998	999	999	999	999	999	999	999	999	999	999	999	.999
	.9995	200^1	200^1	200^1	200^1	200^1	200^1	200^1	200^1	200^1	200^1	200^1	200^1	.9995
3	.0005	$.0^646$	$.0^550$	$.0^344$	.012	.023	.033	.043	.052	.060	.067	.074	.079	.0005
	.001	$.0^519$	$.0^210$	$.0^271$	.018	.030	.042	.053	.063	.072	.079	.086	.093	.001
	.005	$.0^446$	$.0^250$	.021	.041	.060	.077	.092	.104	.115	.124	.132	.138	.005
	.01	$.0^319$	.010	.034	.060	.083	.102	.118	.132	.143	.153	.161	.168	.01
	.025	$.0^212$	.026	.065	.100	.129	.152	.170	.185	.197	.207	.216	.224	.025
	.05	$.0^246$	.052	.108	.152	.185	.210	.230	.246	.259	.270	.279	.287	.05
	.10	.019	.109	.185	.239	.276	.304	.325	.342	.356	.367	.376	.384	.10
	.25	.122	.317	.424	.489	.531	.561	.582	.600	.613	.624	.633	.641	.25
	.50	.585	.881	1.00	1.06	1.10	1.13	1.15	1.16	1.17	1.18	1.19	1.20	.50
	.75	2.02	2.28	2.36	2.39	2.41	2.42	2.43	2.44	2.44	2.44	2.45	2.45	.75
	.90	5.54	5.46	5.39	5.34	5.31	5.28	5.27	5.25	5.24	5.23	5.22	5.22	.90
	.95	10.1	9.55	9.28	9.12	9.01	8.94	8.89	8.85	8.81	8.79	8.76	8.74	.95
	.975	17.4	16.0	15.4	15.1	14.9	14.7	14.6	14.5	14.5	14.4	14.4	14.3	.975
	.99	34.1	30.8	29.5	28.7	28.2	27.9	27.7	27.5	27.3	27.2	27.1	27.1	.99
	.995	55.6	49.8	47.5	46.2	45.4	44.8	44.4	44.1	43.9	43.7	43.5	43.4	.995
	.999	167	149	141	137	135	133	132	131	130	129	129	128	.999
	.9995	266	237	225	218	214	211	209	208	207	206	204	204	.9995

Read $.0^356$ as .00056, 200^1 as 2000, 162^4 as 1620000, etc.

* From *Introduction to Statistical Analysis*, 2d ed., by W. J. Dixon and F. J. Massey, Jr. Copyright, 1957. McGraw-Hill Book Company. Used by permission.

Table A.7   **Percentiles of the $F(\nu_1, \nu_2)$ Distribution with Degrees of Freedom ν_1 for the Numerator and ν_2 for the Denominator** (*Continued*)

Cum. Prop.	15	20	24	30	40	50	60	100	120	200	500	∞	Cum. Prop.	ν_2
.0005	.051	.058	062	.066	.069	.072	.074	.077	.078	.080	.081	.083	.0005	1
.001	.060	.067	.071	.075	.079	.082	.084	.087	.088	.089	.091	.092	.001	
.005	.093	.101	.105	.109	.113	.116	.118	.121	.122	.124	.126	.127	.005	
.01	.115	.124	.128	.132	.137	.139	.141	.145	.146	.148	.150	.151	.01	
.025	.161	.170	.175	.180	.184	.187	.189	.193	.194	.196	.198	.199	.025	
.05	.220	.230	.235	.240	.245	.248	.250	.254	.255	.257	.259	.261	.05	
.10	.325	.336	.342	.347	.353	.356	.358	.362	.364	.366	.368	.370	.10	
.25	.698	.712	.719	.727	.734	.738	.741	.747	.749	.752	.754	.756	.25	
.50	2.09	2.12	2.13	2.15	2.16	2.17	2.17	2.18	2.18	2.19	2.19	2.20	.50	
.75	9.49	9.58	9.63	9.67	9.71	9.74	9.76	9.78	9.80	9.82	9.84	9.85	.75	
.90	61.2	61.7	62.0	62.3	62.5	62.7	62.8	63.0	63.1	63.2	63.3	63.3	.90	
.95	246	248	249	250	251	252	252	253	253	254	254	254	.95	
.975	985	993	997	100^1	101^1	101^1	101^1	101^1	101^1	102^1	102^1	102^1	.975	
.99	616^1	621^1	623^1	626^1	629^1	630^1	631^1	633^1	634^1	635^1	636^1	637^1	.99	
.995	246^2	248^2	249^2	250^2	251^2	252^2	253^2	253^2	254^2	254^2	254^2	255^2	.995	
.999	616^3	621^3	623^3	626^3	629^3	630^3	631^3	633^3	634^3	635^3	636^3	637^3	.999	
.9995	246^4	248^4	249^4	250^4	251^4	252^4	252^4	253^4	253^4	253^4	254^4	254^4	.9995	
.0005	.076	.088	.094	.101	.108	.113	.116	.122	.124	.127	.130	.132	.0005	2
.001	.088	.100	.107	.114	.121	.126	.129	.135	.137	.140	.143	.145	.001	
.005	.130	.143	.150	.157	.165	.169	.173	.179	.181	.184	.187	.189	.005	
.01	.157	.171	.178	.186	.193	.198	.201	.207	.209	.212	.215	.217	.01	
.025	.210	.224	.232	.239	.247	.251	.255	.261	.263	.266	.269	.271	.025	
.05	.272	.286	.294	.302	.309	.314	.317	.324	.326	.329	.332	.334	.05	
.10	.371	.386	.394	.402	.410	.415	.418	.424	.426	.429	.433	.434	.10	
.25	.657	.672	.680	.689	.697	.702	.705	.711	.713	.716	.719	.721	.25	
.50	1.38	1.39	1.40	1.41	1.42	1.43	1.43	1.43	1.43	1.44	1.44	1.44	.50	
.75	3.41	3.43	3.43	3.44	3.45	3.45	3.46	3.47	3.47	3.48	3.48	3.48	.75	
.90	9.42	9.44	9.45	9.46	9.47	9.47	9.47	9.48	9.48	9.49	9.49	9.49	.90	
.95	19.4	19.4	19.5	19.5	19.5	19.5	19.5	19.5	19.5	19.5	19.5	19.5	.95	
.975	39.4	39.4	39.5	39.5	39.5	39.5	39.5	39.5	39.5	39.5	39.5	39.5	.975	
.99	99.4	99.4	99.5	99.5	99.5	99.5	99.5	99.5	99.5	99.5	99.5	99.5	.99	
.995	199	199	199	199	199	199	199	199	199	199	199	200	.995	
.999	999	999	999	999	999	999	999	999	999	999	999	999	.999	
.9995	200^1	200^1	200^1	200^1	200^1	200^1	200^1	200^1	200^1	200^1	200^1	200^1	.9995	
.0005	.093	.109	.117	.127	.136	.143	.147	.156	.158	.162	.166	.169	.0005	3
.001	.107	.123	.132	.142	.152	.158	.162	.171	.173	.177	.181	.184	.001	
.005	.154	.172	.181	.191	.201	.207	.211	.220	.222	.227	.231	.234	.005	
.01	.185	.203	.212	.222	.232	.238	.242	.251	.253	.258	.262	.264	.01	
.025	.241	.259	.269	.279	.289	.295	.299	.308	.310	.314	.318	.321	.025	
.05	.304	.323	.332	.342	.352	.358	.363	.370	.373	.377	.382	.384	.05	
.10	.402	.420	.430	.439	.449	.455	.459	.467	.469	.474	.476	.480	.10	
.25	.658	.675	.684	.693	.702	.708	.711	.719	.721	.724	.728	.730	.25	
.50	1.21	1.23	1.23	1.24	1.25	1.25	1.25	1.26	1.26	1.26	1.27	1.27	.50	
.75	2.46	2.46	2.46	2.47	2.47	2.47	2.47	2.47	2.47	2.47	2.47	2.47	.75	
.90	5.20	5.18	5.18	5.17	5.16	5.15	5.15	5.14	5.14	5.14	5.14	5.13	.90	
.95	8.70	8.66	8.63	8.62	8.59	8.58	8.57	8.55	8.55	8.54	8.53	8.53	.95	
.975	14.3	14.2	14.1	14.1	14.0	14.0	14.0	14.0	13.9	13.9	13.9	13.9	.975	
.99	26.9	26.7	26.6	26.5	26.4	26.4	26.3	26.2	26.2	26.2	26.1	26.1	.99	
.995	43.1	42.8	42.6	42.5	42.3	42.2	42.1	42.0	42.0	41.9	41.9	41.8	.995	
.999	127	126	126	125	125	125	124	124	124	124	124	123	.999	
.9995	203	201	200	199	199	198	198	197	197	197	196	196	.9995	

Table A.7   Percentiles of the $F(\nu_1,\nu_2)$ Distribution with Degrees of Freedom ν_1 for the Numerator and ν_2 for the Denominator (*Continued*)

ν_2	Cum. Prop.	1	2	3	4	5	6	7	8	9	10	11	12	Cum. Prop.
4	.0005	$.0^644$	$.0^350$	$.0^246$	.013	.024	.036	.047	.057	.066	.075	.082	.089	.0005
	.001	$.0^518$	$.0^210$	$.0^273$	.019	.032	.046	.058	.069	.079	.089	.097	.104	.001
	.005	$.0^444$	$.0^250$	.022	.043	.064	.083	.100	.114	.126	.137	.145	.153	.005
	.01	$.0^318$	.010	.035	.063	.088	.109	.127	.143	.156	.167	.176	.185	.01
	.025	$.0^211$	.026	.066	.104	.135	.161	.181	.198	.212	.224	.234	.243	.025
	.05	$.0^244$	.052	.110	.157	.193	.221	.243	.261	.275	.288	.298	.307	.05
	.10	.018	.108	.187	.243	.284	.314	.338	.356	.371	.384	.394	.403	.10
	.25	.117	.309	.418	.484	.528	.560	.583	.601	.615	.627	.637	.645	.25
	.50	.549	.828	.941	1.00	1.04	1.06	1.08	1.09	1.10	1.11	1.12	1.13	.50
	.75	1.81	2.00	2.05	2.06	2.07	2.08	2.08	2.08	2.08	2.08	2.08	2.08	.75
	.90	4.54	4.32	4.19	4.11	4.05	4.01	3.98	3.95	3.94	3.92	3.91	3.90	.90
	.95	7.71	6.94	6.59	6.39	6.26	6.16	6.09	6.04	6.00	5.96	5.94	5.91	.95
	.975	12.2	10.6	9.98	9.60	9.36	9.20	9.07	8.98	8.90	8.84	8.79	8.75	.975
	.99	21.2	18.0	16.7	16.0	15.5	15.2	15.0	14.8	14.7	14.5	14.4	14.4	.99
	.995	31.3	26.3	24.3	23.2	22.5	22.0	21.6	21.4	21.1	21.0	20.8	20.7	.995
	.999	74.1	61.2	56.2	53.4	51.7	50.5	49.7	49.0	48.5	48.0	47.7	47.4	.999
	.9995	106	87.4	80.1	76.1	73.6	71.9	70.6	69.7	68.9	68.3	67.8	67.4	.9995
5	.0005	$.0^643$	$.0^350$	$.0^247$	.014	.025	.038	.050	.061	.070	.081	.089	.096	.0005
	.001	$.0^517$	$.0^210$	$.0^275$	.019	.034	.048	.062	.074	.085	.095	.104	.112	.001
	.005	$.0^443$	$.0^250$	.022	.045	.067	.087	.105	.120	.134	.146	.156	.165	.005
	.01	$.0^317$	.010	.035	.064	.091	.114	.134	.151	.165	.177	.188	.197	.01
	.025	$.0^211$	.025	.067	.107	.140	.167	.189	.208	.223	.236	.248	.257	.025
	.05	$.0^243$	.052	.111	.160	.198	.228	.252	.271	.287	.301	.313	.322	.05
	.10	.017	.108	.188	.247	.290	.322	.347	.367	.383	.397	.408	.418	.10
	.25	.113	.305	.415	.483	.528	.560	.584	.604	.618	.631	.641	.650	.25
	.50	.528	.799	.907	.965	1.00	1.02	1.04	1.05	1.06	1.07	1.08	1.09	.50
	.75	1.69	1.85	1.88	1.89	1.89	1.89	1.89	1.89	1.89	1.89	1.89	1.89	.75
	.90	4.06	3.78	3.62	3.52	3.45	3.40	3.37	3.34	3.32	3.30	3.28	3.27	.90
	.95	6.61	5.79	5.41	5.19	5.05	4.95	4.88	4.82	4.77	4.74	4.71	4.68	.95
	.975	10.0	8.43	7.76	7.39	7.15	6.98	6.85	6.76	6.68	6.62	6.57	6.52	.975
	.99	16.3	13.3	12.1	11.4	11.0	10.7	10.5	10.3	10.2	10.1	9.96	9.89	.99
	.995	22.8	18.3	16.5	15.6	14.9	14.5	14.2	14.0	13.8	13.6	13.5	13.4	.995
	.999	47.2	37.1	33.2	31.1	29.7	28.8	28.2	27.6	27.2	26.9	26.6	26.4	.999
	.9995	63.6	49.8	44.4	41.5	39.7	38.5	37.6	36.9	36.4	35.9	35.6	35.2	.9995
6	.0005	$.0^643$	$.0^350$	$.0^247$	.014	.026	.039	.052	.064	.075	.085	.094	.103	.0005
	.001	$.0^517$	$.0^210$	$.0^275$	.020	.035	.050	.064	.078	.090	.101	.111	.119	.001
	.005	$.0^443$	$.0^250$	.022	.045	.069	.090	.109	.126	.140	.153	.164	.174	.005
	.01	$.0^317$	.010	.036	.066	.094	.118	.139	.157	.172	.186	.197	.207	.01
	.025	$.0^211$	.025	.068	.109	.143	.172	.195	.215	.231	.246	.258	.268	.025
	.05	$.0^243$	.052	.112	.162	.202	.233	.259	.279	.296	.311	.324	.334	.05
	.10	.017	.107	.189	.249	.294	.327	.354	.375	.392	.406	.418	.429	.10
	.25	.111	.302	.413	.481	.524	.561	.586	.606	.622	.635	.645	.654	.25
	.50	.515	.780	.886	.942	.977	1.00	1.02	1.03	1.04	1.05	1.05	1.06	.50
	.75	1.62	1.76	1.78	1.79	1.79	1.78	1.78	1.78	1.77	1.77	1.77	1.77	.75
	.90	3.78	3.46	3.29	3.18	3.11	3.05	3.01	2.98	2.96	2.94	2.92	2.90	.90
	.95	5.99	5.14	4.76	4.53	4.39	4.28	4.21	4.15	4.10	4.06	4.03	4.00	.95
	.975	8.81	7.26	6.60	6.23	5.99	5.82	5.70	5.60	5.52	5.46	5.41	5.37	.975
	.99	13.7	10.9	9.78	9.15	8.75	8.47	8.26	8.10	7.98	7.87	7.79	7.72	.99
	.995	18.6	14.5	12.9	12.0	11.5	11.1	10.8	10.6	10.4	10.2	10.1	10.0	.995
	.999	35.5	27.0	23.7	21.9	20.8	20.0	19.5	19.0	18.7	18.4	18.2	18.0	.999
	.9995	46.1	34.8	30.4	28.1	26.6	25.6	24.9	24.3	23.9	23.5	23.2	23.0	.9995

Table A.7   Percentiles of the $F(\nu_1,\nu_2)$ Distribution with Degrees of Freedom ν for the Numerator and ν_2 for the Denominator (*Continued*)

Cum. Prop.	15	20	24	30	40	50	60	100	120	200	500	∞	Cum. Prop.	ν_2
.0005	.105	.125	.135	.147	.159	.166	.172	.183	.186	.191	.196	.200	.0005	**4**
.001	.121	.141	.152	.163	.176	.183	.188	.200	.202	.208	.213	.217	.001	
.005	.172	.193	.204	.216	.229	.237	.242	.253	.255	.260	.266	.269	.005	
.01	.204	.226	.237	.249	.261	.269	.274	.285	.287	.293	.298	.301	.01	
.025	.263	.284	.296	.308	.320	.327	.332	.342	.346	.351	.356	.359	.025	
.05	.327	.349	.360	.372	.384	.391	.396	.407	.409	.413	.418	.422	.05	
.10	.424	.445	.456	.467	.478	.485	.490	.500	.502	.508	.510	.514	.10	
.25	.664	.683	.692	.702	.712	.718	.722	.731	.733	.737	.740	.743	.25	
.50	1.14	1.15	1.16	1.16	1.17	1.18	1.18	1.18	1.18	1.19	1.19	1.19	.50	
.75	2.08	2.08	2.08	2.08	2.08	2.08	2.08	2.08	2.08	2.08	2.08	2.08	.75	
.90	3.87	3.84	3.83	3.82	3.80	3.80	3.79	3.78	3.78	3.77	3.76	3.76	.90	
.95	5.86	5.80	5.77	5.75	5.72	5.70	5.69	5.66	5.66	5.65	5.64	5.63	.95	
.975	8.66	8.56	8.51	8.46	8.41	8.38	8.36	8.32	8.31	8.29	8.27	8.26	.975	
.99	14.2	14.0	13.9	13.8	13.7	13.7	13.7	13.6	13.6	13.5	13.5	13.5	.99	
.995	20.4	20.2	20.0	19.9	19.8	19.7	19.6	19.5	19.5	19.4	19.4	19.3	.995	
.999	46.8	46.1	45.8	45.4	45.1	44.9	44.7	44.5	44.4	44.3	44.1	44.0	.999	
.9995	66.5	65.5	65.1	64.6	64.1	63.8	63.6	63.2	63.1	62.9	62.7	62.6	.9995	
.0005	.115	.137	.150	.163	.177	.186	.192	.205	.209	.216	.222	.226	.0005	**5**
.001	.132	.155	.167	.181	.195	.204	.210	.223	.227	.233	.239	.244	.001	
.005	.186	.210	.223	.237	.251	.260	.266	.279	.282	.288	.294	.299	.005	
.01	.219	.244	.257	.270	.285	.293	.299	.312	.315	.322	.328	.331	.01	
.025	.280	.304	.317	.330	.344	.353	.359	.370	.374	.380	.386	.390	.025	
.05	.345	.369	.382	.395	.408	.417	.422	.432	.437	.442	.448	.452	.05	
.10	.440	.463	.476	.488	.501	.508	.514	.524	.527	.532	.538	.541	.10	
.25	.669	.690	.700	.711	.722	.728	.732	.741	.743	.748	.752	.755	.25	
.50	1.10	1.11	1.12	1.12	1.13	1.13	1.14	1.14	1.14	1.15	1.15	1.15	.50	
.75	1.89	1.88	1.88	1.88	1.88	1.88	1.87	1.87	1.87	1.87	1.87	1.87	.75	
.90	3.24	3.21	3.19	3.17	3.16	3.15	3.14	3.13	3.12	3.12	3.11	3.10	.90	
.95	4.62	4.56	4.53	4.50	4.46	4.44	4.43	4.41	4.40	4.39	4.37	4.36	.95	
.975	6.43	6.33	6.28	6.23	6.18	6.14	6.12	6.08	6.07	6.05	6.03	6.02	.975	
.99	9.72	9.55	9.47	9.38	9.29	9.24	9.20	9.13	9.11	9.08	9.04	9.02	.99	
.995	13.1	12.9	12.8	12.7	12.5	12.5	12.4	12.3	12.3	12.2	12.2	12.1	.995	
.999	25.9	25.4	25.1	24.9	24.6	24.4	24.3	24.1	24.1	23.9	23.8	23.8	.999	
.9995	34.6	33.9	33.5	33.1	32.7	32.5	32.3	32.1	32.0	31.8	31.7	31.6	.9995	
.0005	.123	.148	.162	.177	.193	.203	.210	.225	.229	.236	.244	.249	.0005	**6**
.001	.141	.166	.180	.195	.211	.222	.229	.243	.247	.255	.262	.267	.001	
.005	.197	.224	.238	.253	.269	.279	.286	.301	.304	.312	.318	.324	.005	
.01	.232	.258	.273	.288	.304	.313	.321	.334	.338	.346	.352	.357	.01	
.025	.293	.320	.334	.349	.364	.375	.381	.394	.398	.405	.412	.415	.025	
.05	.358	.385	.399	.413	.428	.437	.444	.457	.460	.467	.472	.476	.05	
.10	.453	.478	.491	.505	.519	.526	.533	.546	.548	.556	.559	.564	.10	
.25	.675	.696	.707	.718	.729	.736	.741	.751	.753	.758	.762	.765	.25	
.50	1.07	1.08	1.09	1.10	1.10	1.11	1.11	1.11	1.11	1.12	1.12	1.12	.50	
.75	1.76	1.76	1.75	1.75	1.75	1.75	1.74	1.74	1.74	1.74	1.74	1.74	.75	
.90	2.87	2.84	2.82	2.80	2.78	2.77	2.76	2.75	2.74	2.73	2.73	2.72	.90	
.95	3.94	3.87	3.84	3.81	3.77	3.75	3.74	3.71	3.70	3.69	3.68	3.67	.95	
.975	5.27	5.17	5.12	5.07	5.01	4.98	4.96	4.92	4.90	4.88	4.86	4.85	.975	
.99	7.56	7.40	7.31	7.23	7.14	7.09	7.06	6.99	6.97	6.93	6.90	6.88	.99	
.995	9.81	9.59	9.47	9.36	9.24	9.17	9.12	9.03	9.00	8.95	8.91	8.88	.995	
.999	17.6	17.1	16.9	16.7	16.4	16.3	16.2	16.0	16.0	15.9	15.8	15.7	.999	
.9995	22.4	21.9	21.7	21.4	21.1	20.9	20.7	20.5	20.4	20.3	20.2	20.1	.9995	

Table A.7　Percentiles of the $F(\nu_1,\nu_2)$ Distribution with Degrees of Freedom ν_1 for the Numerator and ν_2 for the Denominator (*Continued*)

ν_2	Cum. Prop.	1	2	3	4	5	6	7	8	9	10	11	12	Cum. Prop.
7	.0005	$.0^642$	$.0^550$	$.0^248$	.014	.027	.040	.053	.066	.078	.088	.099	.108	.0005
	.001	$.0^517$	$.0^210$	$.0^276$	.020	.035	.051	.067	.081	.093	.105	.115	.125	.001
	.005	$.0^442$	$.0^250$	.023	.046	.070	.093	.113	.130	.145	.159	.171	.181	.005
	.01	$.0^317$	.010	.036	.067	.096	.121	.143	.162	.178	.192	.205	.216	.01
	.025	$.0^210$	.025	.068	.110	.146	.176	.200	.221	.238	.253	.266	.277	.025
	.05	$.0^242$	.052	.113	.164	.205	.238	.264	.286	.304	.319	.332	.343	.05
	.10	.017	.107	.190	.251	.297	.332	.359	.381	.399	.414	.427	.438	.10
	.25	.110	.300	.412	.481	.528	.562	.588	.608	.624	.637	.649	.658	.25
	.50	.506	.767	.871	.926	.960	.983	1.00	1.01	1.02	1.03	1.04	1.04	.50
	.75	1.57	1.70	1.72	1.72	1.71	1.71	1.70	1.70	1.69	1.69	1.69	1.68	.75
	.90	3.59	3.26	3.07	2.96	2.88	2.83	2.78	2.75	2.72	2.70	2.68	2.67	.90
	.95	5.59	4.74	4.35	4.12	3.97	3.87	3.79	3.73	3.68	3.64	3.60	3.57	.95
	.975	8.07	6.54	5.89	5.52	5.29	5.12	4.99	4.90	4.82	4.76	4.71	4.67	.975
	.99	12.2	9.55	8.45	7.85	7.46	7.19	6.99	6.84	6.72	6.62	6.54	6.47	.99
	.995	16.2	12.4	10.9	10.0	9.52	9.16	8.89	8.68	8.51	8.38	8.27	8.18	.995
	.999	29.2	21.7	18.8	17.2	16.2	15.5	15.0	14.6	14.3	14.1	13.9	13.7	.999
	.9995	37.0	27.2	23.5	21.4	20.2	19.3	18.7	18.2	17.8	17.5	17.2	17.0	.9995
8	.0005	$.0^342$	$.0^550$	$.0^248$	.014	.027	.041	.055	.068	.081	.092	.102	.112	.0005
	.001	$.0^517$	$.0^210$	$.0^276$	.020	.036	.053	.068	.083	.096	.109	.120	.130	.001
	.005	$.0^442$	$.0^250$	.027	.047	.072	.095	.115	.133	.149	.164	.176	.187	.005
	.01	$.0^317$	.010	.036	.068	.097	.123	.146	.166	.183	.198	.211	.222	.01
	.025	$.0^210$	.025	.069	.111	.148	.179	.204	.226	.244	.259	.273	.285	.025
	.05	$.0^242$	.052	.113	.166	.208	.241	.268	.291	.310	.326	.339	.351	.05
	.10	.017	.107	.190	.253	.299	.335	.363	.386	.405	.421	.435	.445	.10
	.25	.109	.298	.411	.481	.529	.563	.589	.610	.627	.640	.654	.661	.25
	.50	.499	.757	.860	.915	.948	.971	.988	1.00	1.01	1.02	1.02	1.03	.50
	.75	1.54	1.66	1.67	1.66	1.66	1.65	1.64	1.64	1.63	1.63	1.63	1.62	.75
	.90	3.46	3.11	2.92	2.81	2.73	2.67	2.62	2.59	2.56	2.54	2.52	2.50	.90
	.95	5.32	4.46	4.07	3.84	3.69	3.58	3.50	3.44	3.39	3.35	3.31	3.28	.95
	.975	7.57	6.06	5.42	5.05	4.82	4.65	4.53	4.43	4.36	4.30	4.24	4.20	.975
	.99	11.3	8.65	7.59	7.01	6.63	6.37	6.18	6.03	5.91	5.81	5.73	5.67	.99
	.995	14.7	11.0	9.60	8.81	8.30	7.95	7.69	7.50	7.34	7.21	7.10	7.01	.995
	.999	25.4	18.5	15.8	14.4	13.5	12.9	12.4	12.0	11.8	11.5	11.4	11.2	.999
	.9995	31.6	22.8	19.4	17.6	16.4	15.7	15.1	14.6	14.3	14.0	13.8	13.6	.9995
9	.0005	$.0^641$	$.0^550$	$.0^248$	.015	.027	.042	.056	.070	.083	.094	.105	.115	.0005
	.001	$.0^517$	$.0^210$	$.0^277$	.021	.037	.054	.070	.085	.099	.112	.123	.134	.001
	.005	$.0^442$	$.0^250$	.023	.047	.073	.096	.117	.136	.153	.168	.181	.192	.005
	.01	$.0^317$	.010	.037	.068	.098	.125	.149	.169	.187	.202	.216	.228	.01
	.025	$.0^210$	.025	.069	.112	.150	.181	.207	.230	.248	.265	.279	.291	.025
	.05	$.0^240$	.052	.113	.167	.210	.244	.272	.296	.315	.331	.345	.358	.05
	.10	.017	.107	.191	.254	.302	.338	.367	.390	.410	.426	.441	.452	.10
	.25	.108	.297	.410	.480	.529	.564	.591	.612	.629	.643	.654	.664	.25
	.50	.494	.749	.852	.906	.939	.962	.978	.990	1.00	1.01	1.01	1.02	.50
	.75	1.51	1.62	1.63	1.63	1.62	1.61	1.60	1.60	1.59	1.59	1.58	1.58	.75
	.90	3.36	3.01	2.81	2.69	2.61	2.55	2.51	2.47	2.44	2.42	2.40	2.38	.90
	.95	5.12	4.26	3.86	3.63	3.48	3.37	3.29	3.23	3.18	3.14	3.10	3.07	.95
	.975	7.21	5.71	5.08	4.72	4.48	4.32	4.20	4.10	4.03	3.96	3.91	3.87	.975
	.99	10.6	8.02	6.99	6.42	6.06	5.80	5.61	5.47	5.35	5.26	5.18	5.11	.99
	.995	13.6	10.1	8.72	7.96	7.47	7.13	6.88	6.69	6.54	6.42	6.31	6.23	.995
	.999	22.9	16.4	13.9	12.6	11.7	11.1	10.7	10.4	10.1	9.89	9.71	9.57	.999
	.9995	28.0	19.9	16.8	15.1	14.1	13.3	12.8	12.4	12.1	11.8	11.6	11.4	.9995

Table A.7 Percentiles of the $F(\nu_1,\nu_2)$ Distribution with Degrees of Freedom ν_1 for the Numerator and ν_2 for the Denominator (*Continued*)

Cum. Prop. \ ν_1	15	20	24	30	40	50	60	100	120	200	500	∞	Cum. Prop.	ν_2
.0005	.130	.157	.172	.188	.206	.217	.225	.242	.246	.255	.263	.268	.0005	7
.001	.148	.176	.191	.208	.225	.237	.245	.261	.266	.274	.282	.288	.001	
.005	.206	.235	.251	.267	.285	.296	.304	.319	.324	.332	.340	.345	.005	
.01	.241	.270	.286	.303	.320	.331	.339	.355	.358	.366	.373	.379	.01	
.025	.304	.333	.348	.364	.381	.392	.399	.413	.418	.426	.433	.437	.025	
.05	.369	.398	.413	.428	.445	.455	.461	.476	.479	.485	.493	.498	.05	
.10	.463	.491	.504	.519	.534	.543	.550	.562	.566	.571	.578	.582	.10	
.25	.679	.702	.713	.725	.737	.745	.749	.760	.762	.767	.772	.775	.25	
.50	1.05	1.07	1.07	1.08	1.08	1.09	1.09	1.10	1.10	1.10	1.10	1.10	.50	
.75	1.68	1.67	1.67	1.66	1.66	1.66	1.65	1.65	1.65	1.65	1.65	1.65	.75	
.90	2.63	2.59	2.58	2.56	2.54	2.52	2.51	2.50	2.49	2.48	2.48	2.47	.90	
.95	3.51	3.44	3.41	3.38	3.34	3.32	3.30	3.27	3.27	3.25	3.24	3.23	.95	
.975	4.57	4.47	4.42	4.36	4.31	4.28	4.25	4.21	4.20	4.18	4.16	4.14	.975	
.99	6.31	6.16	6.07	5.99	5.91	5.86	5.82	5.75	5.74	5.70	5.67	5.65	.99	
.995	7.97	7.75	7.65	7.53	7.42	7.35	7.31	7.22	7.19	7.15	7.10	7.08	.995	
.999	13.3	12.9	12.7	12.5	12.3	12.2	12.1	11.9	11.9	11.8	11.7	11.7	.999	
.9995	16.5	16.0	15.7	15.5	15.2	15.1	15.0	14.7	14.7	14.6	14.5	14.4	.9995	
.0005	.136	.164	.181	.198	.218	.230	.239	.257	.262	.271	.281	.287	.0005	8
.001	.155	.184	.200	.218	.238	.250	.259	.277	.282	.292	.300	.306	.001	
.005	.214	.244	.261	.279	.299	.311	.319	.337	.341	.351	.358	.364	.005	
.01	.250	.281	.297	.315	.334	.346	.354	.372	.376	.385	.392	.398	.01	
.025	.313	.343	.360	.377	.395	.407	.415	.431	.435	.442	.450	.456	.025	
.05	.379	.409	.425	.441	.459	.469	.477	.493	.496	.505	.510	.516	.05	
.10	.472	.500	.515	.531	.547	.556	.563	.578	.581	.588	.595	.599	.10	
.25	.684	.707	.718	.730	.743	.751	.756	.767	.769	.775	.780	.783	.25	
.50	1.04	1.05	1.06	1.07	1.07	1.07	1.08	1.08	1.08	1.09	1.09	1.09	.50	
.75	1.62	1.61	1.60	1.60	1.59	1.59	1.59	1.58	1.58	1.58	1.58	1.58	.75	
.90	2.46	2.42	2.40	2.38	2.36	2.35	2.34	2.32	2.32	2.31	2.30	2.29	.90	
.95	3.22	3.15	3.12	3.08	3.04	3.02	3.01	2.97	2.97	2.95	2.94	2.93	.95	
.975	4.10	4.00	3.95	3.89	3.84	3.81	3.78	3.74	3.73	3.70	3.68	3.67	.975	
.99	5.52	5.36	5.28	5.20	5.12	5.07	5.03	4.96	4.95	4.91	4.88	4.86	.99	
.995	6.81	6.61	6.50	6.40	6.29	6.22	6.18	6.09	6.06	6.02	5.98	5.95	.995	
.999	10.8	10.5	10.3	10.1	9.92	9.80	9.73	9.57	9.54	9.46	9.39	9.34	.999	
.9995	13.1	12.7	12.5	12.2	12.0	11.8	11.8	11.6	11.5	11.4	11.4	11.3	.9995	
.0005	.141	.171	.188	.207	.228	.242	.251	.270	.276	.287	.297	.303	.0005	9
.001	.160	.191	.208	.228	.249	.262	.271	.291	.296	.307	.316	.323	.001	
.005	.220	.253	.271	.290	.310	.324	.332	.351	.356	.366	.376	.382	.005	
.01	.257	.289	.307	.326	.346	.358	.368	.386	.391	.400	.410	.415	.01	
.025	.320	.352	.370	.388	.408	.420	.428	.446	.450	.459	.467	.473	.025	
.05	.386	.418	.435	.452	.471	.483	.490	.508	.510	.518	.526	.532	.05	
.10	.479	.509	.525	.541	.558	.568	.575	.588	.594	.602	.610	.613	.10	
.25	.687	.711	.723	.736	.749	.757	.762	.773	.776	.782	.787	.791	.25	
.50	1.03	1.04	1.05	1.05	1.06	1.06	1.07	1.07	1.07	1.08	1.08	1.08	.50	
.75	1.57	1.56	1.56	1.55	1.55	1.54	1.54	1.53	1.53	1.53	1.53	1.53	.75	
.90	2.34	2.30	2.28	2.25	2.23	2.22	2.21	2.19	2.18	2.17	2.17	2.16	.90	
.95	3.01	2.94	2.90	2.86	2.83	2.80	2.79	2.76	2.75	2.73	2.72	2.71	.95	
.975	3.77	3.67	3.61	3.56	3.51	3.47	3.45	3.40	3.39	3.37	3.35	3.33	.975	
.99	4.96	4.81	4.73	4.65	4.57	4.52	4.48	4.42	4.40	4.36	4.33	4.31	.99	
.995	6.03	5.83	5.73	5.62	5.52	5.45	5.41	5.32	5.30	5.26	5.21	5.19	.995	
.999	9.24	8.90	8.72	8.55	8.37	8.26	8.19	8.04	8.00	7.93	7.86	7.81	.999	
.9995	11.0	10.6	10.4	10.2	9.94	9.80	9.71	9.53	9.49	9.40	9.32	9.26	.9995	

Table A.7   Percentiles of the $F(\nu_1,\nu_2)$ Distribution with Degrees of Freedom ν_1 for the Numerator and ν_2 for the Denominator (*Continued*)

ν_2	Cum. Prop.	1	2	3	4	5	6	7	8	9	10	11	12	Cum. Prop.
10	.0005	$.0^641$	$.0^350$	$.0^249$	.015	.028	.043	.057	.071	.085	.097	.108	.119	.0005
	.001	$.0^517$	$.0^210$	$.0^277$	.021	.037	.054	.071	.087	.101	.114	.126	.137	.001
	.005	$.0^441$	$.0^250$	.023	.048	.073	.098	.119	.139	.156	.171	.185	.197	.005
	.01	$.0^317$	.010	.037	.069	.100	.127	.151	.172	.190	.206	.220	.233	.01
	.025	$.0^210$	.025	.069	.113	.151	.183	.210	.233	.252	.269	.283	.296	.025
	.05	$.0^241$	.052	.114	.168	.211	.246	.275	.299	.319	.336	.351	.363	.05
	.10	.017	.106	.191	.255	.303	.340	.370	.394	.414	.430	.444	.457	.10
	.25	.107	.296	.409	.480	.529	.565	.592	.613	.631	.645	.657	.667	.25
	.50	.490	.743	.845	.899	.932	.954	.971	.983	.992	1.00	1.01	1.01	.50
	.75	1.49	1.60	1.60	1.59	1.59	1.58	1.57	1.56	1.56	1.55	1.55	1.54	.75
	.90	3.28	2.92	2.73	2.61	2.52	2.46	2.41	2.38	2.35	2.32	2.30	2.28	.90
	.95	4.96	4.10	3.71	3.48	3.33	3.22	3.14	3.07	3.02	2.98	2.94	2.91	.95
	.975	6.94	5.46	4.83	4.47	4.24	4.07	3.95	3.85	3.78	3.72	3.66	3.62	.975
	.99	10.0	7.56	6.55	5.99	5.64	5.39	5.20	5.06	4.94	4.85	4.77	4.71	.99
	.995	12.8	9.43	8.08	7.34	6.87	6.54	6.30	6.12	5.97	5.85	5.75	5.66	.995
	.999	21.0	14.9	12.6	11.3	10.5	9.92	9.52	9.20	8.96	8.75	8.58	8.44	.999
	.9995	25.5	17.9	15.0	13.4	12.4	11.8	11.3	10.9	10.6	10.3	10.1	9.93	.9995
11	.0005	$.0^641$	$.0^350$	$.0^249$	.015	.028	.043	.058	.072	.086	.099	.111	.121	.0005
	.001	$.0^516$	$.0^210$	$.0^278$	.021	.038	.055	.072	.088	.103	.116	.129	.140	.001
	.005	$.0^440$	$.0^250$	.023	.048	.074	.099	.121	.141	.158	.174	.188	.200	.005
	.01	$.0^316$	.010	.037	.069	.100	.128	.153	.175	.193	.210	.224	.237	.01
	.025	$.0^210$	.025	.069	.114	.152	.185	.212	.236	.256	.273	.288	.301	.025
	.05	$.0^241$	.052	.114	.168	.212	.248	.278	.302	.323	.340	.355	.368	.05
	.10	.017	.106	.192	.256	.305	.342	.373	.397	.417	.435	.448	.461	.10
	.25	.107	.295	.408	.481	.529	.565	.592	.614	.633	.645	.658	.667	.25
	.50	.486	.739	.840	.893	.926	.948	.964	.977	.986	.994	1.00	1.01	.50
	.75	1.47	1.58	1.58	1.57	1.56	1.55	1.54	1.53	1.53	1.52	1.52	1.51	.75
	.90	3.23	2.86	2.66	2.54	2.45	2.39	2.34	2.30	2.27	2.25	2.23	2.21	.90
	.95	4.84	3.98	3.59	3.36	3.20	3.09	3.01	2.95	2.90	2.85	2.82	2.79	.95
	.975	6.72	5.26	4.63	4.28	4.04	3.88	3.76	3.66	3.59	3.53	3.47	3.43	.975
	.99	9.65	7.21	6.22	5.67	5.32	5.07	4.89	4.74	4.63	4.54	4.46	4.40	.99
	.995	12.2	8.91	7.60	6.88	6.42	6.10	5.86	5.68	5.54	5.42	5.32	5.24	.995
	.999	19.7	13.8	11.6	10.3	9.58	9.05	8.66	8.35	8.12	7.92	7.76	7.62	.999
	.9995	23.6	16.4	13.6	12.2	11.2	10.6	10.1	9.76	9.48	9.24	9.04	8.88	.9995
12	.0005	$.0^641$	$.0^350$	$.0^249$	.015	.028	.044	.058	.073	.087	.101	.113	.124	.0005
	.001	$.0^516$	$.0^210$	$.0^278$	.021	.038	.056	.073	.089	.104	.118	.131	.143	.001
	.005	$.0^439$	$.0^250$	.023	.048	.075	.100	.122	.143	.161	.177	.191	.204	.005
	.01	$.0^316$	.010	.037	.070	.101	.130	.155	.176	.196	.212	.227	.241	.01
	.025	$.0^210$	.025	.070	.114	.153	.186	.214	.238	.259	.276	.292	.305	.025
	.05	$.0^241$	.052	.114	.169	.214	.250	.280	.305	.325	.343	.358	.372	.05
	.10	.016	.106	.192	.257	.306	.344	.375	.400	.420	.438	.452	.466	.10
	.25	.106	.295	.408	.480	.530	.566	.594	.616	.633	.649	.662	.671	.25
	.50	.484	.735	.835	.888	.921	.943	.959	.972	.981	.989	.995	1.00	.50
	.75	1.46	1.56	1.56	1.55	1.54	1.53	1.52	1.51	1.51	1.50	1.50	1.49	.75
	.90	3.18	2.81	2.61	2.48	2.39	2.33	2.28	2.24	2.21	2.19	2.17	2.15	.90
	.95	4.75	3.89	3.49	3.26	3.11	3.00	2.91	2.85	2.80	2.75	2.72	2.69	.95
	.975	6.55	5.10	4.47	4.12	3.89	3.73	3.61	3.51	3.44	3.37	3.32	3.28	.975
	.99	9.33	6.93	5.95	5.41	5.06	4.82	4.64	4.50	4.39	4.30	4.22	4.16	.99
	.995	11.8	8.51	7.23	6.52	6.07	5.76	5.52	5.35	5.20	5.09	4.99	4.91	.995
	.999	18.6	13.0	10.8	9.63	8.89	8.38	8.00	7.71	7.48	7.29	7.14	7.01	.999
	.9995	22.2	15.3	12.7	11.2	10.4	9.74	9.28	8.94	8.66	8.43	8.24	8.08	.9995

Table A.7 Percentiles of the $F(\nu_1, \nu_2)$ Distribution with Degrees of Freedom ν_1 for the Numerator and ν_2 for the Denominator (*Continued*)

Cum. Prop.	ν_1 15	20	24	30	40	50	60	100	120	200	500	∞	Cum. Prop.	ν_2
.0005	.145	.177	.195	.215	.238	.251	.262	.282	.288	.299	.311	.319	.0005	**10**
.001	.164	.197	.216	.236	.258	.272	.282	.303	.309	.321	.331	.338	.001	
.005	.226	.260	.279	.299	.321	.334	.344	.365	.370	.380	.391	.397	.005	
.01	.263	.297	.316	.336	.357	.370	.380	.400	.405	.415	.424	.431	.01	
.025	.327	.360	.379	.398	.419	.431	.441	.459	.464	.474	.483	.488	.025	
.05	.393	.426	.444	.462	.481	.493	.502	.518	.523	.532	.541	.546	.05	
.10	.486	.516	.532	.549	.567	.578	.586	.602	.605	.614	.621	.625	.10	
.25	.691	.714	.727	.740	.754	.762	.767	.779	.782	.788	.793	.797	.25	
.50	1.02	1.03	1.04	1.05	1.05	1.06	1.06	1.06	1.06	1.07	1.07	1.07	.50	
.75	1.53	1.52	1.52	1.51	1.51	1.50	1.50	1.49	1.49	1.49	1.48	1.48	.75	
.90	2.24	2.20	2.18	2.16	2.13	2.12	2.11	2.09	2.08	2.07	2.06	2.06	.90	
.95	2.85	2.77	2.74	2.70	2.66	2.64	2.62	2.59	2.58	2.56	2.55	2.54	.95	
.975	3.52	3.42	3.37	3.31	3.26	3.22	3.20	3.15	3.14	3.12	3.09	3.08	.975	
.99	4.56	4.41	4.33	4.25	4.17	4.12	4.08	4.01	4.00	3.96	3.93	3.91	.99	
.995	5.47	5.27	5.17	5.07	4.97	4.90	4.86	4.77	4.75	4.71	4.67	4.64	.995	
.999	8.13	7.80	7.64	7.47	7.30	7.19	7.12	6.98	6.94	6.87	6.81	6.76	.999	
.9995	9.56	9.16	8.96	8.75	8.54	8.42	8.33	8.16	8.12	8.04	7.96	7.90	.9995	
.0005	.148	.182	.201	.222	.246	.261	.271	.293	.299	.312	.324	.331	.0005	**11**
.001	.168	.202	.222	.243	.266	.282	.292	.313	.320	.332	.343	.353	.001	
.005	.231	.266	.286	.308	.330	.345	.355	.376	.382	.394	.403	.412	.005	
.01	.268	.304	.324	.344	.366	.380	.391	.412	.417	.427	.439	.444	.01	
.025	.332	.368	.386	.407	.429	.442	.450	.472	.476	.485	.495	.503	.025	
.05	.398	.433	.452	.469	.490	.503	.513	.529	.535	.543	.552	.559	.05	
.10	.490	.524	.541	.559	.578	.588	.595	.614	.617	.625	.633	.637	.10	
.25	.694	.719	.730	.744	.758	.767	.773	.780	.788	.794	.799	.803	.25	
.50	1.02	1.03	1.03	1.04	1.05	1.05	1.05	1.06	1.06	1.06	1.06	1.06	.50	
.75	1.50	1.49	1.49	1.48	1.47	1.47	1.47	1.46	1.46	1.46	1.45	1.45	.75	
.90	2.17	2.12	2.10	2.08	2.05	2.04	2.03	2.00	2.00	1.99	1.98	1.97	.90	
.95	2.72	2.65	2.61	2.57	2.53	2.51	2.49	2.46	2.45	2.43	2.42	2.40	.95	
.975	3.33	3.23	3.17	3.12	3.06	3.03	3.00	2.96	2.94	2.92	2.90	2.88	.975	
.99	4.25	4.10	4.02	3.94	3.86	3.81	3.78	3.71	3.69	3.66	3.62	3.60	.99	
.995	5.05	4.86	4.76	4.65	4.55	4.49	4.45	4.36	4.34	4.29	4.25	4.23	.995	
.999	7.32	7.01	6.85	6.68	6.52	6.41	6.35	6.21	6.17	6.10	6.04	6.00	.999	
.9995	8.52	8.14	7.94	7.75	7.57	7.43	7.35	7.18	7.14	7.06	6.98	6.93	.9995	
.0005	.152	.186	.206	.228	.253	.269	.280	.305	.311	.323	.337	.345	.0005	**12**
.001	.172	.207	.228	.250	.275	.291	.302	.326	.332	.344	.357	.365	.001	
.005	.235	.272	.292	.315	.339	.355	.365	.388	.393	.405	.417	.424	.005	
.01	.273	.310	.330	.352	.375	.391	.401	.422	.428	.441	.450	.458	.01	
.025	.337	.374	.394	.416	.437	.450	.461	.481	.487	.498	.508	.514	.025	
.05	.404	.439	.458	.478	.499	.513	.522	.541	.545	.556	.565	.571	.05	
.10	.496	.528	.546	.564	.583	.595	.604	.621	.625	.633	.641	.647	.10	
.25	.695	.721	.734	.748	.762	.771	.777	.789	.792	.799	.804	.808	.25	
.50	1.01	1.02	1.03	1.03	1.04	1.04	1.05	1.05	1.05	1.05	1.06	1.06	.50	
.75	1.48	1.47	1.46	1.45	1.45	1.44	1.44	1.43	1.43	1.43	1.42	1.42	.75	
.90	2.11	2.06	2.04	2.01	1.99	1.97	1.96	1.94	1.93	1.92	1.91	1.90	.90	
.95	2.62	2.54	2.51	2.47	2.43	2.40	2.38	2.35	2.34	2.32	2.31	2.30	.95	
.975	3.18	3.07	3.02	2.96	2.91	2.87	2.85	2.80	2.79	2.76	2.74	2.72	.975	
.99	4.01	3.86	3.78	3.70	3.62	3.57	3.54	3.47	3.45	3.41	3.38	3.36	.99	
.995	4.72	4.53	4.43	4.33	4.23	4.17	4.12	4.04	4.01	3.97	3.93	3.90	.995	
.999	6.71	6.40	6.25	6.09	5.93	5.83	5.76	5.63	5.59	5.52	5.46	5.42	.999	
.9995	7.74	7.37	7.18	7.00	6.80	6.68	6.61	6.45	6.41	6.33	6.25	6.20	.9995	

Table A.7　Percentiles of the $F(\nu_1,\nu_2)$ Distribution with Degrees of Freedom ν_1 for the Numerator and ν_2 for the Denominator (*Continued*)

ν_2	Cum. Prop.	1	2	3	4	5	6	7	8	9	10	11	12	Cum. Prop.
15	.0005	$.0^641$	$.0^350$	$.0^249$	.015	.029	.045	.061	.076	.091	.105	.117	.129	.0005
	.001	$.0^516$	$.0^210$	$.0^279$	.021	.039	.057	.075	.092	.108	.123	.137	.149	.001
	.005	$.0^439$	$.0^250$	.023	.049	.076	.102	.125	.147	.166	.183	.198	.212	.005
	.01	$.0^316$	.010	.037	.070	.103	.132	.158	.181	.202	.219	.235	.249	.01
	.025	$.0^210$	.025	.070	.116	.156	.190	.219	.244	.265	.284	.300	.315	.025
	.05	$.0^241$	.051	.115	.170	.216	.254	.285	.311	.333	.351	.368	.382	.05
	.10	.016	.106	.192	.258	.309	.348	.380	.406	.427	.446	.461	.475	.10
	.25	.105	.293	.407	.480	.531	.568	.596	.618	.637	.652	.667	.676	.25
	.50	.478	.726	.826	.878	.911	.933	.948	.960	.970	.977	.984	.989	.50
	.75	1.43	1.52	1.52	1.51	1.49	1.48	1.47	1.46	1.46	1.45	1.44	1.44	.75
	.90	3.07	2.70	2.49	2.36	2.27	2.21	2.16	2.12	2.09	2.06	2.04	2.02	.90
	.95	4.54	3.68	3.29	3.06	2.90	2.79	2.71	2.64	2.59	2.54	2.51	2.48	.95
	.975	6.20	4.76	4.15	3.80	3.58	3.41	3.29	3.20	3.12	3.06	3.01	2.96	.975
	.99	8.68	6.36	5.42	4.89	4.56	4.32	4.14	4.00	3.89	3.80	3.73	3.67	.99
	.995	10.8	7.70	6.48	5.80	5.37	5.07	4.85	4.67	4.54	4.42	4.33	4.25	.995
	.999	16.6	11.3	9.34	8.25	7.57	7.09	6.74	6.47	6.26	6.08	5.93	5.81	.999
	.9995	19.5	13.2	10.8	9.48	8.66	8.10	7.68	7.36	7.11	6.91	6.75	6.60	.9995
20	.0005	$.0^640$	$.0^350$	$.0^250$	.015	.029	.046	.063	.079	.094	.109	.123	.136	.0005
	.001	$.0^516$	$.0^210$	$.0^279$	.022	.039	.058	.077	.095	.112	.128	.143	.156	.001
	.005	$.0^439$	$.0^250$	.023	.050	.077	.104	.129	.151	.171	.190	.206	.221	.005
	.01	$.0^316$	.010	.037	.071	.105	.135	.162	.187	.208	.227	.244	.259	.01
	.025	$.0^210$	.025	.071	.117	.158	.193	.224	.250	.273	.292	.310	.325	.025
	.05	$.0^240$	.051	.115	.172	.219	.258	.290	.318	.340	.360	.377	.393	.05
	.10	.016	.106	.193	.260	.312	.353	.385	.412	.435	.454	.472	.485	.10
	.25	.104	.292	.407	.480	.531	.569	.598	.622	.641	.656	.671	.681	.25
	.50	.472	.718	.816	.868	.900	.922	.938	.950	.959	.966	.972	.977	.50
	.75	1.40	1.49	1.48	1.47	1.45	1.44	1.43	1.42	1.41	1.40	1.39	1.39	.75
	.90	2.97	2.59	2.38	2.25	2.16	2.09	2.04	2.00	1.96	1.94	1.91	1.89	.90
	.95	4.35	3.49	3.10	2.87	2.71	2.60	2.51	2.45	2.39	2.35	2.31	2.28	.95
	.975	5.87	4.46	3.86	3.51	3.29	3.13	3.01	2.91	2.84	2.77	2.72	2.68	.975
	.99	8.10	5.85	4.94	4.43	4.10	3.87	3.70	3.56	3.46	3.37	3.29	3.23	.99
	.995	9.94	6.99	5.82	5.17	4.76	4.47	4.26	4.09	3.96	3.85	3.76	3.68	.995
	.999	14.8	9.95	8.10	7.10	6.46	6.02	5.69	5.44	5.24	5.08	4.94	4.82	.999
	.9995	17.2	11.4	9.20	8.02	7.28	6.76	6.38	6.08	5.85	5.66	5.51	5.38	.9995
24	.0005	$.0^640$	$.0^350$	$.0^250$	.015	.030	.046	.064	.080	.096	.112	.126	.139	.0005
	.001	$.0^516$	$.0^210$	$.0^279$	.022	.040	.059	.079	.097	.115	.131	.146	.160	.001
	.005	$.0^440$	$.0^250$	.023	.050	.078	.106	.131	.154	.175	.193	.210	.226	.005
	.01	$.0^316$	.010	.038	.072	.106	.137	.165	.189	.211	.231	.249	.264	.01
	.025	$.0^210$	.025	.071	.117	.159	.195	.227	.253	.277	.297	.315	.331	.025
	.05	$.0^240$	.051	.116	.173	.221	.260	.293	.321	.345	.365	.383	.399	.05
	.10	.016	.106	.193	.261	.313	.355	.388	.416	.439	.459	.476	.491	.10
	.25	.104	.291	.406	.480	.532	.570	.600	.623	.643	.659	.671	.684	.25
	.50	.469	.714	.812	.863	.895	.917	.932	.944	.953	.961	.967	.972	.50
	.75	1.39	1.47	1.46	1.44	1.43	1.41	1.40	1.39	1.38	1.38	1.37	1.36	.75
	.90	2.93	2.54	2.33	2.19	2.10	2.04	1.98	1.94	1.91	1.88	1.85	1.83	.90
	.95	4.26	3.40	3.01	2.78	2.62	2.51	2.42	2.36	2.30	2.25	2.21	2.18	.95
	.975	5.72	4.32	3.72	3.38	3.15	2.99	2.87	2.78	2.70	2.64	2.59	2.54	.975
	.99	7.82	5.61	4.72	4.22	3.90	3.67	3.50	3.36	3.26	3.17	3.09	3.03	.99
	.995	9.55	6.66	5.52	4.89	4.49	4.20	3.99	3.83	3.69	3.59	3.50	3.42	.995
	.999	14.0	9.34	7.55	6.59	5.98	5.55	5.23	4.99	4.80	4.64	4.50	4.39	.999
	.9995	16.2	10.6	8.52	7.39	6.68	6.18	5.82	5.54	5.31	5.13	4.98	4.85	.9995

Table A.7   Percentiles of the $F(\nu_1,\nu_2)$ Distribution with Degrees of Freedom ν_1 for the Numerator and ν_2 for the Denominator *(Continued)*

Cum. Prop.	15	20	24	30	40	50	60	100	120	200	500	∞	Cum. Prop.	ν_2
.0005	.159	.197	.220	.244	.272	.290	.303	.330	.339	.353	.368	.377	.0005	**15**
.001	.181	.219	.242	.266	.294	.313	.325	.352	.360	.375	.388	.398	.001	
.005	.246	.286	.308	.333	.360	.377	.389	.415	.422	.435	.448	.457	.005	
.01	.284	.324	.346	.370	.397	.413	.425	.450	.456	.469	.483	.490	.01	
.025	.349	.389	.410	.433	.458	.474	.485	.508	.514	.526	.538	.546	.025	
.05	.416	.454	.474	.496	.519	.535	.545	.565	.571	.581	.592	.600	.05	
.10	.507	.542	.561	.581	.602	.614	.624	.641	.647	.658	.667	.672	.10	
.25	.701	.728	.742	.757	.772	.782	.788	.802	.805	.812	.818	.822	.25	
.50	1.00	1.01	1.02	1.02	1.03	1.03	1.03	1.04	1.04	1.04	1.04	1.05	.50	
.75	1.43	1.41	1.41	1.40	1.39	1.39	1.38	1.38	1.37	1.37	1.36	1.36	.75	
.90	1.97	1.92	1.90	1.87	1.85	1.83	1.82	1.79	1.79	1.77	1.76	1.76	.90	
.95	2.40	2.33	2.29	2.25	2.20	2.18	2.16	2.12	2.11	2.10	2.08	2.07	.95	
.975	2.86	2.76	2.70	2.64	2.59	2.55	2.52	2.47	2.46	2.44	2.41	2.40	.975	
.99	3.52	3.37	3.29	3.21	3.13	3.08	3.05	2.98	2.96	2.92	2.89	2.87	.99	
.995	4.07	3.88	3.79	3.69	3.59	3.52	3.48	3.39	3.37	3.33	3.29	3.26	.995	
.999	5.54	5.25	5.10	4.95	4.80	4.70	4.64	4.51	4.47	4.41	4.35	4.31	.999	
.9995	6.27	5.93	5.75	5.58	5.40	5.29	5.21	5.06	5.02	4.94	4.87	4.83	.9995	
.0005	.169	.211	.235	.263	.295	.316	.331	.364	.375	.391	.408	.422	.0005	**20**
.001	.191	.233	.258	.286	.318	.339	.354	.386	.395	.413	.429	.441	.001	
.005	.258	.301	.327	.354	.385	.405	.419	.448	.457	.474	.490	.500	.005	
.01	.297	.340	.365	.392	.422	.441	.455	.483	.491	.508	.521	.532	.01	
.025	.363	.406	.430	.456	.484	.503	.514	.541	.548	.562	.575	.585	.025	
.05	.430	.471	.493	.518	.544	.562	.572	.595	.603	.617	.629	.637	.05	
.10	.520	.557	.578	.600	.623	.637	.648	.671	.675	.685	.694	.704	.10	
.25	.708	.736	.751	.767	.784	.794	.801	.816	.820	.827	.835	.840	.25	
.50	.989	1.00	1.01	1.01	1.02	1.02	1.02	1.03	1.03	1.03	1.03	1.03	.50	
.75	1.37	1.36	1.35	1.34	1.33	1.33	1.32	1.31	1.31	1.30	1.30	1.29	.75	
.90	1.84	1.79	1.77	1.74	1.71	1.69	1.68	1.65	1.64	1.63	1.62	1.61	.90	
.95	2.20	2.12	2.08	2.04	1.99	1.97	1.95	1.91	1.90	1.88	1.86	1.84	.95	
.975	2.57	2.46	2.41	2.35	2.29	2.25	2.22	2.17	2.16	2.13	2.10	2.09	.975	
.99	3.09	2.94	2.86	2.78	2.69	2.64	2.61	2.54	2.52	2.48	2.44	2.42	.99	
.995	3.50	3.32	3.22	3.12	3.02	2.96	2.92	2.83	2.81	2.76	2.72	2.69	.995	
.999	4.56	4.29	4.15	4.01	3.86	3.77	3.70	3.58	3.54	3.48	3.42	3.38	.999	
.9995	5.07	4.75	4.58	4.42	4.24	4.15	4.07	3.93	3.90	3.82	3.75	3.70	.9995	
.0005	.174	.218	.244	.274	.309	.331	.349	.384	.395	.416	.434	.449	.0005	**24**
.001	.196	.241	.268	.298	.332	.354	.371	.405	.417	.437	.455	.469	.001	
.005	.264	.310	.337	.367	.400	.422	.437	.469	.479	.498	.515	.527	.005	
.01	.304	.350	.376	.405	.437	.459	.473	.505	.513	.529	.546	.558	.01	
.025	.370	.415	.441	.468	.498	.518	.531	.562	.568	.585	.599	.610	.025	
.05	.437	.480	.504	.530	.558	.575	.588	.613	.622	.637	.649	.659	.05	
.10	.527	.566	.588	.611	.635	.651	.662	.685	.691	.701	.715	.723	.10	
.25	.712	.741	.757	.773	.791	.802	.809	.825	.829	.837	.844	.850	.25	
.50	.983	.994	1.00	1.01	1.01	1.02	1.02	1.02	1.02	1.02	1.03	1.03	.50	
.75	1.35	1.33	1.32	1.31	1.30	1.29	1.29	1.28	1.28	1.27	1.27	1.26	.75	
.90	1.78	1.73	1.70	1.67	1.64	1.62	1.61	1.58	1.57	1.56	1.54	1.53	.90	
.95	2.11	2.03	1.98	1.94	1.89	1.86	1.84	1.80	1.79	1.77	1.75	1.73	.95	
.975	2.44	2.33	2.27	2.21	2.15	2.11	2.08	2.02	2.01	1.98	1.95	1.94	.975	
.99	2.89	2.74	2.66	2.58	2.49	2.44	2.40	2.33	2.31	2.27	2.24	2.21	.99	
.995	3.25	3.06	2.97	2.87	2.77	2.70	2.66	2.57	2.55	2.50	2.46	2.43	.995	
.999	4.14	3.87	3.74	3.59	3.45	3.35	3.29	3.16	3.14	3.07	3.01	2.97	.999	
.9995	4.55	4.25	4.09	3.93	3.76	3.66	3.59	3.44	3.41	3.33	3.27	3.22	.9995	

Table A.7   Percentiles of the $F(\nu_1,\nu_2)$ Distribution with Degrees of Freedom ν_1 for the Numerator and ν_2 for the Denominator (*Continued*)

ν_2	Cum. Prop	1	2	3	4	5	6	7	8	9	10	11	12	Cum. Prop.
30	.0005	$.0^640$	$.0^350$	$.0^250$	.015	.030	.047	.065	.082	.098	.114	.129	.143	.0005
	.001	$.0^516$	$.0^210$	$.0^280$	.022	.040	.060	.080	.099	.117	.134	.150	.164	.001
	.005	$.0^440$	$.0^250$	.024	.050	.079	.107	.133	.156	.178	.197	.215	.231	.005
	.01	$.0^316$	.010	.038	.072	.107	.138	.167	.192	.215	.235	.254	.270	.01
	.025	$.0^210$	.025	.071	.118	.161	.197	.229	.257	.281	.302	.321	.337	.025
	.05	$.0^240$	.051	.116	.174	.222	.263	.296	.325	.349	.370	.389	.406	.05
	.10	.016	.106	.193	.262	.315	.357	.391	.420	.443	.464	.481	.497	.10
	.25	.103	.290	.406	.480	.532	.571	.601	.625	.645	.661	.676	.688	.25
	.50	.466	.709	.807	.858	.890	.912	.927	.939	.948	.955	.961	.966	.50
	.75	1.38	1.45	1.44	1.42	1.41	1.39	1.38	1.37	1.36	1.35	1.35	1.35	.75
	.90	2.88	2.49	2.28	2.14	2.05	1.98	1.93	1.88	1.85	1.82	1.79	1.77	.90
	.95	4.17	3.32	2.92	2.69	2.53	2.42	2.33	2.27	2.21	2.16	2.13	2.09	.95
	.975	5.57	4.18	3.59	3.25	3.03	2.87	2.75	2.65	2.57	2.51	2.46	2.41	.975
	.99	7.56	5.39	4.51	4.02	3.70	3.47	3.30	3.17	3.07	2.98	2.91	2.84	.99
	.995	9.18	6.35	5.24	4.62	4.23	3.95	3.74	3.58	3.45	3.34	3.25	3.18	.995
	.999	13.3	8.77	7.05	6.12	5.53	5.12	4.82	4.58	4.39	4.24	4.11	4.00	.999
	.9995	15.2	9.90	7.90	6.82	6.14	5.66	5.31	5.04	4.82	4.65	4.51	4.38	.9995
40	.0005	$.0^640$	$.0^350$	$.0^250$	.016	.030	.048	.066	.084	.100	.117	.132	.147	.0005
	.001	$.0^516$	$.0^210$	$.0^280$	.022	.042	.061	.081	.101	.119	.137	.153	.169	.001
	.005	$.0^440$	$.0^250$	.024	.051	.080	.108	.135	.159	.181	.201	.220	.237	.005
	.01	$.0^316$	.010	.038	.073	.108	.140	.169	.195	.219	.240	.259	.276	.01
	.025	$.0^399$	.025	.071	.119	.162	.199	.232	.260	.285	.307	.327	.344	.025
	.05	$.0^240$	.051	.116	.175	.224	.265	.299	.329	.354	.376	.395	.412	.05
	.10	.016	.106	.194	.263	.317	.360	.394	.424	.448	.469	.488	.504	.10
	.25	.103	.290	.405	.480	.533	.572	.603	.627	.647	.664	.680	.691	.25
	.50	.463	.705	.802	.854	.885	.907	.922	.934	.943	.950	.956	.961	.50
	.75	1.36	1.44	1.42	1.40	1.39	1.37	1.36	1.35	1.34	1.33	1.32	1.31	.75
	.90	2.84	2.44	2.23	2.09	2.00	1.93	1.87	1.83	1.79	1.76	1.73	1.71	.90
	.95	4.08	3.23	2.84	2.61	2.45	2.34	2.25	2.18	2.12	2.08	2.04	2.00	.95
	.975	5.42	4.05	3.46	3.13	2.90	2.74	2.62	2.53	2.45	2.39	2.33	2.29	.975
	.99	7.31	5.18	4.31	3.83	3.51	3.29	3.12	2.99	2.89	2.80	2.73	2.66	.99
	.995	8.83	6.07	4.98	4.37	3.99	3.71	3.51	3.35	3.22	3.12	3.03	2.95	.995
	.999	12.6	8.25	6.60	5.70	5.13	4.73	4.44	4.21	4.02	3.87	3.75	3.64	.999
	.9995	14.4	9.25	7.33	6.30	5.64	5.19	4.85	4.59	4.38	4.21	4.07	3.95	.9995
60	.0005	$.0^640$	$.0^350$	$.0^251$	.016	.031	.048	.067	.085	.103	.120	.136	.152	.0005
	.001	$.0^516$	$.0^210$	$.0^280$	.022	.041	.062	.083	.103	.122	.140	.157	.174	.001
	.005	$.0^440$	$.0^250$	.024	.051	.081	.110	.137	.162	.185	.206	.225	.243	.005
	.01	$.0^316$	.010	.038	.073	.109	.142	.172	.199	.223	.245	.265	.283	.01
	.025	$.0^399$	.025	.071	.120	.163	.202	.235	.264	.290	.313	.333	.351	.025
	.05	$.0^240$	.051	.116	.176	.226	.267	.303	.333	.359	.382	.402	.419	.05
	.10	.016	.106	.194	.264	.318	.362	.398	.428	.453	.475	.493	.510	.10
	.25	.102	.289	.405	.480	.534	.573	.604	.629	.650	.667	.680	.695	.25
	.50	.461	.701	.798	.849	.880	.901	.917	.928	.937	.945	.951	.956	.50
	.75	1.35	1.42	1.41	1.38	1.37	1.35	1.33	1.32	1.31	1.30	1.29	1.29	.75
	.90	2.79	2.39	2.18	2.04	1.95	1.87	1.82	1.77	1.74	1.71	1.68	1.66	.90
	.95	4.00	3.15	2.76	2.53	2.37	2.25	2.17	2.10	2.04	1.99	1.95	1.92	.95
	.975	5.29	3.93	3.34	3.01	2.79	2.63	2.51	2.41	2.33	2.27	2.22	2.17	.975
	.99	7.08	4.98	4.13	3.65	3.34	3.12	2.95	2.82	2.72	2.63	2.56	2.50	.99
	.995	8.49	5.80	4.73	4.14	3.76	3.49	3.29	3.13	3.01	2.90	2.82	2.74	.995
	.999	12.0	7.76	6.17	5.31	4.76	4.37	4.09	3.87	3.69	3.54	3.43	3.31	.999
	.9995	13.6	8.65	6.81	5.82	5.20	4.76	4.44	4.18	3.98	3.82	3.69	3.57	.9995

Table A.7   Percentiles of the $F(\nu_1, \nu_2)$ Distribution with Degrees of Freedom ν_1 for the Numerator and ν_2 for the Denominator (*Continued*)

Cum. Prop.	ν_1 15	20	24	30	40	50	60	100	120	200	500	∞	Cum. Prop.	ν_2
.0005	.179	.226	.254	.287	.325	.350	.369	.410	.420	.444	.467	.483	.0005	30
.001	.202	.250	.278	.311	.348	.373	.391	.431	.442	.465	.488	.503	.001	
.005	.271	.320	.349	.381	.416	.441	.457	.495	.504	.524	.543	.559	.005	
.01	.311	.360	.388	.419	.454	.476	.493	.529	.538	.559	.575	.590	.01	
.025	.378	.426	.453	.482	.515	.535	.551	.585	.592	.610	.625	.639	.025	
.05	.445	.490	.516	.543	.573	.592	.606	.637	.644	.658	.676	.685	.05	
.10	.534	.575	.598	.623	.649	.667	.678	.704	.710	.725	.735	.746	.10	
.25	.716	.746	.763	.780	.798	.810	.818	.835	.839	.848	.856	.862	.25	
.50	.978	.989	.994	1.00	1.01	1.01	1.01	1.02	1.02	1.02	1.02	1.02	.50	
.75	1.32	1.30	1.29	1.28	1.27	1.26	1.26	1.25	1.24	1.24	1.23	1.23	.75	
.90	1.72	1.67	1.64	1.61	1.57	1.55	1.54	1.51	1.50	1.48	1.47	1.46	.90	
.95	2.01	1.93	1.89	1.84	1.79	1.76	1.74	1.70	1.68	1.66	1.64	1.62	.95	
.975	2.31	2.20	2.14	2.07	2.01	1.97	1.94	1.88	1.87	1.84	1.81	1.79	.975	
.99	2.70	2.55	2.47	2.39	2.30	2.25	2.21	2.13	2.11	2.07	2.03	2.01	.99	
.995	3.01	2.82	2.73	2.63	2.52	2.46	2.42	2.32	2.30	2.25	2.21	2.18	.995	
.999	3.75	3.49	3.36	3.22	3.07	2.98	2.92	2.79	2.76	2.69	2.63	2.59	.999	
.9995	4.10	3.80	3.65	3.48	3.32	3.22	3.15	3.00	2.97	2.89	2.82	2.78	.9995	
.0005	.185	.236	.266	.301	.343	.373	.393	.441	.453	.480	.504	.525	.0005	40
.001	.209	.259	.290	.326	.367	.396	.415	.461	.473	.500	.524	.545	.001	
.005	.279	.331	.362	.396	.436	.463	.481	.524	.534	.559	.581	.599	.005	
.01	.319	.371	.401	.435	.473	.498	.516	.556	.567	.592	.613	.628	.01	
.025	.387	.437	.466	.498	.533	.556	.573	.610	.620	.641	.662	.674	.025	
.05	.454	.502	.529	.558	.591	.613	.627	.658	.669	.685	.704	.717	.05	
.10	.542	.585	.609	.636	.664	.683	.696	.724	.731	.747	.762	.772	.10	
.25	.720	.752	.769	.787	.806	.819	.828	.846	.851	.861	.870	.877	.25	
.50	.972	.983	.989	.994	1.00	1.00	1.01	1.01	1.01	1.01	1.02	1.02	.50	
.75	1.30	1.28	1.26	1.25	1.24	1.23	1.22	1.21	1.21	1.20	1.19	1.19	.75	
.90	1.66	1.61	1.57	1.54	1.51	1.48	1.47	1.43	1.42	1.41	1.39	1.38	.90	
.95	1.92	1.84	1.79	1.74	1.69	1.66	1.64	1.59	1.58	1.55	1.53	1.51	.95	
.975	2.18	2.07	2.01	1.94	1.88	1.83	1.80	1.74	1.72	1.69	1.66	1.64	.975	
.99	2.52	2.37	2.29	2.20	2.11	2.06	2.02	1.94	1.92	1.87	1.83	1.80	.99	
.995	2.78	2.60	2.50	2.40	2.30	2.23	2.18	2.09	2.06	2.01	1.96	1.93	.995	
.999	3.40	3.15	3.01	2.87	2.73	2.64	2.57	2.44	2.41	2.34	2.28	2.23	.999	
.9995	3.68	3.39	3.24	3.08	2.92	2.82	2.74	2.60	2.57	2.49	2.41	2.37	.9995	
.0005	.192	.246	.278	.318	.365	.398	.421	.478	.493	.527	.561	.585	.0005	60
.001	.216	.270	.304	.343	.389	.421	.444	.497	.512	.545	.579	.602	.001	
.005	.287	.343	.376	.414	.458	.488	.510	.559	.572	.602	.633	.652	.005	
.01	.328	.383	.416	.453	.495	.524	.545	.592	.604	.633	.658	.679	.01	
.025	.396	.450	.481	.515	.555	.581	.600	.641	.654	.680	.704	.720	.025	
.05	.463	.514	.543	.575	.611	.633	.652	.690	.700	.719	.746	.759	.05	
.10	.550	.596	.622	.650	.682	.703	.717	.750	.758	.776	.793	.806	.10	
.25	.725	.758	.776	.796	.816	.830	.840	.860	.865	.877	.888	.896	.25	
.50	.967	.978	.983	.989	.994	.998	1.00	1.00	1.01	1.01	1.01	1.01	.50	
.75	1.27	1.25	1.24	1.22	1.21	1.20	1.19	1.17	1.17	1.16	1.15	1.15	.75	
.90	1.60	1.54	1.51	1.48	1.44	1.41	1.40	1.36	1.35	1.33	1.31	1.29	.90	
.95	1.84	1.75	1.70	1.65	1.59	1.56	1.53	1.48	1.47	1.44	1.41	1.39	.95	
.975	2.06	1.94	1.88	1.82	1.74	1.70	1.67	1.60	1.58	1.54	1.51	1.48	.975	
.99	2.35	2.20	2.12	2.03	1.94	1.88	1.84	1.75	1.73	1.68	1.63	1.60	.99	
.995	2.57	2.39	2.29	2.19	2.08	2.01	1.96	1.86	1.83	1.78	1.73	1.69	.995	
.999	3.08	2.83	2.69	2.56	2.41	2.31	2.25	2.11	2.09	2.01	1.93	1.89	.999	
.9995	3.30	3.02	2.87	2.71	2.55	2.45	2.38	2.23	2.19	2.11	2.03	1.98	.9995	

Table A.7   Percentiles of the $F(\nu_1, \nu_2)$ Distribution with Degrees of Freedom ν_1 for the Numerator and ν_2 for the Denominator (*Continued*)

ν_2	Cum. Prop.	1	2	3	4	5	6	7	8	9	10	11	12	Cum. Prop.
120	.0005	$.0^640$	$.0^350$	$.0^251$	.016	.031	.049	.067	.087	.105	.123	.140	.156	.0005
	.001	$.0^516$	$.0^210$	$.0^281$	.023	.042	.063	.084	.105	.125	.144	.162	.179	.001
	.005	$.0^439$	$.0^250$	.024	.051	.081	.111	.139	.165	.189	.211	.230	.249	.005
	.01	$.0^316$	.010	.038	.074	.110	.143	.174	.202	.227	.250	.271	.290	.01
	.025	$.0^399$	.025	.072	.120	.165	.204	.238	.268	.295	.318	.340	.359	.025
	.05	$.0^239$	.051	.117	.177	.227	.270	.306	.337	.364	.388	.408	.427	.05
	.10	.016	.105	.194	.265	.320	.365	.401	.432	.458	.480	.500	.518	.10
	.25	.102	.288	.405	.481	.534	.574	.606	.631	.652	.670	.685	.699	.25
	.50	.458	.697	.793	.844	.875	.896	.912	.923	.932	.939	.945	.950	.50
	.75	1.34	1.40	1.39	1.37	1.35	1.33	1.31	1.30	1.29	1.28	1.27	1.26	.75
	.90	2.75	2.35	2.13	1.99	1.90	1.82	1.77	1.72	1.68	1.65	1.62	1.60	.90
	.95	3.92	3.07	2.68	2.45	2.29	2.18	2.09	2.02	1.96	1.91	1.87	1.83	.95
	.975	5.15	3.80	3.23	2.89	2.67	2.52	2.39	2.30	2.22	2.16	2.10	2.05	.975
	.99	6.85	4.79	3.95	3.48	3.17	2.96	2.79	2.66	2.56	2.47	2.40	2.34	.99
	.995	8.18	5.54	4.50	3.92	3.55	3.28	3.09	2.93	2.81	2.71	2.62	2.54	.995
	.999	11.4	7.32	5.79	4.95	4.42	4.04	3.77	3.55	3.38	3.24	3.12	3.02	.999
	.9995	12.8	8.10	6.34	5.39	4.79	4.37	4.07	3.82	3.63	3.47	3.34	3.22	.9995
∞	.0005	$.0^639$	$.0^350$	$.0^251$	.016	.032	.050	.069	.088	.108	.127	.144	.161	.0005
	.001	$.0^516$	$.0^210$	$.0^281$	.023	.042	.063	.085	.107	.128	.148	.167	.185	.001
	.005	$.0^439$	$.0^250$	.024	.052	.082	.113	.141	.168	.193	.216	.236	.256	.005
	.01	$.0^316$	.010	.038	.074	.111	.145	.177	.206	.232	.256	.278	.298	.01
	.025	$.0^398$	.025	.072	.121	.166	.206	.241	.272	.300	.325	.347	.367	.025
	.05	$.0^239$	.051	.117	.178	.229	.273	.310	.342	.369	.394	.417	.436	.05
	.10	.016	.105	.195	.266	.322	.367	.405	.436	.463	.487	.508	.525	.10
	.25	.102	.288	.404	.481	.535	.576	.608	.634	.655	.674	.690	.703	.25
	.50	.455	.693	.789	.839	.870	.891	.907	.918	.927	.934	.939	.945	.50
	.75	1.32	1.39	1.37	1.35	1.33	1.31	1.29	1.28	1.27	1.25	1.24	1.24	.75
	.90	2.71	2.30	2.08	1.94	1.85	1.77	1.72	1.67	1.63	1.60	1.57	1.55	.90
	.95	3.84	3.00	2.60	2.37	2.21	2.10	2.01	1.94	1.88	1.83	1.79	1.75	.95
	.975	5.02	3.69	3.12	2.79	2.57	2.41	2.29	2.19	2.11	2.05	1.99	1.94	.975
	.99	6.63	4.61	3.78	3.32	3.02	2.80	2.64	2.51	2.41	2.32	2.25	2.18	.99
	.995	7.88	5.30	4.28	3.72	3.35	3.09	2.90	2.74	2.62	2.52	2.43	2.36	.995
	.999	10.8	6.91	5.42	4.62	4.10	3.74	3.47	3.27	3.10	2.96	2.84	2.74	.999
	.9995	12.1	7.60	5.91	5.00	4.42	4.02	3.72	3.48	3.30	3.14	3.02	2.90	.9995

For sample sizes larger than, say, 30, a fairly good approximation to the F distribution percentiles can be obtained from

$$\log_{10} F_\alpha(\nu_1, \nu_2) \approx \left(\frac{a}{\sqrt{h - b}} \right) - cg$$

where $h = 2\nu_1\nu_2/(\nu_1 + \nu_2)$, $g = (\nu_2 - \nu_1)/\nu_1\nu_2$, and a, b, c are functions of α given below:

α	.50	.75	.90	.95	.975	.99	.995	.999	.9995
a	0	0.5859	1.1131	1.4287	1.7023	2.0206	2.2373	2.6841	2.8580
b	—	0.58	0.77	0.95	1.14	1.40	1.61	2.09	2.30
c	0.290	0.355	0.527	0.681	0.846	1.073	1.250	1.672	1.857

Table A.7 Percentiles of the $F(\nu_1,\nu_2)$ Distribution with Degrees of Freedom ν_1 for the Numerator and ν_2 for the Denominator (*Continued*)

Cum. Prop.	ν_1 15	20	24	30	40	50	60	100	120	200	500	∞	Cum. Prop.	ν_2
.0005	.199	.256	.293	.338	.390	.429	.458	.524	.543	.578	.614	.676	.0005	120
.001	.223	.282	.319	.363	.415	.453	.480	.542	.568	.595	.631	.691	.001	
.005	.297	.356	.393	.434	.484	.520	.545	.605	.623	.661	.702	.733	.005	
.01	.338	.397	.433	.474	.522	.556	.579	.636	.652	.688	.725	.755	0	
.025	.406	.464	.498	.536	.580	.611	.633	.684	.698	.729	.762	.789	.025	
.05	.473	.527	.559	.594	.634	.661	.682	.727	.740	.767	.785	.819	.05	
.10	.560	.609	.636	.667	.702	.726	.742	.781	.791	.815	.838	.855	.10	
.25	.730	.765	.784	.805	.828	.843	.853	.877	.884	.897	.911	.923	.25	
.50	.961	.972	.978	.983	.989	.992	.994	1.00	1.00	1.00	1.01	1.01	.50	
.75	1.24	1.22	1.21	1.19	1.18	1.17	1.16	1.14	1.13	1.12	1.11	1.10	.75	
.90	1.55	1.48	1.45	1.41	1.37	1.34	1.32	1.27	1.26	1.24	1.21	1.19	.90	
.95	1.75	1.66	1.61	1.55	1.50	1.46	1.43	1.37	1.35	1.32	1.28	1.25	.95	
.975	1.95	1.82	1.76	1.69	1.61	1.56	1.53	1.45	1.43	1.39	1.34	1.31	.975	
.99	2.19	2.03	1.95	1.86	1.76	1.70	1.66	1.56	1.53	1.48	1.42	1.38	.99	
.995	2.37	2.19	2.09	1.98	1.87	1.80	1.75	1.64	1.61	1.54	1.48	1.43	.995	
.999	2.78	2.53	2.40	2.26	2.11	2.02	1.95	1.82	1.76	1.70	1.62	1.54	.999	
.9995	2.96	2.67	2.53	2.38	2.21	2.11	2.01	1.88	1.84	1.75	1.67	1.60	.9995	
.0005	.207	.270	.311	.360	.422	.469	.505	.599	.624	.704	.804	1.00	.0005	∞
.001	.232	.296	.338	.386	.448	.493	.527	.617	.649	.719	.819	1.00	.001	
.005	.307	.372	.412	.460	.518	.559	.592	.671	.699	.762	.843	1.00	.005	
.01	.349	.413	.452	.499	.554	.595	.625	.699	.724	.782	.858	1.00	.01	
.025	.418	.480	.517	.560	.611	.645	.675	.741	.763	.813	.878	1.00	.025	
.05	.484	.543	.577	.617	.663	.694	.720	.781	.797	.840	.896	1.00	.05	
.10	.570	.622	.652	.687	.726	.752	.774	.826	.838	.877	.919	1.00	.10	
.25	.736	.773	.793	.816	.842	.860	.872	.901	.910	.932	.957	1.00	.25	
.50	.956	.967	.972	.978	.983	.987	.989	.993	.994	.997	.999	1.00	.50	
.75	1.22	1.19	1.18	1.16	1.14	1.13	1.12	1.09	1.08	1.07	1.04	1.00	.75	
.90	1.49	1.42	1.38	1.34	1.30	1.26	1.24	1.18	1.17	1.13	1.08	1.00	.90	
.95	1.67	1.57	1.52	1.46	1.39	1.35	1.32	1.24	1.22	1.17	1.11	1.00	.95	
.975	1.83	1.71	1.64	1.57	1.48	1.43	1.39	1.30	1.27	1.21	1.13	1.00	.975	
.99	2.04	1.88	1.79	1.70	1.59	1.52	1.47	1.36	1.32	1.25	1.15	1.00	.99	
.995	2.19	2.00	1.90	1.79	1.67	1.59	1.53	1.40	1.36	1.28	1.17	1.00	.995	
.999	2.51	2.27	2.13	1.99	1.84	1.73	1.66	1.49	1.45	1.34	1.21	1.00	.999	
.9995	2.65	2.37	2.22	2.07	1.91	1.79	1.71	1.53	1.48	1.36	1.22	1.00	.9995	

The values given in this table are abstracted with permission from the following sources:

1. All values for ν_1,ν_2 equal to 50, 100, 200, 500 are from A. Hald, *Statistical Tables and Formulas*, John Wiley & Sons, Inc., New York, 1952.

2. For cumulative proportions .5, .75, .9, .95, .975, .99, .995 most of the values are from M. Merrington and C. M. Thompson, *Biometrika*, vol. 33 (1943), p. 73.

3. For cumulative proportions .999 the values are from C. Colcord and L. S. Deming, *Sankhyā*, vol. 2 (1936), p. 423.

4. For cum. prop. = $\alpha < .5$ the values are the reciprocals of values for $1 - \alpha$ (with ν_1 and ν_2 interchanged). The values in Merrington and Thompson and in Colcord and Deming are to five significant figures, and it is hoped (but not expected) that the reciprocals are correct as given. The values in Hald are to three significant figures, and the reciprocals are probably accurate within one to two digits in the third significant figure except for those values very close to unity, where they may be off four to five digits in the third significant figure.

5. Gaps remaining in the table after using the above sources were filled in by interpolation.

$$\alpha = \frac{(\nu_1/\nu_2)^{\frac{1}{2}\nu_1}}{\beta(\frac{1}{2}\nu_1,\frac{1}{2}\nu_2)} \int_{-\infty}^{F\alpha} F^{\frac{1}{2}\nu_1-1}\left(1+\frac{\nu_1 F}{\nu_2}\right)^{-(\nu_1+\nu_2)/2} dF$$

Table A.8 Tolerance Factors for Normal Distributions (One-sided)*

Factors K such that the probability is γ that at least a proportion $1 - \alpha$ of the distribution will be less than $\bar{X} + Ks$ (or greater than $\bar{X} - Ks$), where $\bar{X}$ and s are estimates of the mean and the standard deviation computed from a sample of size n.

n	$\gamma = 0.75$					$\gamma = 0.90$					$\gamma = 0.95$					$\gamma = 0.99$				
α	0.25	0.10	0.05	0.01	0.001	0.25	0.10	0.05	0.01	0.001	0.25	0.10	0.05	0.01	0.001	0.25	0.10	0.05	0.01	0.001
3	1.464	2.501	3.152	4.396	5.805	2.602	4.258	5.310	7.340	9.651	3.804	6.158	7.655	10.552	13.857					
4	1.256	2.134	2.680	3.726	4.910	1.972	3.187	3.957	5.437	7.128	2.619	4.163	5.145	7.042	9.215					
5	1.152	1.961	2.463	3.421	4.507	1.698	2.742	3.400	4.666	6.112	2.149	3.407	4.202	5.741	7.501					
6	1.087	1.860	2.336	3.243	4.273	1.540	2.494	3.091	4.242	5.556	1.895	3.006	3.707	5.062	6.612	2.849	4.408	5.409	7.334	9.540
7	1.043	1.791	2.250	3.126	4.118	1.435	2.333	2.894	3.972	5.201	1.732	2.755	3.399	4.641	6.061	2.490	3.856	4.730	6.411	8.348
8	1.010	1.740	2.190	3.042	4.008	1.360	2.219	2.755	3.783	4.955	1.617	2.582	3.188	4.353	5.686	2.252	3.496	4.287	5.811	7.566
9	0.984	1.702	2.141	2.977	3.924	1.302	2.133	2.649	3.641	4.772	1.532	2.454	3.031	4.143	5.414	2.085	3.242	3.971	5.389	7.014
10	0.964	1.671	2.103	2.927	3.858	1.257	2.065	2.568	3.532	4.629	1.465	2.355	2.911	3.981	5.203	1.954	3.048	3.739	5.075	6.603
11	0.947	1.646	2.073	2.885	3.804	1.219	2.012	2.503	3.444	4.515	1.411	2.275	2.815	3.852	5.036	1.854	2.897	3.557	4.828	6.284
12	0.933	1.624	2.048	2.851	3.760	1.188	1.966	2.448	3.371	4.420	1.366	2.210	2.736	3.747	4.900	1.771	2.773	3.410	4.633	6.032
13	0.919	1.606	2.026	2.822	3.722	1.162	1.928	2.403	3.310	4.341	1.329	2.155	2.670	3.659	4.787	1.702	2.677	3.290	4.472	5.826
14	0.909	1.591	2.007	2.796	3.690	1.139	1.895	2.363	3.257	4.274	1.296	2.108	2.614	3.585	4.690	1.645	2.592	3.189	4.336	5.651
15	0.899	1.577	1.991	2.776	3.661	1.119	1.866	2.329	3.212	4.215	1.268	2.068	2.566	3.520	4.607	1.596	2.521	3.102	4.224	5.507
16	0.891	1.566	1.977	2.756	3.637	1.101	1.842	2.299	3.172	4.164	1.242	2.032	2.523	3.463	4.534	1.553	2.458	3.028	4.124	5.374
17	0.883	1.554	1.964	2.739	3.615	1.085	1.820	2.272	3.136	4.118	1.220	2.001	2.486	3.415	4.471	1.514	2.405	2.962	4.038	5.268
18	0.876	1.544	1.951	2.723	3.595	1.071	1.800	2.249	3.106	4.078	1.200	1.974	2.453	3.370	4.415	1.481	2.357	2.906	3.961	5.167
19	0.870	1.536	1.942	2.710	3.577	1.058	1.781	2.228	3.078	4.041	1.183	1.949	2.423	3.331	4.364	1.450	2.315	2.855	3.893	5.078
20	0.865	1.528	1.933	2.697	3.561	1.046	1.765	2.208	3.052	4.009	1.167	1.926	2.396	3.295	4.319	1.424	2.275	2.807	3.832	5.003
21	0.859	1.520	1.923	2.686	3.545	1.035	1.750	2.190	3.028	3.979	1.152	1.905	2.371	3.262	4.276	1.397	2.241	2.768	3.776	4.932
22	0.854	1.514	1.916	2.675	3.532	1.025	1.736	2.174	3.007	3.952	1.138	1.887	2.350	3.233	4.238	1.376	2.208	2.729	3.727	4.866
23	0.849	1.508	1.907	2.665	3.520	1.016	1.724	2.159	2.987	3.927	1.126	1.869	2.329	3.206	4.204	1.355	2.179	2.693	3.680	4.806
24	0.845	1.502	1.901	2.656	3.509	1.007	1.712	2.145	2.969	3.904	1.114	1.853	2.309	3.181	4.171	1.336	2.154	2.663	3.638	4.755
25	0.842	1.496	1.895	2.647	3.497	0.999	1.702	2.132	2.952	3.882	1.103	1.838	2.292	3.158	4.143	1.319	2.129	2.632	3.601	4.706
30	0.825	1.475	1.869	2.613	3.454	0.966	1.657	2.080	2.884	3.794	1.059	1.778	2.220	3.064	4.022	1.249	2.029	2.516	3.446	4.508
35	0.812	1.458	1.849	2.588	3.421	0.942	1.623	2.041	2.833	3.730	1.025	1.732	2.166	2.994	3.934	1.195	1.957	2.431	3.334	4.364
40	0.803	1.445	1.834	2.568	3.395	0.923	1.598	2.010	2.793	3.679	0.999	1.697	2.126	2.941	3.866	1.154	1.902	2.365	3.250	4.255
45	0.795	1.435	1.821	2.552	3.375	0.908	1.577	1.986	2.762	3.638	0.978	1.669	2.092	2.897	3.811	1.122	1.857	2.313	3.181	4.168
50	0.788	1.426	1.811	2.538	3.358	0.894	1.560	1.965	2.735	3.604	0.961	1.646	2.065	2.863	3.766	1.096	1.821	2.296	3.124	4.096

* Reproduced from "Tables for One-sided Statistical Tolerance Limits," by Gerald J. Lieberman, *Industrial Quality Control*, vol. XIV, no. 10, p. 8, April, 1958, with the permission of the author and journal.

Table A.9 Tolerance Factors for Normal Distributions (Two-sided)*

N	γ = 0.75 P=0.75	0.90	0.95	0.99	0.999	γ = 0.90 P=0.75	0.90	0.95	0.99	0.999	γ = 0.95 P=0.75	0.90	0.95	0.99	0.999	γ = 0.99 P=0.75	0.90	0.95	0.99	0.999
2	4.498	6.301	7.414	9.531	11.920	11.407	15.978	18.800	24.167	30.227	22.858	32.019	37.674	48.430	60.573	114.363	160.193	188.491	242.300	303.054
3	2.501	3.538	4.187	5.431	6.844	4.132	5.847	6.919	8.974	11.309	5.922	8.380	9.916	12.861	16.208	13.378	18.930	22.401	29.055	36.616
4	2.035	2.892	3.431	4.471	5.657	2.932	4.166	4.943	6.440	8.149	3.779	5.369	6.370	8.299	10.502	6.614	9.398	11.150	14.527	18.383
5	1.825	2.599	3.088	4.033	5.117	2.454	3.494	4.152	5.423	6.879	3.002	4.275	5.079	6.634	8.415	4.643	6.612	7.855	10.260	13.015
6	1.704	2.429	2.889	3.779	4.802	2.196	3.131	3.723	4.870	6.188	2.604	3.712	4.414	5.775	7.337	3.743	5.337	6.345	8.301	10.548
7	1.624	2.318	2.757	3.611	4.593	2.034	2.902	3.452	4.521	5.750	2.361	3.369	4.007	5.248	6.676	3.233	4.613	5.488	7.187	9.142
8	1.568	2.238	2.663	3.491	4.444	1.921	2.743	3.264	4.278	5.446	2.197	3.136	3.732	4.891	6.226	2.905	4.147	4.936	6.468	8.234
9	1.525	2.178	2.593	3.400	4.330	1.839	2.626	3.125	4.098	5.220	2.078	2.967	3.532	4.631	5.899	2.677	3.822	4.550	5.966	7.600
10	1.492	2.131	2.537	3.328	4.241	1.775	2.535	3.018	3.959	5.046	1.987	2.839	3.379	4.433	5.649	2.508	3.582	4.265	5.594	7.129
11	1.465	2.093	2.493	3.271	4.169	1.724	2.463	2.933	3.849	4.906	1.916	2.737	3.259	4.277	5.452	2.378	3.397	4.045	5.308	6.766
12	1.443	2.062	2.456	3.223	4.110	1.683	2.404	2.863	3.758	4.792	1.858	2.655	3.162	4.150	5.291	2.274	3.250	3.870	5.079	6.477
13	1.425	2.036	2.424	3.183	4.059	1.648	2.355	2.805	3.682	4.697	1.810	2.587	3.081	4.044	5.158	2.190	3.130	3.727	4.893	6.240
14	1.409	2.013	2.398	3.148	4.016	1.619	2.314	2.756	3.618	4.615	1.770	2.529	3.012	3.955	5.045	2.120	3.029	3.608	4.737	6.043
15	1.395	1.994	2.375	3.118	3.979	1.594	2.278	2.713	3.562	4.545	1.735	2.480	2.954	3.878	4.949	2.060	2.945	3.507	4.605	5.876
16	1.383	1.977	2.355	3.092	3.946	1.572	2.246	2.676	3.514	4.484	1.705	2.437	2.903	3.812	4.865	2.009	2.872	3.421	4.492	5.732
17	1.372	1.962	2.337	3.069	3.917	1.552	2.219	2.643	3.471	4.430	1.679	2.400	2.858	3.754	4.791	1.965	2.808	3.345	4.393	5.607
18	1.363	1.948	2.321	3.048	3.891	1.535	2.194	2.614	3.433	4.382	1.655	2.366	2.819	3.702	4.725	1.926	2.753	3.279	4.307	5.497
19	1.355	1.936	2.307	3.030	3.867	1.520	2.172	2.588	3.399	4.339	1.635	2.337	2.784	3.656	4.667	1.891	2.703	3.221	4.230	5.399
20	1.347	1.925	2.294	3.013	3.846	1.506	2.152	2.564	3.368	4.300	1.616	2.310	2.752	3.615	4.614	1.860	2.659	3.168	4.161	5.312
21	1.340	1.915	2.282	2.998	3.827	1.493	2.135	2.543	3.340	4.264	1.599	2.286	2.723	3.577	4.567	1.833	2.620	3.121	4.100	5.234
22	1.334	1.906	2.271	2.984	3.809	1.482	2.118	2.524	3.315	4.232	1.584	2.264	2.697	3.543	4.523	1.808	2.584	3.078	4.044	5.163
23	1.328	1.898	2.261	2.971	3.793	1.471	2.103	2.506	3.292	4.203	1.570	2.244	2.673	3.512	4.484	1.785	2.551	3.040	3.993	5.098
24	1.322	1.891	2.252	2.959	3.778	1.462	2.089	2.489	3.270	4.176	1.557	2.225	2.651	3.483	4.447	1.764	2.522	3.004	3.947	5.039
25	1.317	1.883	2.244	2.948	3.764	1.453	2.077	2.474	3.251	4.151	1.545	2.208	2.631	3.457	4.413	1.745	2.494	2.972	3.904	4.985
26	1.313	1.877	2.236	2.938	3.751	1.444	2.065	2.460	3.232	4.127	1.534	2.193	2.612	3.432	4.382	1.727	2.460	2.941	3.865	4.935
27	1.309	1.871	2.229	2.929	3.740	1.437	2.054	2.447	3.215	4.106	1.523	2.178	2.595	3.409	4.353	1.711	2.446	2.914	3.828	4.888
30	1.297	1.855	2.210	2.904	3.708	1.417	2.025	2.413	3.170	4.049	1.497	2.140	2.549	3.350	4.278	1.668	2.385	2.841	3.733	4.768
35	1.283	1.834	2.185	2.871	3.667	1.390	1.988	2.368	3.112	3.974	1.462	2.090	2.490	3.272	4.179	1.613	2.306	2.748	3.611	4.611
40	1.271	1.818	2.166	2.846	3.635	1.370	1.959	2.334	3.066	3.917	1.435	2.052	2.445	3.213	4.104	1.571	2.247	2.677	3.518	4.493
45	1.262	1.805	2.150	2.826	3.609	1.354	1.935	2.306	3.030	3.871	1.414	2.021	2.408	3.165	4.042	1.539	2.200	2.621	3.444	4.399
50	1.255	1.794	2.138	2.809	3.588	1.340	1.916	2.284	3.001	3.833	1.396	1.996	2.379	3.126	3.993	1.512	2.162	2.576	3.385	4.323
55	1.249	1.785	2.127	2.795	3.571	1.329	1.901	2.265	2.976	3.801	1.382	1.976	2.354	3.094	3.951	1.490	2.130	2.538	3.335	4.260
60	1.243	1.778	2.118	2.784	3.556	1.320	1.887	2.248	2.955	3.774	1.369	1.958	2.333	3.066	3.916	1.471	2.103	2.506	3.293	4.206

*From *Introduction to Statistical Analysis*, 2d ed., by W. J. Dixon and F. J. Massey, Jr., Copyright, 1957. McGraw-Hill Book Company. Used by permission.

Table A.9 Tolerance Factors for Normal Distributions (Two-sided)* *(Continued)*

N \ P	γ = 0.75					γ = 0.90					γ = 0.95					γ = 0.99				
	0.75	0.90	0.95	0.99	0.999	0.75	0.90	0.95	0.99	0.999	0.75	0.90	0.95	0.99	0.999	0.75	0.90	0.95	0.99	0.999
65	1.239	1.771	2.110	2.773	3.543	1.312	1.875	2.235	2.937	3.751	1.359	1.943	2.315	3.042	3.886	1.455	2.080	2.478	3.257	4.160
70	1.235	1.765	2.104	2.764	3.531	1.304	1.865	2.222	2.920	3.730	1.349	1.929	2.299	3.021	3.859	1.440	2.060	2.454	3.225	4.120
75	1.231	1.760	2.098	2.757	3.521	1.298	1.856	2.211	2.906	3.712	1.341	1.917	2.285	3.002	3.835	1.428	2.042	2.433	3.197	4.084
80	1.228	1.756	2.092	2.749	3.512	1.292	1.848	2.202	2.894	3.696	1.334	1.907	2.272	2.986	3.814	1.417	2.026	2.414	3.173	4.053
85	1.225	1.752	2.087	2.743	3.504	1.287	1.841	2.193	2.882	3.682	1.327	1.897	2.261	2.971	3.795	1.407	2.012	2.397	3.150	4.024
90	1.223	1.748	2.083	2.737	3.497	1.283	1.834	2.185	2.872	3.669	1.321	1.889	2.251	2.958	3.778	1.398	1.999	2.382	3.130	3.999
95	1.220	1.745	2.079	2.732	3.490	1.278	1.828	2.178	2.863	3.657	1.315	1.881	2.241	2.945	3.763	1.390	1.987	2.368	3.112	3.976
100	1.218	1.742	2.075	2.727	3.484	1.275	1.822	2.172	2.854	3.646	1.311	1.874	2.233	2.934	3.748	1.383	1.977	2.355	3.096	3.954
110	1.214	1.736	2.069	2.719	3.473	1.268	1.813	2.160	2.839	3.626	1.302	1.861	2.218	2.915	3.723	1.369	1.958	2.333	3.066	3.917
120	1.211	1.732	2.063	2.712	3.464	1.262	1.804	2.150	2.826	3.610	1.294	1.850	2.205	2.898	3.702	1.358	1.942	2.314	3.041	3.885
130	1.208	1.728	2.059	2.705	3.456	1.257	1.797	2.141	2.814	3.595	1.288	1.841	2.194	2.883	3.683	1.349	1.928	2.298	3.019	3.857
140	1.206	1.724	2.054	2.700	3.449	1.252	1.791	2.134	2.804	3.582	1.282	1.833	2.184	2.870	3.666	1.340	1.916	2.283	3.000	3.833
150	1.204	1.721	2.051	2.695	3.443	1.248	1.785	2.127	2.795	3.571	1.277	1.825	2.175	2.859	3.652	1.332	1.905	2.270	2.983	3.811
160	1.202	1.718	2.047	2.691	3.437	1.245	1.780	2.121	2.787	3.561	1.272	1.819	2.167	2.848	3.638	1.326	1.896	2.259	2.968	3.792
170	1.200	1.716	2.044	2.687	3.432	1.242	1.775	2.116	2.780	3.552	1.268	1.813	2.160	2.839	3.627	1.320	1.887	2.248	2.955	3.774
180	1.198	1.713	2.042	2.683	3.427	1.239	1.771	2.111	2.774	3.543	1.264	1.808	2.154	2.831	3.616	1.314	1.879	2.239	2.942	3.759
190	1.197	1.711	2.039	2.680	3.423	1.236	1.767	2.106	2.768	3.536	1.261	1.803	2.148	2.823	3.606	1.309	1.872	2.230	2.931	3.744
200	1.195	1.709	2.037	2.677	3.419	1.234	1.764	2.102	2.762	3.529	1.258	1.798	2.143	2.816	3.597	1.304	1.865	2.222	2.921	3.731
250	1.190	1.702	2.028	2.665	3.404	1.224	1.750	2.085	2.740	3.501	1.245	1.780	2.121	2.788	3.561	1.286	1.839	2.191	2.880	3.678
300	1.186	1.696	2.021	2.656	3.393	1.217	1.740	2.073	2.725	3.481	1.236	1.767	2.106	2.767	3.535	1.273	1.820	2.169	2.850	3.641
400	1.181	1.688	2.012	2.644	3.378	1.207	1.726	2.057	2.703	3.452	1.223	1.749	2.084	2.739	3.499	1.255	1.794	2.138	2.809	3.589
500	1.177	1.683	2.006	2.636	3.368	1.201	1.717	2.046	2.689	3.434	1.215	1.737	2.070	2.721	3.475	1.243	1.777	2.117	2.783	3.555
600	1.175	1.680	2.002	2.631	3.360	1.196	1.710	2.038	2.678	3.421	1.209	1.729	2.060	2.707	3.458	1.234	1.764	2.102	2.763	3.530
700	1.173	1.677	1.998	2.626	3.355	1.192	1.705	2.032	2.670	3.411	1.204	1.722	2.052	2.697	3.445	1.227	1.755	2.091	2.748	3.511
800	1.171	1.675	1.996	2.623	3.350	1.189	1.701	2.027	2.663	3.402	1.201	1.717	2.046	2.688	3.434	1.222	1.747	2.082	2.736	3.495
900	1.170	1.673	1.993	2.620	3.347	1.187	1.697	2.023	2.658	3.396	1.198	1.712	2.040	2.682	3.426	1.218	1.741	2.075	2.726	3.483
1000	1.169	1.671	1.992	2.617	3.344	1.185	1.695	2.019	2.654	3.390	1.195	1.709	2.036	2.676	3.418	1.214	1.736	2.068	2.718	3.472
∞	1.150	1.645	1.960	2.576	3.291	1.150	1.645	1.960	2.576	3.291	1.150	1.645	1.960	2.576	3.291	1.150	1.645	1.960	2.576	3.291

APPENDIX B: CHARTS

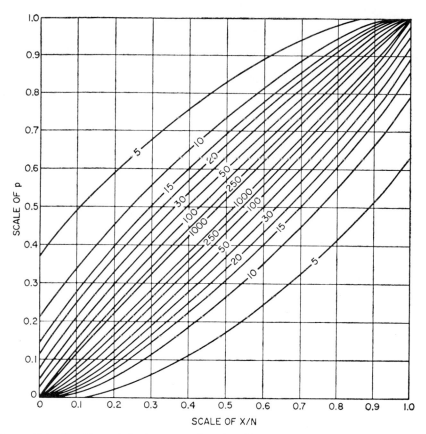

CHART B.1   Confidence belts for proportions (confidence coefficient 0.80).   (*From Introduction to Statistical Analysis, 2d ed., by W. J. Dixon and F. J. Massey, Jr. Copyright, 1957. McGraw-Hill Book Company. Used by permission.*)

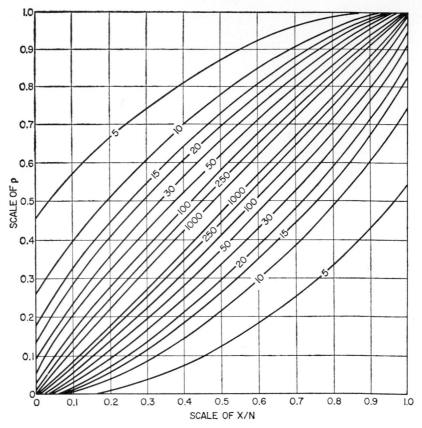

CHART B.2   Confidence belts for proportions (confidence coefficient 0.90).   (*From Introduction to Statistical Analysis, 2d ed., by W. J. Dixon and F. J. Massey, Jr.   Copyright, 1957.   McGraw-Hill Book Company.   Used by permission.*)

LOWER CONFIDENCE LIMIT ON R

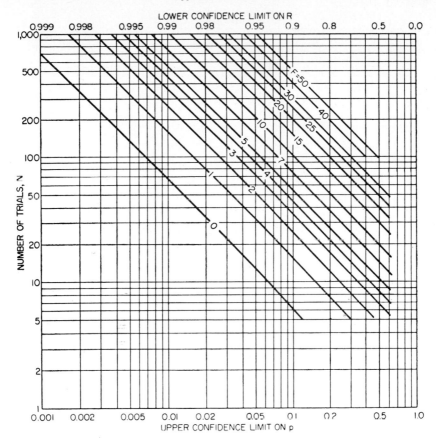

CHART B.3 Confidence limits on reliability. Upper confidence limit on unreliability (1 minus lower confidence limit on reliability), number of trials N, observed failures F, confidence coefficient $\gamma = 0.50$. (*David K. Lloyd and Myron Lipow, Reliability: Management, Methods and Mathematics.* © 1962, by permission of Prentice-Hall, Inc., Englewood Cliffs, New Jersey.)

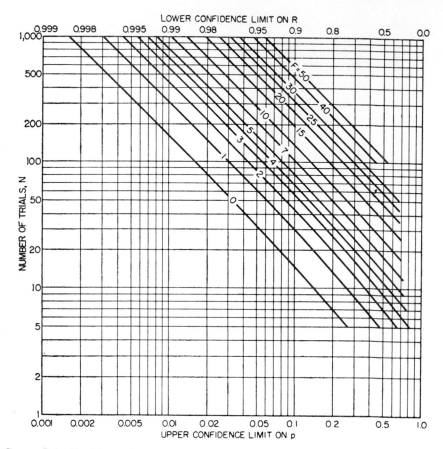

CHART B.4   Confidence limits on reliability. Upper confidence limit on unreliability (1 minus lower confidence limit on reliability), number of trials N, observed failures F, confidence coefficient $\gamma = 0.80$. (*David K. Lloyd and Myron Lipow, Reliability: Management, Methods and Mathematics.* © *1962, by permission of Prentice-Hall, Inc., Englewood Cliffs, New Jersey.*)

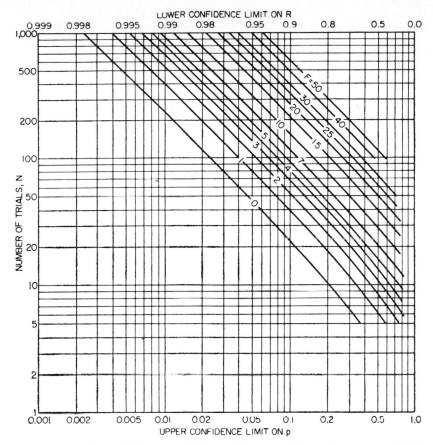

CHART B.5 Confidence limits on reliability. Upper confidence limit on unreliability (1 minus lower confidence limit on reliability), number of trials N, observed failures F, confidence coefficient $\gamma = 0.90$. (*David K. Lloyd and Myron Lipow, Reliability: Management, Methods and Mathematics.* © *1962, by permission of Prentice-Hall, Inc., Englewood Cliffs, New Jersey.*)

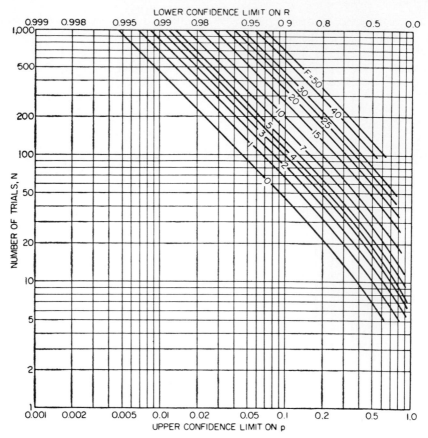

CHART B.6   Confidence limits on reliability.   Upper confidence limit on unreliability (1 minus lower confidence limit on reliability), number of trials N, observed failures F, confidence coefficient $\gamma = 0.95$.   (*David K. Lloyd and Myron Lipow, Reliability: Management, Methods and Mathematics.* © 1962, *by permission of Prentice-Hall, Inc., Englewood Cliffs, New Jersey.*)

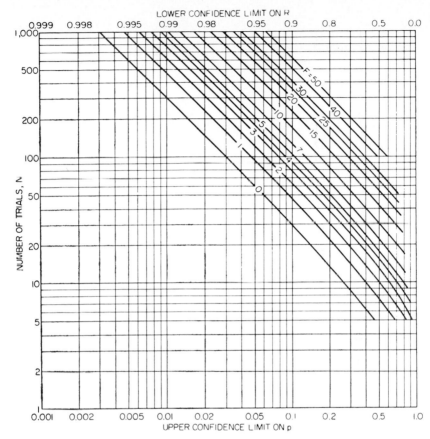

CHART B.7   Confidence limits on reliability.   Upper confidence limit on unreliability (1 minus lower confidence limit on reliability), number of trials N, observed failures F, confidence coefficient $\gamma = 0.99$.   (*David K. Lloyd and Myron Lipow, Reliability: Management, Methods and Mathematics.*   © 1962, *by permission of Prentice-Hall, Inc., Englewood Cliffs, New Jersey.*)

Index